AF769970

FATIGUE AND TRIBOLOGICAL PROPERTIES OF PLASTICS AND ELASTOMERS

FATIGUE AND TRIBOLOGICAL PROPERTIES OF PLASTICS AND ELASTOMERS

Second Edition

Laurence W. McKeen

ELSEVIER

Amsterdam • Boston • Heidelberg • London • New York • Oxford
Paris • San Diego • San Francisco • Sydney • Tokyo
William Andrew is an imprint of Elsevier

William Andrew is an imprint of Elsevier
The Boulevard, Langford Lane, Oxford OX5 1GB, UK
30 Corporate Drive, Suite 400, Burlington, MA 01803, USA

First edition 1994
Second edition 2010

Copyright © 2010, Laurence W. McKeen. Published by Elsevier Inc. All rights reserved

The right of Laurence W. McKeen to be identified as the author of this work has been asserted in accordance with the Copyright, Designs and Patents Act 1988

No part of this publication may be reproduced or transmitted in any form or by any means, electronic or mechanical, including photocopying, recording, or any information storage and retrieval system, without permission in writing from the publisher. Details on how to seek permission, further information about the Publisher's permissions policies and our arrangement with organizations such as the Copyright Clearance Center and the Copyright Licensing Agency, can be found at our website: www.elsevier.com/permissions

This book and the individual contributions contained in it are protected under copyright by the Publisher (other than as may be noted herein).

Notices

Knowledge and best practice in this field are constantly changing. As new research and experience broaden our understanding, changes in research methods, professional practices, or medical treatment may become necessary.

Practitioners and researchers must always rely on their own experience and knowledge in evaluating and using any information, methods, compounds, or experiments described herein. In using such information or methods they should be mindful of their own safety and the safety of others, including parties for whom they have a professional responsibility.

To the fullest extent of the law, neither the Publisher nor the authors, contributors, or editors, assume any liability for any injury and/or damage to persons or property as a matter of products liability, negligence or otherwise, or from any use or operation of any methods, products, instructions, or ideas contained in the material herein.

British Library Cataloguing in Publication Data
A catalogue record for this book is available from the British Library

Library of Congress Cataloging-in-Publication Data
A catalogue record for this book is available from the Library of Congress

ISBN: 978-0-08-096450-8

For information on all Elsevier publications
visit our website at elsevierdirect.com

Printed and bound in United States of America

10 11 12 11 10 9 8 7 6 5 4 3 2 1

Working together to grow
libraries in developing countries

www.elsevier.com | www.bookaid.org | www.sabre.org

ELSEVIER BOOK AID International Sabre Foundation

Contents

Preface

This book is an update to an "authorless" work by the same title. The first edition was published in the early 1990s. A lot has changed in the field since then and a lot has not changed. There are new plastic materials. There has been a huge turnover in ownership of the plastic producing companies. There has been a lot of consolidation, which of course means discontinued products. This update is much more extensive than the usual "next edition."

It has been reorganized from a polymer chemistry point of view. Plastics of similar polymer types are grouped into nine chapters. Each of these chapters includes an introduction with a brief explanation of the chemistry of the polymers used in the plastics.

An extensive introduction has been added as three chapters. The initial chapter focuses on fatigue, what it is, how it is measured, and how data is presented. The second chapter focuses on tribology properties. The field of tribology is extensive, so this chapter focuses primarily on the measures included in the data portion of this book. The third chapter covers polymer chemistry and plastics composition.

Chapters 4–12 are a databank that serves as an evaluation of fatigue and tribology performance of plastics. Each of these chapters is split into two sections, one each for fatigue properties and tribology properties. Several hundred uniform graphs for more than 45 generic families of plastics are contained in these chapters.

The data in each chapter is generally organized by manufacturer and their product number. Most of the fatigue data is in graphical form. While there are a lot of graphical tribology charts, there are many more tables of tribology properties.

Some data from the first edition has been removed. Removed data includes discontinued products, product names, and manufacturers have been updated.

Laurence W. McKeen
2009

1 Introduction to Fatigue and Tribology of Plastics and Elastomers

1.1 Introduction to Fatigue

There are two recently published books on the properties of engineering plastics in this series. The *Effect of Temperature and Other Factors on Plastics*[1] discusses the general mechanical properties of plastics. The mechanical properties as a function of temperature, humidity, and other factors are presented in graphs or tables. That work includes hundreds of graphs of stress versus strain, modulus versus temperature, impact strength versus temperature, etc. Time was not a factor in that book. *The Effect of Creep and Other Time Related Factors on Plastics*[2] discusses the long-term behavior of plastics when exposed to constant stresses or strains for long periods of time. This book adds another two layers of plastics performance criteria, fatigue, and tribology.

This book provides graphical multipoint data and tabular data on fatigue and tribological properties of plastics and elastomers. This first chapter deals with the types of stress and an introduction to fatigue. Tribology is discussed in Chapter 2. The chemistry of plastics follows in Chapter 3. The remaining chapters contain the data.

The idea of fatigue is very simple. If an object is subjected to a stress or deformation, and it is repeated, the object becomes weaker. This weakening of plastic material is called fatigue and occurs when the material is subject to alternating stresses over a long period of time.

1.2 Types of Stress

As noted in Section 1.1, fatigue occurs as a result of rapidly changing stress or strain. Stress and strain can be applied in a number of ways. *Normal stress* (σ) is the ratio of the applied force (F) over the cross-sectional area (A) as shown in Equation 1.1 and Figure 1.1.

$$\sigma = \frac{F}{A} \tag{1.1}$$

1.2.1 Tensile and Compressive Stress

When the applied force is directed away from the part, as shown in Figure 1.1, it is a *tensile force* inducing a *tensile stress*. This is also called a *normal stress* as it is applied perpendicularly. When the force is applied toward the part, it is a *compressive force* inducing a *compressive stress*.

1.2.2 Shear Stress

A *shear stress* (τ) is defined as a stress which is applied parallel or tangential to a face of a material as shown in Figure 1.2. The shear force is applied parallel to the cross-sectional area "A".

Shear stress is also expressed as force per unit area as in Equation 1.2.

$$\tau = \frac{F}{A} \tag{1.2}$$

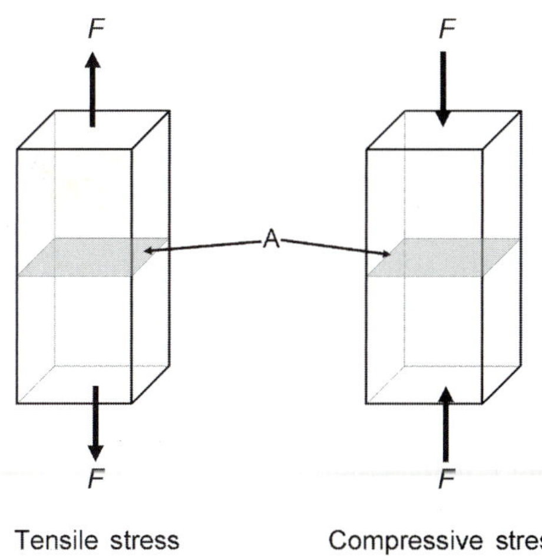

Figure 1.1 Illustration of tensile stress and compressive stress.

Copyright © 2010 Laurence W. McKeen. All rights reserved.

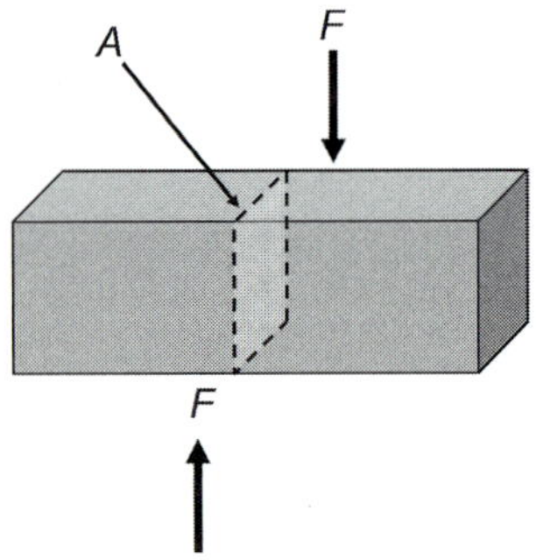

Figure 1.2 Illustration of shear stress.

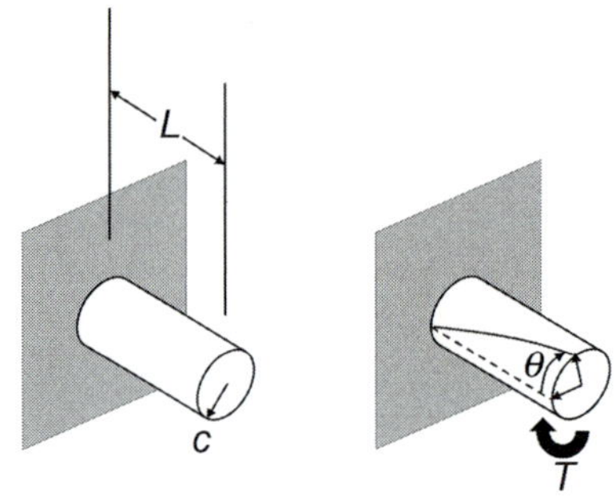

Figure 1.3 Illustration of torsional stress.

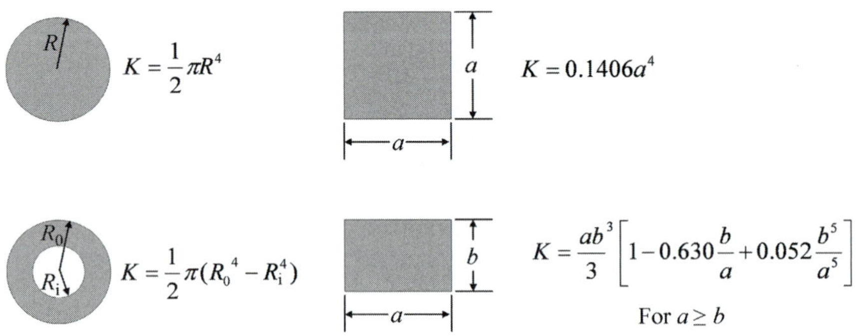

Figure 1.4 Torsional constants for rods or beams of common geometries.

1.2.3 Torsional Stress

Torsional stress (τ) occurs when a part such as a rod for a shaft is twisted as in Figure 1.3. This is also a shear stress, but the stress is variable and depends how far the point of interest is from the center of the shaft. The equation describing torsional stress is shown in Equation 1.3.

$$\tau = \frac{Tc}{K} \tag{1.3}$$

In this equation, T is the torque and c is the distance from the center of the shaft or rod. K is a *torsional constant* that is dependent on the geometry of the shaft, rod, or beam. The torque (T) is further defined by Equation 1.4, in which θ is the angle of twist, G is the modulus of rigidity (material dependent), and L is the length.

$$T = \frac{\theta K G}{L} \tag{1.4}$$

The torsional constant (K) is dependent upon geometry and the formulas for several geometries are shown in Figure 1.4. Additional formulas for torsional constant are published.[3]

1.2.4 Flexural or Bending Stress

Bending stress or *flexural stress* commonly occurs in two instances, shown in Figure 1.5. One is called a simply supported structural beam bending and the other is called cantilever bending. For the simply supported structural beam, the upper surface of the bending beam is in compression and the bottom surface is in tension. The neutral axis (NA) is a region of zero stress. The bending stress (σ) is defined by Equation 1.5. M is the bending moment, which is calculated by multiplying a force by the distance between that point of interest and the force. c is the distance from the neutral axis (N.A. in Figure 1.5) and I is the moment of inertia. The cantilevered beam configuration is also shown in Figure 1.5 and has a similar formula. The formulas for M, c, and I can be complex, depending on the exact configuration and beam shape, but many are published.[3]

$$\sigma = \frac{Mc}{I} \tag{1.5}$$

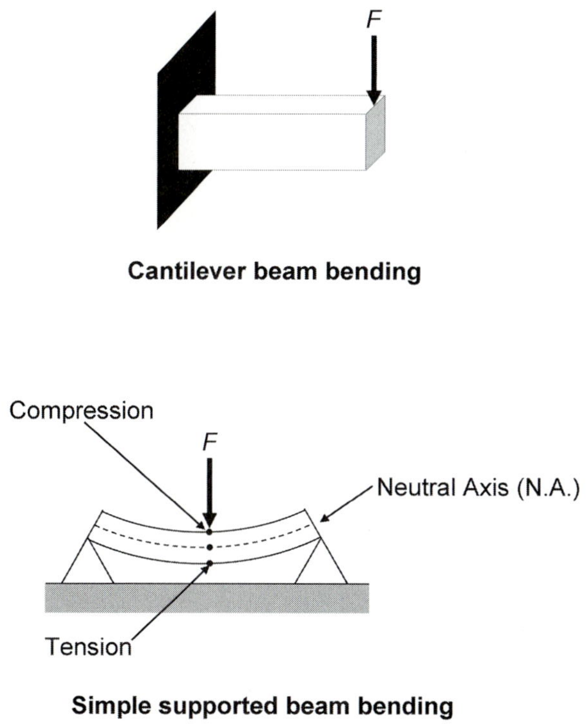

Figure 1.5 Illustration of flexural or bending stress.

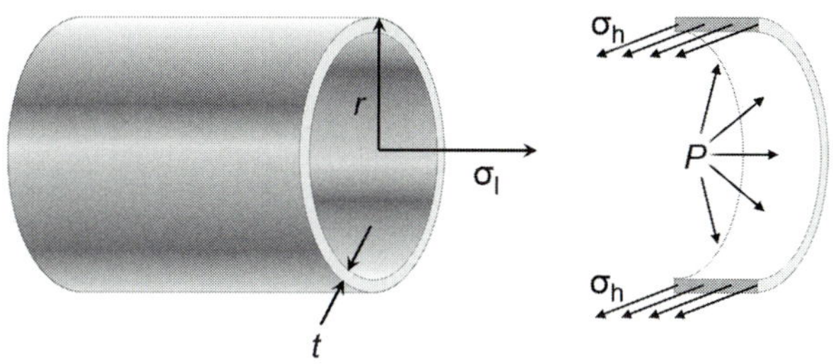

Figure 1.6 Illustration of hoop stress.

1.2.5 Hoop Stress

Hoop stress (σ_h) is mechanical stress defined for rotationally symmetric objects such as pipe or tubing. The real-world view of hoop stress is the tension applied to the iron bands, or hoops, of a wooden barrel. It is the result of forces acting circumferentially. Figure 1.6 shows stresses caused by pressure (P) inside a cylindrical vessel. The hoop stress is indicated in the right-hand side of Figure 1.6 that shows a segment of the pipe.

The classic equation for hoop stress created by an internal pressure on a thin wall cylindrical pressure vessel is given in Equation 1.6.

$$\sigma_h = \frac{Pr}{t} \tag{1.6}$$

where P = the internal pressure, t = the wall thickness, and r = the radius of the cylinder.

The SI unit for P is the Pascal (Pa), while t and r are in meters (m).

If the pipe is closed on the ends, any force applied to them by internal pressure will induce an *axial or longitudinal stress* (σ_l) on the same pipe wall. The longitudinal stress under the same conditions of Figure 1.6 is given in Equation 1.7.

$$\sigma_l = \frac{\sigma_h}{2} \tag{1.7}$$

There could also be a radial stress especially when the pipe walls are thick, but thin walled sections often have negligibly small *radial stress* (σ_r). The stress in radial direction at a point in the tube or cylinder wall is shown in Equation 1.8.

$$\sigma_r = \left(\frac{a^2 P}{b^2 - a^2}\right)\left(1 - \frac{b^2}{r^2}\right) \tag{1.8}$$

where P = internal pressure in the tube or cylinder, a = internal radius of tube or cylinder, b = external radius of tube or cylinder, r = radius to point in tube where radial stress is calculated.

Often the stresses in the pipe are combined into a measure called *equivalent stress*. This is determined using the Von Mises equivalent stress formula which is shown in Equation 1.9.

$$\sigma_e = \sqrt{\sigma_l^2 + \sigma_h^2 - \sigma_l \sigma_h + 3\tau_c^2} \tag{1.9}$$

where σ_l = longitudinal stress, σ_h = hoop stress, and τ_c = tangential shear stress (from material flowing through the pipe).

Failure by fracture in cylindrical vessels is dominated by the hoop stress in the absence of other external loads as it is the largest principal stress. Failure by yielding is affected by an equivalent stress that includes hoop stress and longitudinal stress. The equivalent stress can also include tangential shear stress and radial stress when present.

1.3 Fatigue Testing

There are many machines that have been designed to put a periodic stress or strain on a test coupon or specimen. While the details of these machines vary, they really fall into similar designs. This section will first present several basic fatigue test machine designs. Machines can be designed to put a cycling stress or a strain on the test coupon. The strain is a fixed displacement (% or mm/mm) and the stress is a pressure (MPa).

1.3.1 Tensile Eccentric Fatigue Machine

Many of the machines apply the stress or strain based on a circular drive mechanism and so they are called *eccentric machines*. One such machine for tensile and compressive testing is shown in Figure 1.7. This machine may compress and extend a test specimen repeatedly (Figure 1.8).

The stress and strain in eccentric machines vary in a sinusoidal manner as depicted in Figure 1.9. This shows the change in stress or strain versus time. There are several descriptive parameters noted on this figure that are useful in specifying or describing the test conditions.

The terms and symbols are:

L = Cycle, one full oscillation of the loading (stress or strain), almost always assumed to be constant

f = Cycle frequency; number of cycles per unit time in Hz (1/s)

N = Number of cycles

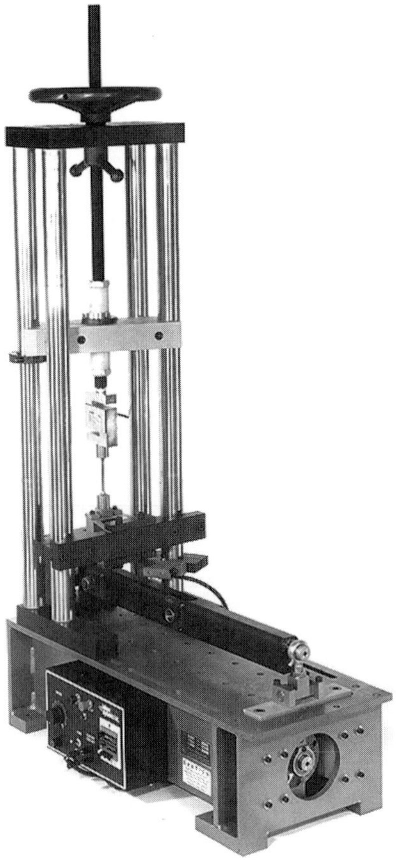

Figure 1.8 Photograph of an eccentric machine for tensile and compressive oscillation fatigue tests (photo courtesy of Fatigue Dynamics, Inc.).

σ_o = maximum stress, highest absolute stress value

σ_u = minimum stress, lowest absolute stress value

σ_m = mean stress = $0.5\,(\sigma_o + \sigma_u)$

σ_a = stress amplitude = $\pm 0.5\,(\sigma_o - \sigma_u)$

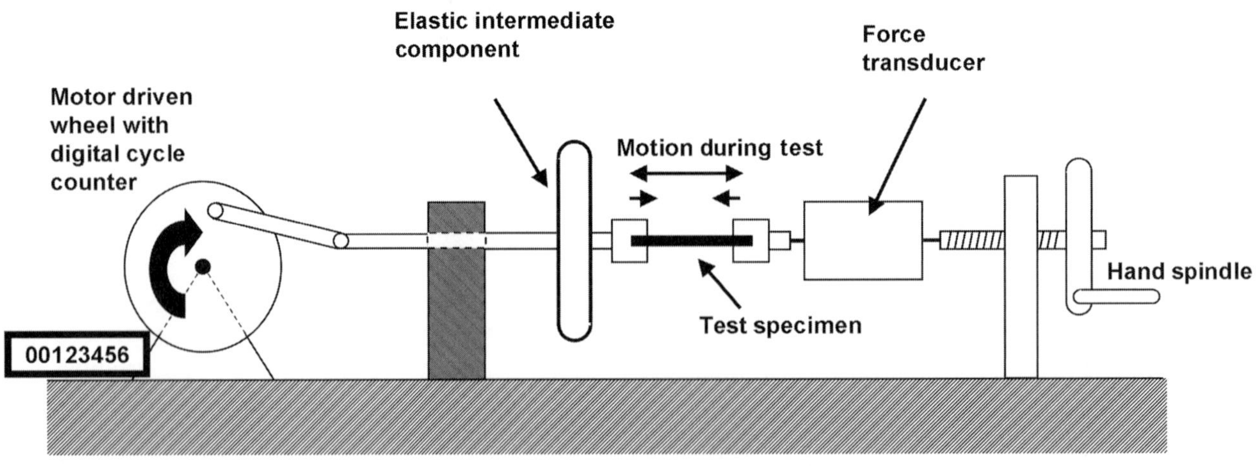

Figure 1.7 Illustration of an eccentric machine for tensile and compressive oscillation fatigue tests.

ε_o = maximum strain (displacement), highest absolute strain value

ε_u = minimum strain (displacement), lowest absolute strain value

ε_m = mean strain (displacement) = 0.5 ($\varepsilon_o + \varepsilon_u$)

ε_a = strain (displacement) amplitude = ±0.5 ($\varepsilon_o - \varepsilon_u$)

The *mean stress*, σ_m, or the *mean strain*, ε_m, is not always zero. A range of values is possible as shown in Figure 1.10. Curves A, D, and F are most common testing conditions. The simplest is the reversed stress cycle, Curve D. This is a sine wave where the maximum stress and minimum stress magnitudes are equal except that they differ by a negative sign.

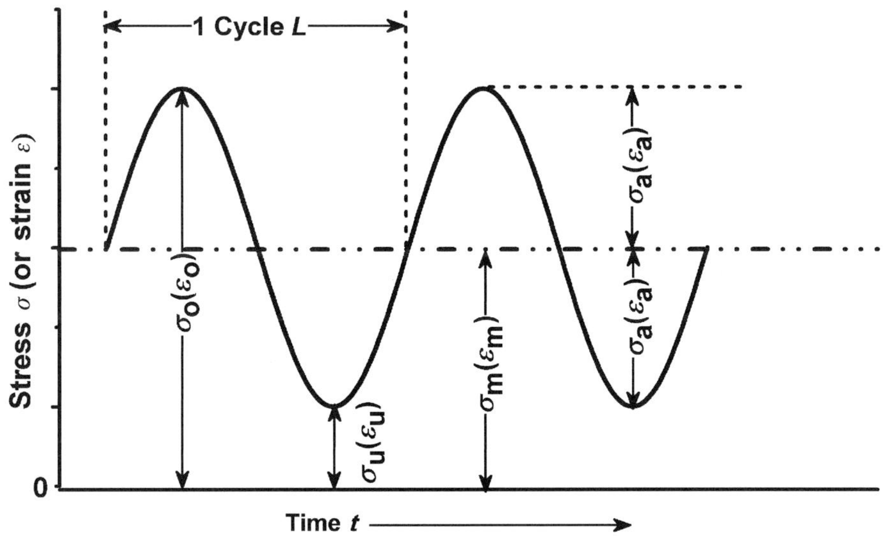

Figure 1.9 Illustration of the cyclic nature of the stress or strain with terms and symbols induced by eccentric tests machines.

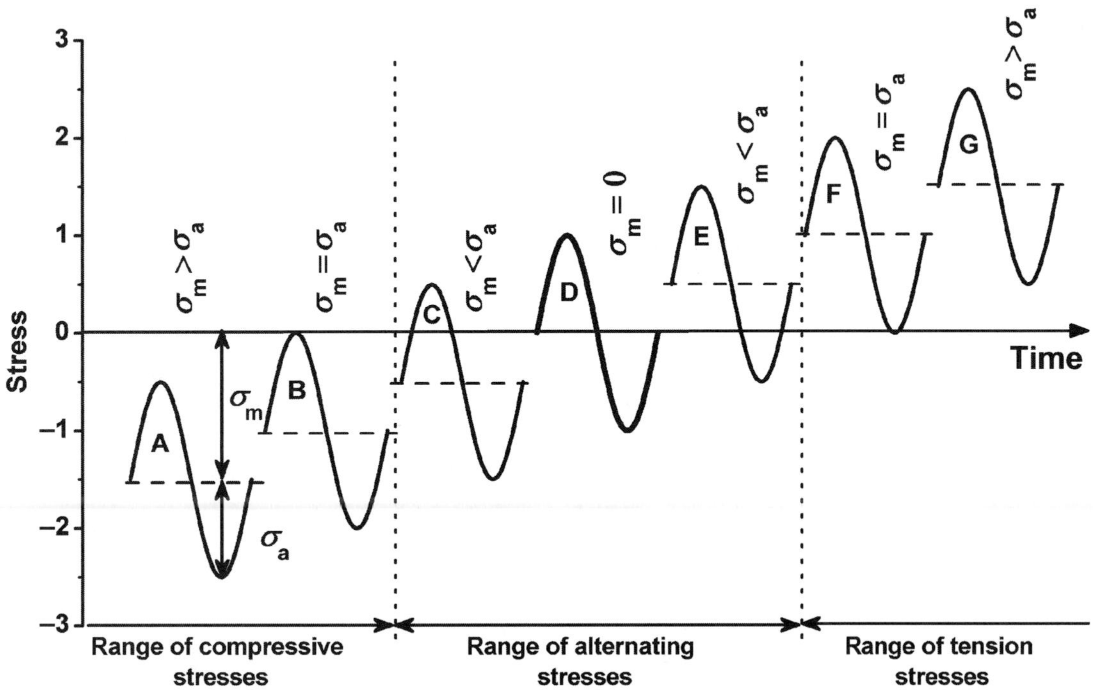

Figure 1.10 Illustration of the cyclic nature of the stress or strain and the ranges of mean stress offset (σ_m).

A real-world example of this type of stress cycle would be in an axle, in which every half turn the stress on a point would be reversed. The most common types of cycle found in engineering applications are the other curves where the maximum and minimum stresses are asymmetric, not equal and opposite. This type of stress cycle is called repeated stress cycle.

The *stroke set* on the rotating wheel on the eccentric unit controls the strain/stress amplitude for the oscillation test. The mean stress is set using the hand spindle shown in Figures 1.7 and 1.8. The cycle frequency is controlled by the rotational speed of the wheel. The frequency is often kept relatively low to minimize sample heating during the test. The mean and minimum stress can be set by adjusting the fixed clamping device. The stress amplitude may decrease during the test, which is caused by relaxation and heating. Correcting stress amplitude for this decrease required increasing the eccentric stroke when the machine is turned off. To avoid any interruption to the test, an elastic intermediate component is incorporated in the test setup as shown in the figure, which considerably reduces the stress reduction, since its spring travel is greater than that of the plastic. This allows the machine to operate with quasiconstant stress values.

1.3.1.1 Fatigue Coupons

The test specimens are usually molded bars or rods that are further machined to specific shapes and configurations. ASTM International, originally known as the American Society for Testing and Materials (ASTM), is one organization that defines standard tests; its standards are the well-known ASTM standards. ASTM E606 describes fatigue specimens as shown in Figures 1.11 and 1.12. Figure 1.11 shows specimens that are made from molded sheet or bars. The test area is primarily in the center of these pieces. Specimen (a) in Figure 1.11 has a rectangular cross section while specimen (b) is circular.

Specimens made from molded rods are shown in Figure 1.12. The rods in this figure do not show the

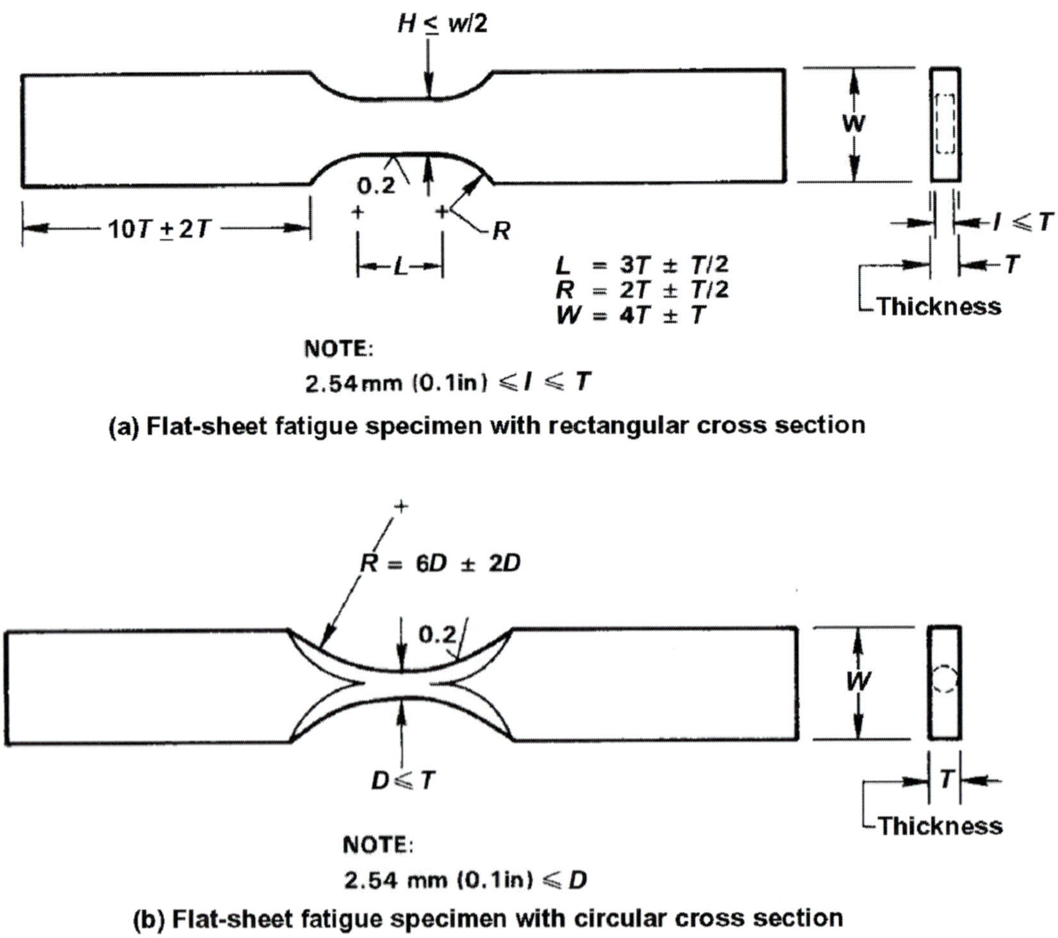

Figure 1.11 Illustration of typical molded flat sheet fatigue testing specimens.

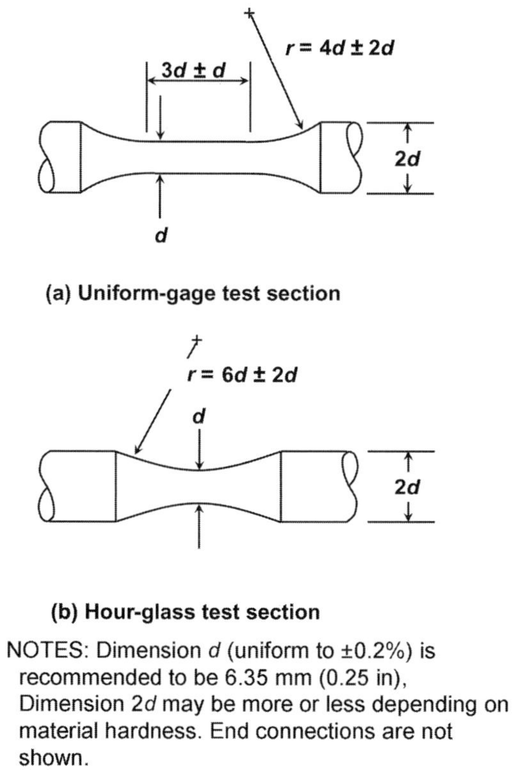

Figure 1.12 Illustration of typical molded rod fatigue testing specimens.

clapping options, of which there are many, to secure the specimen to the test machine.

1.3.1.2 Fatigue Testing Method

Usually a minimum of six identical testing specimens are made for testing. A specimen is tested first at the highest stress or strain amplitude. It is tested until it fails (breaks). The stress/strain amplitude is recorded along with the cycles it took to fail. Because of the variability in the test, measurements are usually replicated a second or third time at the same stress/strain amplitude. Next the stress/strain is reduced and the test is run till failure, which of course takes longer. The reduction in stress or strain continues until failure does not occur in 10^6–10^7 cycles.

A specimen may fail to break even with at the highest stress or strain. In cases such as these, it is necessary to specify the number of cycles up to a particular level of material damage (e.g., 20% stress reduction if strain is controlled, or 20% increase in strain if stress is controlled) instead of the number of cycles to failure.

The temperature of the specimen is monitored while testing as the specimen may heat up during the test. This process is referred to as *hysteretic heating*. Temperature is measured at the surface either by thermocouples that are attached to the surface or by noncontact infrared thermometers. Figure 1.13 and

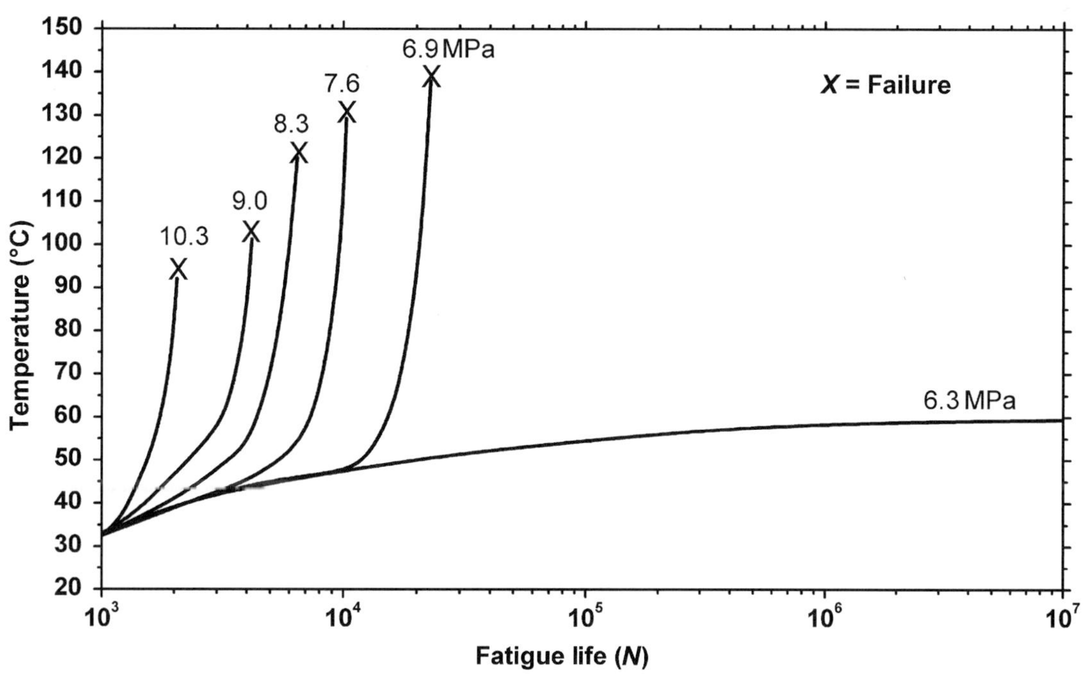

Figure 1.13 Measured temperature of PTFE samples undergoing fatigue testing at various constant stress levels at 30 Hz.

Table 1.1 show examples of heating in polytetrafluoroethylene (PTFE) during fatigue testing.

The data shown in Figure 1.13 and Table 1.1 do show the effect of temperature rise, and that it is most significant as the material being tested approaches failure.

Frequency also affects fatigue testing because it also contributes to hysteretic temperature rise. An example of this is shown in Figure 1.14. The significance of this curve is explained in a later section.

Most fatigue tests are conducted at room temperature with a cycle frequency, f, of 7 Hz, but the cycle frequency may be adjusted to minimize temperature rise and reduce testing time. Fatigue tests in all three loading ranges (compressive, alternating, and tension) can be conducted on the eccentric testing machines. These machines impart a constant strain. The measured imparted stress amplitude may become smaller with an increasing number of cycles if the specimen relaxes and heats up.

Table 1.1 Measured Temperature at Failure of PTFE Samples Undergoing Fatigue Testing at Various Constant Stress Levels at 30 Hz[4]

Stress (MPa)	*N* (Cycles)	Temperature (°C)
10.3	2×10^3	100
9.0	4×10^3	115
8.3	6.1×10^3	125
7.6	9.5×10^3	130
6.9	1.9×10^4	141
6.3	1×10^7	60

The plotting and analysis of the data are discussed in a later section of this chapter.

1.3.2 Flexural Eccentric Fatigue Machine

An eccentric machine for flexural fatigue testing is shown in Figures 1.15 and 1.16. The stroke on this type of flexural unit imposes a constant bending radius on the specimen during the fatigue test at the axis of rotation in the figure. The guide springs under the right-hand clamping unit permit the specimen to move in the longitudinal direction which reduces the additional tensile forces that would otherwise develop during bending.

1.3.3 Cantilevered Beam Eccentric Flexural Fatigue Machine

The cantilevered beam flexural fatigue machine is similar to the machine shown in Figure 1.17 except that the test specimen is fixed and immovable at one end. ASTM D671 describes this test. Test specimen

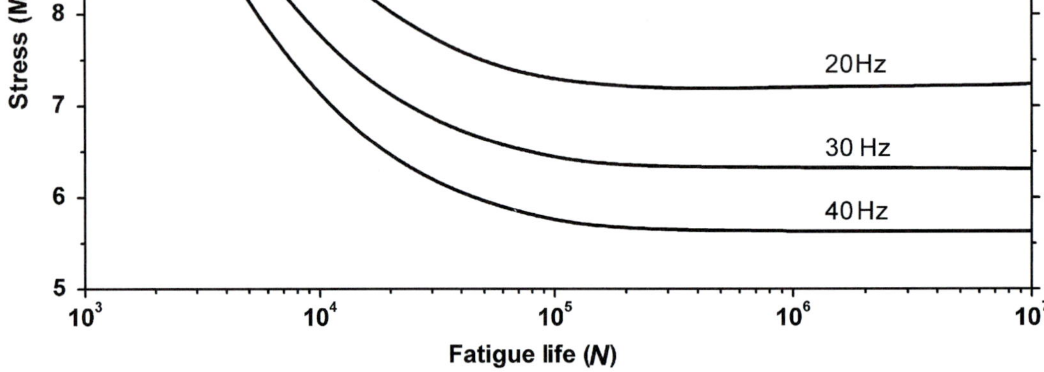

Figure 1.14 The effect of testing frequency on the fatigue properties of PTFE.

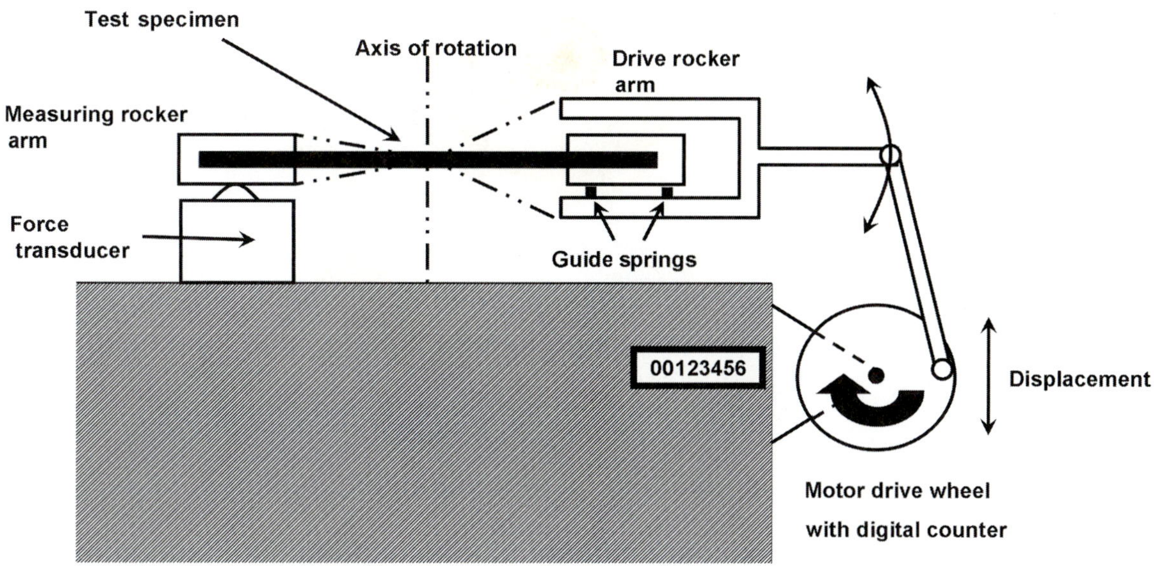

Figure 1.15 Illustration of an eccentric machine for flexural oscillation fatigue tests.

Figure 1.16 Photograph of an eccentric machine for flexural oscillation fatigue tests (photo courtesy of Fatigue Dynamics, Inc.).

is supported as a cantilevered beam and is subjected to an alternating force at one end as shown in Figure 1.18. The alternating applied stress and the cycles to failure are recorded. The "odd" triangular shape of the test specimen is designed to produce a constant stress along the length of the test section of the specimen. The machine that is used to perform this test is shown in Figures 1.18 and 1.19.

1.3.4 Servohydraulic, Electrohydraulic, or Pulsator Fatigue Testing Machines

Servohydraulic, electrohydraulic, or pulsator fatigue testing machines do not use an eccentric wheel to apply cyclic stress and strain. These machines use a computer-controlled hydraulic drive or pulsator to apply the varying stress or strain to the test specimen. This is particularly important because some real-world cycle modes have stress level, and frequency varies randomly. Wave forms do not need to be limited to sine waves either. A real-world example of this situation would be the simulation of the function of automobile shocks, where the frequency magnitude of imperfections in the road will produce varying minimum and maximum stresses. Figure 1.20 shows a picture of a servohydraulic machine. These machines may apply compressive, tensile, or flexural loads. The pulsator in Figure 1.20 is located at the top. The machines are usually run in a force- or stress-controlled manner. This particular example includes an environmental chamber which can be used to control the temperature, humidity, and atmosphere that the fatigue test will take place in.

This machine can be run in a pulsating bending or flexing mode by utilizing a testing stage such as that shown in Figure 1.21. The specimen is supported by roller bearings and is subjected to three-point bending. The bearings minimize the generated tensile stresses.

1.3.5 MIT Flex Life Machine

The MIT Flex Test is used to measure the ability of plastic films to withstand fatigue from flexing. This test method is described in ASTM Standard

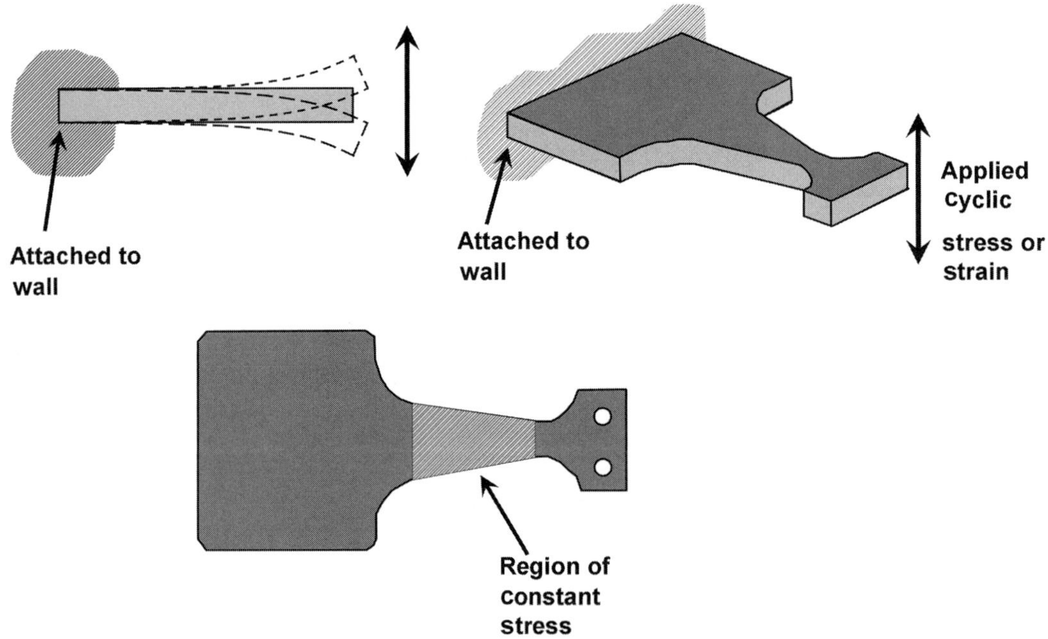

Figure 1.17 Illustration of cantilevered fatigue testing specimens per ASTM D671.

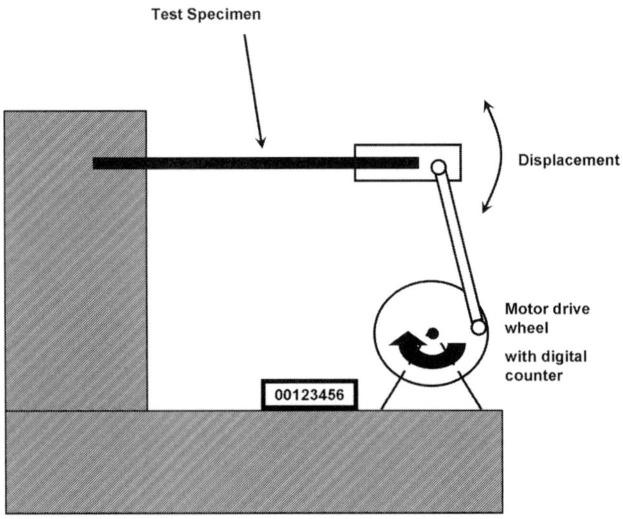

Figure 1.18 Diagram of a cantilevered fatigue testing machine.

Figure 1.19 Photograph of a cantilevered fatigue testing machine (photo courtesy of Fatigue Dynamics, Inc.).

D-2176-69, which is the standard method for testing the endurance of paper with the MIT test apparatus. A diagram of the flex held in the MIT flex tester is shown in Figure 1.22. One end of the plastic test film is clamped in a holder that rotates through 270° very rapidly. The other end is pulled with a constant stress. Even though the ASTM standard describes a paper test, it can be applied to any thin film plastic. It is often used in the evaluation of wire plastic insulation. This test may also help provide insight into the effect of tensioning on life. Flex testing is occasionally performed after exposing the plastics to heat and/or chemicals in order to simulate in use exposure conditions. The flex life is the number of cycles before the film breaks.

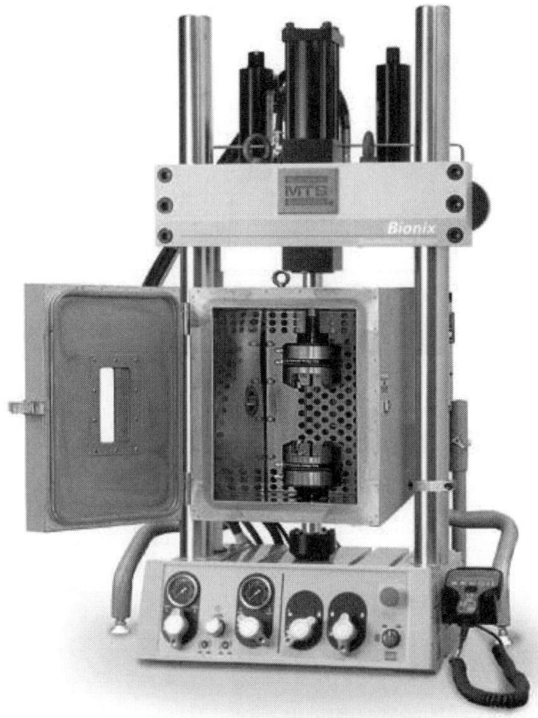

Figure 1.20 Photograph of a servohydraulic fatigue testing machine with environmental chamber (photo courtesy of MTS Systems Corporation © 2009).

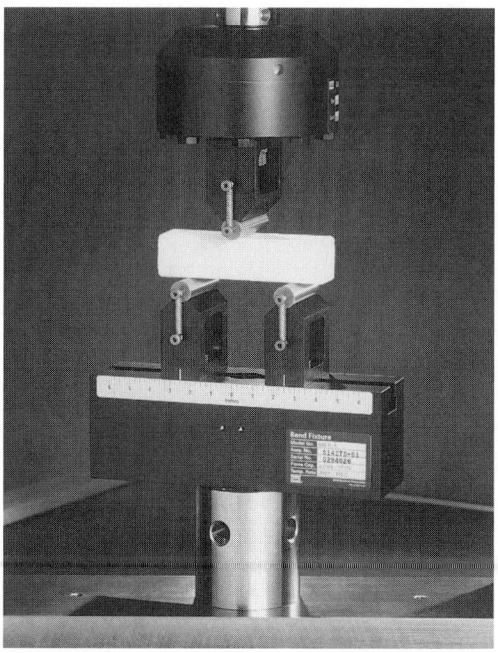

Figure 1.21 Photograph of a flexural test rig used in a servohydraulic fatigue testing machine (photo courtesy of MTS Systems Corporation © 2009).

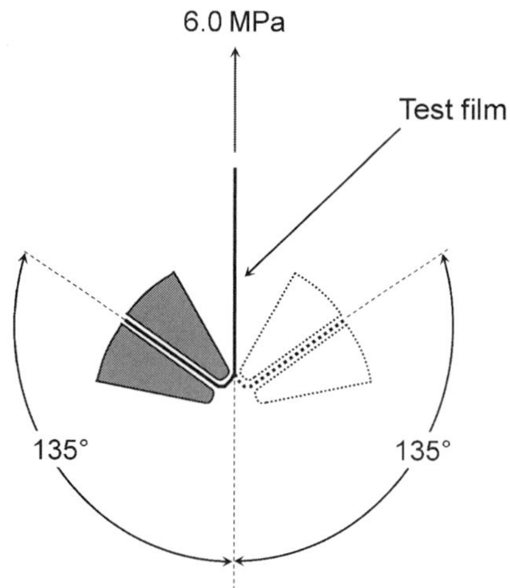

Figure 1.22 Illustration of mode of operation of the MIT Flex Life tester.

1.3.6 Fatigue and Fracture Standards

There are many testing or standard agencies that have standards concerning fatigue and fracture. Some of them include:

- ASTM—ASTM International, originally known as the American Society for Testing and Materials
- ISO—ISO (International Organization for Standardization)
- DIN—Deutsches Institut für Normung.-German Institute for Standardization
- ANSI—American National Standards Institute
- JIS—Japanese Industrial Standards
- SAE—Society of Automotive Engineers

Tables 1.2–1.5 list many, but not all, of the test standards. The individual organizations should be contacted for the details of these tests.

1.4 Understanding Fatigue Testing Data

This section develops an understanding of what happens to a specimen during fatigue testing and describes various ways of reporting fatigue testing results.

Table 1.2 ASTM Fatigue Related Standards

Standard Designation	Standard Title
E467-08	Standard Practice for Verification of Constant Amplitude Dynamic Forces in an Axial Fatigue Testing System
E1942-98(2004)	Standard Guide for Evaluating Data Acquisition Systems Used in Cyclic Fatigue and Fracture Mechanics Testing
E2208-02	Standard Guide for Evaluating Non-Contacting Optical Strain Measurement Systems
E2443-05	Standard Guide for Verifying Computer-Generated Test Results Through the Use of Standard Data Sets
E647-08	Standard Test Method for Measurement of Fatigue Crack Growth Rates
E1457-07e1	Standard Test Method for Measurement of Creep Crack Growth Times in Metals
E1681-03(2008)	Standard Test Method for Determining a Threshold Stress Intensity Factor for Environment-Assisted Cracking of Metallic Materials
E466-07	Standard Practice for Conducting Force Controlled Constant Amplitude Axial Fatigue Tests of Metallic Materials
E468-90(2004)e1	Standard Practice for Presentation of Constant Amplitude Fatigue Test Results for Metallic Materials
E606-04e1	Standard Practice for Strain-Controlled Fatigue Testing
E1922-04	Standard Test Method for Translaminar Fracture Toughness of Laminated and Pultruded Polymer Matrix Composite Materials
E2207-08	Standard Practice for Strain-Controlled Axial-Torsional Fatigue Testing with Thin-Walled Tubular Specimens
E2244-05	Standard Test Method for In-Plane Length Measurements of Thin, Reflecting Films Using an Optical Interferometer
E2245-05	Standard Test Method for Residual Strain Measurements of Thin, Reflecting Films Using an Optical Interferometer
E2246-05	Standard Test Method for Strain Gradient Measurements of Thin, Reflecting Films Using an Optical Interferometer
E2368-04e1	Standard Practice for Strain Controlled Thermomechanical Fatigue Testing
E2444-05e1	Terminology Relating to Measurements Taken on Thin, Reflecting Films
E399-08	Standard Test Method for Linear-Elastic Plane-Strain Fracture Toughness $K\ Ic$ of Metallic Materials
E561-05e1	Standard Test Method for K–R Curve Determination
E740-03	Standard Practice for Fracture Testing with Surface-Crack Tension Specimens
E1221-06	Standard Test Method for Determining Plane-Strain Crack-Arrest Fracture Toughness, K_{Ia}, of Ferritic Steels
E1290-08	Standard Test Method for Crack-Tip Opening Displacement (CTOD) Fracture Toughness Measurement
E1820-08a	Standard Test Method for Measurement of Fracture Toughness
E1921-08ae1	Standard Test Method for Determination of Reference Temperature, T_o, for Ferritic Steels in the Transition Range
E2472-06	Standard Test Method for Determination of Resistance to Stable Crack Extension Under Low-Constraint Conditions
E338-03	Standard Test Method of Sharp-Notch Tension Testing of High-Strength Sheet Materials
E436-03(2008)	Standard Test Method for Drop-Weight Tear Tests of Ferritic Steels
E602-03	Standard Test Method for Sharp-Notch Tension Testing with Cylindrical Specimens

Table 1.2 (Continued)

Standard Designation	Standard Title
E1304-97(2002)	Standard Test Method for Plane-Strain (Chevron-Notch) Fracture Toughness of Metallic Materials
E1823-07a	Standard Terminology Relating to Fatigue and Fracture Testing
E739-91(2004)e1	Standard Practice for Statistical Analysis of Linear or Linearized Stress-Life (S–N) and Strain-Life (ε–N) Fatigue Data
E1049-85(2005)	Standard Practices for Cycle Counting in Fatigue Analysis
D2176-97a(2007)	Standard Test Method for Folding Endurance of Paper by the M.I.T. Tester

Table 1.3 ISO Fatigue Related Standards

Standard Designation	Standard Title
ISO 1099:2006	Metallic materials—Fatigue testing—Axial force-controlled method
ISO 1143:1975	Metals—Rotating bar bending fatigue testing
ISO 12106:2003	Metallic materials—Fatigue testing—Axial-strain-controlled method
ISO 12107:2003	Metallic materials—Fatigue testing—Statistical planning and analysis of data
ISO 12108:2002	Metallic materials—Fatigue testing—Fatigue crack growth method
ISO 13003:2003	Fibre-reinforced plastics—Determination of fatigue properties under cyclic loading conditions
ISO 1352:1977	Steel—Torsional stress fatigue testing
ISO 24999:2008	Flexible cellular polymeric materials—Determination of fatigue by a constant-strain procedure
ISO 27727:2008	Rubber, vulcanized—Measurement of fatigue crack growth rate
ISO 3800:1993	Threaded fasteners—Axial load fatigue testing—Test methods and evaluation of results
ISO 4664-1:2005	Rubber, vulcanized or thermoplastic—Determination of dynamic properties—Part 1: General guidance
ISO 4666-1:1982	Rubber, vulcanized—Determination of temperature rise and resistance to fatigue in flexometer testing—Part 1: Basic principles
ISO 4666-3:1982	Rubber, vulcanized—Determination of temperature rise and resistance to fatigue in flexometer testing—Part 3: Compression flexometer
ISO 4666-4:2007	Rubber, vulcanized—Determination of temperature rise and resistance to fatigue in flexometer testing—Part 4: Constant-stress flexometer
ISO 4965:1979	Axial load fatigue testing machines—Dynamic force calibration—Strain gauge technique
ISO 5999:2007	Flexible cellular polymeric materials—Polyurethane foam for load-bearing applications excluding carpet underlay—Specification
ISO 1099:2006	Metallic materials—Fatigue testing—Axial force-controlled method
ISO 1143:1975	Metals—Rotating bar bending fatigue testing
ISO 12106:2003	Metallic materials—Fatigue testing—Axial-strain-controlled method
ISO 12107:2003	Metallic materials—Fatigue testing—Statistical planning and analysis of data
ISO 12108:2002	Metallic materials—Fatigue testing—Fatigue crack growth method
ISO 13003:2003	Fibre-reinforced plastics—Determination of fatigue properties under cyclic loading conditions
ISO 1352:1977	Steel—Torsional stress fatigue testing

(Continued)

Table 1.3 (Continued)

Standard Designation	Standard Title
ISO 24999:2008	Flexible cellular polymeric materials—Determination of fatigue by a constant-strain procedure
ISO 27727:2008	Rubber, vulcanized—Measurement of fatigue crack growth rate
ISO 3800:1993	Threaded fasteners—Axial load fatigue testing—Test methods and evaluation of results
ISO 4664-1:2005	Rubber, vulcanized or thermoplastic—Determination of dynamic properties—Part 1: General guidance
ISO 4666-1:1982	Rubber, vulcanized—Determination of temperature rise and resistance to fatigue in flexometer testing—Part 1: Basic principles
ISO 4666-3:1982	Rubber, vulcanized—Determination of temperature rise and resistance to fatigue in flexometer testing—Part 3: Compression flexometer
ISO 4666-4:2007	Rubber, vulcanized—Determination of temperature rise and resistance to fatigue in flexometer testing—Part 4: Constant-stress flexometer
ISO 4965:1979	Axial load fatigue testing machines—Dynamic force calibration—Strain gauge technique
ISO 5999:2007	Flexible cellular polymeric materials—Polyurethane foam for load-bearing applications excluding carpet underlay—Specification
ISO 6943:2007	Rubber, vulcanized—Determination of tension fatigue
ISO/AWI 12108	Metallic materials—Fatigue testing—Fatigue crack growth method
ISO/CD 1143	Metallic materials—Rotating bar bending fatigue testing
ISO/CD 12107	Metallic materials—Fatigue testing—Statistical planning and analysis of data
ISO/CD 1352	Metallic materials—Torsional stress fatigue testing
ISO/DIS 12111	Metallic materials—Fatigue testing—Strain-controlled thermomechanical fatigue testing method

Table 1.4 DIN Fatigue Related Standards

Standard Designation	Standard Title
DIN 50113:1982	Testing of metals; Rotating bar bending fatigue test
DIN 50142:1982	Testing of metallic materials; Flat bending fatigue test
DIN EN ISO 5999	Polymeric materials, cellular flexible—Polyurethane foam for load-bearing applications excluding carpet underlay

Table 1.5 JIS Fatigue Related Standards

Standard Designation	Standard Title
K 6265:2001	Rubber, vulcanized and thermoplastic—Determination of temperature rise and resistance to fatigue in flexometer testing
K 7082:1993	Testing method for complete reversed plane bending fatigue of carbon fibre reinforced plastics
K 7083:1993	Testing method for constant-load amplitude tension-tension fatigue of carbon fibre reinforced plastics
K 7118:1995	General rules for testing fatigue of rigid plastics
K 7119:1972	Testing method of flexural fatigue of rigid plastics by plane bending

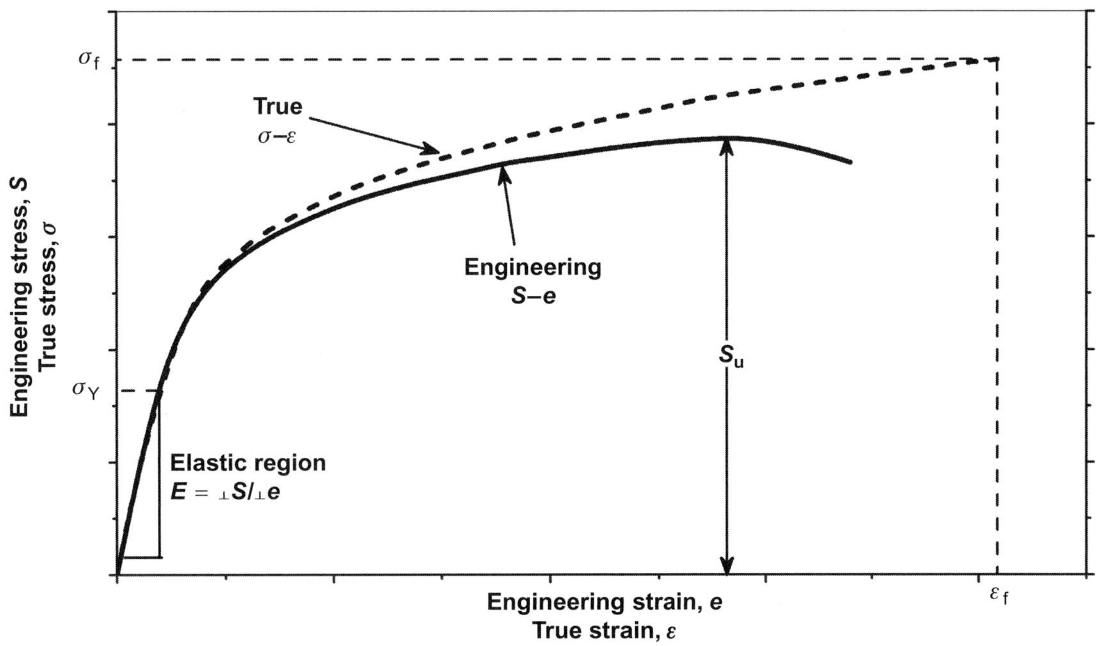

Figure 1.23 Typical monotonic stress–strain curve.

1.4.1 Monotonic Stress–Strain Behavior

Monotonic stress–strain curves such as the one shown in Figure 1.23 are very common. They are used to obtain design parameters for limiting stresses on structures and components subjected to static loading. They are frequently measured at a series of temperatures or strain rates as shown in an earlier book of this series.[1]

The engineering stress–strain curve shown in the figure is obtained by means of a tension test, in which a specimen is subjected to a continually increasing, monotonic load. The elongation of the specimen is measured, and engineering stress and strain values are derived as described below.

The measured engineering stress is the average stress in the specimen, and is given by:

$$S = \frac{P}{A_0} \tag{1.10}$$

where P = applied load, A_0 = unloaded cross-sectional area of the specimen. The engineering strain is the average linear strain obtained from:

$$e = \frac{l - l_0}{l_0} \tag{1.11}$$

where l_0 = original unstrained specimen length and A_0 = strained length.

Keeping in mind that when the specimen is stretched, the cross-sectional area changes, the true stress–strain curve, shown in black, is calculated using the instantaneous length and cross-sectional area, instead of average values. The true stress, σ, is calculated from Equation 1.12 and is always larger than the engineering stress.

$$\sigma = \frac{P}{A} \tag{1.12}$$

where P = applied load and A = true cross-sectional area of the specimen.

The true strain is also calculated by Equation 1.13.

$$\varepsilon = \ln \frac{l}{l_0} \tag{1.13}$$

where l = instantaneous length of the specimen and l_0 = original length of the specimen.

The stress–strain curve ends (at the ultimate tensile strength S_u), when the specimen fails, either by breaking or yielding. If it fails by yielding the specimen necks, it thins nonuniformly as shown in Figure 1.24.

Until failure occurs, true stress and strain are related to engineering stress and strain by Equations 1.14 and 1.15.

$$\varepsilon = \ln(1 + e) \tag{1.14}$$

$$\sigma = S(1 + e) \tag{1.15}$$

Also shown in Figure 1.18 is the true fracture strength which is the true stress at final fracture, and is calculated by Equation 1.16.

$$\sigma_f = \frac{P_f}{A_f} \tag{1.16}$$

where P_f = load at fracture and A_f = measured cross-sectional area at fracture.

The true fracture strain is the true strain at final fracture, and is calculated by:

$$\varepsilon_f = \ln\frac{A_0}{A_f} = \ln\frac{1}{1 - RA} \tag{1.17}$$

where $RA = (A_0 - A_f)/A_0$ (the reduction in cross-sectional area of the specimen).

Figure 1.23 has a region labeled the *elastic region*. This region of the stress–strain curve is linear. In this region, ideally, if the stress is removed, the strain returns back to zero. The deformation is completely reversible. The *modulus of elasticity* or *Young's modulus* is defined by the slope of the stress–strain curve in the elastic region. The end point of the linear elastic region is called the *yield point* or *elastic limit*. The stress at the yield point is called the *yield stress*, σ_Y. The rest of the stress–strain curve beyond the elastic region is called the *plastic region*. The *total true strain* is calculated from the equations above. The true stress–strain curve shown in Figure 1.23 can be approximately modeled by Equation 1.18:

$$\varepsilon_t = \frac{\sigma}{E} + \left(\frac{\sigma}{K}\right)^{1/n} \tag{1.18}$$

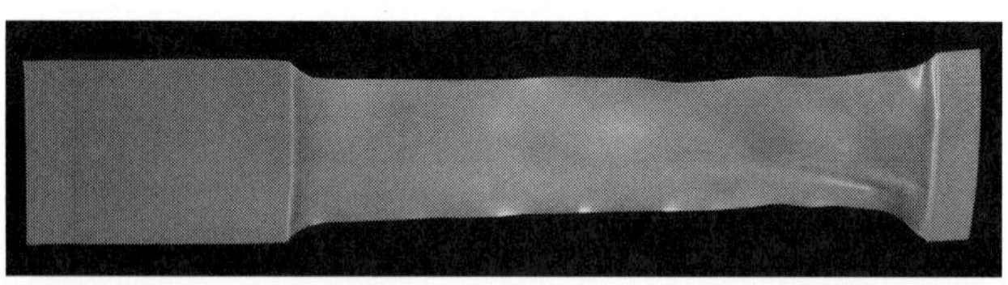

Figure 1.24 Necking in a plastic specimen at failure.

Considered in the next section is what happens when this measurement is reversed cyclically.

1.4.2 Cyclic Stress–Strain Behavior

When the stress–strain measurement shown in Figure 1.23 is reversed at a point after the yield stress, σ_Y, but before failure, ε_f, the stress–strain relationship will initially follow a line with a slope equivalent to the elastic modulus E. This is illustrated by segment A–B in Figure 1.25. If the process were stopped at point B, the length of the specimen does not fully recover to its initial value. However, in this particular example, the specimen is then subjected to a compressive load to σ_{max} to point C in Figure 1.25.

If the loading process shown in Figure 1.25 is reversed again from $-\sigma_{max}$ to $+\sigma_{max}$, then a hysteresis loop will result such as that shown in Figure 1.26. The hysteresis loop defines a single fatigue cycle in the strain-life method. After a number of cycles, the hysteresis loop stabilizes. The stability occurs normally in less than 10% of the total life. The hysteresis loop is

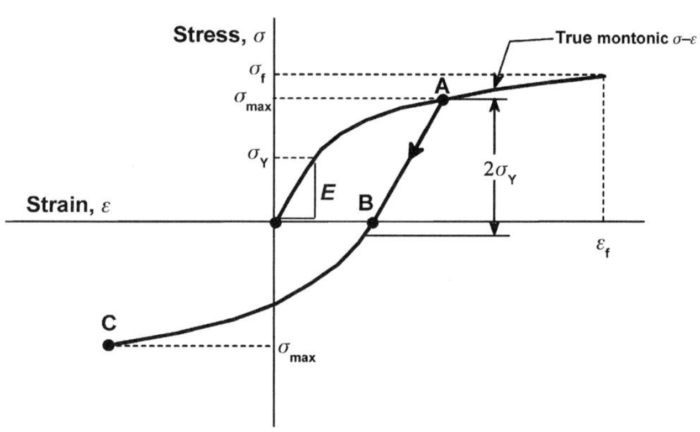

Figure 1.25 Stress–strain behavior after a reversal.

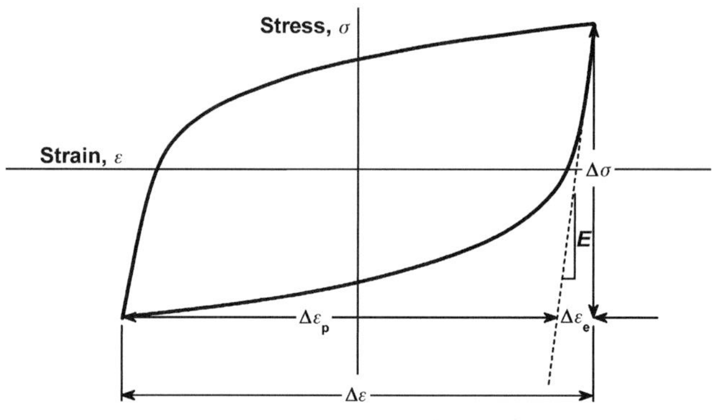

Figure 1.26 Stress–strain behavior of a single fatigue cycle.

often characterized by its stress range, $\Delta\sigma$, and strain range, $\Delta\varepsilon$. The strain range may be split into an elastic part, $\Delta\varepsilon_e$ and a plastic part, $\Delta\varepsilon_p$.

When subjected to strain-controlled cyclic loading, the stress–strain response of a material can change depending upon the number of applied cycles. In plastics, the maximum stress generally decreases with the increase in the number of cycles. The test is typically run to failure of the specimen or some maximum number of cycles, often 1×10^7 cycles.

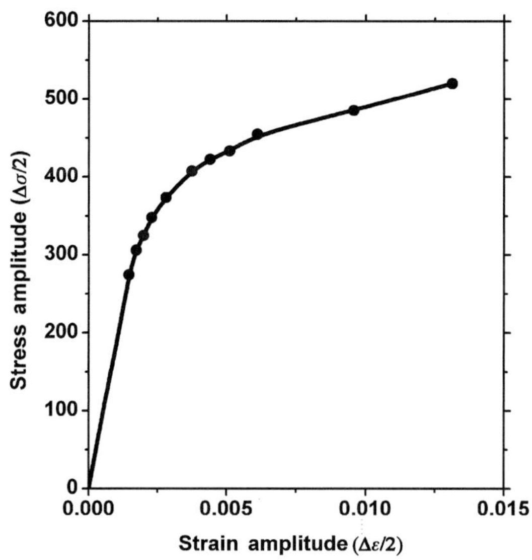

Figure 1.27 A cyclic stress strain curve.

The test is often run at a series of different strain ranges (or strain amplitudes) on new specimens. Each strain range tested will have a corresponding stress range that is measured. This data can be plotted as shown in Figure 1.27 and is called a cyclic stress–strain curve.

The cyclic stress–strain curve is different from the initial behavior that is measured in a traditional tensile test. A power function, Equation 1.19, may be fitted to this curve to obtain three material properties.

$$\frac{\Delta\varepsilon}{2} = \frac{\Delta\sigma}{2E} + \left(\frac{\Delta\sigma}{2K'}\right)^{1/n'} \quad (1.19)$$

where K' = cyclic strength coefficient, n' = cyclic strain hardening exponent, and E = elastic modulus.

1.4.3 Strain-Life Behavior

The cyclic stress–strain measurement can be run until the specimen fails or a maximum number of cycles have been made. These measurements are done with machines that control the strain. The strain range is controlled and the corresponding stress range and fatigue life are measured. When a series of these cyclic stress–strain measurements (to failure) are done at different strain levels, the data may be plotted as shown in Figure 1.28. The data are usually plotted on a log–log plot, with reversals

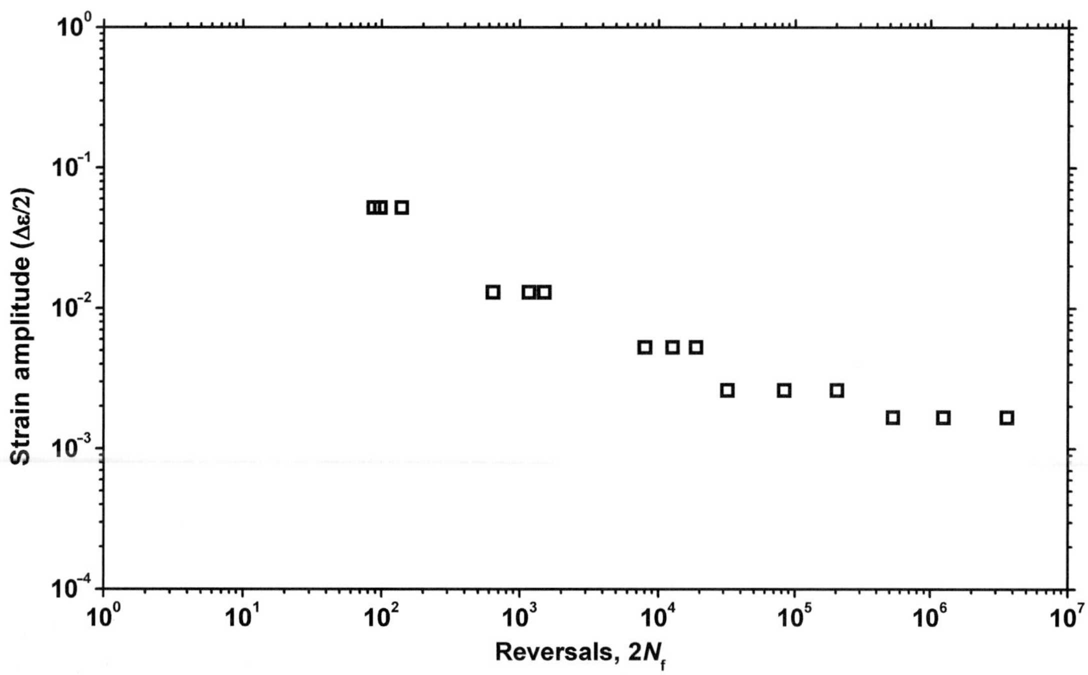

Figure 1.28 A strain-life plot.

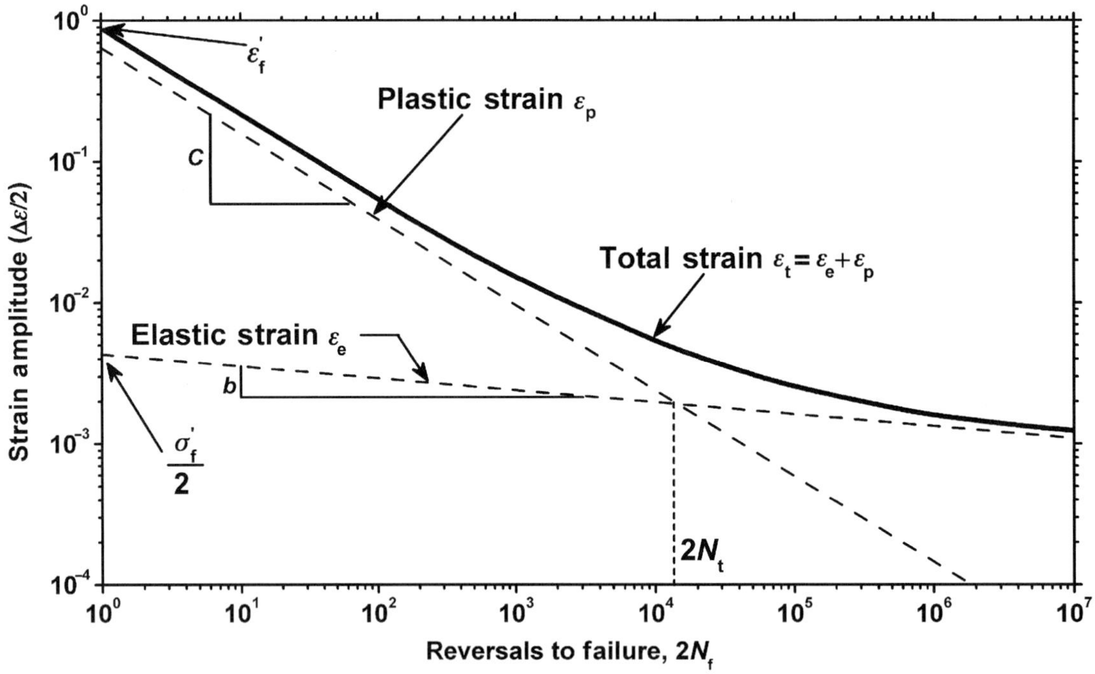

Figure 1.29 A strain life curve modeled.

(note: 2 reversals = 1 cycle) or cycles to failure on the *X*-axis and strain amplitude on the *Y*-axis. As can be deduced from this plot, the data are usually run in duplicate or triplicate at each set strain amplitude.

Separate researchers had noticed that the lower cycle data points could be fit by a straight line and the higher cycle points could be fit by separate straight lines as shown in Figure 1.29.[4–6]

The equation developed for the high-cycle straight line on the log–log strain life plot corresponds to elastic material behavior of the material. The equation developed, shown in Equation 1.20, defines two material parameters.[7]

$$\varepsilon_e = \frac{\sigma_a}{E} = \frac{\sigma'_f}{E}(2N_f)^b \qquad (1.20)$$

where ε_e = the elastic component of the cyclic strain amplitude, E = elastic modulus, σ_a = cyclic stress amplitude, σ'_f = called the fatigue strength coefficient, N_f = number of cycles to failure, and b = called the fatigue strength exponent.

The equation developed for the low-cycle straight line on the log–log strain life plot corresponds to plastic material behavior of the material. The equation developed, shown in Equation 1.21, defines two material parameters.[4,5]

$$\varepsilon_p = \varepsilon'_f(2N_f)^c \qquad (1.21)$$

where ε_p = the plastic component of the cyclic strain amplitude, ε'_f = called the fatigue ductility coefficient, N_f = number of cycles to failure, and c = called the fatigue ductility exponent.

The complete strain-life curve, ε_t, is the sum of the elastic and plastic components, Equation 1.22.

$$\varepsilon_t = \varepsilon_e + \varepsilon_p = \frac{\sigma'_f}{E}(2N_f)^b + \varepsilon'_f(2N_f)^c \qquad (1.22)$$

All of these are summarized in Figure 1.29.

One additional parameter shown on this graph is the transition life, $2N_t$. This represents the life at which the elastic and plastic strain ranges are equivalent and can be expressed by Equation 1.23. The transition life provides an accepted demarcation between low-cycle and high-cycle fatigue regimes.

$$2N_t = \left(\frac{\varepsilon'_f E}{\sigma'_f}\right)^{1/(b-c)} \qquad (1.23)$$

While fatigue data collected in the laboratory are generated using a fully reversed stress cycle, actual loading applications usually involve a nonzero mean stress. The mean stress can be tensile, zero or compressive and it effects the strain-life curve as shown schematically in Figure 1.30. Mean stress has its largest effects in the high-cycle regime. Compressive means extend life and tensile means reduce it.

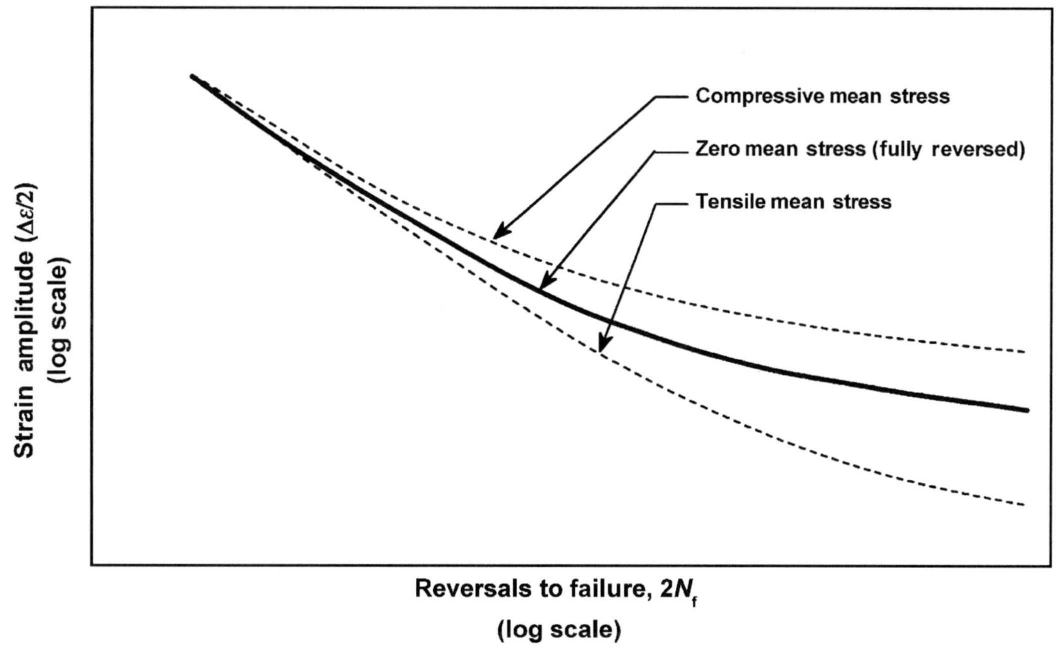

Figure 1.30 The effect of mean stress on the strain-life curve.

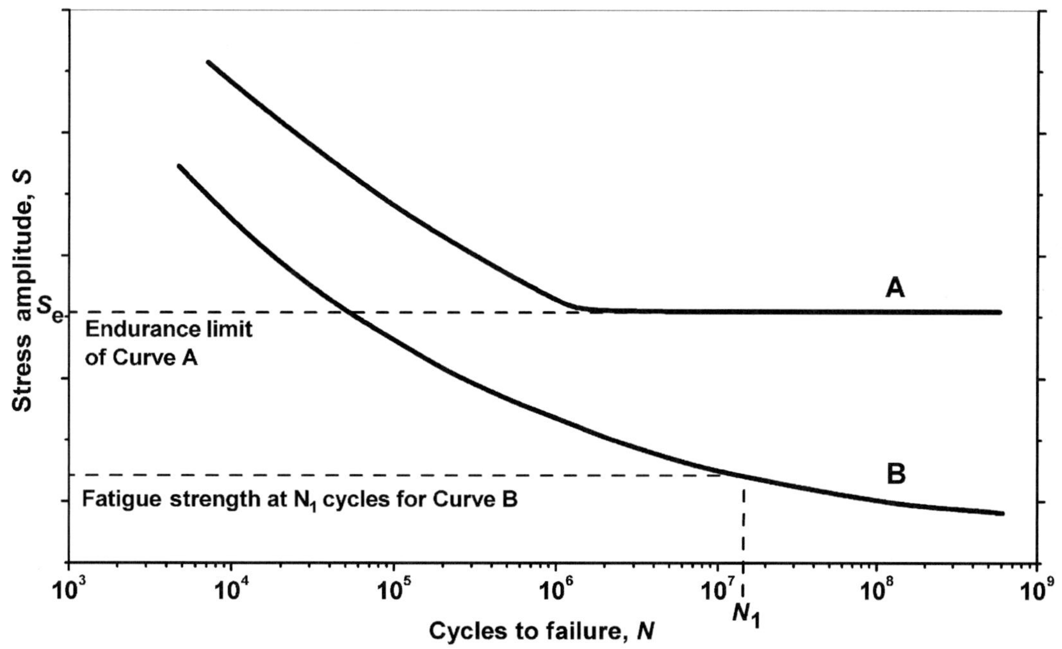

Figure 1.31 Two typical stress-life curves.

1.4.4 Stress-Life Behavior

The most common published fatigue data chart is the stress-life curve which is commonly called an *S–N* curve or a Wöhler[8] curve. This is a graph of the magnitude of a cyclical stress (*S*), linear or log scale, against the cycles to failure (*N*) on a log scale. The cyclic measurement is made under constant oscillatory load amplitude. It is generally applied in high-cycle regimes, where the strain-life behavior is used in low-cycle regimes. Figure 1.31 shows two generic *S–N* curves. Curve A in this figure shows a *fatigue limit*. If the material is loaded below the fatigue limit, it will not fail, regardless of the number of fatigue cycles it experiences. Many materials do not behave in this manner and their *S–N* curve will look more like Curve B in Figure 1.31. *Fatigue strength* is noted on this curve and is defined as the stress amplitude

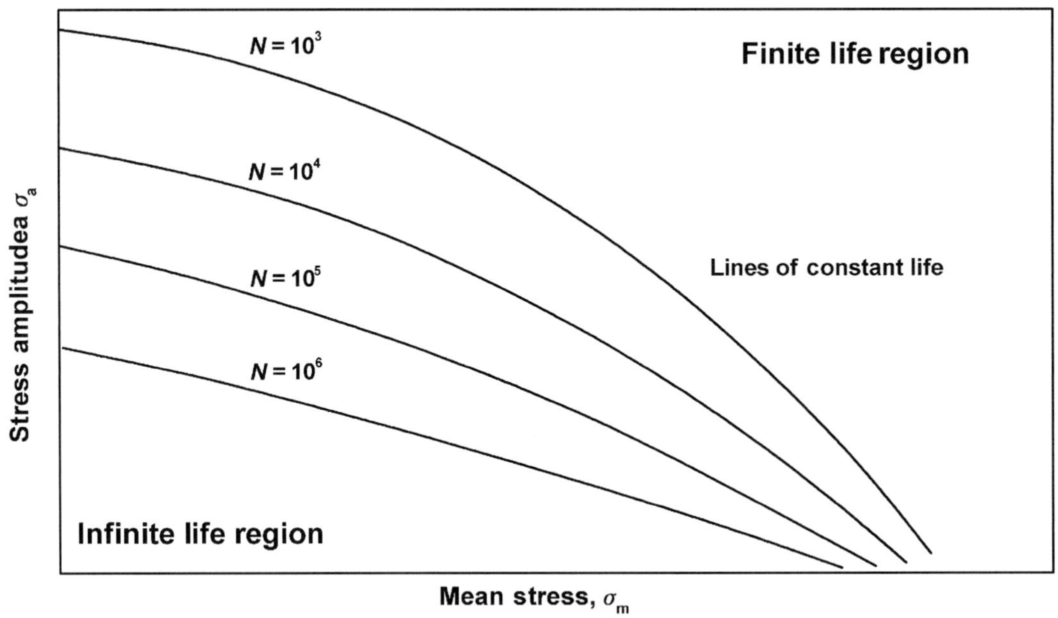

Figure 1.32 A generic Haigh diagram.

at which failure occurs for a given number of cycles. Inversely, *fatigue life* is the number of cycles required for a material to fail at a given stress amplitude.

For those fatigue tested specimens that survive the test through the maximum specified cycle limit without failure, the fatigue damage may still be estimated. The short-term properties may be measured on these specimens (e.g. tensile strength and elongation at break). The ratio of the short-term properties of new (untested) specimens to those of the tested specimens constitutes a measure of the damage suffered in the fatigue test.

Most *S–N* curves are run at zero mean stress. When the fatigue tests are run at a nonzero mean stress, a different plot called a Haigh diagram is often made. The Haigh diagram, as shown in Figure 1.32, plots the mean stress on the *X*-axis versus the stress amplitude on the *Y*-axis. A family of curves is typical with lines drawn at a given life. The region under the lowest curve is called the infinite life region. The finite life region is the region above the curves.

1.5 The Fatigue Process

Failure by fatigue always involves cracking.[9,10] The process may be simplified into three steps:

1. Crack initiation or nucleation
2. Crack growth or propagation
3. Final fracture

1.5.1 Crack Initiation

The initial crack occurs in this stage. The crack may be caused by:

1. cyclic loading
2. surface scratches induced during handling or tooling of the material
3. a defect introduced during manufacture, such as during casting or molding
4. mechanical impact
5. thermal shock, thermal expansion, or contraction
6. chemical attack (such as pitting or corrosion)

1.5.2 Crack Growth or Propagation

Once a crack has started, it continues to grow as a result of continuous applied stresses present under the influence of cyclic loading. If the crack grows to a critical length, then fracture of the component will occur. The rate of the crack growth before it reaches the critical length directly influences fatigue life. Fortunately a mathematical model known as Paris' Law[11] provides a way to predict the crack growth rate.

The stress intensity factor, *K*, is used in fracture mechanics to accurately predict the stress intensity near the tip of a crack in an item caused by a load applied someplace on that item or by residual

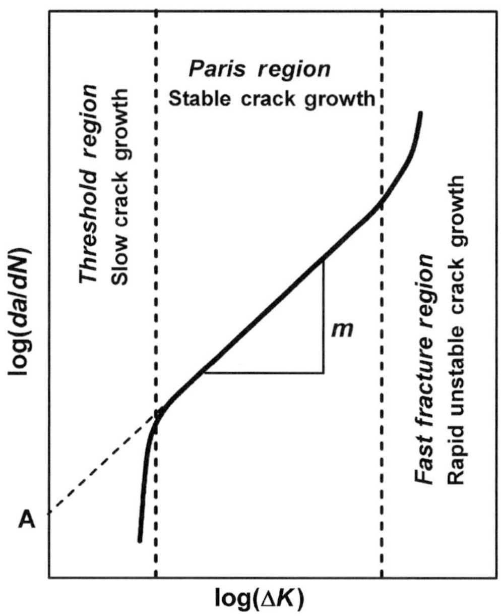

Figure 1.33 A crack growth graph showing three regions.

stresses. The magnitude of K depends on sample geometry, the size and location of the crack, and the magnitude and the distribution of loads on the material. Equation 1.24 shows the calculation of the stress intensity factor.

$$K = Y\sigma\sqrt{\pi a} \tag{1.24}$$

where Y = dimensionless parameter used to account for geometry, σ = uniform tensile stress perpendicular to the plane of the crack, and a = the crack size.

Stress intensity factors have been tabulated for thousands of part and crack geometries.[12]

Paris proposed that the *stress intensity factor range*, ΔK, characterizes subcritical crack growth under fatigue loading, because he found that plots of crack growth rate versus stress intensity factor range gave straight lines on log–log scales. The stress intensity factor range is defined by Equation 1.25.

$$\Delta K = Y\Delta\sigma\sqrt{\pi a} \tag{1.25}$$

The equation of that line is shown in Equation 1.26, where C and m are constants for a given material. Equation 1.26 can be rearranged to remove the logs, giving Equation 1.27.

$$\log\left(\frac{da}{dN}\right) = m \bullet \log(\Delta K) + \log C \tag{1.26}$$

$$\frac{da}{dN} = C\Delta K^m = C(Y\Delta\sigma\sqrt{\pi a})^m \tag{1.27}$$

Integrating this equation from zero to the number of cycles which caused fast fracture, or from initial and final crack size gives Equation 1.28, which became known as the *Paris Law*.

$$\int_0^{N_f} dN = \int_{a_i}^{a_f} \frac{da}{CY^m\Delta\sigma^m(\pi a)^{m/2}} \tag{1.28}$$

It was later realized that the *Paris Law* applied to growth rates in a particular range as shown in Figure 1.33. This figure, a fatigue crack growth rate curve, plots the fatigue crack growth rate against the stress intensity factor range. The lower crack growth rate region is called the *threshold regime*. The *higher growth rate regime* occurs where values of maximum stress intensity in the fatigue cycle and failure approach rapidly. A more detailed information is available in the literature.[13]

1.5.3 Failure

As the crack grows, there is less material available to withstand the applied stress or strain. Failure occurs when the material that has not been affected by the crack cannot withstand the applied stress. This stage happens very quickly. Failure in materials is often classified as *ductile* or *brittle*. Brittle failure occurs in some metals, which experience little or no plastic deformation prior to fracture. Ductile failure shows observable plastic deformation prior to fracture. At times materials behave in a transitional manner—partially ductile/brittle.

Fatigue failure is often classified into two types: high-cycle fatigue and low-cycle fatigue. High-cycle failure is generally classified as failure above 10^4 cycles. In high-cycle fatigue situations, material performance is commonly characterized by the *S–N* curve described in the previous section.

Where the stress is high enough for plastic deformation to occur leading to failure in less than 10^4 cycles, low-cycle fatigue is usually characterized by the Coffin–Manson relation[14,15] given in Equation 1.29.

$$\frac{\Delta\varepsilon_p}{2} = \varepsilon_f'(2N)^c \tag{1.29}$$

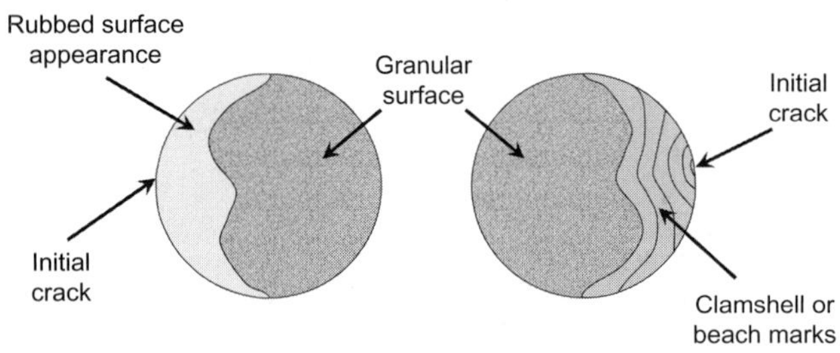

Figure 1.34 A diagram showing the surface of a fatigue fracture.

where $\Delta\varepsilon_p/2$ = the plastic strain amplitude, ε_f' = the fatigue ductility coefficient, the failure strain for a single reversal, $2N$ = the number of reversals to failure, and c = the fatigue ductility exponent.

Examination of the fracture site of material failed by fatigue often shows two distinct regions. One region is smooth or burnished as a result of the rubbing of the bottom and top of the crack during the cyclic action of the stress or strain. The second region appears granular due to the rapid failure of the material. These may be seen in Figure 1.34. The rough, granular surface indicates brittle failure, while the smooth surface represents crack propagation.

Often features of a fatigue fracture are visible, such as beach marks or clamshell marks and striations. Beach marks or clamshell marks may be seen in fatigue failures of materials that are not in continuous use. They may be used for a period of time, allowed to rest and then used again. Striations are thought to be steps in crack propagation. Thousands of striations may be found within each beach mark.

1.6 Factors That Affect Fatigue Life

The following factors are known to affect fatigue life:

- Cyclic stress state—Stress amplitude, mean stress, biaxiality, in-phase or out-of-phase shear stress, and load sequence.
- Geometry—Notches and variation in cross section throughout a part lead to stress concentrations where fatigue cracks can begin.
- Surface quality—Surface roughness can cause microscopic stress concentrations that lower the fatigue strength.
- Material type—Behavior during cyclic loading, varies widely for different materials and is the basis for the data portion of this book.
- Residual stresses—Molding, cutting, machining, and other manufacturing processes involving heat or deformation can produce high levels of tensile residual stress, which decreases the fatigue strength.
- Size and distribution of internal defects—Defects such as gas porosity shrinkage voids can significantly reduce fatigue strength.
- Direction of loading—For nonisotropic materials such as fiber reinforced plastics, fatigue strength depends on the direction of the principal stress.
- Environment—Environmental conditions can cause erosion, oxidation, degradation, and environmental or solvent stress cracking which all affect fatigue life.
- Temperature—Higher temperatures generally decrease fatigue strength.

1.7 Design Against Fatigue

Design against fatigue failure requires thorough education and experience in structural engineering, mechanical engineering, and materials science.[16] This subject is beyond the objectives of this book. To dependably design against fatigue failure, one needs a thorough engineering education and years of experience in engineering and materials science. There are three principal approaches to life assurance for mechanical parts:

1. Design to keep stress below threshold of the material's fatigue limit (sometimes called the infinite lifetime concept). This depends on having enough fatigue data which this book aims to provide.
2. Design for a fixed life and plan to replace the part with a new one much like car manufacturers do with their maintenance schedules. This is sometimes called "a so-called *lifed* part," finite lifetime concept,[17] or "safe-life" design practice.
3. Plan to inspect the part periodically for cracks and replace the part once an observed crack exceeds a critical length. This approach usually requires an accurate prediction of the rate of fatigue crack growth. This sometimes is referred to as damage tolerant design[18] or "retirement-for-cause."

There are other strategies to deal with the factors that accelerate fatigue listed in Section 1.6. Perhaps most important, apart from the material of manufacture, is paying particular attention to the manufacturing process. The aim is to minimize internal and surface defects that concentrate stresses. One can also engineer temperature control and environmental exposure.

1.8 Summary

This chapter has provided a general summary of fatigue concepts, measurement techniques or methods, data presentation, and theory. It was meant to be introductory only and additional details should be obtained from the literature cited in this chapter.[19–21] Chapters 4–12 contain hundreds of plots of fatigue-related data on hundreds of different plastics.

References

1. McKeen LW. *The effect of temperature and other factors on plastics, plastics design library*. Norwich, NY, William Andrew Publishing; 2008.
2. McKeen LW. *The effect of creep and other time related factors on plastics, plastics design library*. Norwich, NY, William Andrew Publishing; 2009.
3. Pilkey WD. *Formulas for stress, strain, and structural matrices*. Hobaken, NJ, 2nd ed. John Wiley & Sons; 2005 pp. 63–76. Online version available at: http://www.knovel.com/knovel2/Toc.jsp?BookID=1429&VerticalID=0.
4. Basquin OH. The exponential law of endurance tests. In: *American Society for Testing and Materials Proceedings*, Vol. 10; 1910.
5. Coffin LF Jr., A study of the effects of cyclic thermal stresses on a ductile metal. New York, NY, *Trans ASME* 1954;**76**:931–50.
6. Manson SS. Behavior of materials under conditions of thermal stress. *Heat Transfer Symposium*, University of Michigan Engineering Research Institute; 1953.
7. Riddell MN, Koo GP, O'Toole JL. *Polym Eng Sci* 1966;**6**:363.
8. Day L. *Biographical dictionary of the history of technology*. London, Routledge; 1995. p. 765.
9. Hertzberg RW, Manson J. *Fatigue of engineering plastics*. New York, NY, Academic Press; 1980.
10. Moalli J. *Plastics failure—analysis and prevention*. Norwich, NY, William Andrew Publishing/Plastics Design Library; 2001, Online version available at: http://knovel.com/web/portal/browse/display?_EXT_KNOVEL_DISPLAY_bookid=382&VerticalID=0.
11. Paris P, Erdogan F. A critical analysis of crack propagation laws. *J Basic Eng. Trans Am Soc Mech Eng* 1963;December:528−534.
12. Murakami Y. *Stress intensity factors handbook*. Oxford, UK, Elsevier Science Ltd; 2003.
13. Suresh S. *Fatigue of materials*. 2nd ed. Cambridge, England: Cambridge University Press; 1998.
14. Coffin LF Jr. *Trans ASME* 1954;**76**:931–50.
15. Manson SS. NACA, TN 2933; 1953.
16. Schijve J. *Fatigue of structures and materials*. 2nd ed. Netherlands: Springer; 2009, pp. 559–586.
17. Ashby MF, Bréchet Y. Materials selection for a finite life time. *Adv Eng Mater* 2002;**4**(6):35–341.
18. Handbook for Damage Tolerant Design. Online handbook by U.S. *Air Force Research Laboratory*. http://www.afgrow.net/applications/DTDHandbook/default.aspx.
19. Jansen J. *Fatigue of plastics*. http://www.4spe.org/online-store/fatigue-plastics; 2006.
20. Harris B. *Fatigue in composites*. Boca Raton, FL, CRC Press; 2003.
21. Lee Y-L, et al. *Fatigue testing and analysis: theory and practice*. Elsevier Butterworth-Heinemann; 2005.

2 Introduction to the Tribology of Plastics and Elastomers

The first chapter of this book was an introduction to fatigue. This chapter is a brief introduction to tribology. There are many texts that deal with this subject in much more detail.[1,2] Tribology is the science and technology of surfaces in contact with each other and therefore covers friction, lubrication, and wear. Tribological properties are most often of concern when the materials are used in bearing applications. Especially when engineering plastics are used for bearing materials, they must have a suitable combination of mechanical and tribological properties under the conditions experienced in use. The three main performance areas that need to be examined for bearing are friction, wear, and limiting PV. However, tribological properties often need to be considered in other applications.

2.1 Friction

Frictional force is not always intuitive. This is apparent when one considers two blocks on a plate as shown in Figure 2.1.

The blocks are of equal mass and surface finish. The block on the right has twice the surface contact area of the other. An equal vertical force (F_n) is applied to each block. Both blocks are made to slide by the application of an equal horizontal force (F). A frictional force (F_s) resists the sliding motion. What most people will find surprising is that the frictional force (F_s) will be the same for both blocks even though the surface contact area is different. The frictional force (F_s) depends only on the vertical applied force (F_n) and is described by Equation 2.1.

$$F_s = \mu \times F_n \tag{2.1}$$

where μ = the coefficient of friction.

The coefficient of friction is a parameter that depends on the combination of block and plate materials. It is approximately 0.5 for many material combinations, but fortunately not for all materials. As it turns out, the coefficient of friction is constant only under a given set of conditions. It can vary with velocity and temperature. There are actually two coefficients of friction for each material pair. The *static coefficient of friction* (μ_s) is determined from the force that is just enough to start the block moving. Once the block is moving, *the dynamic coefficient of friction* (μ_d) is determined from the force that is just enough to keep the block moving. Dynamic coefficient of friction is sometimes called *kinetic coefficient of friction.*

The sliding surfaces do not contact completely over the expected contact area. Even the smoothest surface is "rough" at a microscopic scale as shown in Figure 2.2. At the junction between the two surfaces, the materials only touch over small patches, called "asperities." The asperities support the load and deform (especially for plastics, elastically or plastically) to reach an equilibrium. When the apparent contact area is measured or calculated, it is not the real contact area in tribological terms. The apparent contact area is much larger than the true contact area.

When movement of the block occurs, the asperities rub against one another creating a natural resistance

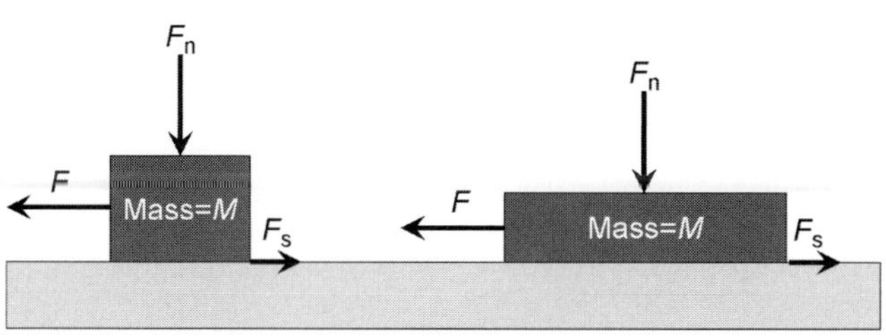

Figure 2.1 Illustration of basic frictional forces.

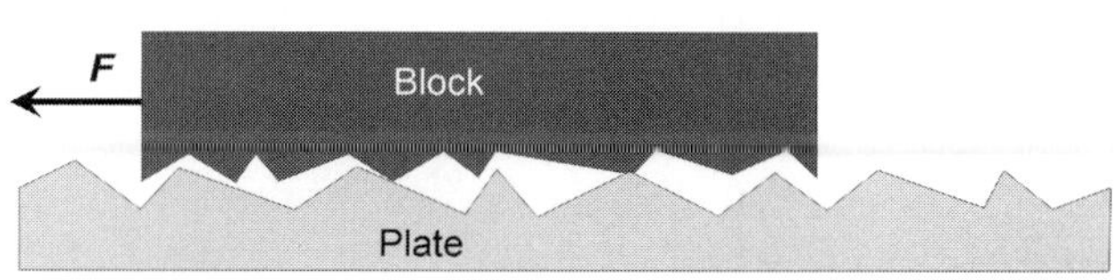

Figure 2.2 Illustration of the contact between block and plate on a microscopic scale.

Copyright © 2010 Laurence W. McKeen. All rights reserved.

to movement as they slide over and deform one another. This resistance to the movement is the frictional force (F_s) as defined by Equation 2.1.

Frictional properties of plastics differ markedly from those of metals. The coefficients of friction vary with applied load, velocity, and temperature.

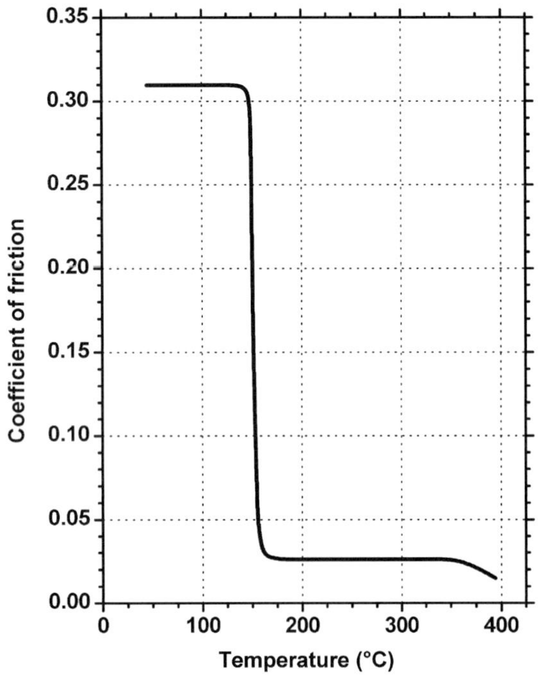

Figure 2.3 The coefficient of friction varies with temperature for Vespel® SP-21 (against mild carbon steel).

Figure 2.3 shows an example of the temperature dependence of the coefficient of friction for a polyimide, Vespel® SP-21.

The rigidity of even the highly reinforced resins is low compared to that of metals; therefore, plastics do not exactly behave according to the classic laws of friction. Metal to plastic friction is characterized by adhesion and deformation of the plastic, resulting in frictional forces that are more dependent on velocity rather than load. In thermoplastics, friction generally decreases as load increases. Figure 2.4 shows the dependence of the coefficient of friction of Teflon® PTFE as a function of both velocity (sliding speed) and load (pressure).

A unique characteristic of most thermoplastics is that the static coefficient of friction is typically less than the dynamic coefficient of friction. This accounts for the slip/stick sliding motion associated with many plastics on metal and with plastics on plastics.

2.2 Lubrication

Lubrication is the common approach to reducing friction. The lubricant resides between the two surfaces as shown in Figure 2.5. When an incompressible solid or liquid lubricant is inserted between the two contacting surfaces, it tends to fill the gaps between the asperities (as shown in Figure 2.5). It acts as a

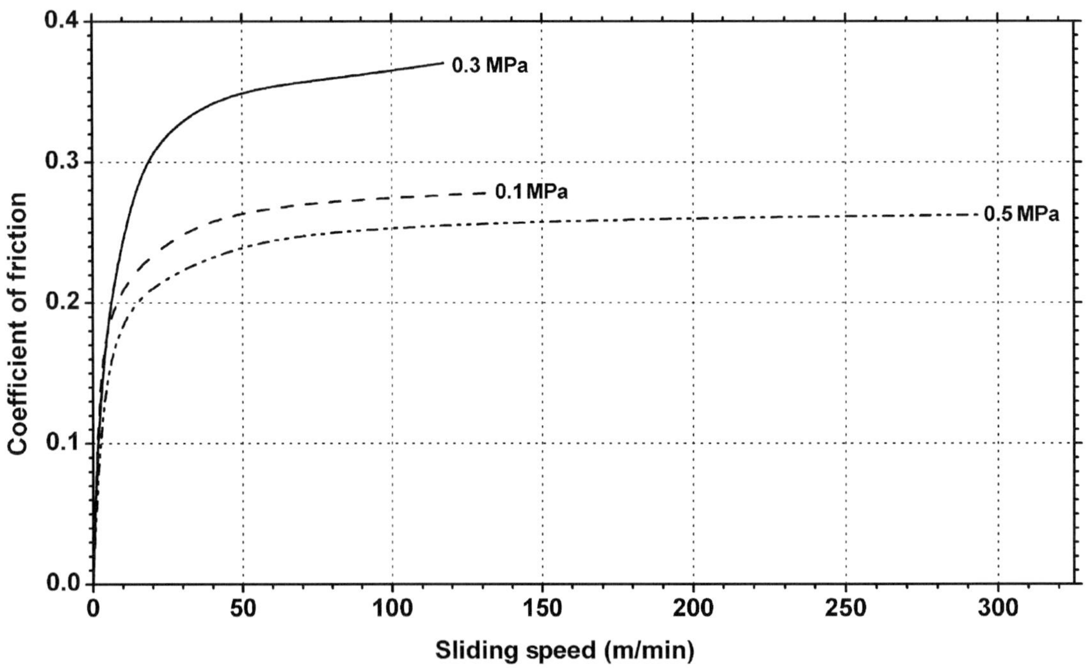

Figure 2.4 The coefficient of friction varies with sliding speed and pressure for Teflon® PTFE (against mild carbon steel).

fluid bearing surface and allows smoother movement of the two materials. This gives a greatly reduced frictional force than for unlubricated movement. Traditional fluid lubricants are oils, but it is also possible to use solid lubricants such as graphite, molybdenum disulphide (MoS_2), or PTFE. The solid lubricants are often dispersed in oil or water.

Lubrication is classified as *partial* when there is still some contact between the block and the plate. The coefficient of friction for partial lubrication is generally between 0.01 and 0.1. When there is a total separation of the block and the plate by a layer of lubricant it is called *fully hydrodynamic*. The coefficient of friction for fully hydrodynamic lubrication is usually between 0.001 and 0.01. Lubrication is common and effective even for materials with a very low coefficient of friction, such as Teflon® PTFE because the lubricant layer not only greatly reduces the coefficient of friction but also reduces the surface damage caused by the asperities rubbing together. Figure 2.6 shows the effect of lubrication on Vespel® SP-21, which is a polyimide plastic that contains 15% graphite for internal lubrication. This figure plots the coefficient of friction versus time of the Vespel® SP-21 rubbing against AISI 1080 carbon steel. The system is lubricated with oil at the start, but then the flow of oil between the two surfaces is shut off after 1 hour. As can be seen, the coefficient of friction starts to rise and continues to do so slowly. When the lubricant is completely gone, the coefficient of friction rises to the level characteristic of this internally lubricated material.

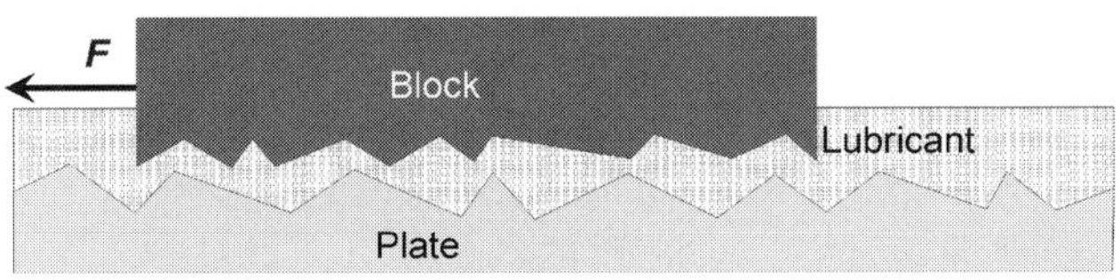

Figure 2.5 Illustration of lubricant between block and plate on a microscopic scale.

2.3 Wear and Erosion

Wear is defined as the removal of material from a solid surface as a result of friction or impact. Considering the frictional block and plate model in Figure 3.1, it is easy to imagine that the constant movement of the asperities over one another will lead to material removal as the asperities are ground down. Wear occurs and material is lost from both surfaces, even if one is much harder than the other.

2.3.1 Classification of Wear

One may envision many contact scenarios that lead to wear. Wear has been classified in various ways. One possible classification is based on the fundamental motion that is causing the removal of

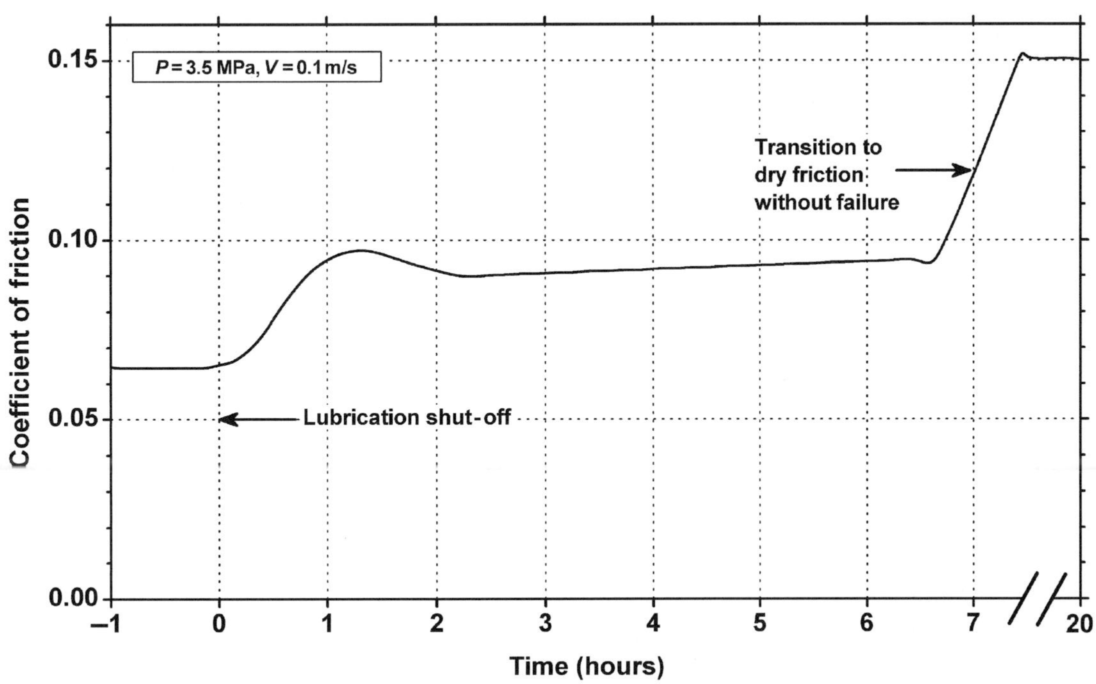

Figure 2.6 The effect of lubrication on the coefficient of friction of Vespel® SP-21.

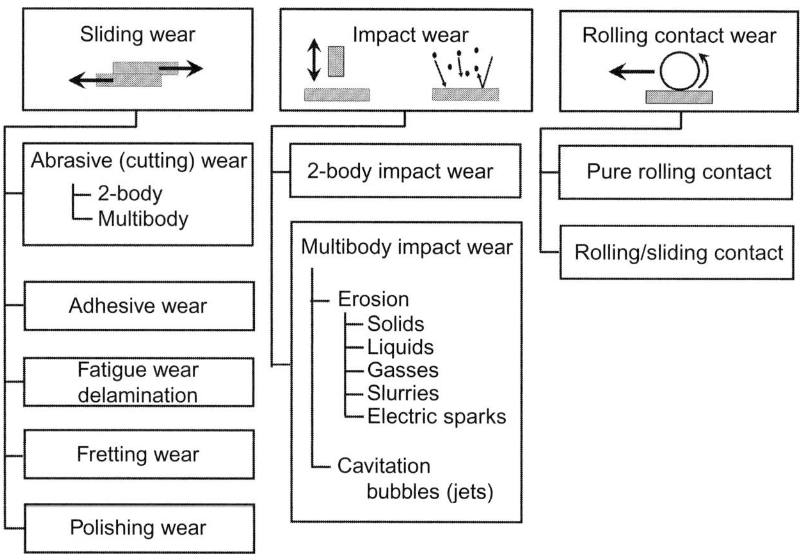

Figure 2.7 Major categories of wear classified by the type of relative motion encountered.[3]

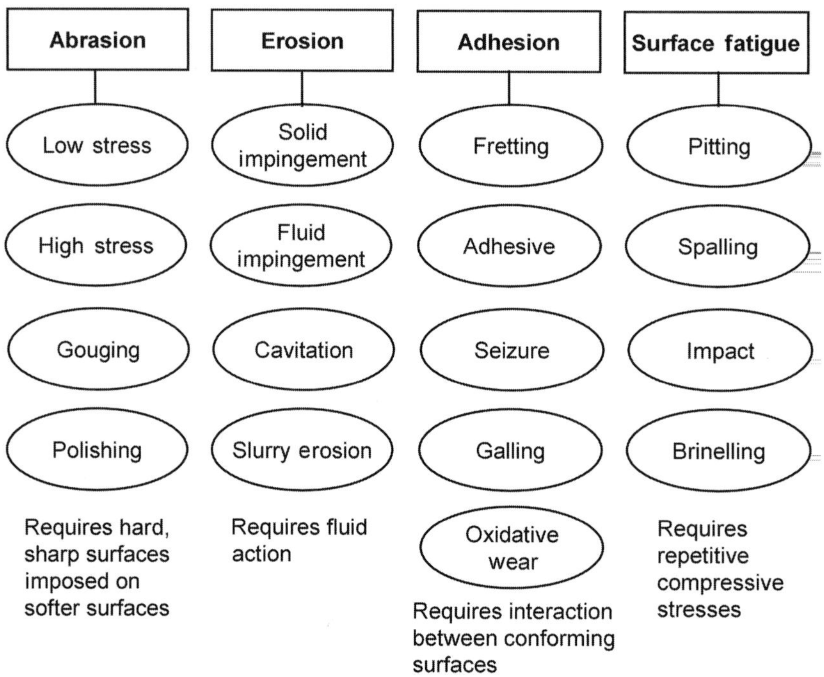

Figure 2.8 Major categories of wear classified by mechanism.[4]

the material. Many schemes based on motion have been proposed, but none is universally accepted. As an example, one particular scheme has been proposed by Blau.[3] This scheme, shown in Figure 2.7, puts wear processes into one of three categories based on the type of motion producing the wear. The details of each of these processes may be found in the original reference. However, processes that only displace material and not remove it such as galling, scuffing, and scoring are not considered wear, but are considered surface damage. When applying this scheme to real-world wear, problems can get complicated because often more than one process may be the cause of the wear observed.

Budinski[4] classifies wear processes into four categories based on mechanism, i.e., abrasion, erosion, adhesion, and surface fatigue as shown in Figure 2.8. The reference has the details of these mechanisms.

A distinction between erosion and abrasion should be noted. This is because testing is quite different.

Solid particle impingement (erosion) refers to the striking and rebounding of solid particles from the surface. The particles transfer energy to the surface during that strike and rebound. That depends upon the particle velocity, angle of strike, and particle mass. Fluid impingement and cavitation are usually classified as erosion but the process is different.

The most common wear mechanism for thermoplastics is adhesive wear. Adhesive wear occurs when opposing/mating surfaces slide against each other, and fragments of one surface pull off and adhere to the other. The adhesive forces between the polymer and the counterpart are sufficient to inhibit sliding at the original interface. The harder of the opposing surfaces scrapes or abrades away the mating part. The adhesive junctions which form at the real points of contact rupture within the polymer itself and a layer of polymer is deposited on the counterpart. The counterpart surface can become effectively smoother leading to a reduction in the rate of wear. PTFE has been shown to be very effective at forming such a transfer film.[5] The tribological properties of most neat polymers (without additives) are relatively poor since adhesion of most polymeric transfer layers to metal counterparts is very weak.

2.3.2 Characterizing Wear

Even though there are many wear mechanisms or processes, wear is usually characterized by several parameters. These will be discussed in the following sections.

2.3.2.1 Wear Rate

Wear can be characterized in several ways. It is often reported as the removal of material on a volume, weight, or depth (thickness) basis. The rate of wear would also relate the amount of material removed to a variable such as time, cycles, or distance. Tests for making these measurements in the laboratory are discussed in the next section.

2.3.2.2 Wear Factor

The wear resistance of materials can be predicted from an experimentally determined *wear factor*. The wear factor is derived from an equation relating the volume of material removed by wear in a given time per unit of load and surface velocity. The general equation is given in Equation 2.2 and the special case of a flat surface is given in Equation 2.3.

$$W = K \times F \times V \times T \tag{2.2}$$

where W = wear volume (cm^3), K = wear factor (cm^3-min/m-kg-h), F = load (kg), V = velocity (m/min), and T = time (hours).

For flat surfaces:

$$X = K \times P \times V \times T \tag{2.3}$$

where X = wear depth (cm), K = wear factor (cm-min/m-MPa-h), P = pressure (MPa), V = velocity (m/min), and T = time (hours).

The time, velocity, pressure, and depth units are often different so the wear factor units need to be carefully matched.

Once a K wear factor is established, it can be used to calculate wear rates of such components as bearings and gears. However, the engineer must bear in mind that the wear rate of the plastic is affected by test PV, plastic material finish, part geometry, ambient temperature, mating surface finish, mating surface hardness, and mating surface thermal conductivity. But at a given PV condition, a lower wear rate factor also indicates lower wear rate. Nonetheless, as a relative measure of one material versus another under the same operating conditions, K factors have proved to be highly reliable.

The wear factor is temperature dependent. One example of this is shown in Figure 2.9 for a particularly thermally stable material, Vespel®.

Materials can rise in temperature during use even though the general environment is not a high-temperature one. The PV multiplier also represents the work done per unit area per unit time at the contact surface. A part of this work may then be transformed into heat. The amount of heat generated proportional to the PV value times the coefficient of friction, μ, as shown in Equation 2.4.

$$\text{Heat} \approx PV \times \mu \tag{2.4}$$

The actual temperature rise will also depend on the thermal conductivity and heat capacity of the materials. If the thermoplastic is sensitive to temperature change (typified by low heat deflection temperature or low melting point), this frictional heating may cause the polymer to soften or even melt. This means the wear mode (thus the wear rate) and part shape are changed rapidly, and the wear part can no longer function adequately due to large dimensional deformation. High friction coefficient is often indicative of such a softening of the thermoplastic part.

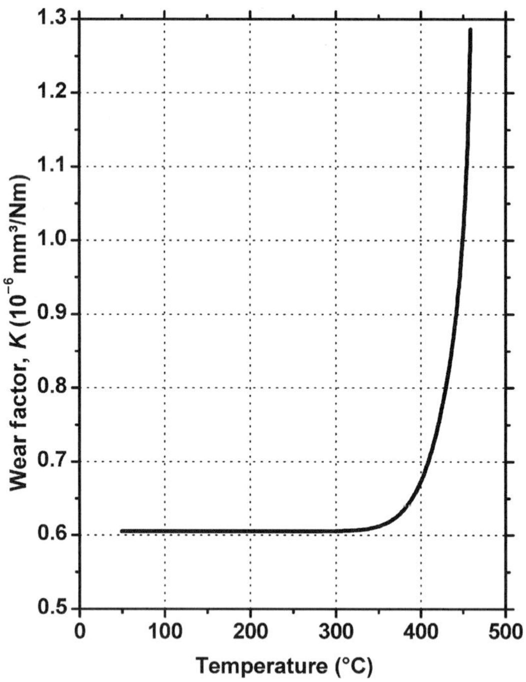

Figure 2.9 The temperature dependence of wear factor for Vespel® SP-21 in unlubricated operation against mild carbon steel.

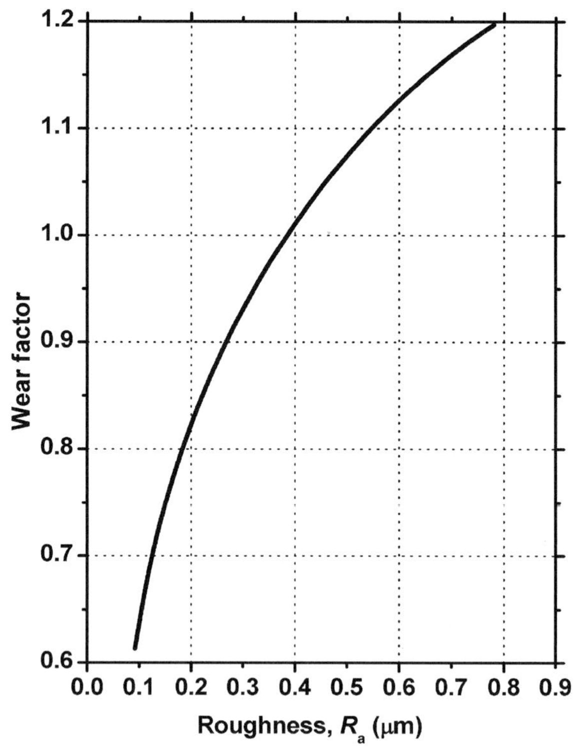

Figure 2.10 The dependence of wear factor for Vespel® SP-21 in unlubricated operation against mild carbon steel of different roughness levels.

The temperature at which wear increases dramatically is called the *wear transition temperature*. This is sometimes reported in vacuum or inert atmosphere and in air. For instance, the Vespel® products used in several of the preceding charts have a wear transition temperature in the range of 482–538°C in vacuum or inert gases, and 371–399°C in air. Figure 2.9 shows the wear factor of Vespel® bearings is essentially constant over a wide range of temperature but then it rises quite rapidly and from the chart one would say the wear transition temperature is someplace between 350°C and 400°C.

Mating materials can have different degrees of smoothness, or conversely roughness. There are devices that characterize roughness, where higher roughness values mean the surface is rougher. The effect of the roughness on the mating surface on the wear factor of Vespel® SP-21 is shown in Figure 2.10.

The hardness of the mating material can also affect the wear factor. This is shown in Figure 2.11 for Vespel® SP-21.

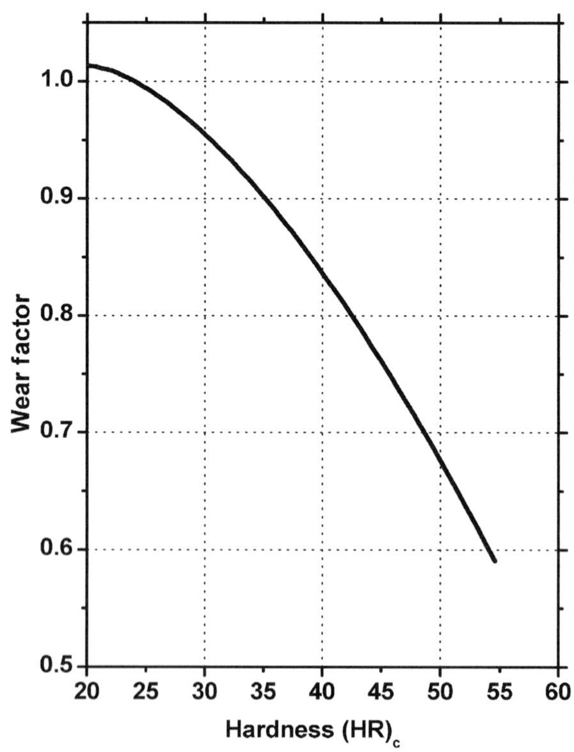

Figure 2.11 The dependence of wear factor for Vespel® SP-21 in unlubricated operation against materials of different hardness.

2.3.2.3 PV Limit

In addition to the wear factor and coefficient of friction, another key parameter that is often used to select a material for parts requiring excellent resistance to the effects of wear is the PV *limit*. PV is the product of load or pressure (*P*) and sliding velocity (*V*). By definition, the PV limit is simply a PV multiplier above which the material can no longer function as a wear part due to softening, melting, and deformation. But in reality, the PV limit remains more a concept than a clear-cut number that one can determine experimentally.

In a bearing application, the PV limit for a material is the product of limiting bearing pressure MPa (psi) and peripheral velocity m/min (fpm), or bearing pressure and limiting velocity, in a given dynamic system. It describes a critical, easily recognizable change in the bearing performance of the material in the given system. When the PV limit is exceeded, one of the following manifestations may occur:

1. Melting
2. Cold flow or creep
3. Unstable friction
4. Transition from mild to severe wear

PV limit is generally related to rubbing surface temperature limit. As such, PV limit decreases with increasing ambient temperature. The PV limits determined on any given tester geometry and ambient temperature can rank materials, but translation of test PV limits to other geometries is difficult.

As long as the mechanical strength of the material is not exceeded, the temperature of the surface is generally the most important factor in determining PV limit. Therefore, anything that affects surface temperature will also affect the PV limit of the material. The following factors are known to affect the PV limit:

1. Coefficient of friction
2. Thermal conductivity of both mating materials
3. Lubrication
4. Ambient temperature
5. Running clearance
6. Hardness
7. Surface finish of mating materials

2.4 Tribology Testing

There are perhaps a hundred designs of machines called tribometers that may be used to measure wear and coefficient of friction.[6] There are so many designs

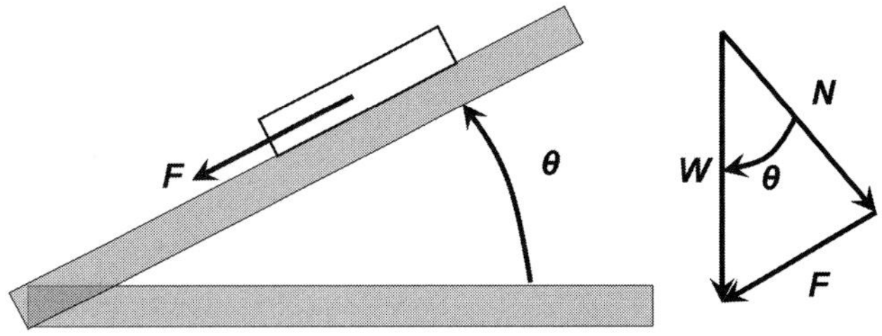

Figure 2.12 An inclined plane may be used to determine the static coefficient of friction.

because one tried to simulate the expected endues environment and conditions as close as possible. Only a few of these tests and machines will be discussed here. These will be the most common ones.

2.4.1 Testing for Friction

Most tribometers used to study wear will also measure coefficient of friction. But there are a few tests aimed just at coefficient of friction tests. As mentioned in Section 2.1, there are two coefficients of friction that can be measured. The static coefficient of friction (μ_s) is found from the force that is just enough to start the block moving. Once the block is moving, it is possible to measure the dynamic coefficient of friction (μ_d) from the force that is just enough to keep the block moving.

One simple way to measure the static coefficient of friction is to place a block of steel on a plaque of test material. The plaque is lifted on one end creating an inclined plane that is tilted higher and higher until the block starts to move as shown in Figure 2.12. The angle of tilt can be used to resolve the forces to calculate the static coefficient of friction as defined in Equation 2.5.

$$\tan\theta = \frac{F}{N} = \mu_s \tag{2.5}$$

For plastic film and sheeting, the most common test is ASTM D1894-08 Standard Test Method for Static and Kinetic Coefficients of Friction of Plastic Film and Sheeting that is capable of measuring both the static and dynamic coefficients of friction (Figure 2.13).

Laboratory testing for friction and wear is often carried out using a motorized tribometer and there are various standardized and nonstandardized test methods available.

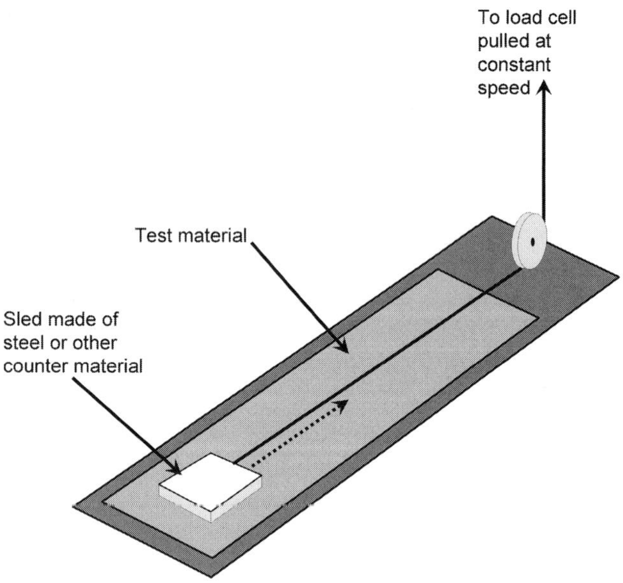

Figure 2.13 Diagram of an attachment for a load frame (such as an Instron®) used to measure static and kinetic coefficients of friction of plastic film and sheeting.

Friction in actual applications is very difficult to predict because there are:

1. a wide range of surface combinations
2. a wide range of lubrication methods and materials
3. the nonlinear relationship between the contact pressure, speed, and the coefficient of friction
4. the effect of temperature rise due to frictional heating

2.4.2 Wear and Abrasion Tests

2.4.2.1 Thrust Washer Abrasion Testing

One of the most common abrasion tests is called the Thrust Washer Abrasion Test and is described by ASTM standard, D3702-94(2004) Standard Test Method for Wear Rate and Coefficient of Friction of Materials in Self-Lubricated Rubbing Contact Using a Thrust Washer Testing Machine. The machine can provide a large amount of quality and detailed information about wear of materials. Two of the manufacturers of the machine are the Falex Corporation and Plint Tribology Products.

The machine tests a precision-machined washer shown in Figure 2.14. The opposing surface is a ring. The test plastic is usually a ring and the thrust washer is usually a steel mating material. In these tests, the stationary ring is mounted in an antifriction bearing equipped with a torque transducer. The moving specimen (thrust washer), which is mounted in the upper sample holder, presses against the stationary specimen.

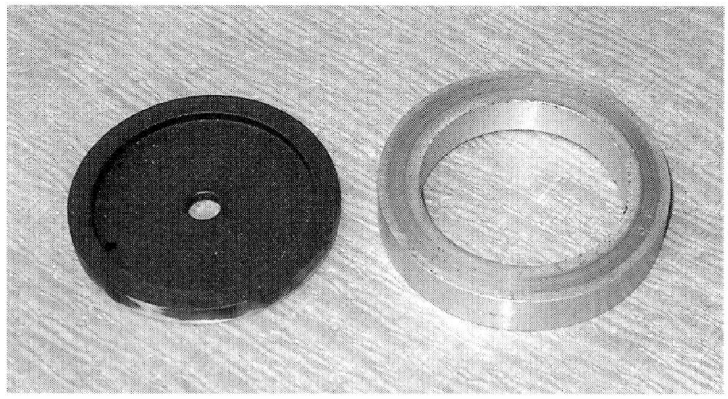

Figure 2.14 Thrust washer (on left) and ring test specimens.

The test specimens are loaded into the test machine as shown in Figure 2.15.

There are a number of selectable variables for this test:

1. The load pressing the washer and ring together (kg/cm^2)
2. The speed or RPM from which the velocity can be calculated (m/min)
3. The ring holder temperature (or the environmental chamber)
4. The presence and temperature of lubricant if chosen
5. The time or number of rotations the experiment runs

The machine monitors the torque applied to the ring by friction. The machine can be set to run a specific number of revolutions or for a specific time. There is usually a break-in period in which the measured torque varies wildly so the experiment must be run until the measured torque stabilizes.

The tests, to be considered valid, are run until an equilibrium condition is reached. After the experiment, the wear depth and weight loss can be measured.

From this data, an array of wear rate and wear factor can be calculated as described in Section 2.3.2. The thrust washer test generates wear information

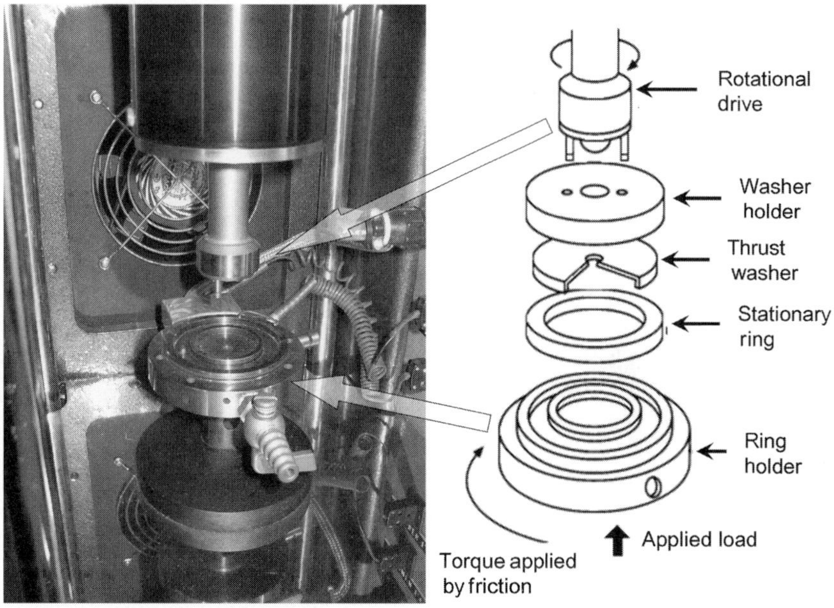

Figure 2.15 Thrust washer test equipment.

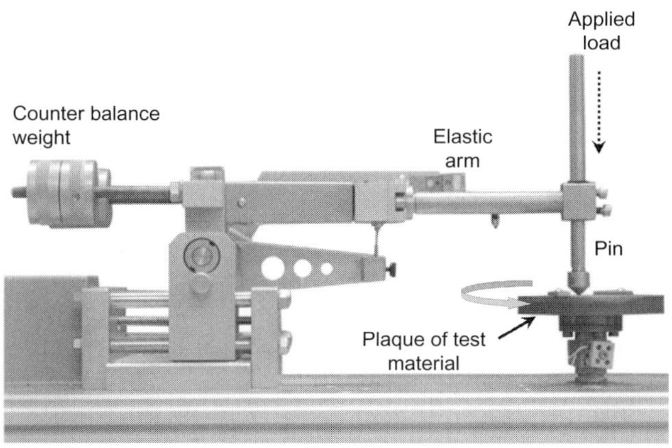

Figure 2.16 Pin-on-disk tribometer equipment (photo courtesy of Nanovea Corporation).

for a material based on area contact, not line or point contact as needed in some bearing applications.

The thrust washer test is capable of generating wear data for plastic against metal, plastic against plastic, or against virtually any mating surface. Testing can be done for a wide temperature range and/or submerged in various fluids.

2.4.2.2 Pin-on-Disk Abrasion Testing

The Pin-on-Disk Tribometer, shown in Figure 2.16, consists of a flat, pin or sphere which is attached to a stiff elastic arm that is weighted down onto a test sample with a precisely known weight. The sample is rotated at a selected speed. The elastic arm ensures a nearly fixed contact point and a stable position in the friction track formed by the pin on the sample. The kinetic friction coefficient is determined during the test by measuring the deflection of the elastic arm, or by direct measurement of the change in torque by a sensor located at the pivot point of the arm. Wear rates for the pin and the disk are calculated from the volume or weight of material removed during the test. Figure 2.17 shows the track and wear debris on a test plaque. With this machine, one can control test parameters such as speed, contact pressure (hence PV), and time. With the right environmental chamber, one can also control and measure the effect of humidity, temperature, and atmospheric composition. The pin-on-disk measurement is usually done per ASTM G99-05 Standard Test Method for Wear Testing with a pin-on-disk apparatus.

2.4.2.3 Linear Reciprocating Abrasion Testing

The pin-on-disk tribometer can be modified by replacing the rotating disk motor with a one directional reciprocating table as shown in Figure 2.18. This arrangement reproduces the reciprocating motion typical in many real-world mechanisms. The configuration of the mating surface to the test plaque on the reciprocating table can be point, line, or

Figure 2.17 Pin-on-disk wear track (photo courtesy of Nanovea Corporation).

Figure 2.18 Reciprocating tribometer equipment (photo courtesy of Nanovea Corporation).

surface contact. The sample is moved at a controlled speed. The elastic arm ensures a nearly fixed contact point and a stable position in the friction track formed on the sample. The static friction coefficient is determined during the test by measuring the deflection of the elastic arm during each change in direction. Wear rates are calculated from the volume or weight of material removed during the test. The reciprocating abrasion measurement is usually done per ASTM G133-05e1 Standard Test Method for Linearly Reciprocating Ball-on-Flat Sliding Wear.

2.4.2.4 Taber Abraser

The Taber Abraser has been used to characterize wear for a long time. It is a standard ASTM

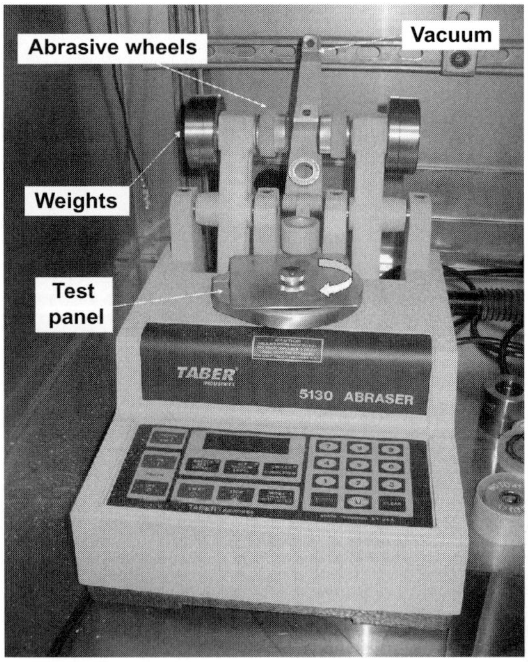

Figure 2.19 A photo of the Taber Abraser.

test, ASTM D1044-08 Standard Test Method for Resistance of Transparent Plastics to Surface Abrasion. This historically has been an industry favorite, because it is inexpensive and easy to do. A test panel has a hole in the center. It is mounted in the Taber Abraser that is shown in Figure 2.19.

Weight is selected and added, as are the types of abrasive wheels. The wheel and weight assembly is lowered onto the test panels, which then rotates allowing the panel to be abraded by the wheels. A vacuum removes abraded debris. The panels are rotated for a given number of cycles, typically one thousand. By measuring the depth of the worn area or by weighing before and after test and knowing the number of cycles, the wear rate can be calculated in terms of thickness loss or weight loss per 1000 cycles. This measurement may also be converted to volume loss per 1000 cycles by geometric calculation using the measured density of the test material.

There are a couple of problems with this test. First, as the plastic is abraded, it tends to fill in the porosity of the abrading wheels. This makes them less efficient at abrading. The abrading wheels need to be cleaned or redressed every 100–200 cycles. This is especially true for materials containing perfluoropolymers. Secondly, the test has poor reproducibility. Comparisons of materials should be restricted to testing in only one laboratory. Inter laboratory comparison should use rankings of coatings

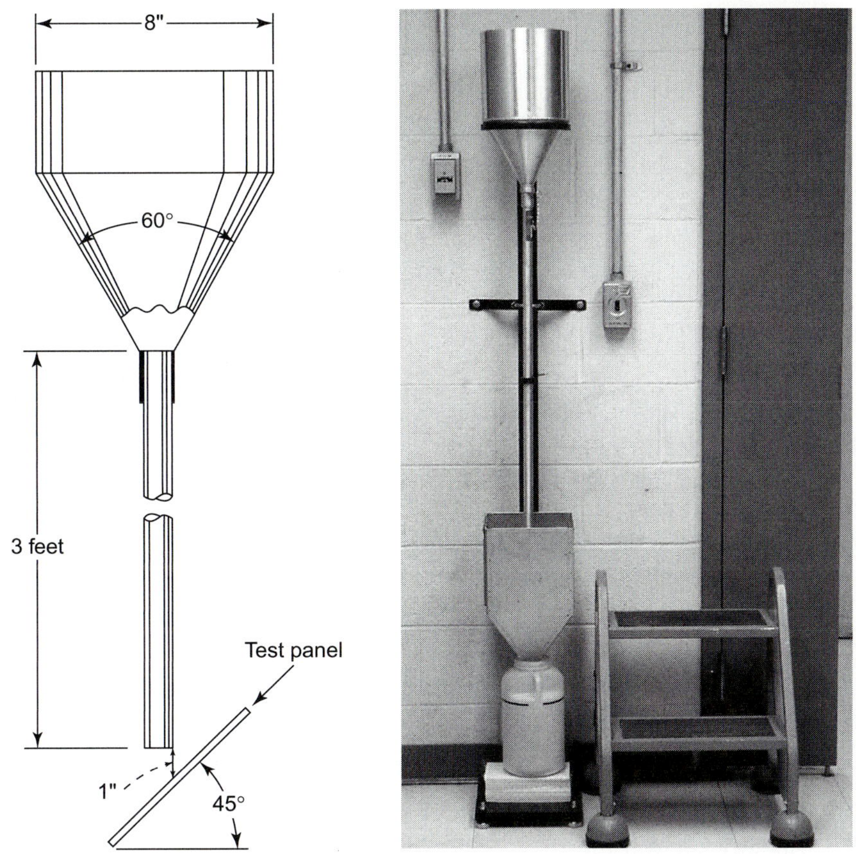

Figure 2.20 Falling abrasive abrasion/erosion tester.

in place of numerical values. The substrate disk must be very flat.

2.4.3 Erosion Tests

Erosion tests are usually based on dry material or slurries. They generally use gravity or pressure to force particles against the test plaque.

2.4.3.1 Falling Abrasive/Erosion Test

A simple, inexpensive reproducible abrasion/erosion test is the falling abrasive test described in ASTM D968-93(2001) Standard Test Methods for Abrasion Resistance of Organic Coatings by Falling Abrasive. Known weights or volume of sand, gravel, aluminum oxide, or silicon carbide are poured on a panel from a given height through a funnel and tube as shown in Figure 2.20. The panel is positioned at a 45° angle. The abrasive is collected for reuse. Abrasive such as aluminum oxide can be reused many times. When desired, after many tests, the fines can be removed by sieving. The change in weight per unit weight of abrasive is used to report abrasion/erosion rates.

2.4.3.2 Slurry Erosion Tests

There are a number of slurry erosion test machines discussed in the literature.[7,8] Many of these pump slurry at high velocity at the surface of a test plaque. A simpler test is described in ASTM G75-07 Standard Test Method for Determination of Slurry Abrasivity (Miller Number) and Slurry Abrasion Response of Materials (SAR Number). The relative effect of slurry abrasivity is determined by measuring the mass loss of a block plastic elastomer after it has been driven in a reciprocating motion in a trough containing the slurry. A direct load is applied to the test plaque. The interior of the trough has a flat-bottomed or truncated "V" shape trough that forces the slurry particles to the reciprocating path taken by test specimen. The slurry may be of any material of interest such as sand in water or other liquid. The Miller Number Machine lifts slightly the test block and delays momentarily at the end of each stroke to allow time for fresh slurry material to flow back into the wear path. The test consists of measuring the mass loss of a part per unit time.

2.4.4 Standard Tests

Table 2.1 lists many, but not all, of the testing standards issued by ASTM and ISO.

2.5 Wear-Resistant Additives

Even among plastic materials with excellent natural lubricity, wear characteristics between two thermoplastics differ greatly. When an application calls for plastic on plastic, dissimilar polymers should be used and incorporated with one or more wear-resistant additives. Reinforcements such as glass, carbon, and aramid fibers enhance wear resistance by increasing the thermal conductivity and creep resistance, thus improving the LPV and working PV of the part.

PTFE has the lowest coefficient of friction of any internal lubricant. Its particles shear during operation to form a lubricous film on the part surface. Often referred to as the best lubricant for metal mating surfaces, PTFE modifies the mating surface after an initial break-in period. PTFE goes an extra step in lessening wear and fatigue failure by actually cushioning shock. What is most important about PTFE is its distribution throughout the thermoplastic compound. PTFE has a typical optimum loading of 15% in amorphous thermoplastic resins and 20% in crystalline resins. However, there is a price performance limit at which PTFE can actually begin to demonstrate diminishing returns.

MoS_2, otherwise known as *moly*, is a solid lubricant usually used in nylon and other composites to reduce wear rates and increase PV limits. Acting as a nucleating agent, MoS_2 creates a better wearing surface by changing the structure of nylons to become more crystalline, creating a harder and more wear-resistant surface. MoS_2 will not lower the coefficients of friction like other modifiers, and its use is therefore confined to nylons where it has this crystallizing effect on the nylon molecular structure.

MoS_2 also has a high affinity for metal. Once attracted to the metal, it fills the metal's microscopic pores, making the metal surface slippery. This makes MoS_2 the ideal lubricant for applications in which nylon wears against metal, such as industrial bushings, cam components, and ball joints. Two added benefits occur during molding: fast injection molding times which lower per part costs; and less and more uniform shrinkage.

Graphite's unique chemical lattice structure allows its molecules to slide easily over one another with little friction. This is especially true in an aqueous environment and makes graphite powder an ideal lubricant for many underwater applications such as water meter housings, impellers, and valve seals.

Silicone or polysiloxane fluid is a migratory lubricant. A particular silicone fluid is chosen that is compatible enough with the base resin to allow compounding, yet incompatible enough to migrate to the surface of the compound to continuously regenerate the wear surface.

Perfluoropolyether (PFPE) synthetic oil marketed by DuPont under the trademark Fluoroguard® is an internal lubricant that imparts improved wear and low friction properties like silicone or polysiloxane fluids.

Silicone resin offers engineers several unique advantages based on its ability to be both a boundary lubricant and an alloying partner with the base resin. Silicone acts as a boundary lubricant because silicone moves or migrates to the surface of a part over time, by both diffusion as a result of random molecular movement, and by its exclusion from the resin matrix which is a result of migration. As a partial alloying material with the base resin, silicone remains in the component over its service lifetime, but because silicone is incompatible enough, the silicone is constantly moving from the matrix to the surface. This continuous secretion eases friction and wear at start-up and when high-speed lubricity is necessary. Silicone is excellent for start-up, high-speed, and low-pressure wear applications such as keyboard keycap receptacles and high-speed printer components.

Silicone fluid is available in a wide range of viscosities. The lower the viscosity, the more fluid the additive is, and the quicker it will migrate to the surface and provide lubrication. This is particularly important in wear applications that require numerous start and stop actions. However, if the additive's viscosity is too low, the silicone can vaporize during processing, or migrate too quickly from the molded part.

Silicone and PTFE will work together to create a high-temperature grease which will create better wearing characteristics and lower friction, particularly at high speeds and during start-ups. When used together, PTFE acts as a thickening agent as well as

Table 2.1 ASTM and ISO Tribology Related Standards

Standard Designation	Standard Title
ASTM D968-05e1	Standard Test Methods for Abrasion Resistance of Organic Coatings by Falling Abrasive
ASTM D1044-08	Standard Test Method for Resistance of Transparent Plastics to Surface Abrasion
ASTM D1242-95a	Standard Test Methods for Resistance of Plastic Materials to Abrasion (Withdrawn 2004)
ASTM D1894-08	Standard Test Method for Static and Kinetic Coefficients of Friction of Plastic Film and Sheeting
ASTM D2670-95(2004)	Standard Test Method for Measuring Wear Properties of Fluid Lubricants (Falex Pin and Vee Block Method)
ASTM D2981-94(2003)	Standard Test Method for Wear Life of Solid Film Lubricants in Oscillating Motion
ASTM D3233-93(2003)	Standard Test Methods for Measurement of Extreme Pressure Properties of Fluid Lubricants (Falex Pin and Vee Block Methods)
ASTM D3702-94(2004)	Standard Test Method for Wear Rate and Coefficient of Friction of Materials in Self-Lubricated Rubbing Contact Using a Thrust Washer Testing Machine
ASTM D4060-07	Standard Test Method for Abrasion Resistance of Organic Coatings by the Taber Abraser
ASTM D4172-94(2004)e1	Standard Test Method for Wear Preventive Characteristics of Lubricating Fluid (Four-Ball Method)
ASTM D4175-09	Standard Terminology Relating to Petroleum, Petroleum Products, and Lubricants
ASTM D5001-08	Standard Test Method for Measurement of Lubricity of Aviation Turbine Fuels by the Ball-on-Cylinder Lubricity Evaluator (BOCLE)
ASTM D5183-05	Standard Test Method for Determination of the Coefficient of Friction of Lubricants Using the Four-Ball Wear Test Machine
ASTM D5707–05	Standard Test Method for Measuring Friction and Wear Properties of Lubricating Grease Using a High-Frequency, Linear-Oscillation (SRV) Test Machine
ASTM D6425-05	Standard Test Method for Measuring Friction and Wear Properties of Extreme Pressure (EP) Lubricating Oils Using SRV Test Machine
ASTM G40-05	Standard Terminology Relating to Wear and Erosion
ASTM G65-04	Standard Test Method for Measuring Abrasion Using the Dry Sand/Rubber Wheel Apparatus
ASTM G75-07	Standard Test Method for Determination of Slurry Abrasivity (Miller Number) and Slurry Abrasion Response of Materials (SAR Number)
ASTM G77-05e1	Standard Test Method for Ranking Resistance of Materials to Sliding Wear Using Block-on-Ring Wear Test
ASTM G98-02	Standard Test Method for Galling Resistance of Materials
ASTM G99-05	Standard Test Method for Wear Testing with a Pin-on-Disk Apparatus
ASTM G132-96(2007)	Standard Test Method for Pin Abrasion Testing
ASTM G133-05e1	Standard Test Method for Linearly Reciprocating Ball-on-Flat Sliding Wear
ASTM G143-03(2004)	Standard Test Method for Measurement of Web/Roller Friction Characteristics
ASTM G171-03	Standard Test Method for Scratch Hardness of Materials Using a Diamond Stylus
ASTM G174-04	Standard Test Method for Measuring Abrasion Resistance of Materials by Abrasive Loop Contact
ISO 6601:2002	Plastics—Friction and wear by sliding—Identification of test parameters
ISO 9352:1995	Plastics—Determination of resistance to wear by abrasive wheels

an extreme pressure additive to make the grease at the surface. Because the silicone is constantly moving to the surface, this provides the added lubricity necessary during start-ups and at high speeds. Since failure at high speeds is more dependent on wear than failure at low speeds, and because the benefits of the silicone/PTFE synergy are most evident at these higher speeds, this combination should not be considered for low-speed components. In these cases, usually a PTFE-only compound is needed.

Glass fibers are mainly added to resins to improve both short-term mechanical and thermal performance properties, particularly strength, creep resistance, hardness, and heat distortion. Wear resistance can also be improved with the addition of glass fibers, but the improvement is directly correlated to the efficiency of the glass sizing system which bonds the resins and fibers together. Glass reinforcement results in a marked improvement of the resins limiting PV by enhancing creep resistance, thermal conductivity, and heat distortion.

Glass fiber reinforcement often leads to increased coefficient of friction and mating surface wear. This can be counteracted with the addition of an internal lubricant.

Carbon fibers are added to engineering resins to produce high-strength, heat distortion temperatures, and modulus as well as creep and fatigue resistance. Often referred to as the perfect additive for wear and friction resins, carbon fibers also greatly increase thermal conductivity and lower coefficients of friction and wear rates. In fact, the strengthened compound may have lower friction coefficients than the base resin.

Carbon fibers should be considered as replacements or alternatives for glass fiber when wear and friction are not sufficiently addressed in glass fiber reinforced components. Unlike glass, carbon is a softer and less abrasive fiber. It will not score the surface of iron or steel. Most resins which are reinforced with 10% or more of carbon fibers will dissipate static electricity and overcome problems with static buildup on moving parts. This can be extremely important for business machines, textile equipment, and other electronic components.

Aromatic polyamide fiber, commonly known as aramid fiber or Kevlar®, is one of the latest wear-resistant additives to be used in thermoplastic composites. Unlike the traditional fiber reinforcements of glass and carbon, aramid is the softest and least abrasive fiber. This is a major advantage in wear applications, particularly if the mating surface is sensitive to abrasion.

2.6 Summary

Published multipoint tribology data is limited and it is included in Chapters 4–12. Tabular data is more extensive.

References

1. Davis JR. *Surface engineering for corrosion and wear resistance. Maney Publishing*. (pp. 43–86) Online version available at: http://knovel.com/web/portal/browse/display?_EXT_KNOVEL_DISPLAY_bookid=1283&VerticalID=0; 2001.
2. Ludema KC. *Friction, wear, lubrication: a textbook in tribology*. Boca Raton, FL, CRC Press; 1996.
3. Blau PJ. Wear testing. In: Davis JR, editor. *Metals handbook desk edition*. 2nd ed. Cleveland, OH, ASM International; 1998. pp. 1342–47.
4. Budinski KG. Wear modes. *Surface engineering for wear resistance*. Prentice Hall, 1988. pp. 15–43.
5. Jintang G. Tribochemical effects in formation of polymer transfer film. *Wear* 2000;**245**(1–2):100–6.
6. Budinski KG. *Guide to friction, wear and erosion testing*. West Conshohocken, PA, ASTM International; 2007.
7. Hawthorne HM. Some coriolis slurry erosion test developments. *Tribol Int* 2002;**35**(10):625–30.
8. Fang Q, et al. Erosion of ceramic materials by a sand/water slurry jet. *Wear* 1999;**224**:183–93.

3 Introduction to Plastics and Polymers

The most basic component of plastic and elastomer materials is polymers. The word polymer is derived from the Greek term for "many parts." Polymers are large molecules comprised of many repeat units, called monomers that have been chemically bonded into long chains. Since World War II, the chemical industry has developed a large quantity of synthetic polymers to satisfy the materials needs for a diverse range of products, including paints, coatings, fibers, films, elastomers, and structural plastics. Literally thousands of materials can be called "plastics," although the term today is typically reserved for polymeric materials, excluding fibers, which can be molded or formed into solid or semi-solid objects. The subject of this chapter includes polymerization chemistry and the different types of polymers and how they can differ from each other. Since plastics are rarely "neat", reinforcement, fillers, and additives are reviewed. A basic understanding of plastic and polymer chemistry will make the discussion of fatigue and tribology of specific plastics easier to understand and it also provides a basis for the introductions of the plastic families in later chapters. This chapter is taken from *The Effect of Temperature and Other Factors on Plastics* book, but it has been refocused on fatigue properties.

3.1 Polymerization

Polymerization is the process of chemically bonding monomer building blocks to form large molecules. Commercial polymer molecules are usually thousands of repeat units long. Polymerization can proceed by one of several methods. The two most common methods are called addition and condensation polymerization.

In addition polymerization, a chain reaction adds new monomer units to the growing polymer molecule one at a time through double or triple bonds in the monomer. Each new monomer unit creates an active site for the next attachment. The net result is shown in Figure 3.1. Many of the plastics discussed in later chapters of this book are formed in this manner. Some of the plastics made by addition polymerization include polyethylene, polyvinyl chloride (PVC), acrylics, polystyrene, and polyoxymethylene (acetal).

The other common method is condensation polymerization in which the reaction between monomer units and the growing polymer chain end group releases a small molecule, often water as shown in Figure 3.2. This reversible reaction will reach equilibrium and halt unless this small molecular by-product is removed. Polyesters and polyamides are among the plastics made by this process.

Understanding the polymerization process used to make a particular plastic gives insight into the nature of the plastic. For example, plastics made via condensation polymerization, in which water is released, can degrade when exposed to water at high temperature. Polyesters such as polyethylene terephthalate

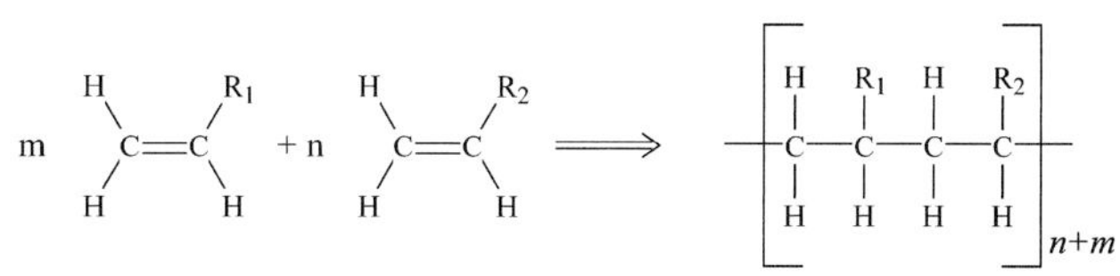

Figure 3.1 Addition polymerization.

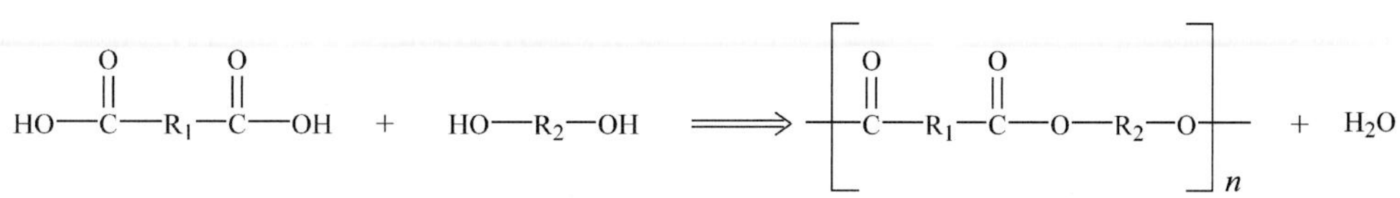

Figure 3.2 Condensation polymerization.

Copyright © 2010 Laurence W. McKeen. All rights reserved.

(PET) can degrade by a process called hydrolysis when exposed to acidic, basic, or even some neutral environments severing the polymer chains. As a result the polymer's properties are degraded.

3.2 Copolymers

A copolymer is a polymer formed when two (or more) different types of monomer are linked in the same polymer chain, as opposed to a homopolymer where only one monomer is used. If exactly three monomers are used, it is called a terpolymer.

Monomers are only occasionally symmetric; the molecular arrangement is the same no matter which end of the monomer molecule you are looking at. The arrangement of the monomers in a copolymer can be head-to-tail, head-to-head, or tail-to-tail. Since a copolymer consists of at least two types of repeating units, copolymers can be classified based on how these units are arranged along the chain. These classifications include:

- Alternating copolymer
- Random copolymer (statistical copolymer)
- Block copolymer
- Graft copolymer.

When the two monomers are arranged in an alternating fashion, the polymer is called, of course, an alternating copolymer:

–A–B–A–B–A–B–A–B–A–B–A–B–A–B–A–B–A–B–
Alternating copolymer

In the following examples A and B are different monomers. Keep in mind the A and B do not have to be present in a one-to-one ratio. In a random copolymer, the two monomers may follow in any order:

–A–A–B–A–B–B–A–B–A–A–B–B–B–A–B–A–A–
Random copolymer

In a block copolymer, all of one type of monomer is grouped together, and all of the other are grouped together. A block copolymer can be thought of as two homopolymers joined together at the ends:

–A–A–A–A–A–A–A–A–A–A–B–B–B–B–B–B–B–B–
Block copolymer

A polymer that consists of large grouped blocks of each of the monomers is also considered a block copolymer:

–A–A–A–A–A–A–B–B–B–B–B–B–A–A–A–A–A–
Block copolymer

When chains of a polymer made of monomer B are grafted onto a polymer chain of monomer A we have a graft copolymer:

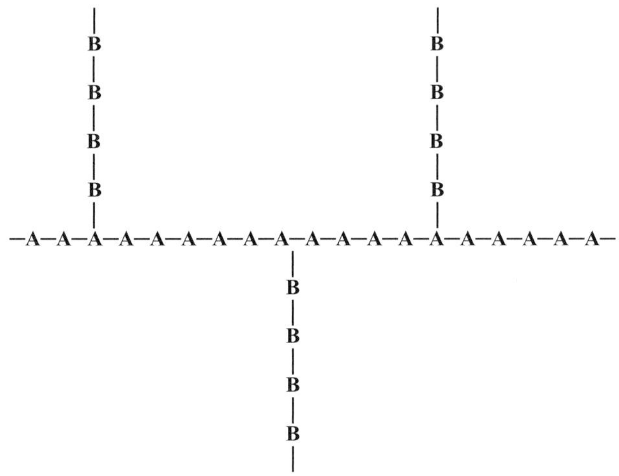

Branched/Grafted copolymer

High-impact polystyrene, or HIPS, is a graft copolymer. It is a polystyrene backbone with chains of polybutadiene grafted onto the backbone. The polystyrene gives the material strength, but the rubbery polybutadiene chains give it resilience to make it less brittle.

3.3 Linear, Branched and Cross-linked Polymers

Some polymers are linear, a long chain of connected monomers. Polyethylene, PVC, Nylon 66, and polymethyl methacrylate are some linear commercial examples found in this book. Branched polymers can be visualized as a linear polymer with side chains of the same polymer attached to the main chain. While the branches may in turn be branched, they do not connect to another polymer chain. The ends of the branches are not connected to anything. Cross-linked polymer, sometimes called network polymer, is one in which different chains are connected. Essentially the branches are connected to different polymer chains on the ends. These three polymer structures are shown in Figure 3.3.

A higher amount of cross-linking in plastics generally leads to higher fatigue crack propagation rates. This is shown in Figure 3.4. This figure shows the fatigue crack propagation rate increases as the amount of cross-linking increases in polymethyl methacrylate. The data to the left side of the plot has the highest amount of cross-linking. The uncross-linked data on the far right has the best performance.

3.4 Molecular Weight

A polymer's molecular weight is the sum of the atomic weights of individual atoms that comprise a molecule. It indicates the *average* length of the bulk resin's polymer chains. All polymer molecules of a particular grade do not all have the exact same molecular weight. There is a range or distribution of molecular weights. The average molecular weight can be determined by several means, but this subject is beyond the scope of this book. Low-molecular-weight polyethylene chains have backbones as small as 1000 carbon atoms long. Ultrahigh molecular weight polyethylene chains can have 500,000 carbon atoms along their length. Many plastics are

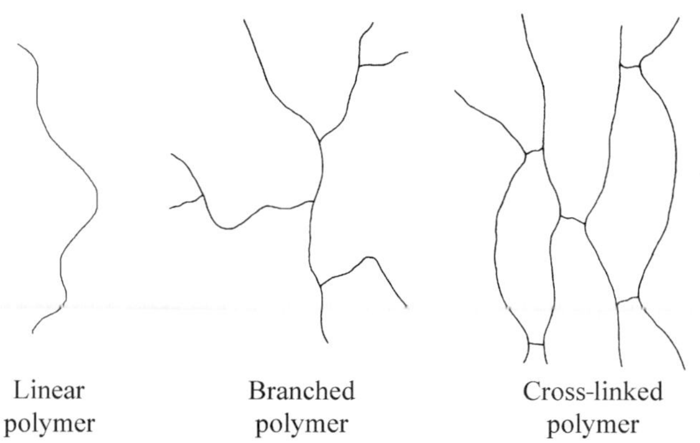

Figure 3.3 Linear, branched, and cross-linked polymers.

available in a variety of chain lengths, or different molecular weight grades. These resins can also be classified indirectly by a viscosity value, rather than molecular weight. Within a resin family, such as polycarbonate, higher molecular weight grades have higher melt viscosities. For example, in the viscosity test for polycarbonate, the melt flow rate ranges from approximately 4 g/10 min for the highest molecular weight, standard grades to more than 60 g/10 min for lowest molecular weight, high flow, specialty grades.

Selecting the correct molecular weight for your injection molding application generally involves a balance between filling ease and material performance. If your application has thin-walled sections, a lower molecular weight/lower viscosity grade offers better flow. For normal wall thicknesses, these resins also offer faster mold cycle times and fewer molded in stresses. The stiffer flowing, high-molecular-weight resins offer the ultimate material performance, being tougher and more resistant to chemical and environmental attack.

Molecular weight of the polymers that are used in engineering plastics affects fatigue lifetimes. While it is not always known exactly what the molecular weights are, as mentioned above higher flowing

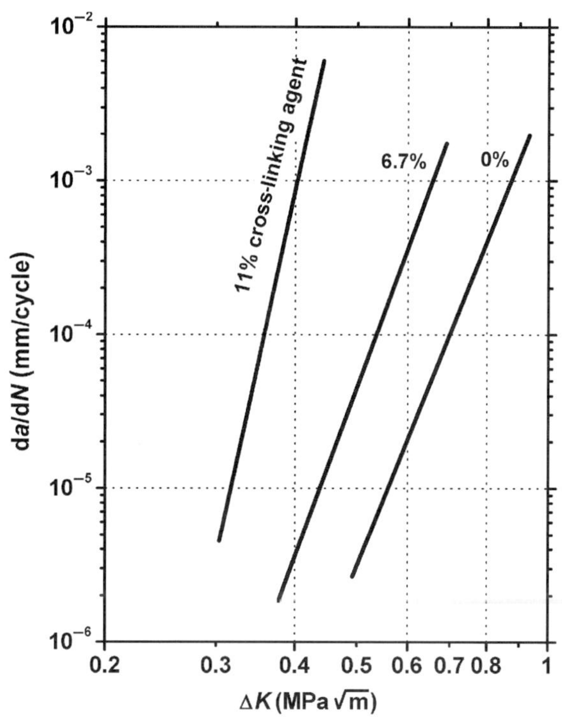

Figure 3.4 Fatigue crack propagation in polystyrene with different amounts of cross-linking agent.

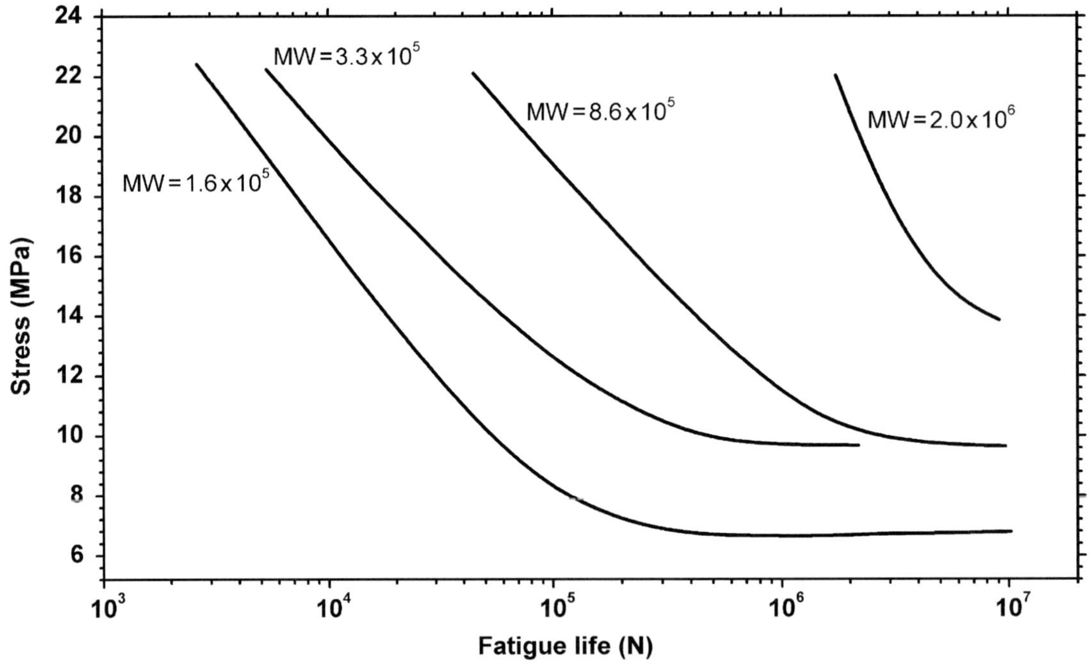

Figure 3.5 The effect of molecular weight of polystyrene on fatigue life.

plastics of a given series of products generally are lower molecular weight polymers. In general, higher molecular weight provides improved resistance to cyclic fatigue damage. This is demonstrated in Figures 3.5 and 3.6. Figure 3.5 shows the effect of molecular weight of polystyrene on fatigue life. Figure 3.6 shows the effect of molecular weight on fatigue crack propagation rates of the same polymer, polystyrene.

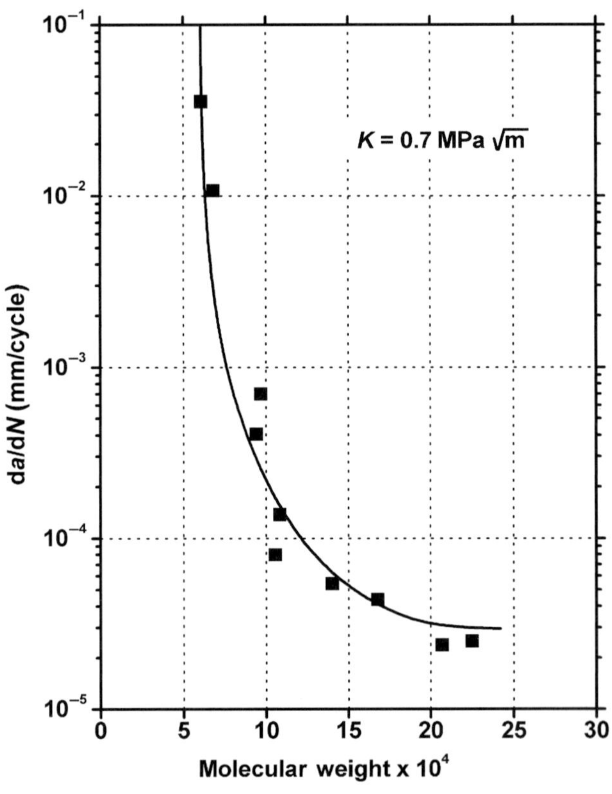

Figure 3.6 The effect of molecular weight of polystyrene on fatigue crack propagation.

3.5 Thermosets versus Thermoplastics

A plastic falls into one of two broad categories depending on its response to heat: thermoplastics and thermosets. Thermoplastics soften and melt when heated and harden when cooled. Because of this behavior, these resins can be injection molded, extruded or formed via other molding techniques. This behavior also allows production scrap runners and trimmings, to be reground and reused.

Unlike thermoplastics, thermosets react chemically to form cross-links, as described earlier that limit chain movement. This network of polymer chains tends to degrade, rather than soften, when exposed to excessive heat. Until recently, thermosets could not be remelted and reused after initial curing. Recent advances in recycling have provided new methods for remelting and reusing thermoset materials.

3.6 Crystalline versus Amorphous

Thermoplastics are further classified by their crystallinity, or the degree of order within the polymer's overall structure. As a crystalline resin cools from the melt, polymer chains fold or align into highly ordered crystalline structures as shown in Figure 3.7.

Some plastics can be completely amorphous or crystalline. Often plastics specifications will report what percent of it is crystalline as a percent, such as 73% crystallinity. Generally, polymer chains with bulky side groups cannot form crystalline regions. The degree of crystallinity depends upon both the polymer and the processing technique. Some polymers such as polyethylene crystallize quickly and reach high levels of crystallinity. Others, such as PET polyester, require slow cooling to crystallize. If cooled quickly, PET polyester remains amorphous in the final product.

Crystalline and amorphous plastics have several characteristic differences. Amorphous polymers do not have a sharp melting point, but do have what is called a glass transition temperature, T_g. A glass transition temperature is the temperature at which a polymer changes from hard and brittle to soft and pliable. The force to generate flow in amorphous materials diminishes slowly as the temperature rises above the glass transition temperature. In crystalline resins, the force requirements diminish quickly as the material is heated above its crystalline melt temperature. Because of these easier flow characteristics, crystalline resins have an advantage in filling thin-walled sections of a mold. Crystalline resins generally have superior chemical resistance, greater stability at elevated temperatures, and better creep resistance. Amorphous plastics typically have better impact strength, less mold shrinkage, and less final part warping than crystalline materials. End use requirements usually dictate whether an amorphous or crystalline resin is preferred.

It is generally accepted that crystalline polymers are more fatigue crack propagation resistant than amorphous polymers and so they should also have improved fatigue life. Figure 3.8 shows the effects of crystallinity in PTFE on the fatigue life. In this particular example the thermal history affects the crystallinity of the PTFE, with samples that were cooled more slowly after molding possessing higher levels of crystallinity. This means that not only does it matter which plastic is used to make a part, but the way a part is made may also impact the fatigue performance.

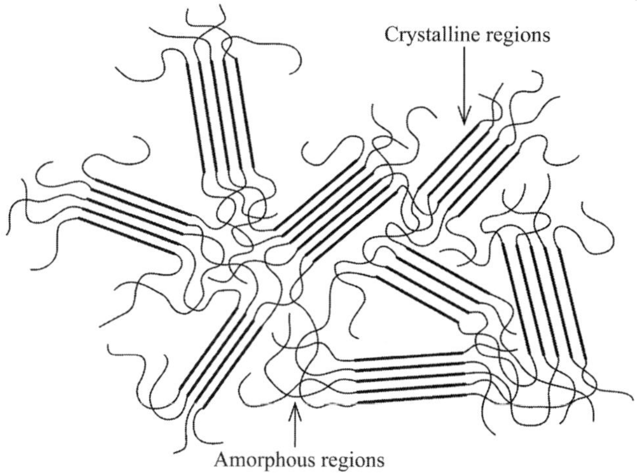

Figure 3.7 Many plastics have crystalline and amorphous regions.

Similarly, thermal history may also affect fatigue life. This is shown in Figure 3.9 which shows the difference heat treatment makes on the fatigue life of polycaproamide. In this case the upper curve has been heat treated for 1 hour at 180°C in oil. The oil keeps oxygen away from the heated polymer. Had the heat treatment been done in air, a different result might be expected.

3.7 Blends

Polymers can often be blended. Occasionally, blended polymers have properties that exceed those of either of the constituents. For instance, blends of polycarbonate resin and PET polyester, originally created to improve the chemical resistance of polycarbonate, actually have fatigue resistance and low-temperature impact resistance superior to either of the individual polymers.

Sometimes a material is needed that has some of the properties of one polymer, and some of the properties of another. Instead of going back into the lab and trying to synthesize a brand new polymer with all the properties wanted, two polymers can be melted together to form a blend, which will hopefully have some properties of both.

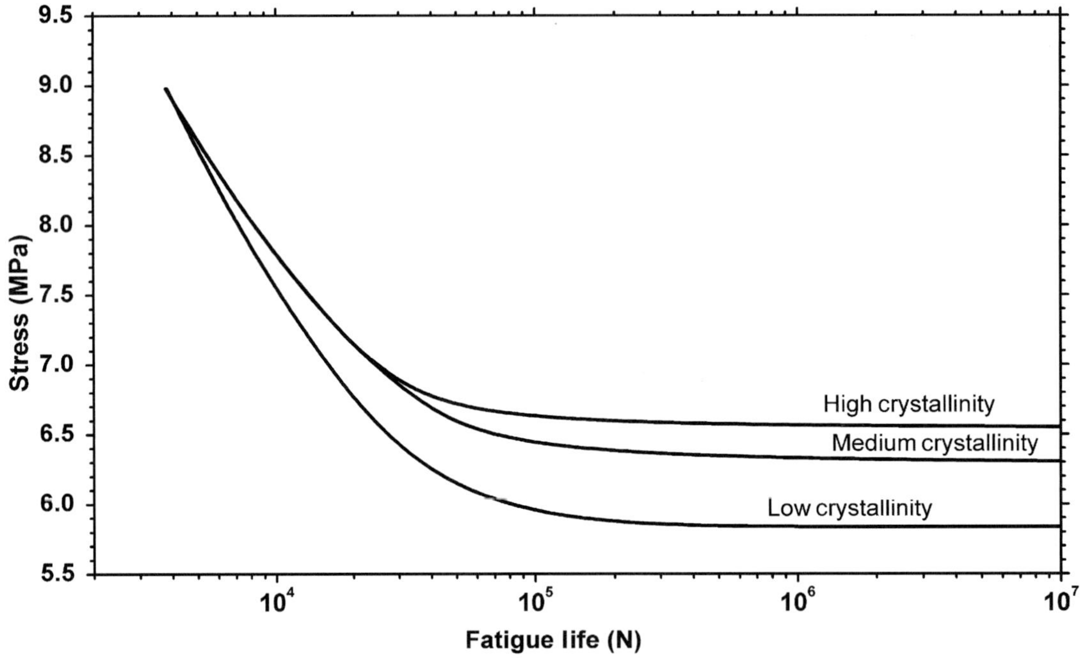

Figure 3.8 Fatigue life of PTFE with different crystallinity levels.

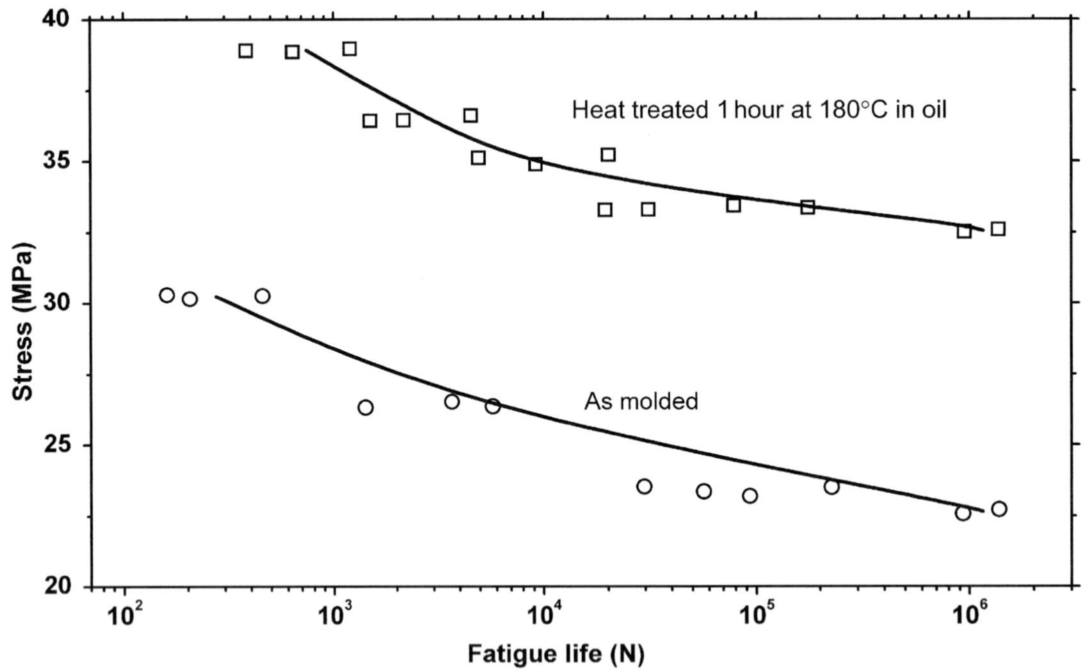

Figure 3.9 The effect of heat treatment on the fatigue life of polycaproamide (Nylon 6).

Two polymers that do actually mix well are polystyrene and polyphenylene oxide. A few other examples of polymer pairs that will blend are:

- PET with polybutylene terephthalate
- Polymethyl methacrylate with polyvinylidene fluoride

Phase-separated mixtures are obtained when one tries to mix most polymers. But strangely enough, the phase-separated materials often turn out to be rather useful. They are called immiscible blends.

Polystyrene and polybutadiene are immiscible. When polystyrene is mixed with a small amount of polybutadiene, the two polymers do not blend.

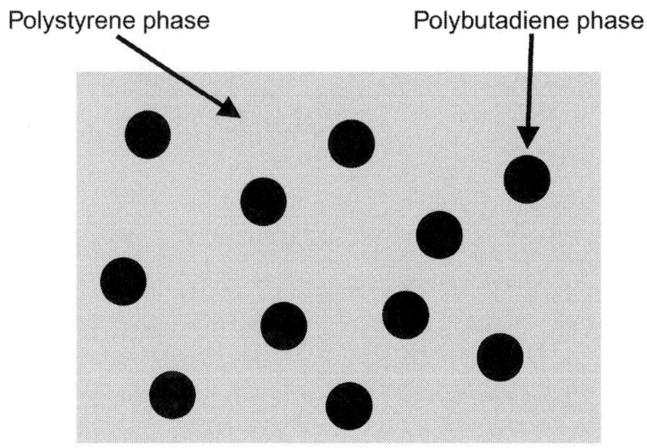

Figure 3.10 Immiscible blend of polystyrene and polybutadiene.

The polybutadiene separates from the polystyrene into little spherical blobs. If this mixture is viewed under a high-power microscope something that looks like the picture in Figure 3.10 would be seen.

Multiphase polymer blends are of major economic importance in the polymer industry. The most common examples involve the impact modification of a thermoplastic by the microdispersion of a rubber into a brittle polymer matrix. Most commercial blends consist of two polymers combined with small amounts of a third, compatibilizing polymer, typically a block or graft copolymer.

Multiphase polymer blends can be easier to process than a single polymer with similar properties. The possible blends from a given set of polymers offer many more physical properties than do the individual polymers. This approach has shown some success but becomes cumbersome when more than a few components are involved.

Blending two or more polymers offers yet another method of tailoring resins to a specific application. Because blends are only physical mixtures, the resulting polymer usually has physical and mechanical properties that lie somewhere between the values of its constituent materials. For instance, an automotive bumper made from a blend of polycarbonate resin and thermoplastic polyurethane elastomer gains rigidity from the polycarbonate resin and retains most of the flexibility and paintability of the polyurethane elastomer. For business machine housings, a blend of polycarbonate and acrylonitrile butadiene styrene (ABS) resins offers the enhanced performance of polycarbonate flame retardance and UV stability at a lower cost.

Additional information on the subject of polymer blends is available in the literature.[1–3]

3.8 Elastomers

Elastomers are a class of polymeric materials that can be repeatedly stretched to over twice the original length with little or no permanent deformation. Elastomers can be made of either thermoplastic or thermoset materials and generally are tested and categorized differently than rigid materials. They are commonly selected according to their hardness and energy absorption characteristics, properties rarely considered in rigid thermoplastics. Elastomers are found in numerous applications, such as automotive bumpers and industrial hoses.

3.9 Additives

Additives encompass a wide range of substances that aid processing or add value to the final product.[4,5] Found in virtually all plastics, most additives are incorporated into a resin family by the supplier as part of a proprietary package. For example, you can choose standard polycarbonate resin grades with additives for improved internal mold release, UV stabilization, and flame retardance; or nylon grades with additives to improve impact performance.

Additives often determine the success or failure of a resin or system in a particular application. Many common additives are discussed in the following sections. Except for reinforcement fillers, most additives are added in very small amounts.

3.9.1 Fillers, Reinforcement, Composites

Reinforcing fillers can be added in large amounts. Some plastics may contain as much as 60% reinforcing fillers. Often, fibrous materials, such as glass or carbon fibers, are added to resins to create reinforced grades with enhanced properties. For example, adding 30% short glass fibers by weight to Nylon 6 improves creep resistance and increases stiffness by 300%. These glass reinforced plastics usually suffer some loss of impact strength and ultimate elongation, and are more prone to warping because of the relatively large difference in mold shrinkage between the flow and cross-flow directions.

Plastics with nonfibrous fillers such as glass spheres or mineral powders generally exhibit higher stiffness characteristics than unfilled resins, but not as high as fiber reinforced grades. Resins with particulate fillers are less likely to warp and show a decrease in mold shrinkage. Particulate fillers typically reduce shrinkage by a percentage roughly equal to the volume percentage of filler in the polymer, an advantage in tight tolerance molding.

Often reinforced plastics are called *composites*. Often, the plastic material containing the reinforcement is referred to as the matrix. One can envision a number of ways different reinforcing materials might be arranged in a composite. Many of these arrangements are shown in Figure 3.11.

In general fiber reinforcement improves the fatigue strength of a plastic over its unreinforced analog. This is clearly demonstrated in Figure 3.12. It generally follows that carbon fibers enhance the performance over glass fibers at equal loading.

The effect of particulate reinforcement on fatigue properties is less clear and has not been studied as much as fibrous reinforcement.

3.9.2 Combustion Modifiers, Fire, Flame Retardants and Smoke Suppressants

Combustion modifiers are added to polymers to help retard the resulting parts from burning. Generally required for electrical and medical housing applications, combustion modifiers and their amounts vary with the inherent flammability of the base polymer. Polymers designed for these applications often are rated using an Underwriters Laboratories rating

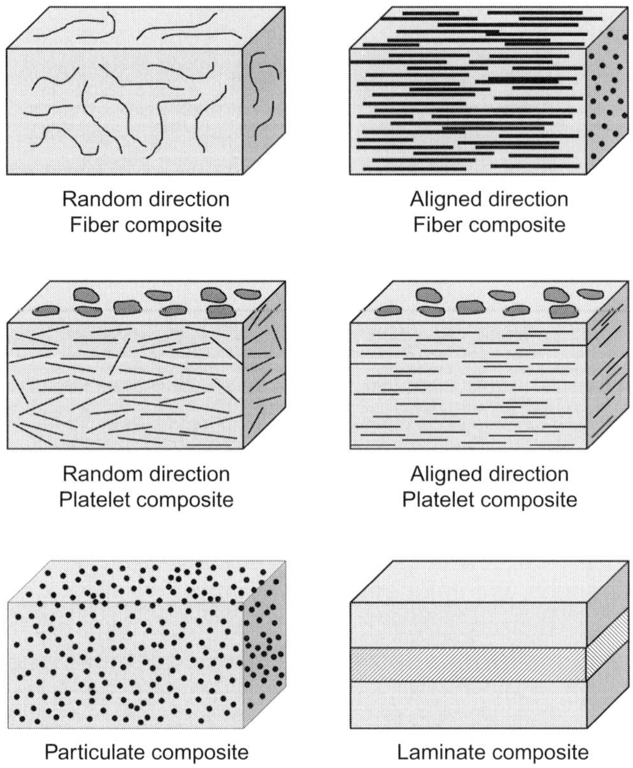

Figure 3.11 Several types of composite materials.

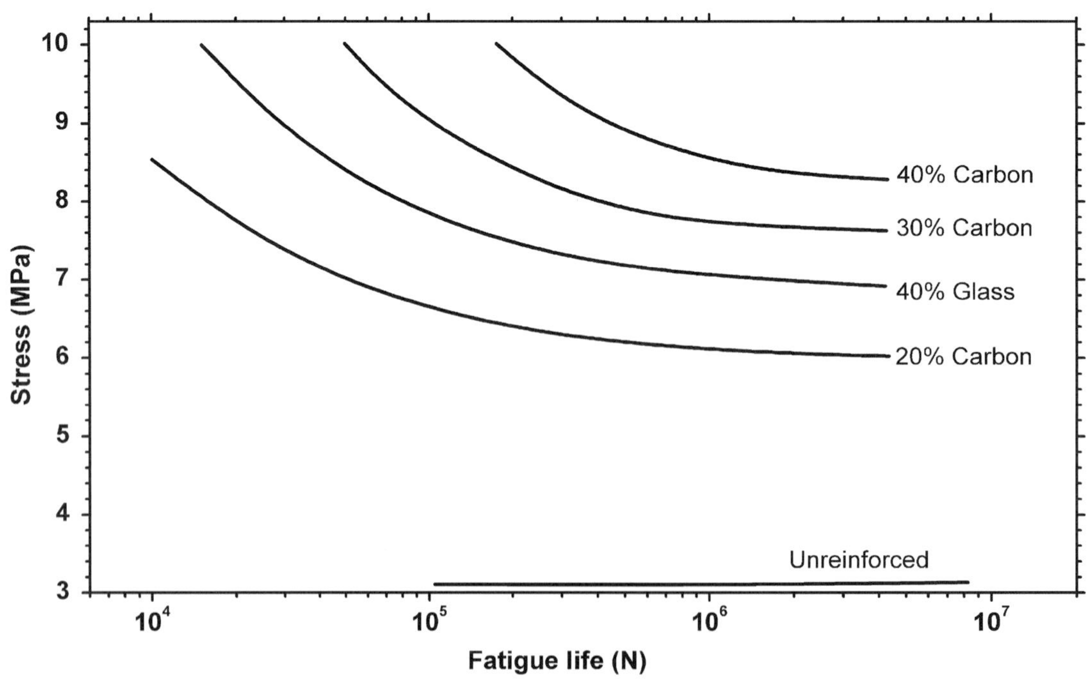

Figure 3.12 Fatigue life comparison of carbon and glass fiber reinforced nylon.

system. Use these ratings for comparison purposes only, as they may not accurately represent the hazard present under actual fire conditions.

3.9.3 Release Agents and Antiblocking Agents

External release agents are lubricants, liquids or powders, which coat a mold cavity to facilitate part removal. Internal release agents can accomplish the same purpose. The identity of the release agent is rarely disclosed, but frequently they are fine fluoropolymer powders, called micropowders, silicone resins, or waxes.

3.9.4 Lubricants and Slip Agents, Tribology Additives

As discussion of many additives used to improve slip, dry lubrication and abrasion resistance were discussed in Chapter 2. They are summarized again here.

- PFPE synthetic oil marketed under the trademark Fluoroguard® is an internal lubricant that imparts improved wear and low-friction properties.
- PTFE imparts the lowest coefficient of friction of any internal lubricant.
- Silicone acts as a boundary lubricant because it migrates to the surface of the plastic over time.
- Molybdenum disulfide, commonly called *moly* is a solid lubricant often used in bearing applications.
- Graphite is a solid lubricant used like molybdenum disulfide.
- Carbon fiber improves mechanical and thermal performance which leads to higher PV limits. Carbon fiber reinforced plastics may have lower coefficient of friction than the base resin. Carbon is softer and less abrasive than glass fiber and some carbon fiber compounds can dissipate static electricity.
- Aramid Fiber, Kevlar® being one, is softer and less abrasive than carbon or glass fiber, this additive is most commonly used for reduction in the wear of the mating surface, especially softer materials.

3.9.5 Catalysts

Catalysts, substances that initiate or change the rate of a chemical reaction, do not undergo a permanent change in composition or become part of the molecular structure of the final product. Occasionally used to describe a setting agent, hardener, curing agent, promoter, etc., they are added in minute quantities, typically less than 1%.

3.9.6 Impact Modifiers and Tougheners

Many plastics do not have sufficient impact resistance for the use for which they are intended. Rather than change to a different type of plastic, they can be impact modified in order to fulfill the performance in use requirements. Addition of modifiers called impact modifiers or tougheners can significantly improve impact resistance. This is one of the most important additives. There are many suppliers and chemical types of these modifiers.

General-purpose impact modification is a very low level of impact modification. It improves room temperature impact strength but does not take into account any requirements for low-temperature (below 0°C) impact strength. For most of these types of applications only low levels of impact modifier will be required (<10%).

Low-temperature impact strength is required for applications that require a certain level of low-temperature flexibility and resistance to break. This is, for example, the case for many applications in the appliance area. For this purpose modifier levels between 5% and 15% of mostly reactive modifiers will be necessary. Reactive modifiers can bond chemically to the base polymer.

Super tough impact strength may be required for applications that should not lead to a failure of the part even if hit at low temperatures (−30°C to −40°C) under high speed. This requirement can only be fulfilled with high levels (20–25%) of reactive impact modifier with low glass transition temperature.

Figure 3.13 shows the effect of one toughener on the Izod performance of a common Nylon 6 plastic. The toughener used in this graph is DuPont's Fusabond®N MN-493D. The graph shows the improvement in notched Izod performance versus temperature with differing levels of toughener additive. As shown in this figure, the performance can be dramatically improved.

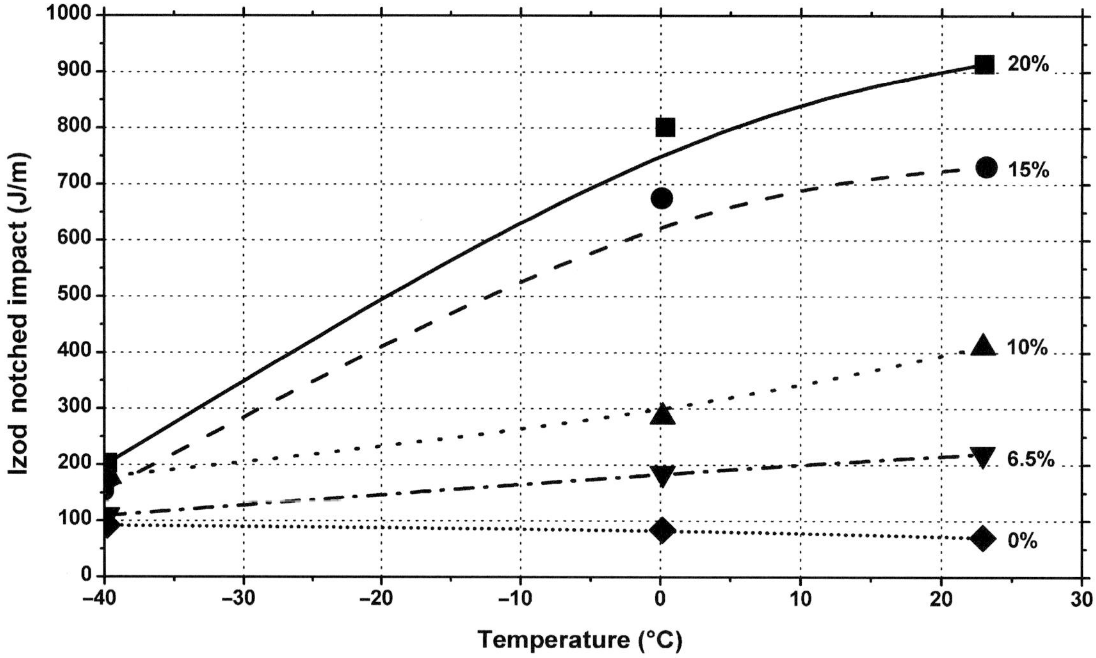

Figure 3.13 Notched Izod of BASF Ultramid® B-3Nylon 6 modified with various levels of Fusabond® N NM-493D toughener.

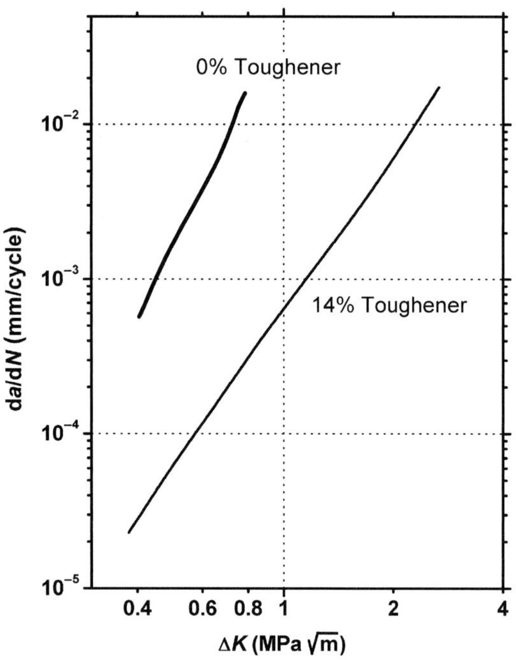

Figure 3.14 The effect of adding methacrylate–butadiene styrene rubber (MBS) toughener to PVC on the fatigue crack propagation rate.

In general toughened plastics are more fatigue crack propagation resistant than the corresponding untoughened analogs. This is graphically demonstrated in Figure 3.14 which compares the fatigue crack propagation rates of toughened and untoughened PVC.

3.9.7 UV Stabilizers

Sunshine and its UV radiation have a deteriorating effect on many polymers. UV stabilizers play an important role in plastics for external uses by counteracting the effects of the sun. UV stabilizers are used in plastic items such as greenhouse film, outdoor furniture, and automotive plastic parts. The amounts added are very small, generally less than 1%.

3.9.8 Antistatic Agents

Antistatic additives are capable of modifying properties of plastics in such a way that they become antistatic, conductive, and/or improve electromagnetic interference shielding (EMI). Carbon fibers, conductive carbon powders, and other electrically conductive materials are used for this purpose.

3.9.9 Plasticizers

Plasticizers are added to help maintain flexibility in a plastic. Various phthalates are commonly used for this purpose. Since they are small molecules they may extract or leach out of the plastic causing a loss of flexibility with time. Just as purposely added

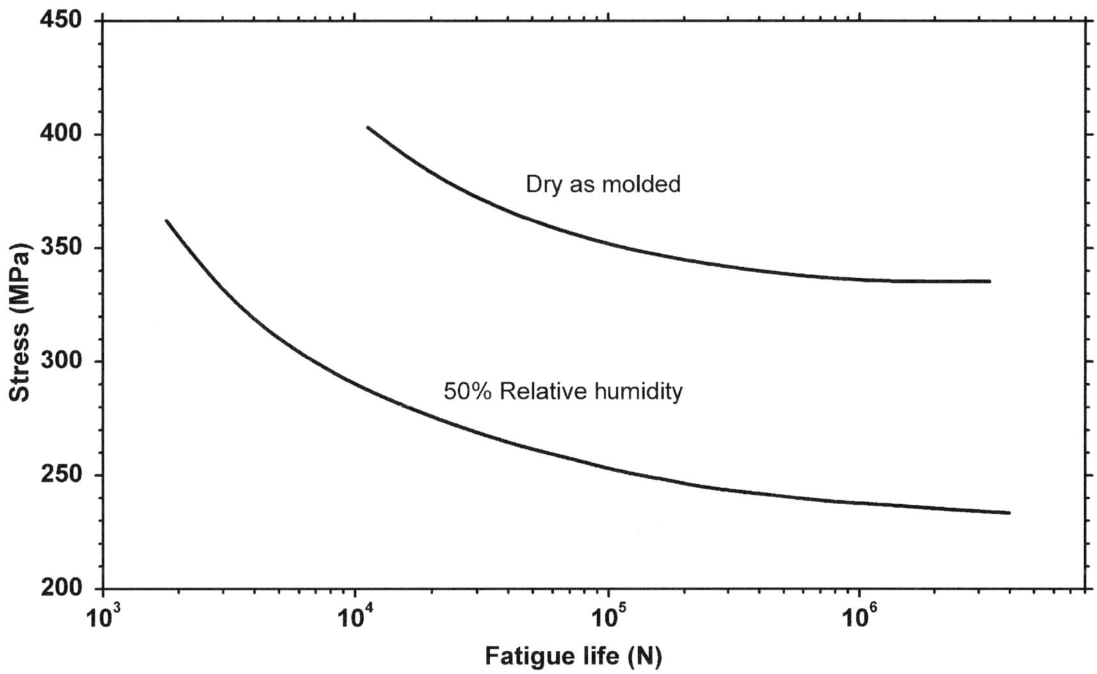

Figure 3.15 The effect of plasticization by water absorption on the flexural fatigue of a polyamide (nylon) resin.

small molecules may leach out, small molecules from the environment may be absorbed by the plastic and act like a plasticizer as shown in Figure 3.15.

3.9.10 Pigments, Extenders, Dyes, Mica

Pigments are added to give a plastic color, but they may also affect the physical properties. Extenders are usually cheap materials added to reduce the cost of plastic resins. Dyes are colorants chemically different than pigments. Mica is a special pigment added to impact sparkle or metallic appearance.

3.9.11 Coupling Agents

The purpose of adding fillers is either to lower the cost of the polymer, make it tougher or stiffer or make it flame retardant so that it does not burn when it is ignited. Often the addition of the filler will reduce the elongation at break, the flexibility and in many cases the toughness of the polymer because the fillers are added at very high levels. One reason for the degradation of properties is that the fillers in most cases are not compatible with the polymers. The addition of coupling agents can improve the compatibility of the filler with the polymer. As a result the polymer will like the filler more, the filler will adhere better to the polymer matrix and the properties of the final mixture (e.g., elongation, flexibility) will be enhanced.

3.9.12 Thermal Stabilizers

One of the limiting factors in the use of plastics at high temperatures is their tendency to not only become softer but also thermally degrade. Thermal degradation can present an upper limit to the service temperature of plastics. Thermal degradation can occur at temperatures much lower than those at which mechanical failure is likely to occur. Plastics can be protected from thermal degradation by incorporating stabilizers into them. Stabilizers can work in a variety of ways but discussion of these mechanisms are beyond the purpose of this book.

There are other additives used in plastics, but the ones discussed above are the most common.

3.10 Summary

These first three chapters provide a good basis for analyzing the fatigue and tribology data that follow in the next chapters. If the particular plastic manufacturer's grade is not found in these chapters, one may find a composition similar from another manufacturer. If the composition is nearly the same, one should expect that data to be representative.

References

1. Utracki LA. *Polymer blends handbook,* vol 1–2. Springer-Verlag. Online version available at: http://www.knovel.com/knovel2/Toc.jsp?BookID=1117&VerticalID=0: 2002.
2. Utracki LA. *Commercial polymer blends*. Springer-Verlag. Online version available at: http://www.knovel.com/knovel2/Toc.jsp?BookID=878&VerticalID=0: 1998.
3. Utracki LA. *Encyclopaedic dictionary of commercial polymer blends*. ChemTec Publishing, 1994. Online version available at: http://www.knovel.com/knovel2/Toc.jsp?BookID=285&VerticalID=0: 1994.
4. Flick EW. *Plastics additives – an industrial guide.* 2nd ed. William Andrew Publishing/Noyes. Online version available at: http://www.knovel.com/knovel2/Toc.jsp?BookID=353&VerticalID=0: 1993.
5. Pritchard G. *Plastics additives – an A–Z reference*. Springer-Verlag. Online version available at: http://www.knovel.com/knovel2/Toc.jsp?BookID=335&VerticalID=0: 1998.

4 Styrenic Plastics

4.1 Background

This chapter on styrenic plastics covers a broad class of polymeric materials of which an important part is styrene. Styrene, also known as vinyl benzene, is an organic compound with the chemical formula $C_6H_5CH = CH_2$. Its structure is shown in Figure 4.1.

It is used as a monomer to make plastics such as polystyrene, acrylonitrile butadiene styrene (ABS), styrene acrylonitrile (SAN), and the other polymers in this chapter.

4.1.1 Polystyrene

Polystyrene is the simplest plastic based on styrene. Its structure is shown in Figure 4.2.

Pure solid polystyrene is a colorless, hard plastic with limited flexibility. Polystyrene can be transparent or can be made in various colors. It is economical and is used for producing plastic model assembly kits, plastic cutlery, CD "jewel" cases, and many other objects where a fairly rigid, economical plastic is desired.

Polystyrene's most common use, however, is expanded polystyrene (EPS). EPS is produced from a mixture of about 5–10% gaseous blowing agent (most commonly pentane or carbon dioxide) and 90–95% polystyrene by weight. The solid plastic beads are expanded into foam through the use of heat (usually steam). The heating is carried out in a large vessel holding 200–2000 liters. An agitator is used to keep the beads from fusing together. The expanded beads are lighter than unexpanded beads so they are forced to the top of the vessel and removed. This expansion process lowers the density of the beads to 3% of their original value and yields a smooth-skinned, closed cell structure. Next, the preexpanded beads are usually "aged" for at least 24 hours in mesh storage silos. This allows air to diffuse into the beads, cooling them, and making them harder. These expanded beads are excellent for detailed molding. Extruded polystyrene (XPS), which is different from EPS, is commonly known by the trade name Styrofoam™. All these foams are not of interest in this book.

One of the most important plastics is high-impact polystyrene, or HIPS. This is a polystyrene matrix that is imbedded with an impact modifier, which is basically a rubber like polymer such as polybutadiene. This is shown in Figure 4.3.

4.1.2 Acrylonitrile Styrene Acrylate

ASA is the acronym for acrylate rubber–modified styrene–acrylonitrile copolymer. ASA is a terpolymer that can be produced by either a reaction process of all three monomers or a graft process. ASA is usually made

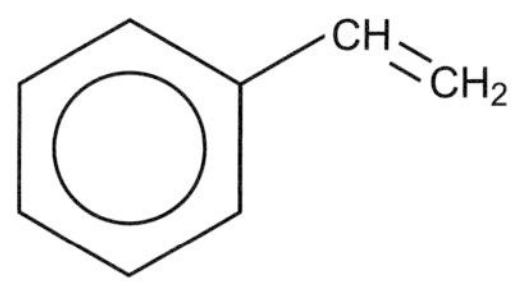

Figure 4.1 Chemical structure of styrene.

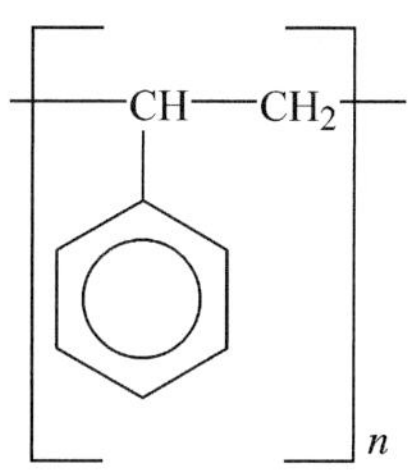

Figure 4.2 Chemical structure of polystyrene.

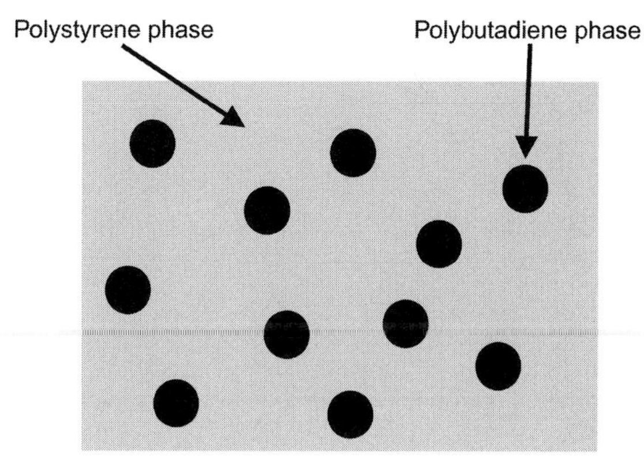

Figure 4.3 The structure of HIPS.

Copyright © 2010 Laurence W. McKeen. All rights reserved.

by introducing a grafted acrylic ester elastomer during the copolymerization of styrene and acrylonitrile, known as SAN. SAN is described in the next section of this chapter. The finely divided elastomer powder is uniformly distributed and grafted to the SAN molecular chains. The outstanding weatherability of ASA is due to the acrylic ester elastomer. ASA polymers are amorphous plastics, which have mechanical properties similar to those of the ABS resins described in Section 4.1.4. However, the ASA properties are far less affected by outdoor weathering.

ASA resins are available in natural, off-white, and a broad range of standard and custom-matched colors. ASA resins can be compounded with other polymers to make alloys and compounds that benefit from ASA's weather resistance. ASA is used in many products including lawn and garden equipment, sporting goods, automotive exterior parts, safety helmets, and building materials.

4.1.3 Styrene Acrylonitrile

Styrene and acrylonitrile monomers can be copolymerized to form a random, amorphous copolymer that has good weatherability, stress crack resistance, and barrier properties. The copolymer is called styrene acrylonitrile or SAN. The SAN copolymer generally contains 70–80% styrene and 20–30% acrylonitrile. It is a simple random copolymer. This monomer combination provides higher strength, rigidity, and chemical resistance than polystyrene, but it is not quite as clear as crystal polystyrene and its appearance tends to discolor more quickly. The general structure is shown in Figure 4.4.

SAN is used for household goods and tableware, in cosmetics packaging, sanitary and toiletry articles as well as for writing materials and office supplies.

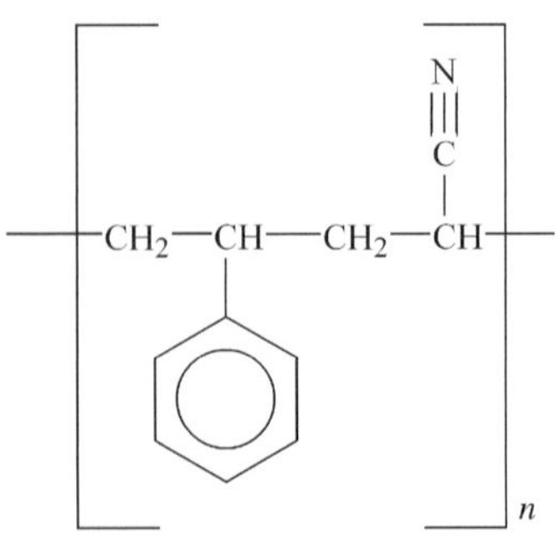

Figure 4.4 Chemical structure of SAN.

4.1.4 Acrylonitrile Butadiene Styrene

Acrylonitrile butadiene styrene, or ABS, is a common thermoplastic used to make light, rigid, molded products such as pipe, automotive body parts, wheel covers, enclosures, and protective head gear.

SAN copolymers have been available since the 1940s and while its increased toughness over styrene made it suitable for many applications, its limitations led to the introduction of a rubber, butadiene, as a third monomer producing the range of materials popularly referred to as ABS plastics. These became available in the 1950s and the availability of these plastics and ease of processing led ABS to become one of the most popular of the engineering polymers.

The chemical structures of the monomers are shown in Figure 4.5. The proportions of the monomers typically range from 15% to 35% acrylonitrile, 5% to 30% butadiene and 40% to 60% styrene. It can be found as a graft copolymer, in which SAN polymer is formed in a polymerization system in the presence of polybutadiene rubber latex; the final product is a complex mixture consisting of SAN copolymer, a graft polymer of styrene acrylonitrile and polybutadiene and some free polybutadiene rubber.

4.1.5 Methyl Methacrylate Acrylonitrile Butadiene Styrene

Methyl methacrylate acrylonitrile butadiene styrene, or MABS, is a newer modification of ABS. It is sometimes called transparent ABS, a copolymer of methyl methacrylate, acrylonitrile, butadiene, and styrene (MABS). Key properties of MABS are excellent transparency, high-impact strength, and good chemical resistance. This is an exceptional combination of properties for an impact-modified thermoplastic. MABS can be used to create particularly brilliant visual effects such as very deep colors, pearly or sparkle effects. It is easy to process and can also be printed upon.

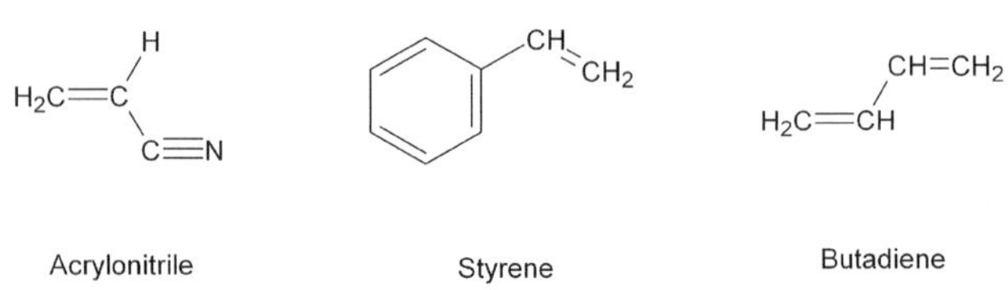

Figure 4.5 Chemical structure of ABS raw materials.

4.1.6 Styrene Maleic Anhydride

Copolymerization of styrene with maleic anhydride creates a copolymer called styrene maleic anhydride (SMA). This reaction is shown in Figure 4.6. SMA has a higher glass transition temperature than polystyrene and is chemically reactive because of active functional groups. Thus, SMA polymers are often used in blends or composites where interaction or reaction of the maleic anhydride provides for desirable interfacial effects. The anhydride reaction with primary amines is particularly potent.

4.1.7 Styrenic Block Copolymers

Styrenic block copolymer, or SBC, is a commercially important thermoplastic elastomer. The polymer is made of three separate polymeric blocks (see Section 3.2 for an explanation of block copolymers). At one end is a hard polystyrene block, in the middle a long polybutadiene (or other elastomeric) block, followed by a second hard block of polystyrene. These blocks are immiscible, so they form discrete domains of polystyrene within a polybutadiene matrix. The separate domains are chemically connected. This is shown in Figure 4.7, where one might notice that this looks a lot like HIPS, except that the continuous phase and hard discrete phase are switched in SBC and the domains are connected. One additional property of interest is that some SBCs blend well with general-purpose polystyrene, allowing customization of properties.

4.1.8 Styrenic Blends

While the number of styrenic blends might seem limitless, compatibility and morphology limit blend types. Styrenic blends are numerous but most are limited to only a couple of types. The most important blend is ABS and polycarbonate (PC). Next in importance is ABS and polyamide (or nylon, PA). Polystyrene and polyethylene are often used in expandable foams. Polystyrene and polyphenylene ether (PPE or PPO) are commercially important blends, which are covered in a later chapter. The other classes of the styrenic blends are not major product lines but can be very important in some applications.

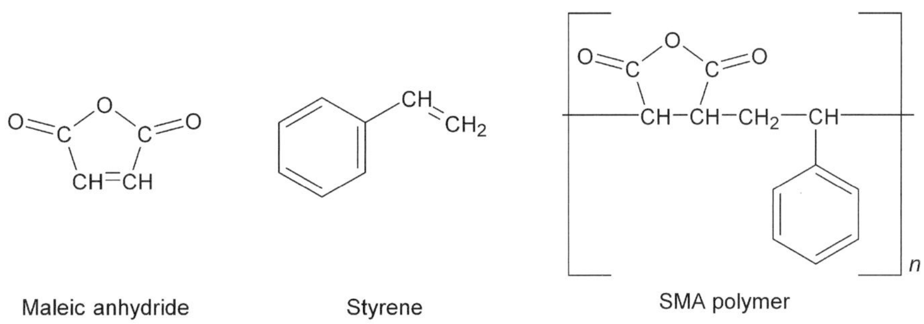

Figure 4.6 The production of SMA.

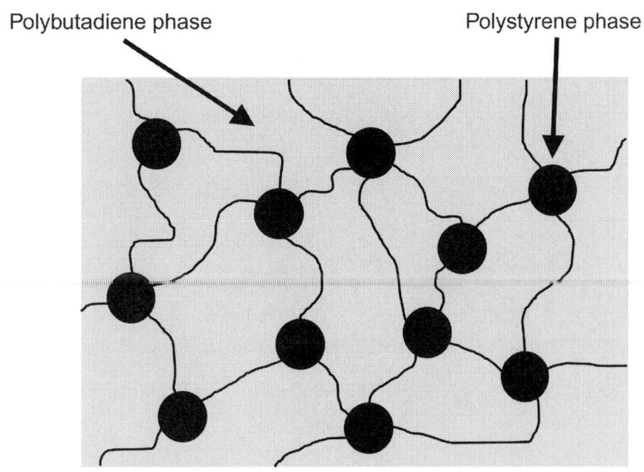

Figure 4.7 The "microscopic" structure of SBC.

4.2 Polystyrene

4.2.1 Fatigue Data

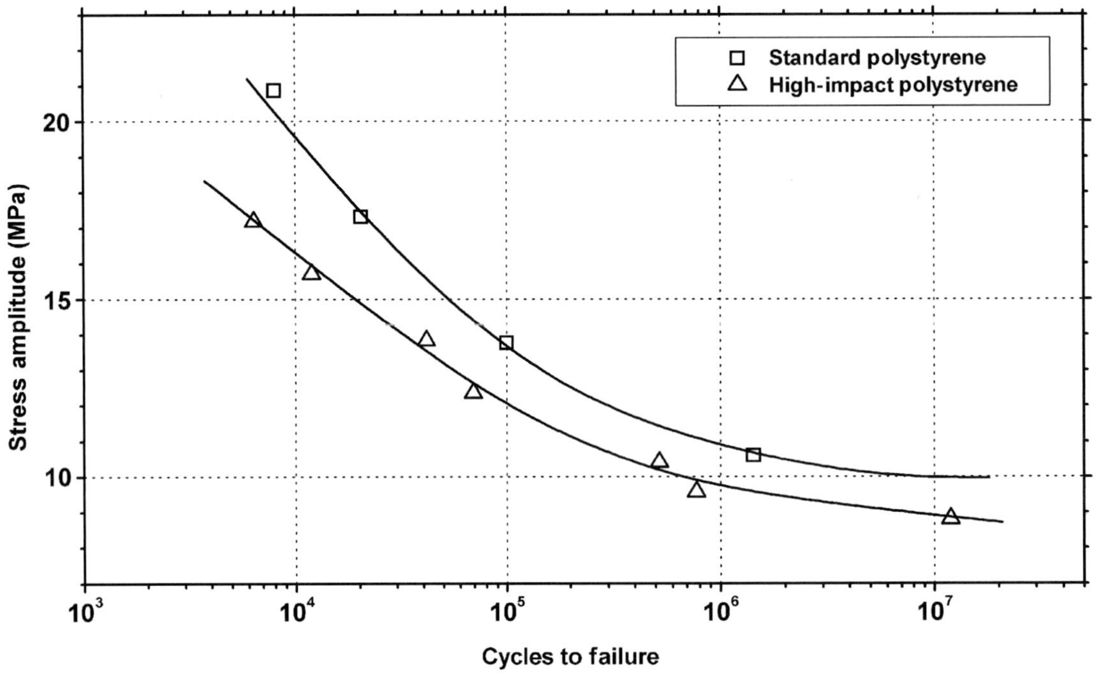

Figure 4.8 Stress amplitude vs. cycles to failure for HIPS and standard polystyrene.

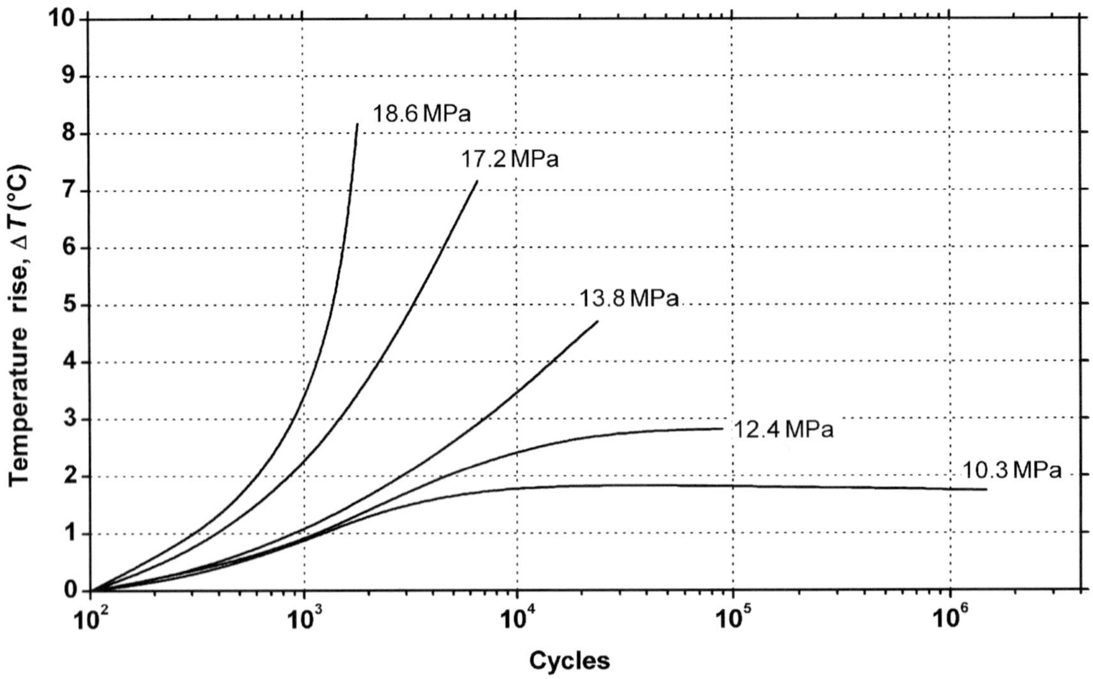

Figure 4.9 Test specimen temperature rise HIPS vs. the number of fatigue cycles at several different stress amplitudes.

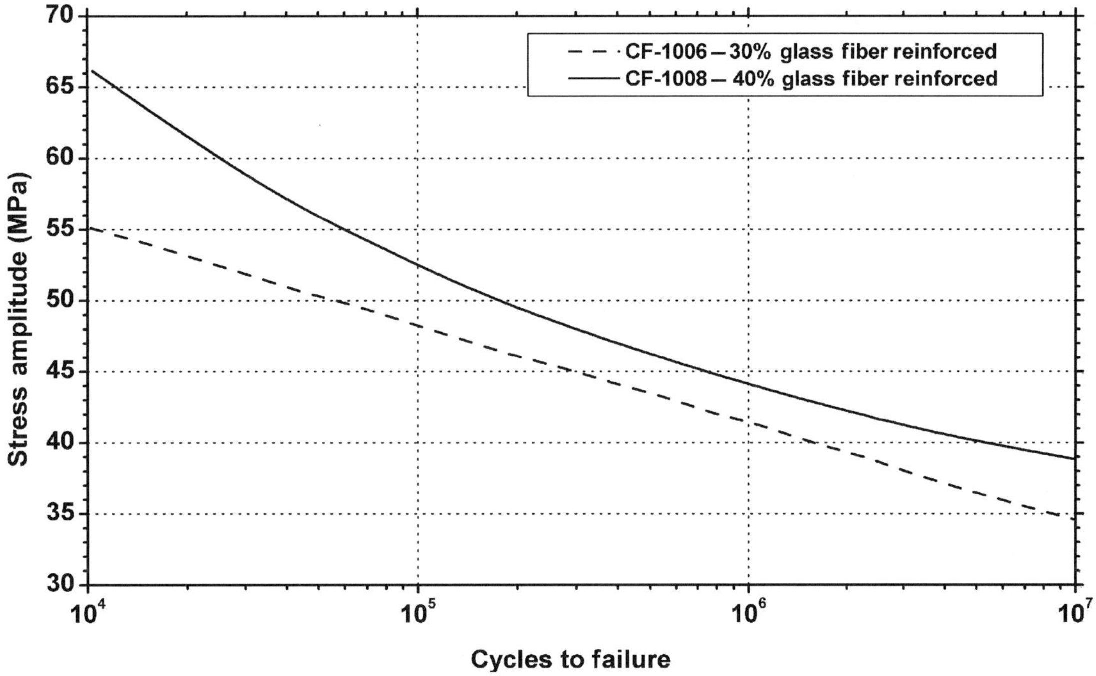

Figure 4.10 Flexural stress amplitude vs. cycles to failure for SABIC Innovative Plastics Thermocomp® glass fiber reinforced polystyrene.

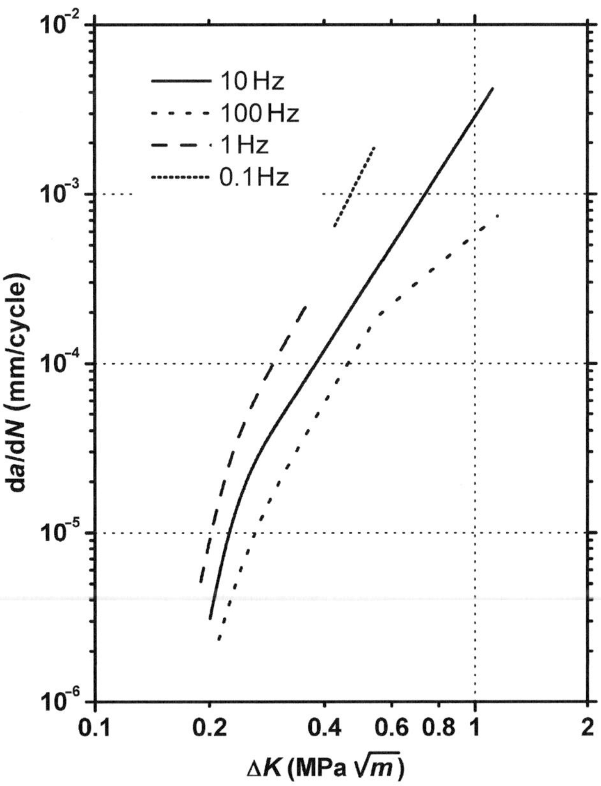

Figure 4.11 Fatigue crack propagation rates of polystyrene at different test frequencies.

4.3 Acrylonitrile Styrene Acrylate

4.3.1 Fatigue Data

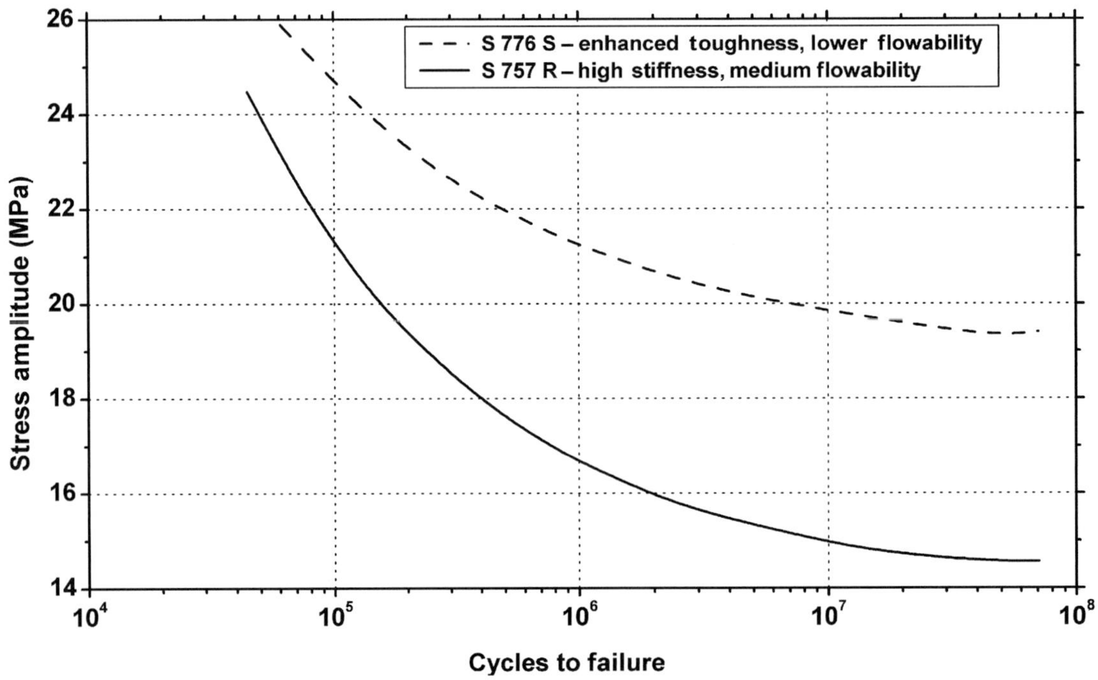

Figure 4.12 Flexural stress amplitude vs. cycles to failure for two BASF Luran® ASA plastics.

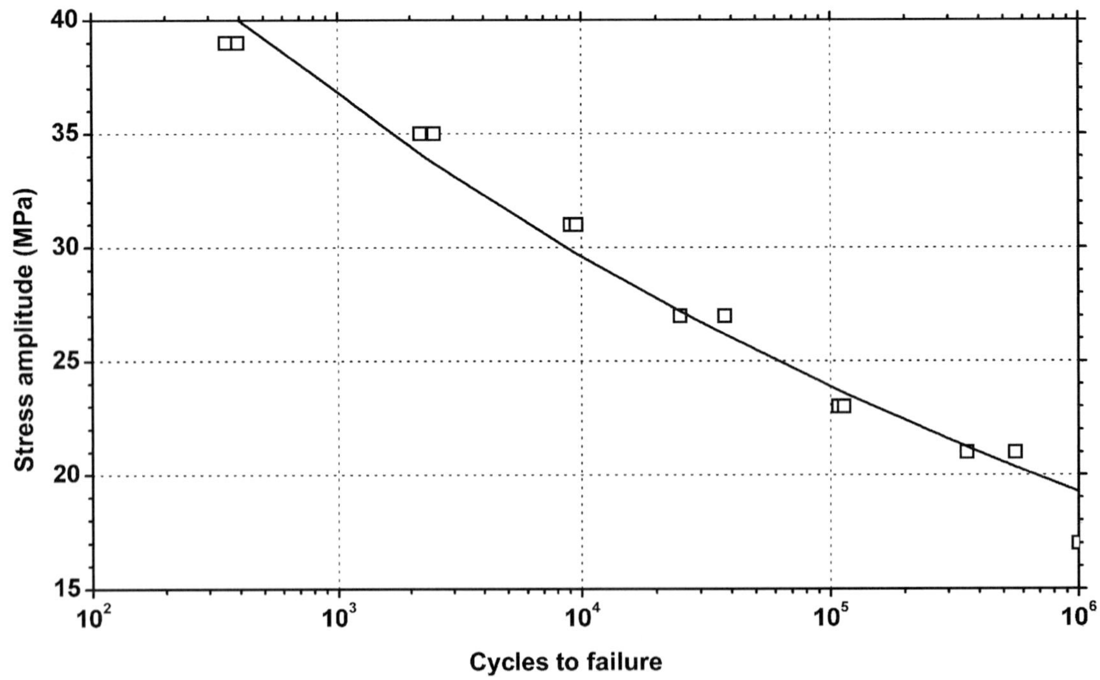

Figure 4.13 Tensile stress amplitude vs. cycles to failure for SABIC Innovative Plastics Geloy® CR7010—sheet extrusion, high-impact ASA.

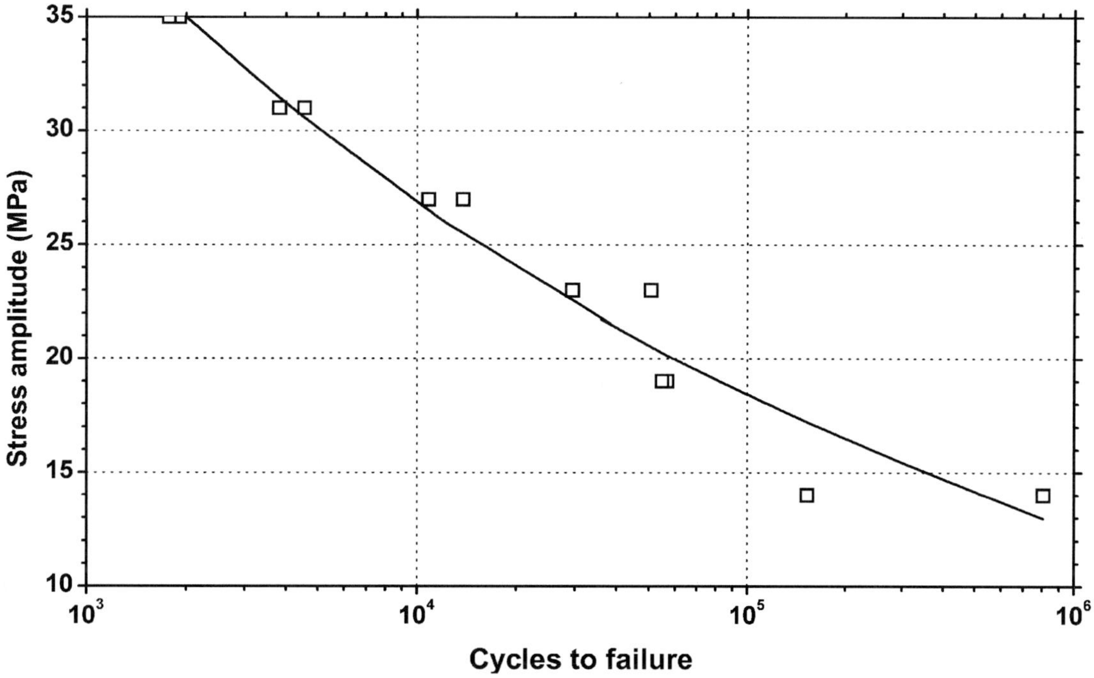

Figure 4.14 Tensile stress amplitude vs. cycles to failure for SABIC Innovative Plastics Geloy® CR7020 ASA for sheet coextrusion over ABS.

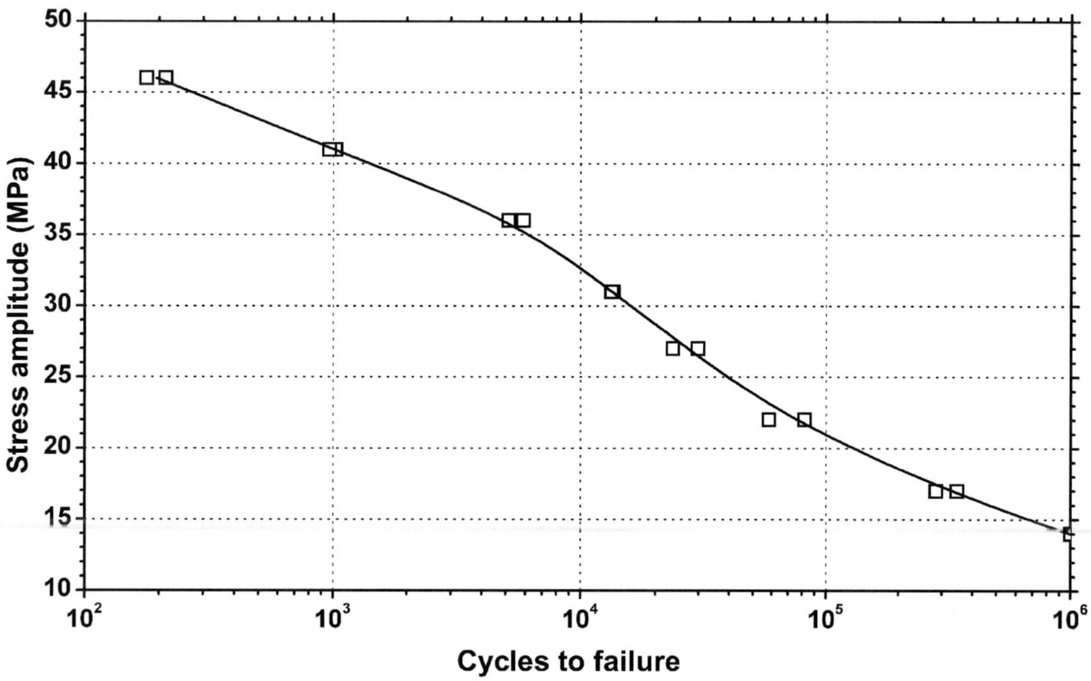

Figure 4.15 Tensile stress amplitude vs. cycles to failure for SABIC Innovative Plastics Geloy® CR7510—high heat, automotive exterior ASA.

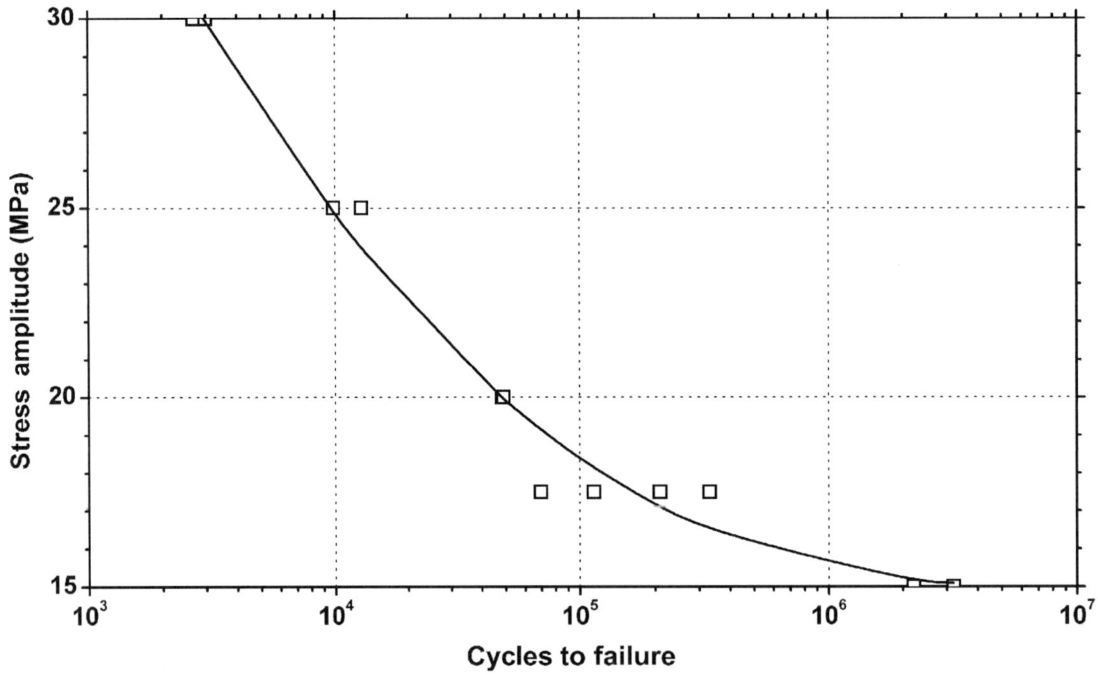

Figure 4.16 Tensile stress amplitude vs. cycles to failure for SABIC Innovative Plastics Geloy® CR7520—high-impact, good flow automotive exterior ASA.

4.4 Styrene Acrylonitrile

4.4.1 Fatigue Data

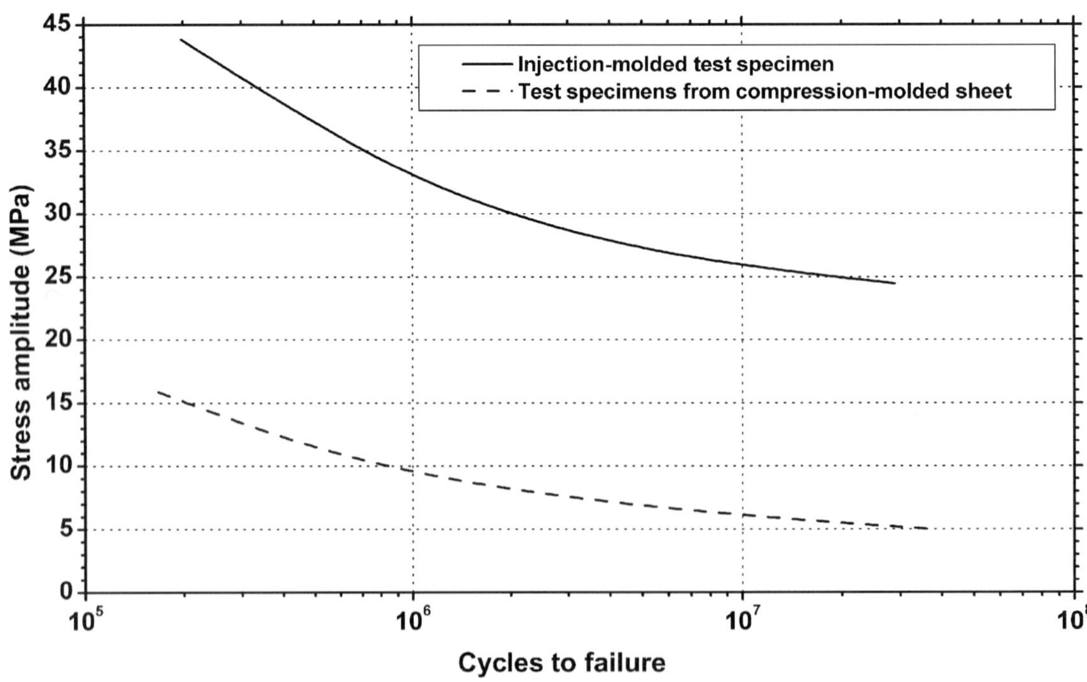

Figure 4.17 Flexural stress amplitude vs. cycles to failure for BASF Luran® 368 R—general-purpose SAN.

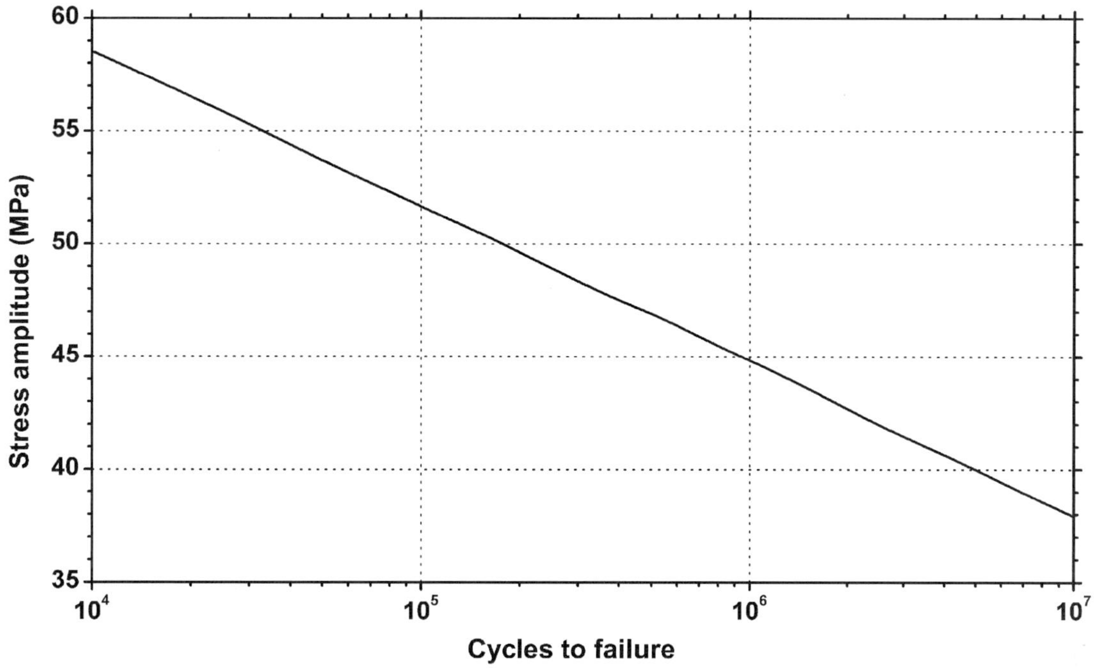

Figure 4.18 Flexural stress amplitude vs. cycles to failure for SABIC Innovative Plastics Thermocomp® BF-1006—30% glass fiber filled SAN.

4.5 Acrylonitrile Butadiene Styrene

4.5.1 Fatigue Data

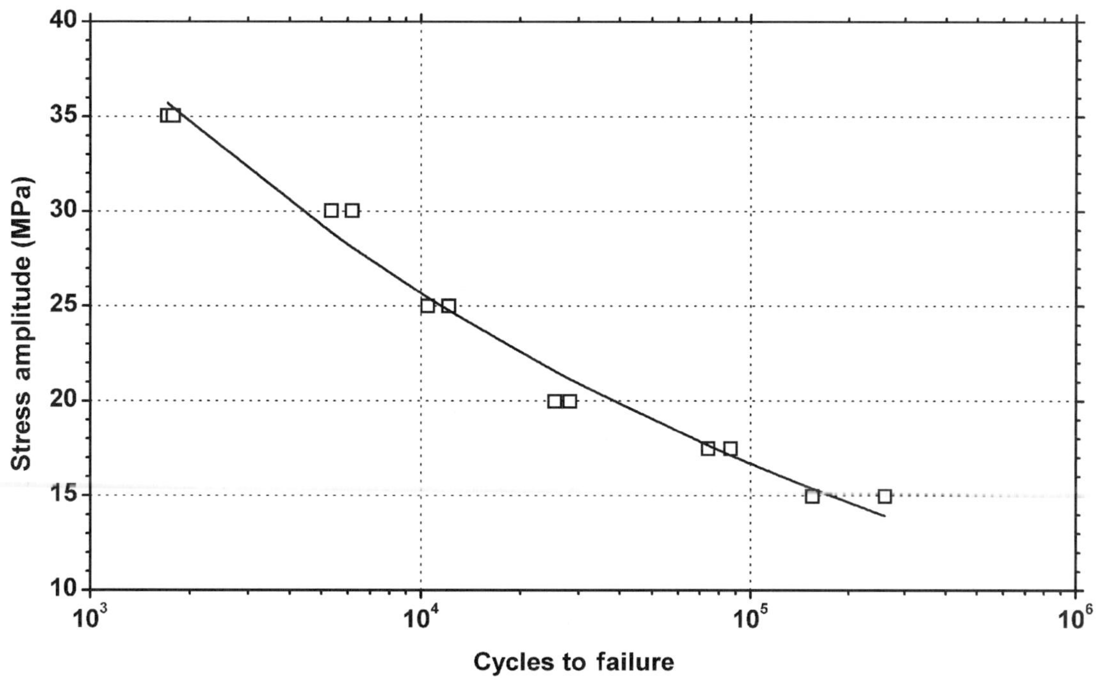

Figure 4.19 Tensile stress amplitude vs. cycles to failure for SABIC Innovative Plastics Cycolac® G-100 ABS.

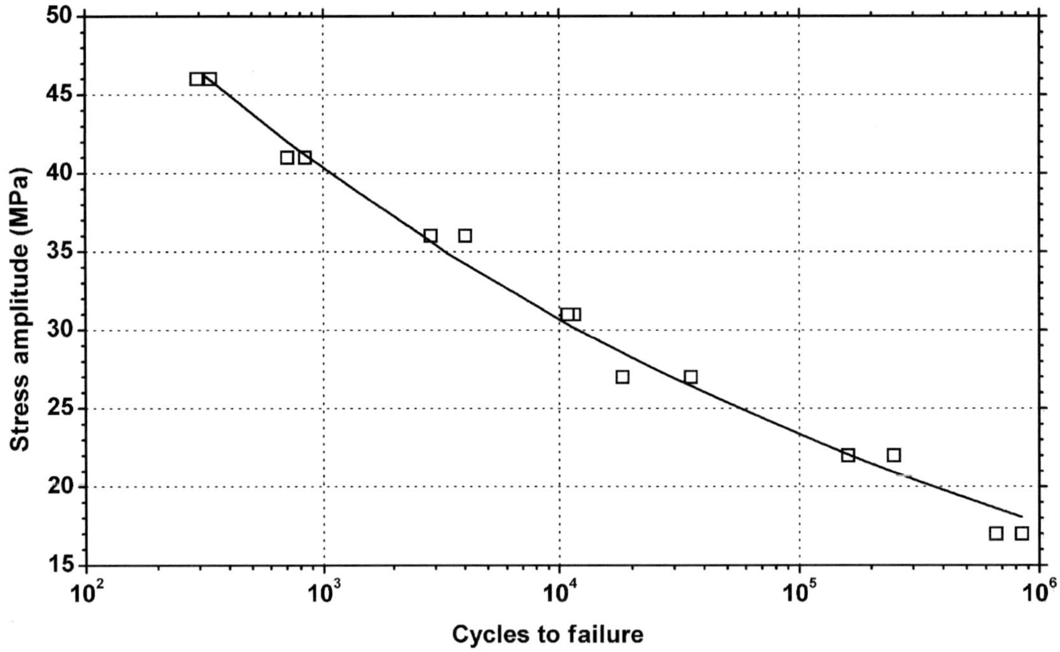

Figure 4.20 Tensile stress amplitude vs. cycles to failure for SABIC Innovative Plastics Cycolac® BDT5510 ABS.

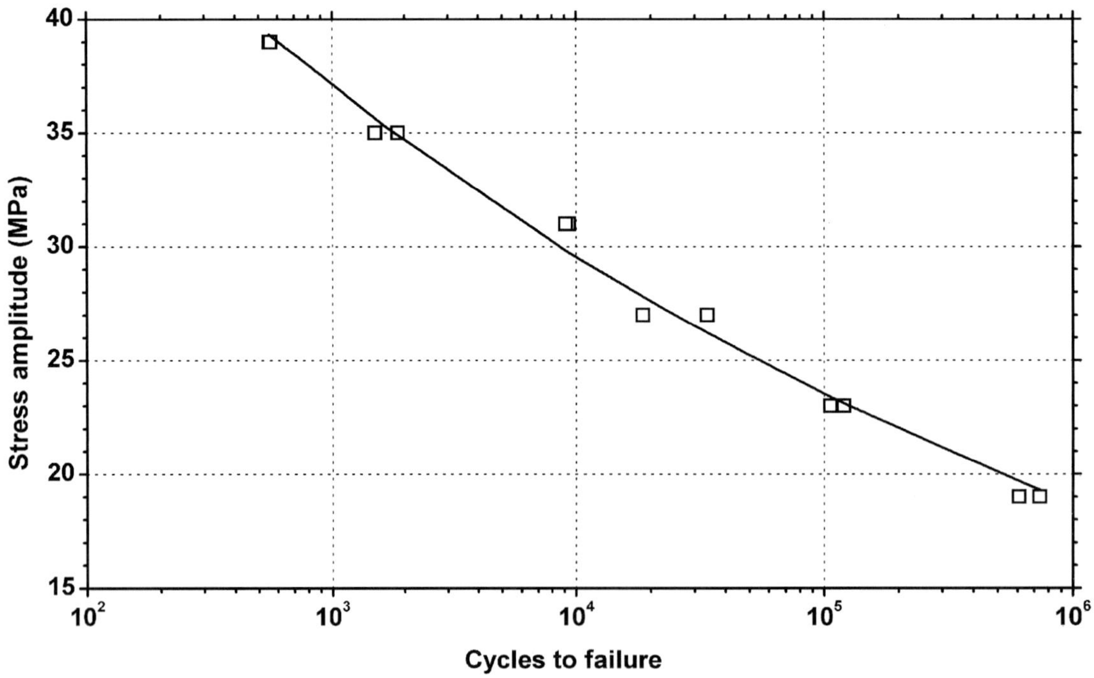

Figure 4.21 Tensile stress amplitude vs. cycles to failure for SABIC Innovative Plastics Cycolac® BDT6500—ABS for automotive interior applications, low gloss, color concentratable.

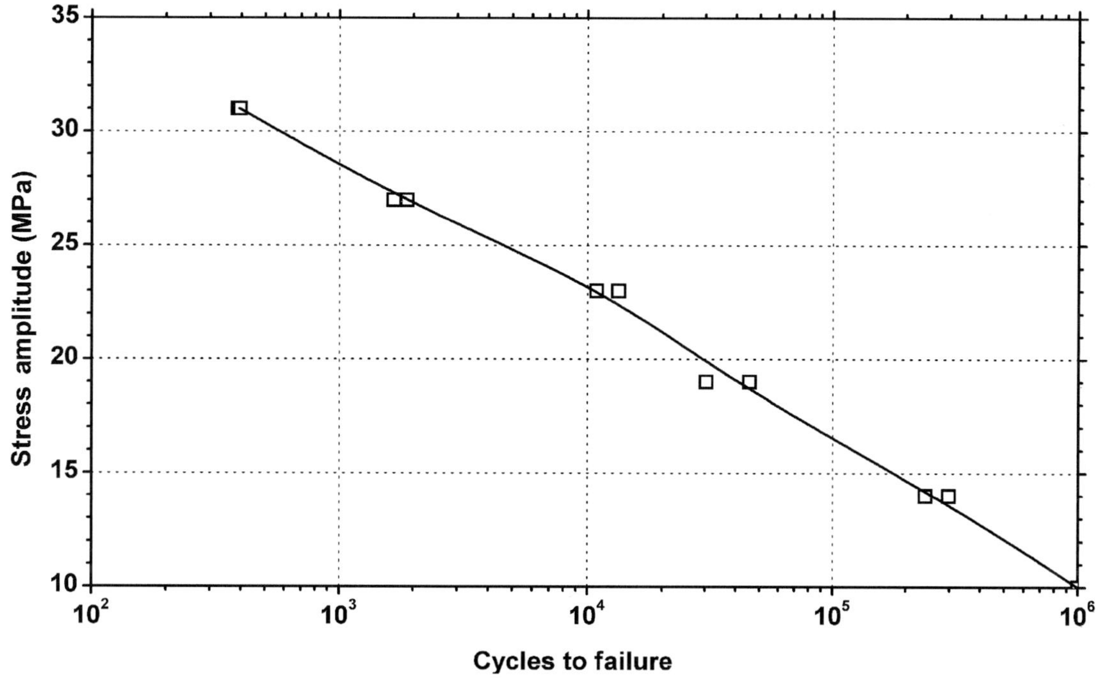

Figure 4.22 Tensile stress amplitude vs. cycles to failure for SABIC Innovative Plastics Cycolac® CGA—ABS for extrusion and thermoforming.

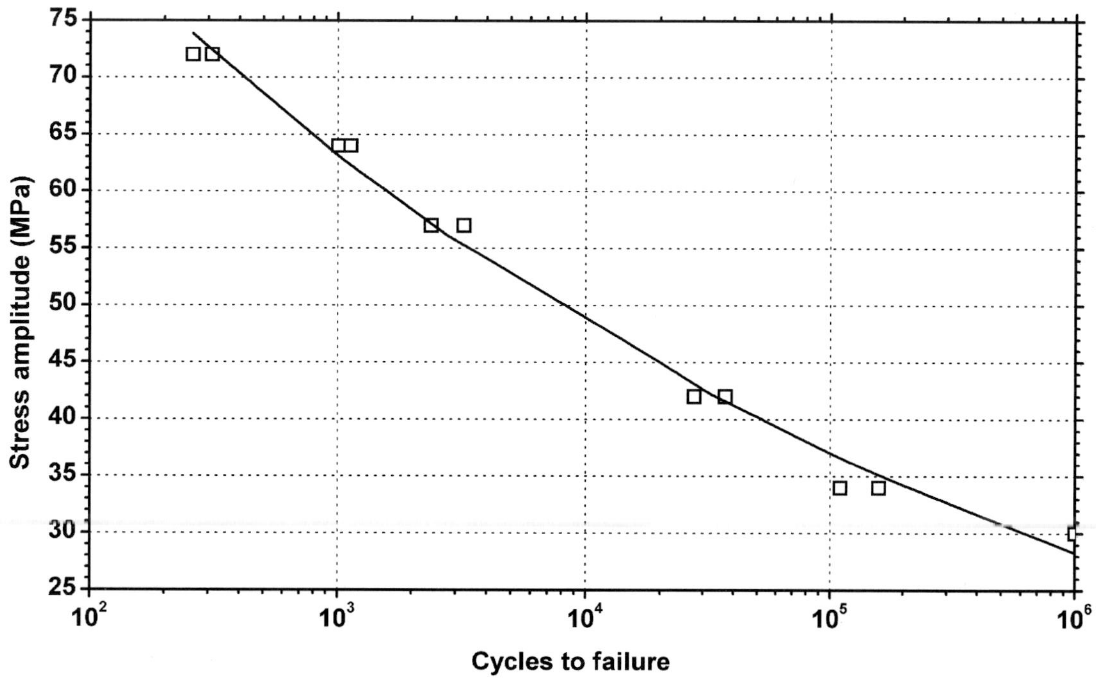

Figure 4.23 Tensile stress amplitude vs. cycles to failure for SABIC Innovative Plastics Cycolac® CGF20—20% glass fiber filled, high flow ABS.

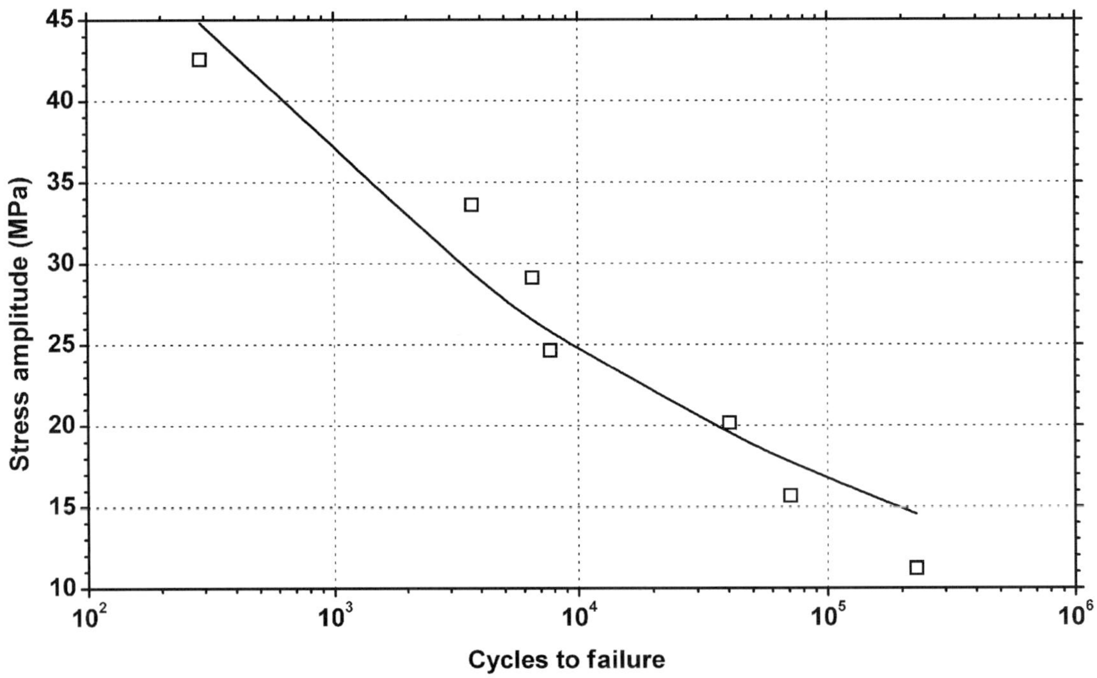

Figure 4.24 Tensile stress amplitude vs. cycles to failure for SABIC Innovative Plastics Cycolac® CTR52—clear ABS.

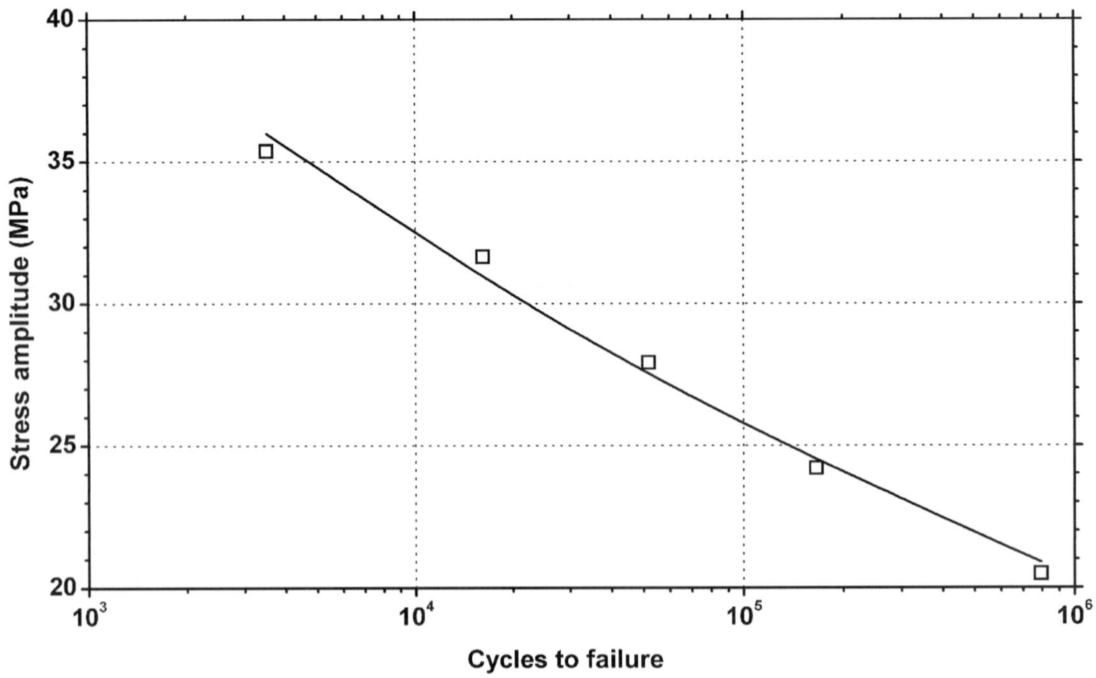

Figure 4.25 Tensile stress amplitude vs. cycles to failure for SABIC Innovative Plastics Cycolac® EX39—ABS with highest impact extrusion for sheet and blow molding.

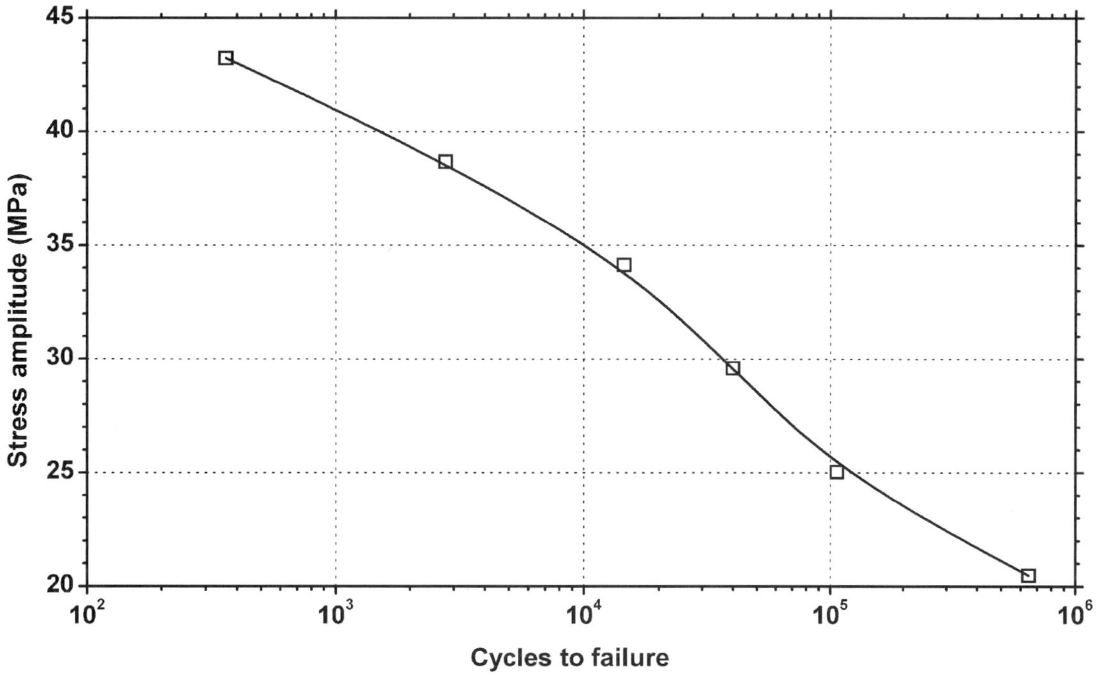

Figure 4.26 Tensile stress amplitude vs. cycles to failure for SABIC Innovative Plastics Cycolac® EX58—high-impact ABS for sheet extrusion and blow molding.

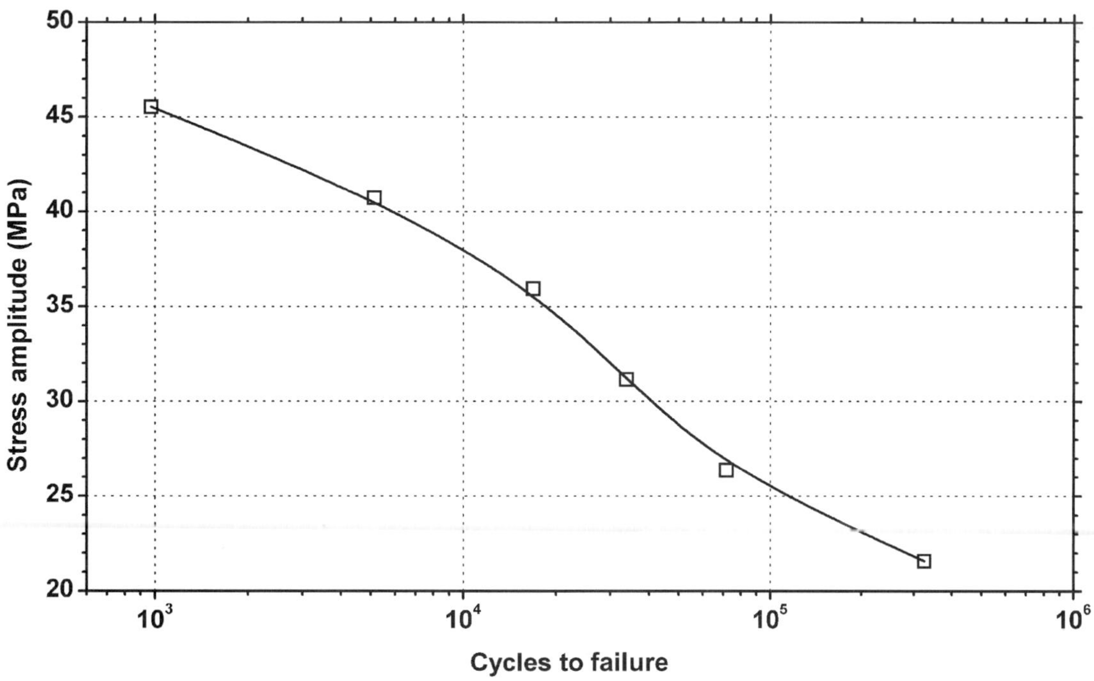

Figure 4.27 Tensile stress amplitude vs. cycles to failure for SABIC Innovative Plastics Cycolac® EX75—multipurpose, extrusion ABS.

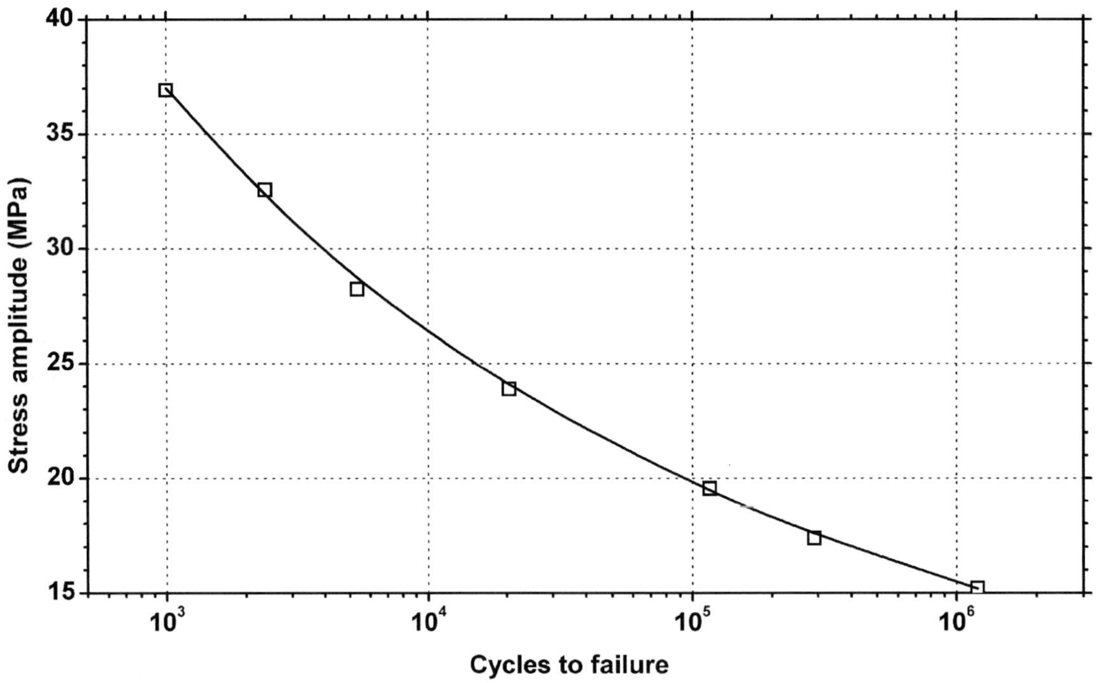

Figure 4.28 Tensile stress amplitude vs. cycles to failure for SABIC Innovative Plastics Cycolac® FR15—flame retardant ABS.

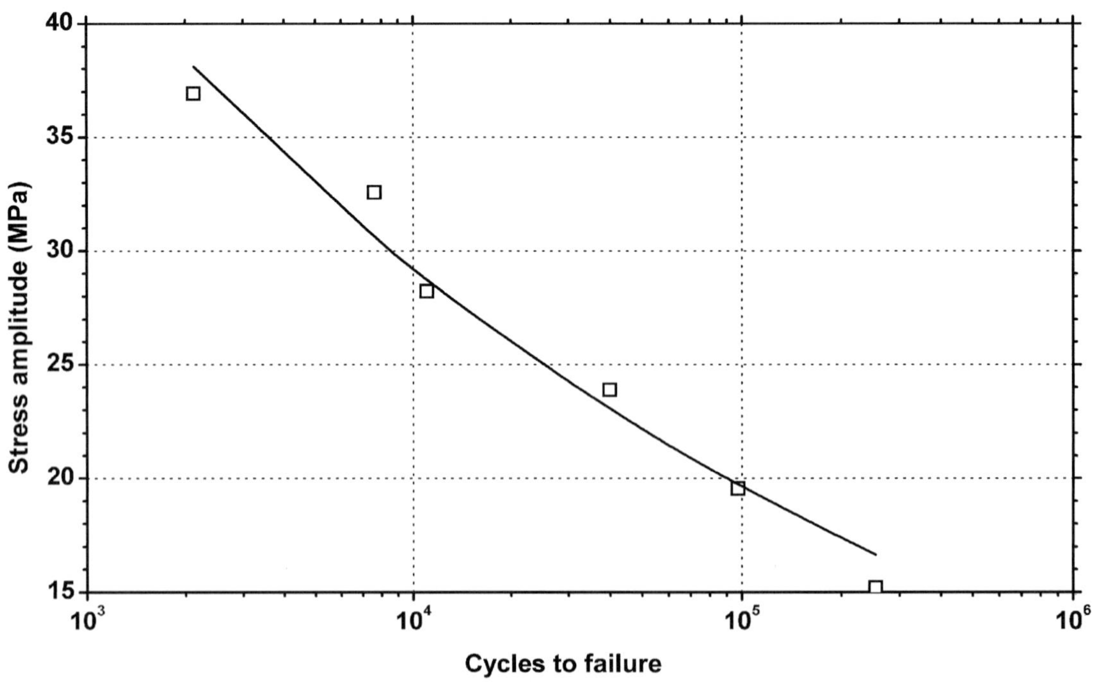

Figure 4.29 Tensile stress amplitude vs. cycles to failure for SABIC Innovative Plastics Cycolac® FR23—flame retardant ABS.

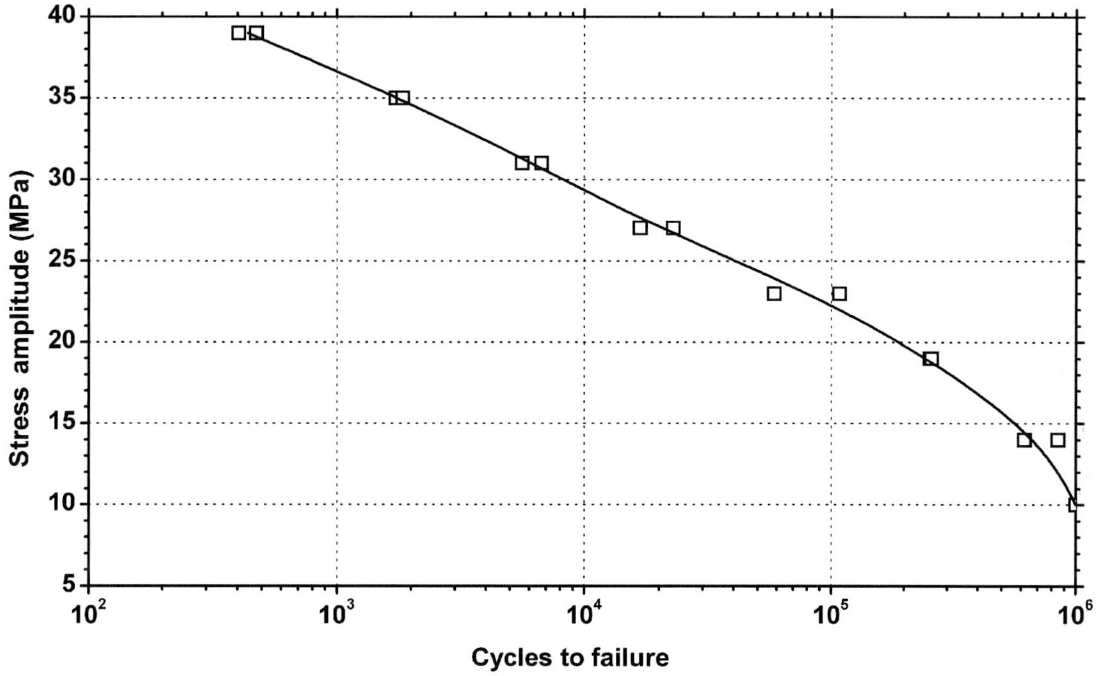

Figure 4.30 Tensile stress amplitude vs. cycles to failure for SABIC Innovative Plastics Cycolac® KJB—medium-impact ABS with wide processing range, UL94 rated V-0 ABS.

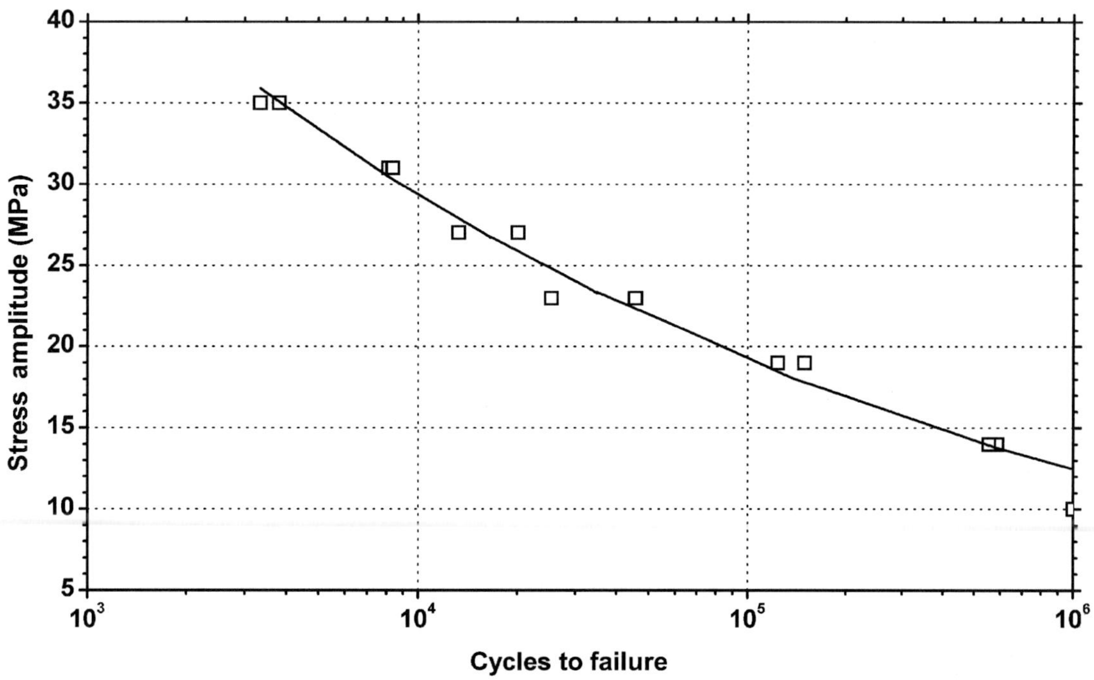

Figure 4.31 Tensile stress amplitude vs. cycles to failure for SABIC Innovative Plastics Cycolac® LDA—ABS for pipe extrusion.

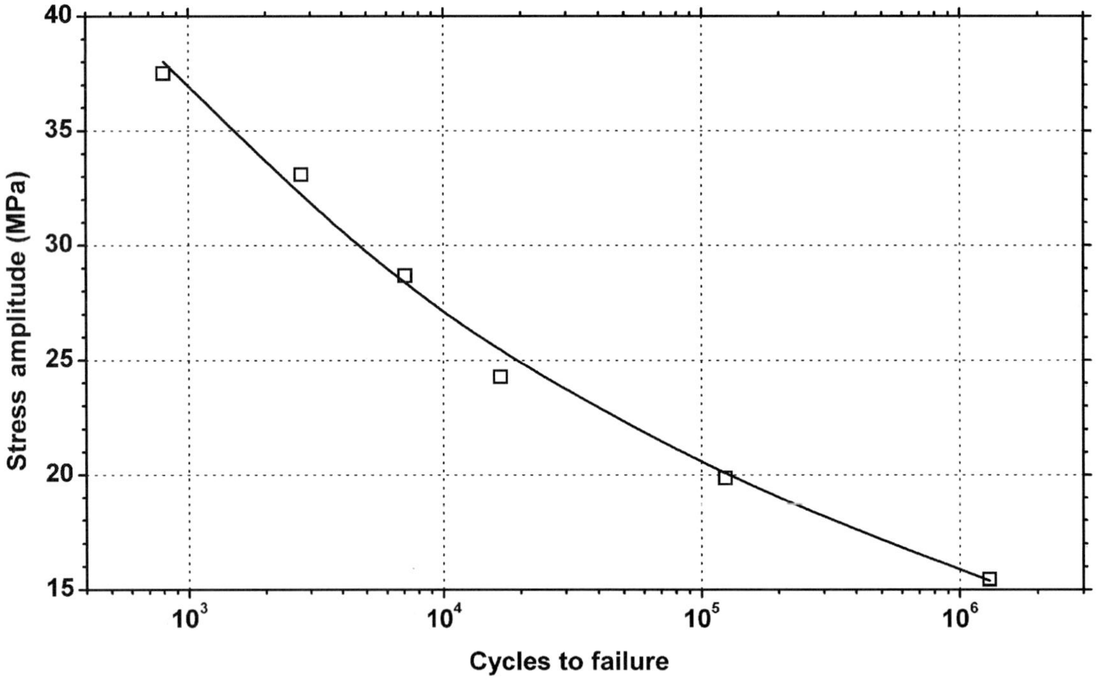

Figure 4.32 Tensile stress amplitude vs. cycles to failure for SABIC Innovative Plastics Cycolac® MG38F—very high-impact ABS, with toughness and rigidity. FDA compliant.

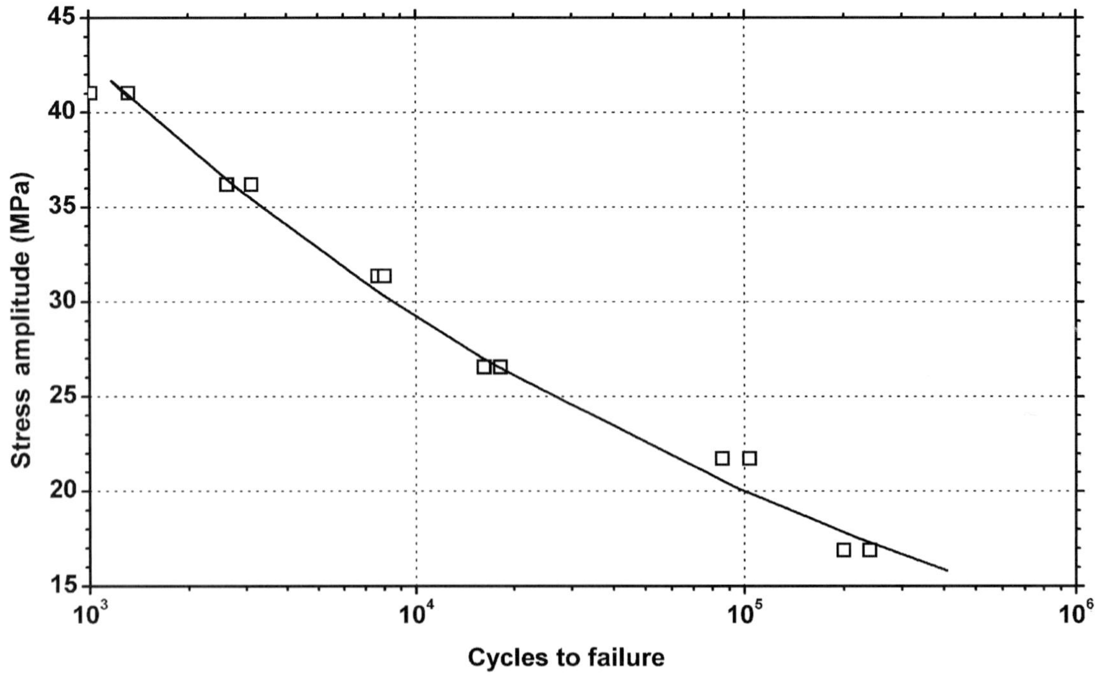

Figure 4.33 Tensile stress amplitude vs. cycles to failure for SABIC Innovative Plastics Cycolac® MG47—multipurpose ABS for injection molding.

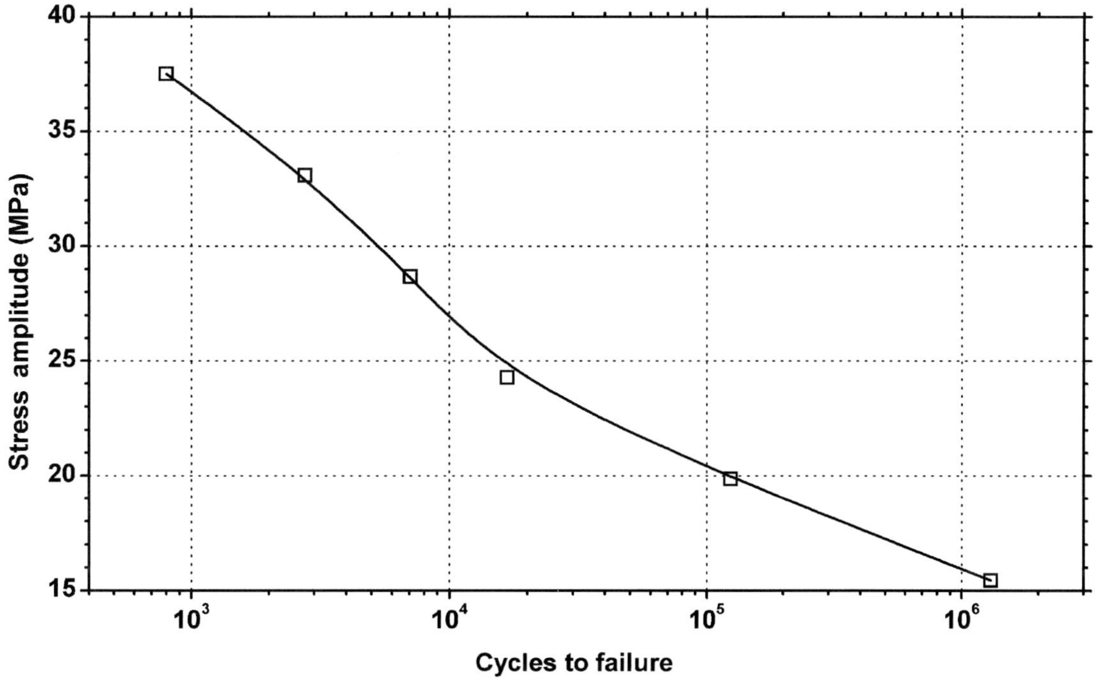

Figure 4.34 Tensile stress amplitude vs. cycles to failure for SABIC Innovative Plastics Cycolac® MGABS01 ABS.

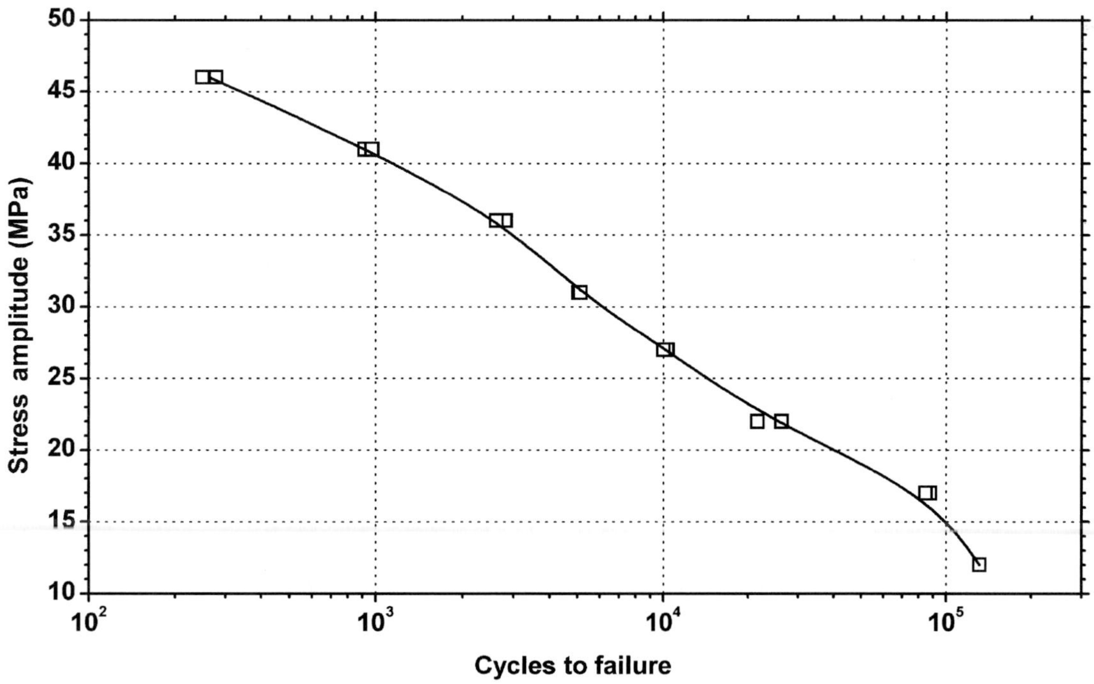

Figure 4.35 Tensile stress amplitude vs. cycles to failure for SABIC Innovative Plastics Cycolac® MGX53GP—general-purpose ABS, MAGIX™ visual effect technology.

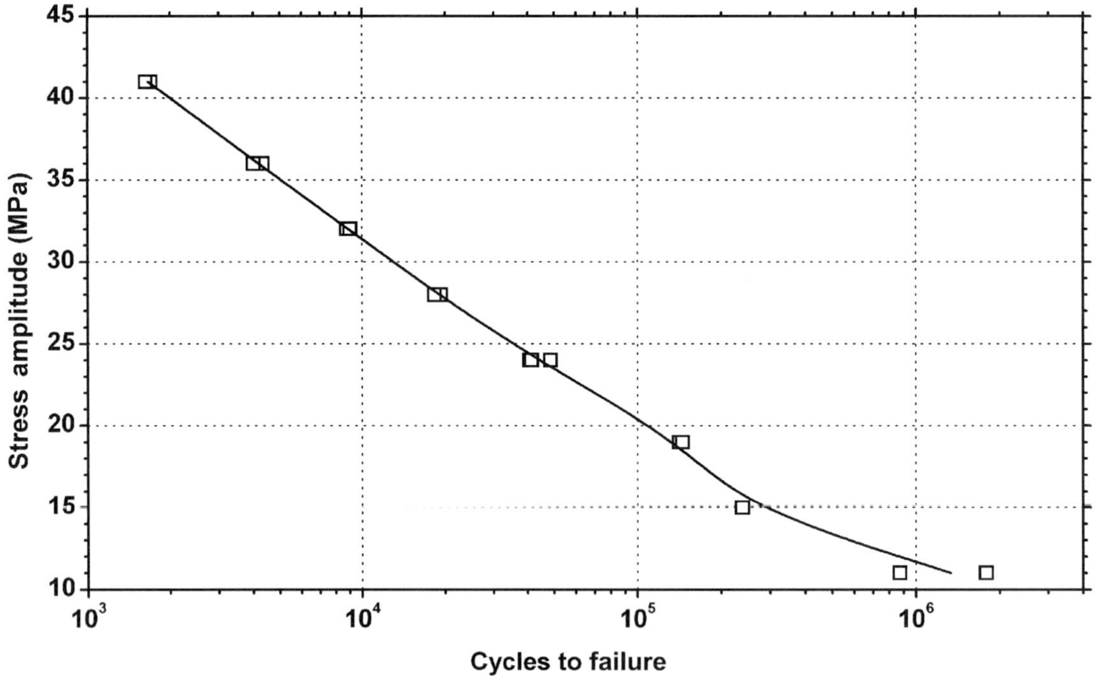

Figure 4.36 Tensile stress amplitude vs. cycles to failure for SABIC Innovative Plastics Cycolac® X11—high heat resistant ABS for automotive.

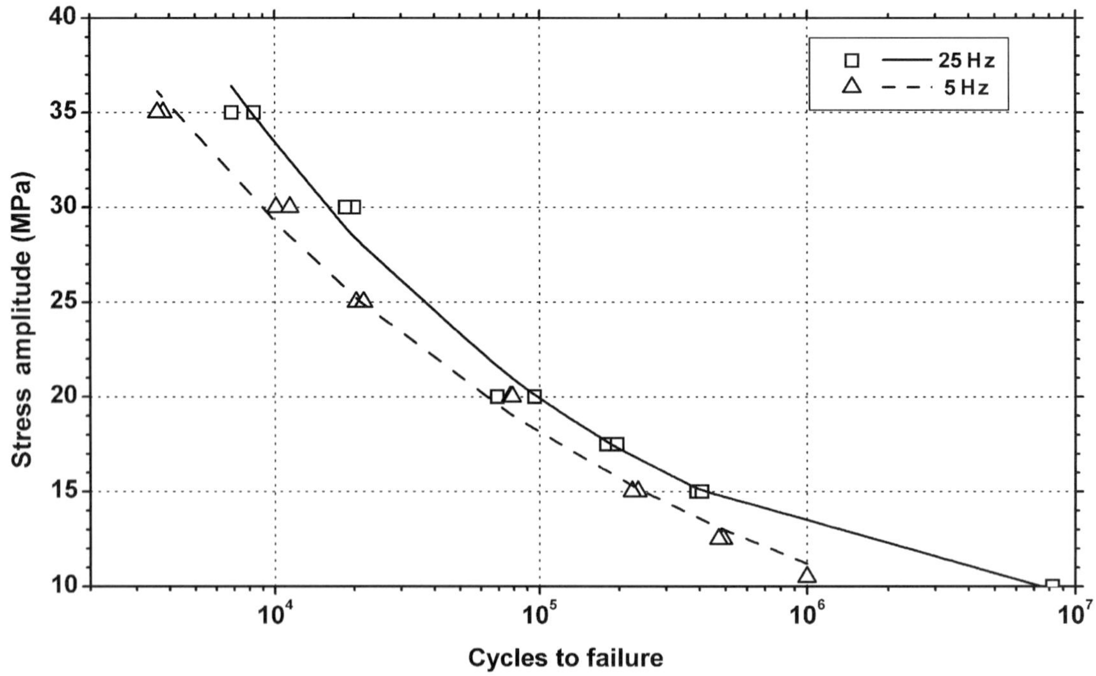

Figure 4.37 Tensile stress amplitude vs. cycles to failure for SABIC Innovative Plastics Cycolac® X37—high heat, injection molding ABS.

4.6 Styrenic Blends

4.6.1 Fatigue Data

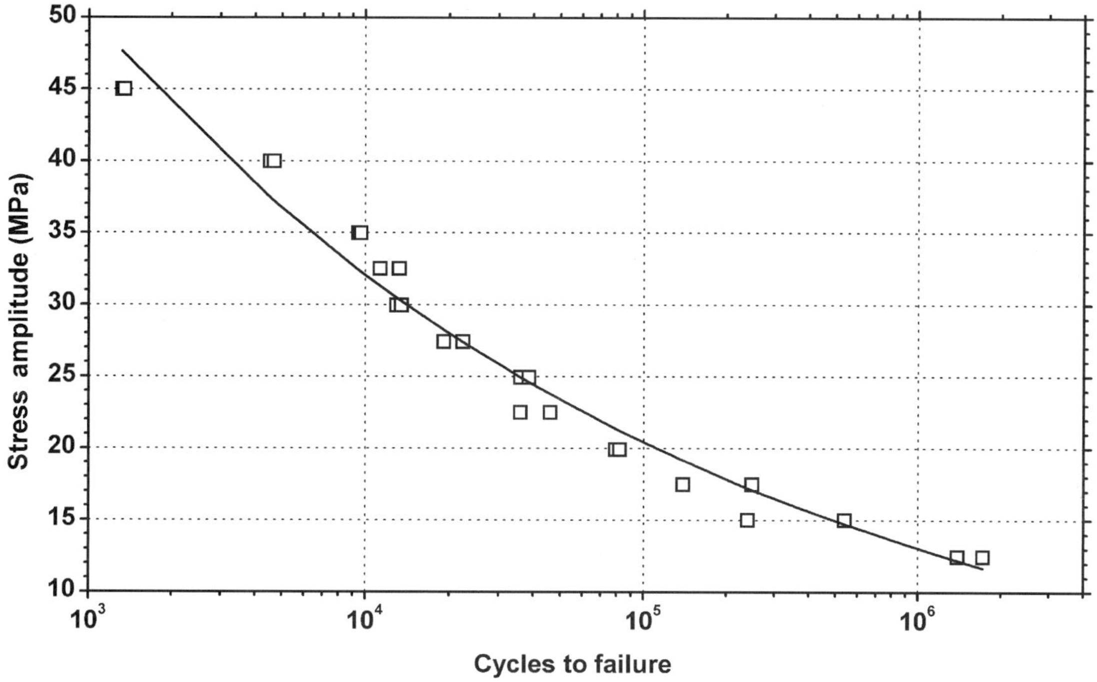

Figure 4.38 Tensile stress amplitude vs. cycles to failure for SABIC Innovative Plastics Cycoloy® C1100—high-impact ABS/PC blend.

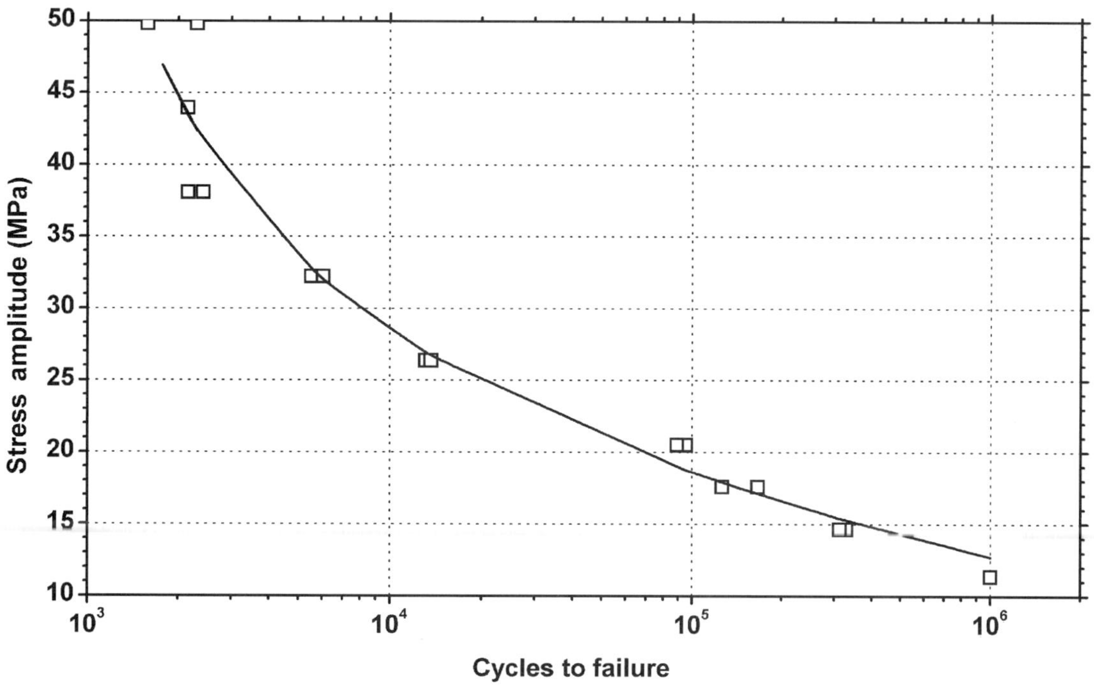

Figure 4.39 Tensile stress amplitude vs. cycles to failure for SABIC Innovative Plastics Geloy® XP4020R—weatherable, up to 20% postindustrial recycle ABS/PC blend.

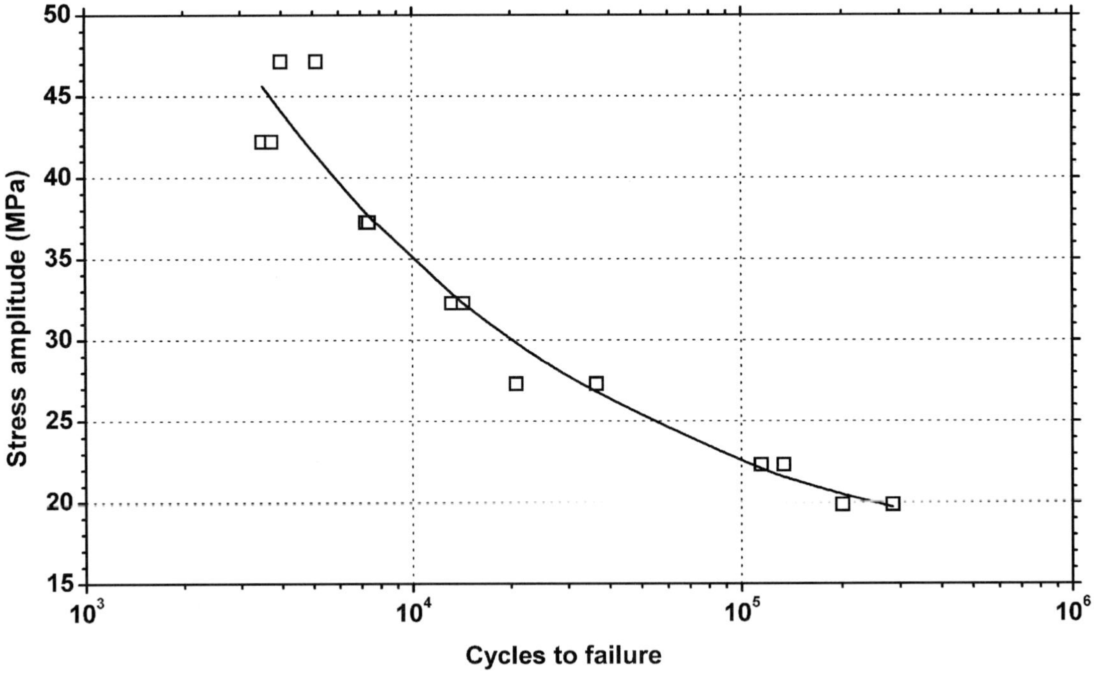

Figure 4.40 Tensile stress amplitude vs. cycles to failure for SABIC Innovative Plastics Geloy® XP4025—weatherable, injection molding ABS/PC blend.

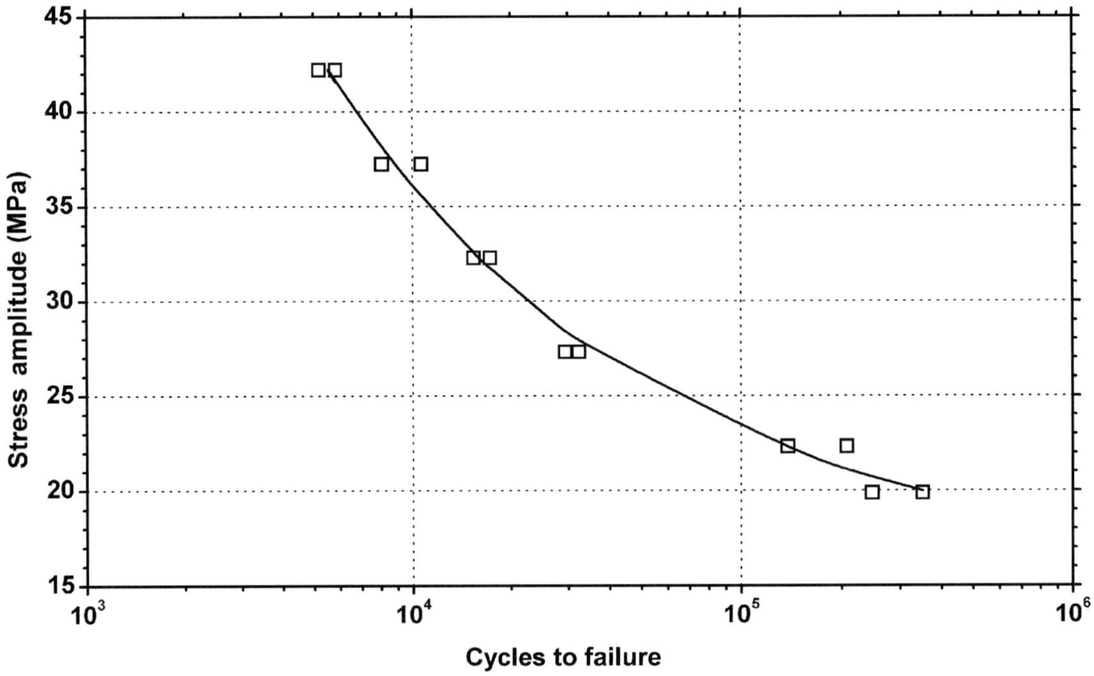

Figure 4.41 Tensile stress amplitude vs. cycles to failure for SABIC Innovative Plastics Geloy® XP4034—weatherable, injection molding ABS/PC blend.

4.6.2 Tribology Data

Table 4.1 Taber Abrasion SABIC Innovative Plastics Cycoloy® PC ABS Plastics

Product Code and Description	mg/1000 revolutions
C1100—High impact	79
C1100HF—High impact, high flow	81
C1200—High impact/high heat resistance	63
C1200HF—High impact/high heat resistance, high flow	63
C1204HF—High impact/high heat resistance, high flow food grade	63
C2100	62
C2100HF	62
C2800—Nonchlorinated and nonbrominated flame retardant	72
C2950—Nonchlorinated and nonbrominated flame retardant, improved heat	54
C3100	55
C3600	54
C3650—Nonchlorinated and nonbrominated flame retardant	54
C6200—Nonchlorinated and nonbrominated flame retardant	82
CU6800—Nonchlorinated and nonbrominated flame retardant, good flow	15
CX5430—General-purpose, good weld line strength	70
FXC630xy	82
FXC810xy	63
LG9000—Low gloss and UV stable	82

5 Polyether Plastics

5.1 Background

This chapter covers polymers in which the most important linking group is the ether moiety, which is –O–.

5.1.1 Polyoxymethylene (or Acetal Homopolymer)

Acetal polymers, also known as polyoxymethylene (POM) or polyacetal, are formaldehyde-based thermoplastics that have been commercially available since the 1960s. Polyformaldehyde is thermally unstable. It decomposes on heating to yield formaldehyde gas. Two methods of stabilizing polyformaldehyde for use as an engineering polymer were developed and introduced by DuPont in 1959 and Celanese in 1962 (now Ticona).

DuPont's method for making polyacetal yields a homopolymer through the condensation reaction of polyformaldehyde and acetic acid (or acetic anhydride). The acetic acid puts acetate groups (CH_3COO–) on the ends of the polymer as shown in Figure 5.1, which provide thermal protection against decomposition to formaldehyde.

Further stabilization of acetal polymers also includes the addition of antioxidants and acid scavengers. Polyacetals are subject to oxidative and acidic degradation, which leads to molecular weight decline. Once the chain of the homopolymer is ruptured by such an attack, the exposed polyformaldehyde ends may decompose to formaldehyde and acetic acid.

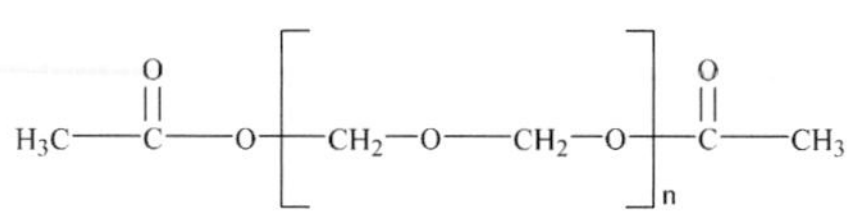

Figure 5.1 Chemical structure of acetal homopolymer.

5.1.2 Polyoxymethylene Copolymer (POM-Co or Acetal Copolymer)

The Celanese route for the production of polyacetal yields a more stable copolymer product via the reaction of trioxane, a cyclic trimer of formaldehyde, and a cyclic ether, such as ethylene oxide or 1,3-dioxolane. The structures of these monomers are shown in Figure 5.2. The polymer structure is given in Figure 5.3.

The improved thermal and chemical stability of the copolymer versus the homopolymer is a result of randomly distributed oxyethylene groups, which is circled in Figure 5.5. All polyacetals are subject to oxidative and acidic degradation, which leads to molecular weight reduction. Degradation of the copolymer ceases, however, when one of the randomly distributed oxyethylene linkages is reached. These groups offer stability to oxidative, thermal, acidic and alkaline attack. The raw copolymer is hydrolyzed to an oxyethylene end cap to provide thermally stable polyacetal copolymer.

The copolymer is also more stable than the homopolymer in an alkaline environment. Its oxyethylene end cap is stable in the presence of strong bases. The acetate end cap of the homopolymer, however, is readily hydrolyzed in the presence of alkalis, causing significant polymer degradation.

The homopolymer is more crystalline than the copolymer. The homopolymer provides better mechanical properties, except for elongation. The oxyethylene groups of the copolymer provide improved long-term chemical and environmental

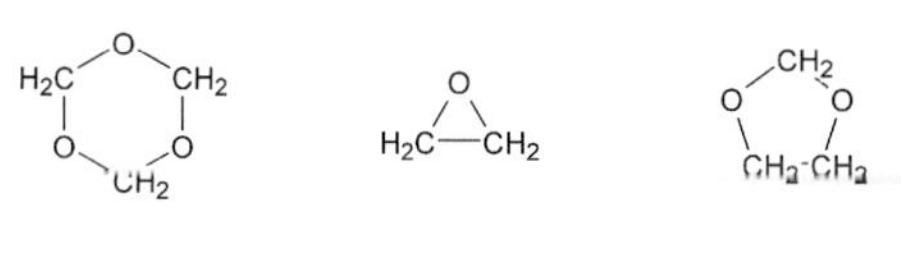

Figure 5.2 Chemical structure of POM-Co monomers.

Copyright © 2010 Laurence W. McKeen. All rights reserved.

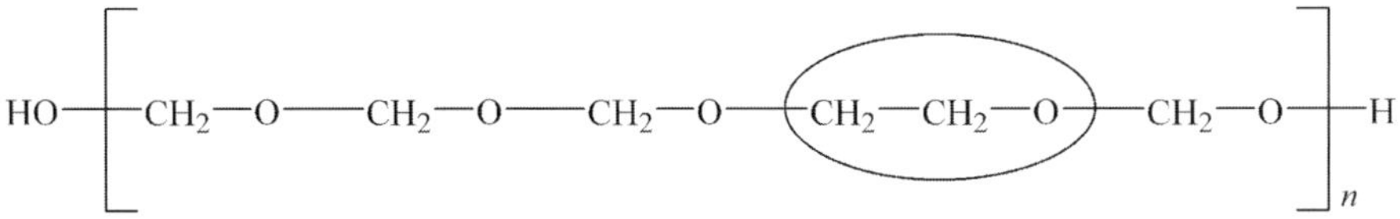

Figure 5.3 Chemical structure of acetal copolymer.

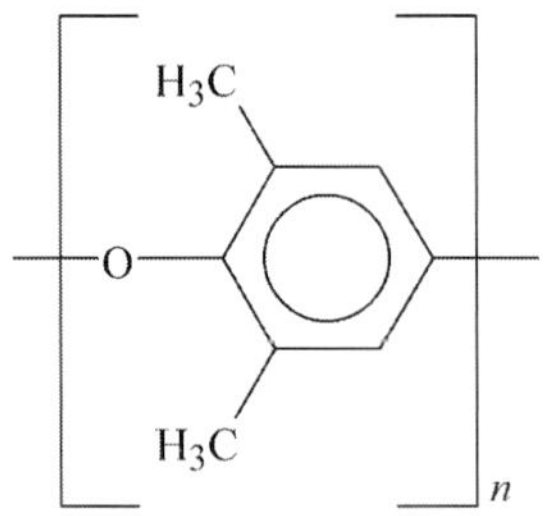

Figure 5.4 Chemical structure of PPE or PPO.

stability. The copolymer's chemical stability results in better retention of mechanical properties over an extended product life.

Acetal polymers have been particularly successful in replacing cast and stamped metal parts due to their toughness, abrasion resistance and ability to withstand prolonged stresses with minimal creep. Polyacetals are inherently self-lubricating. Their lubricity allows the incorporation of polyacetal in a variety of metal-to-polymer and polymer-to-polymer interface applications such as bearings, gears and switch plungers. These properties have permitted the material to meet a wide range of market requirements.

The properties of polyacetals can be summarized as follows:

- Excellent wear resistance
- Very good strength and stiffness
- Good heat resistance
- Excellent chemical resistance
- Opaque
- Moderate to high price
- Somewhat restricted processing

5.1.3 Modified Polyphenylene Ether/Polyphenylene Oxides

Polyphenylene ether (PPE) plastics are also referred to as polyphenylene oxide (PPO). The structure of the polymer is shown in Figure 5.4.

The PPE materials are always blended or alloyed with other plastics, so they are called modified PPE or PPO. PPE is compatible with polystyrene (PS) and is usually blended with high-impact PS over a wide range of ratios. Because both PPE and PS plastics are hydrophobic, the alloys have very low water absorption rates and high dimensional stability. They exhibit excellent dielectric properties over a wide range of frequencies and temperatures. PPE/PS alloys are supplied in flame-retardant, filled and reinforced, and structural foam molding grades. PPE can also be alloyed with polyamide (nylon) plastics to provide increased resistance to organic chemicals and better high-temperature performance.

End uses include automotive electrical applications, water pump impellers, HVAC equipment, solar heating systems, packaging and circuit breakers.

Graphs of the creep properties of the polyether-based plastics are in the following sections.

5.2 Acetals–POM Homopolymer

5.2.1 Fatigue Data

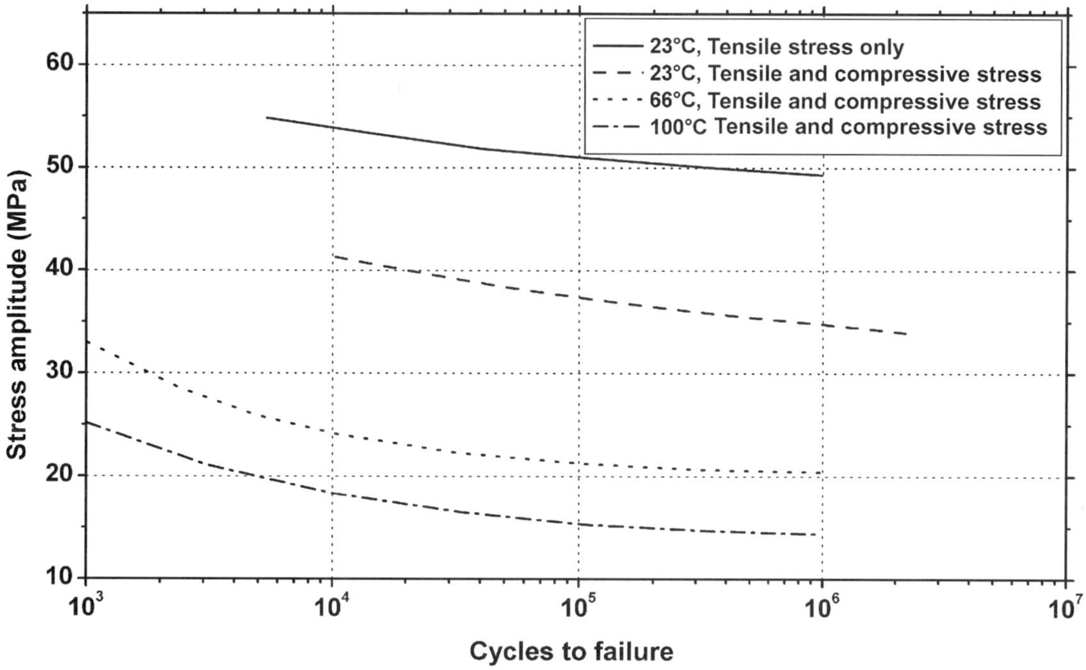

Figure 5.5 Stress amplitude vs. cycles to failure at various temperatures for DuPont Engineering Polymers Delrin® 500—unfilled medium viscosity POM.

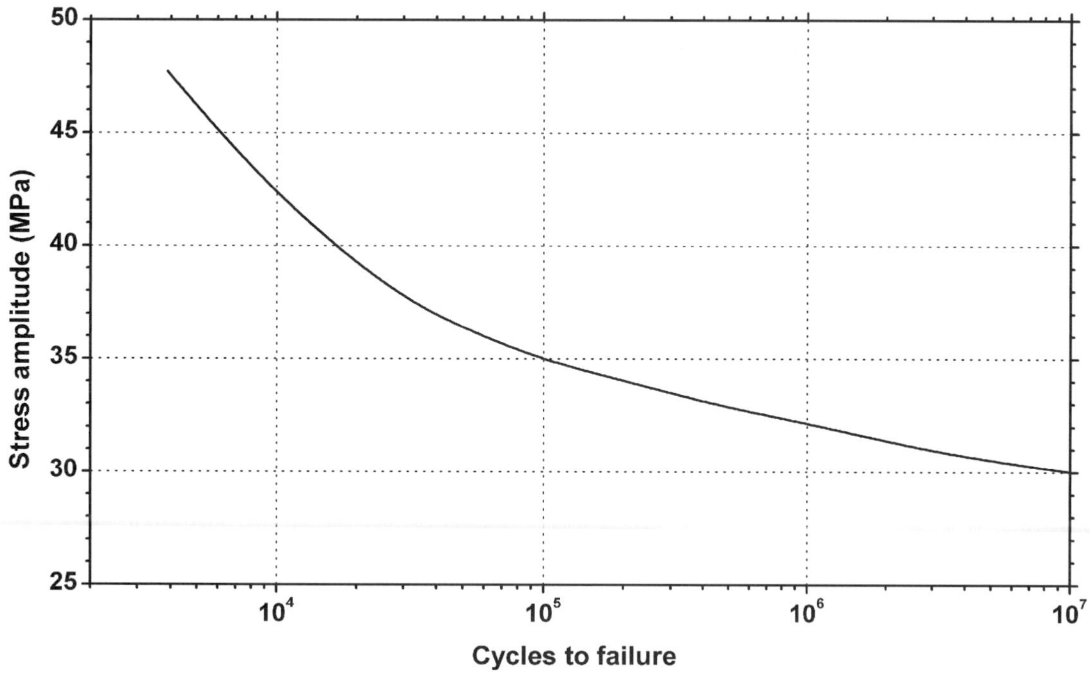

Figure 5.6 Flexural stress amplitude vs. cycles to failure at various temperatures for DuPont Engineering Polymers Delrin® 100, 500 and 900—unfilled POM.

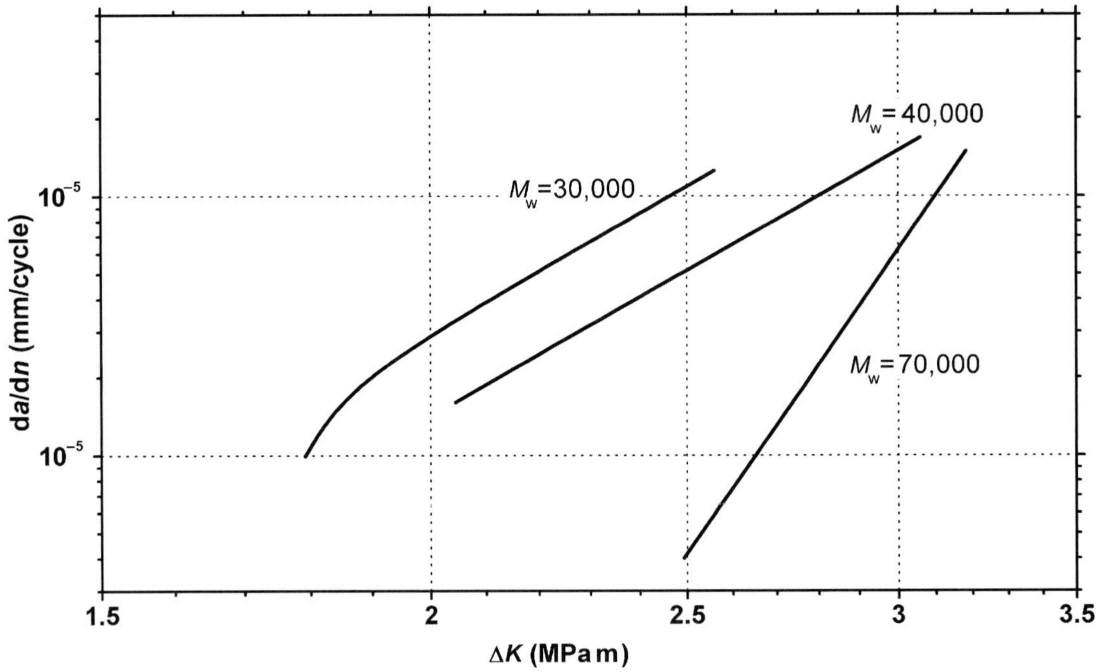

Figure 5.7 Fatigue crack propagation vs. stress intensity factor of various molecular weights of generic unfilled POM.

5.2.2 Tribology Data

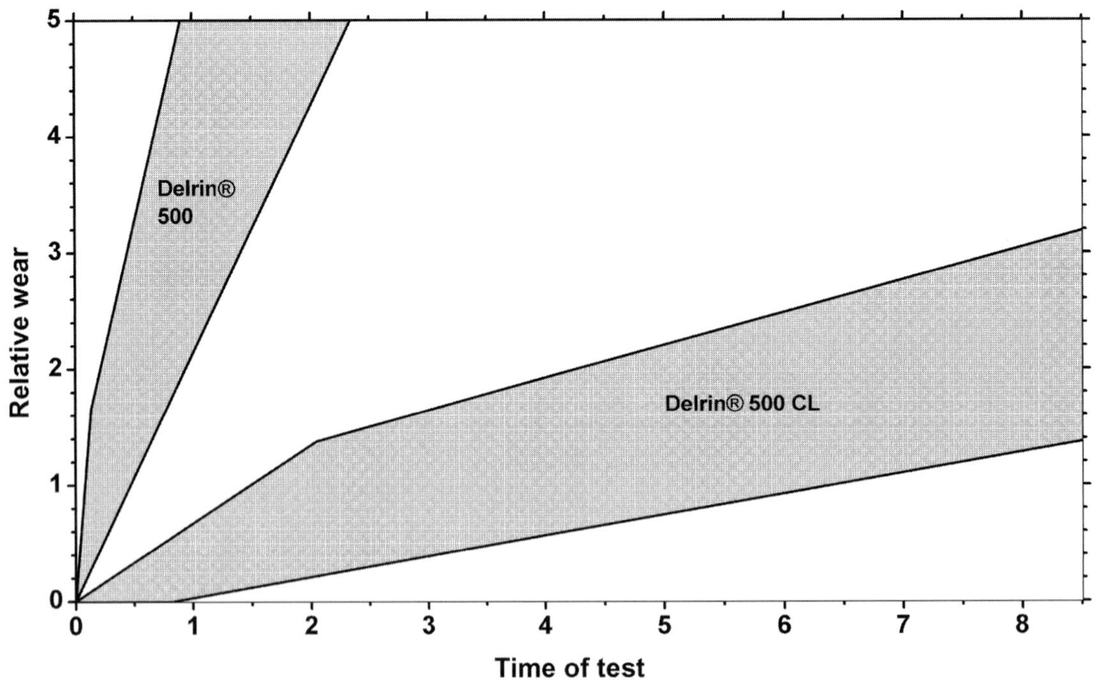

Figure 5.8 Wear against mild steel in a thrust washer test for DuPont Engineering Polymers Delrin® 500 and 500 CL (internally lubricated) medium viscosity POM resins (nonlubricated, $P = 0.04$ MPa, $V = 0.95$ m/s).

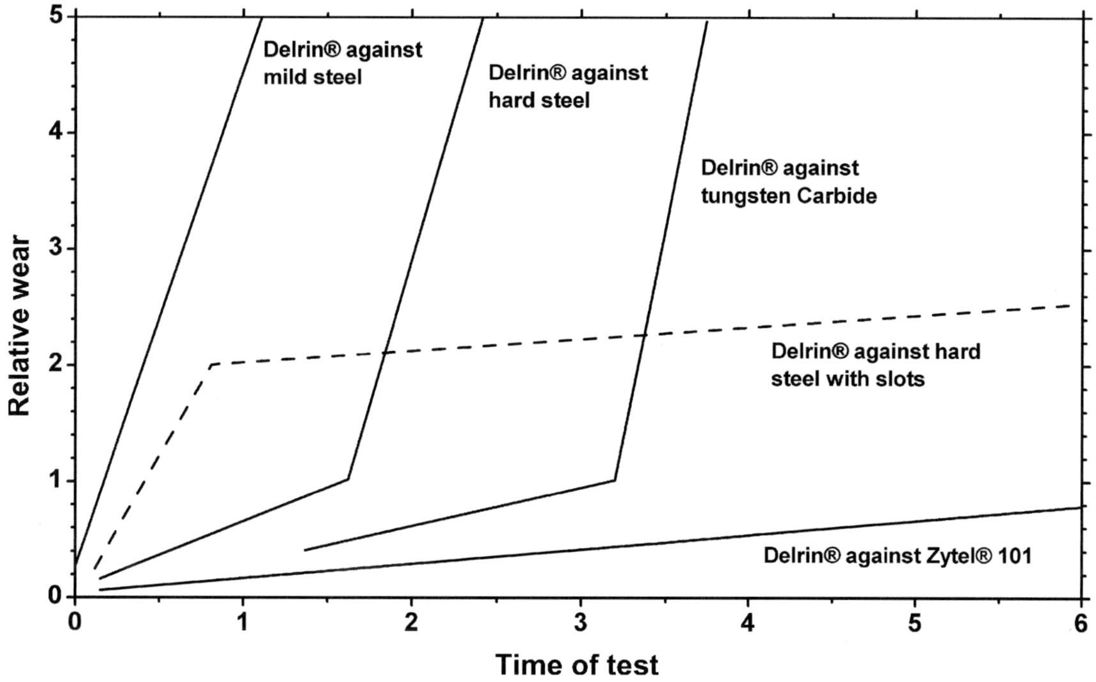

Figure 5.9 Wear against various materials of DuPont Engineering Polymers Delrin® 100, 500 and 900—unfilled POM.

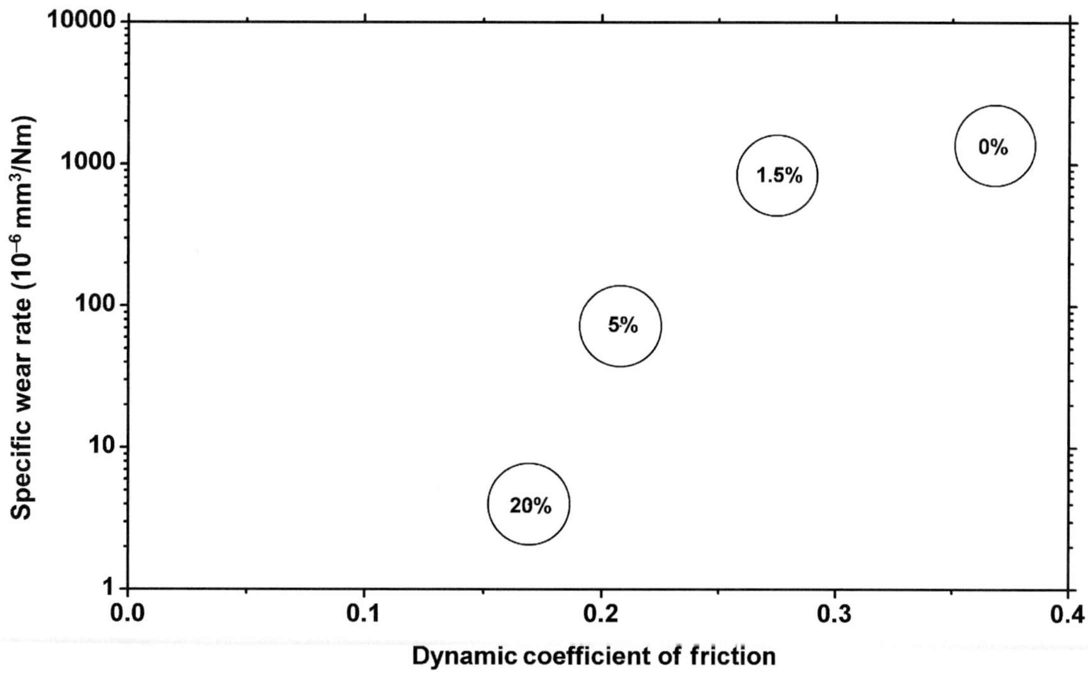

Figure 5.10 The effect of Teflon® PTFE levels in DuPont Engineering Polymers Delrin® on wear rate and dynamic coefficient of friction.

Table 5.1 Wear Properties of RTP Company RTP 800 TFE 20 DEL—POM Homopolymer with PTFE 20% vs. 1018 C Steel

PV (KPa m/s)	Load (N)	Speed (m/s)	Wear Factor × 10^{-8} (mm^3/N m)	Dynamic Coefficient of Friction
70	1.80	0.25	321	0.30
175	2.25	0.50	440	0.40
350	2.25	1.00	609	0.57

Table 5.2 Coefficient of Friction of Several DuPont Engineering Polymers Delrin® POM Plastics (Thrust Washer Test, Nonlubricated, 23°C; $P = 2.1$ MPa; $V = 3$ m/min)

Material/Counter Material	Static Coefficient of Friction	Dynamic Coefficient of Friction
Delrin 100, 500, 900 on Steel	0.20	0.35
Delrin 500CL on Steel	0.10	0.20
Delrin AF on Steel	0.08	0.14
Delrin 500 on Delrin 500	0.30	0.40
Delrin 500 on Zytel 101	0.10	0.20

Table 5.3 Characteristics, Specific Wear Rate and Dynamic Coefficient of Friction of Various Grades of DELRIN® Designed for Friction and Wear Applications

Delrin Grade	Specific Wear Rate 10^{-6} mm^3/N·m		Dynamic Coefficient of Friction	
	Against Itself	Against 100 Cr_6 Steel	Against Itself	Against 100 Cr_6 Steel
100P	1150	12	0.38	0.27
500P	1500	12	0.35	0.33
500CL	1200	4	0.36	0.27
500AF	40	2	0.22	0.20
500AL	22	6	0.16	0.18
520MP	5	3	0.19	0.18
900P	1400	12	0.33	0.33
900SP	20		0.10	

1. Surface and countersurface are consisting of the same grade of DELRIN®. The specific wear rate was measured at low speed (0.084 m/s) with a pressure of 0.624 MPa in a reciprocating motion (total sliding distance: 1.52 km), and the coefficient of friction was measured at a similar speed (0.08 m/s) with a pressure of 0.196 MPa, also in reciprocating motion.

2. Surface roughness Ra (μm): 0.10; hardness HRB: 93. The specific wear rate was measured at low speed (0.084 m/s) with a pressure of 0.624 MPa in a reciprocating motion (total sliding distance: 4.25 km); the coefficient of friction was measured at a high speed (0.5 m/s) with a load of 10 N in a sliding motion (Block-on-Ring).

5.3 Acetals–POM-Co

5.3.1 Fatigue Data

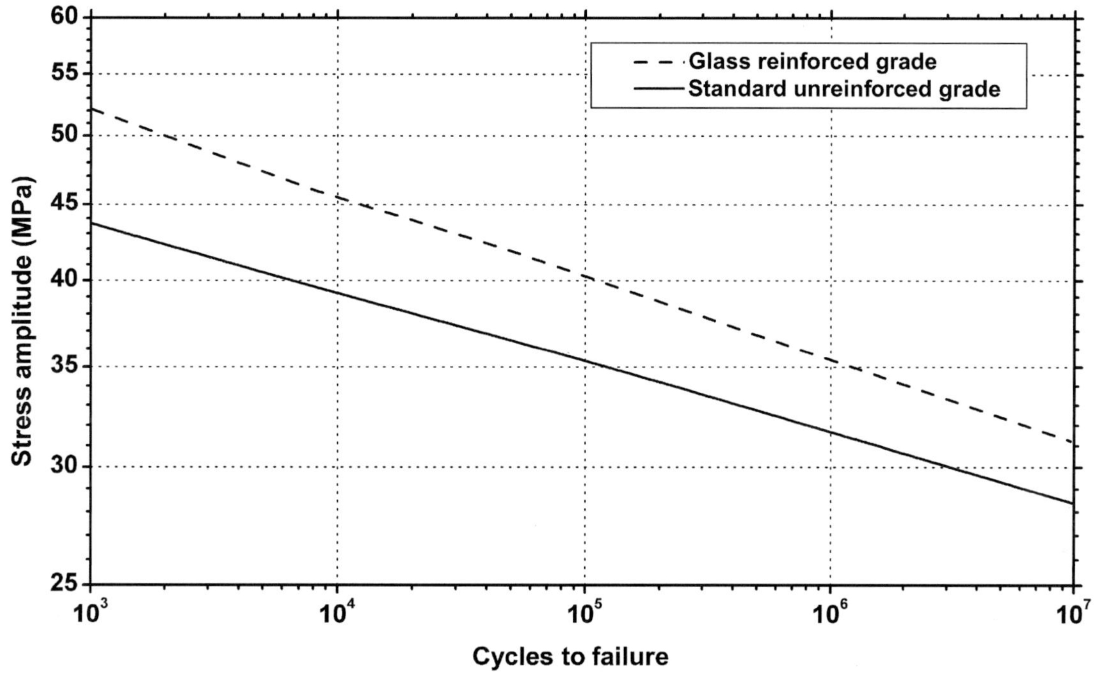

Figure 5.11 Flexural stress amplitude vs. cycles to failure for Ticona Celcon® POM-Co plastics.

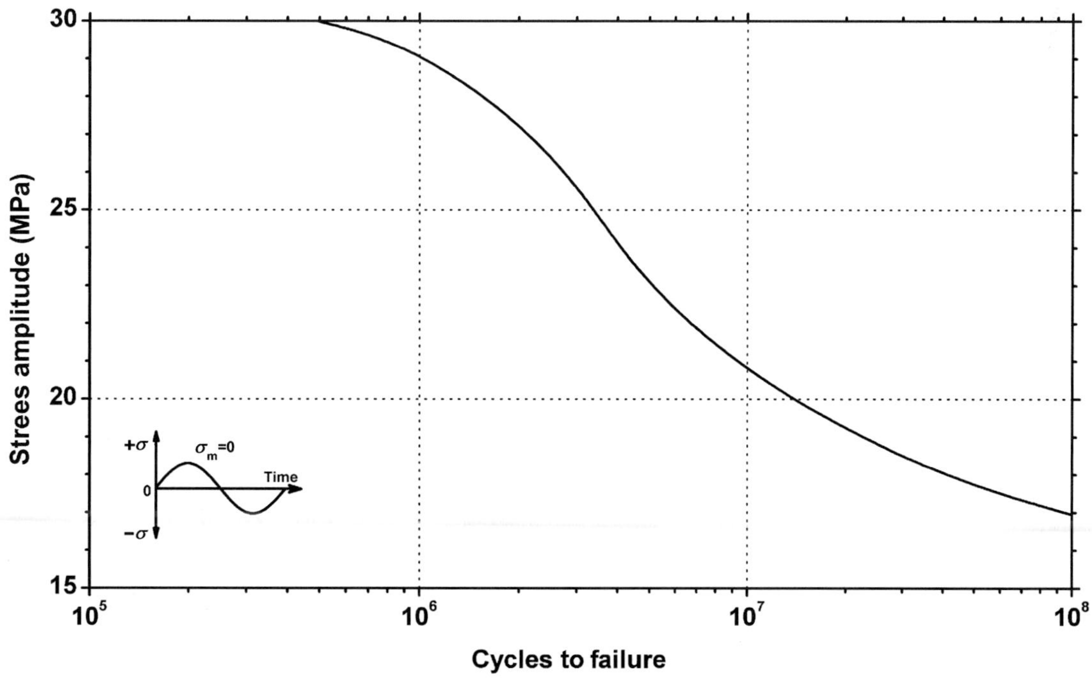

Figure 5.12 Fluctuating stress amplitude vs. cycles to failure at 23°C and 10 Hz for Ticona Hostaform® C 9021—Standard Injection Molding Grade POM-Co.

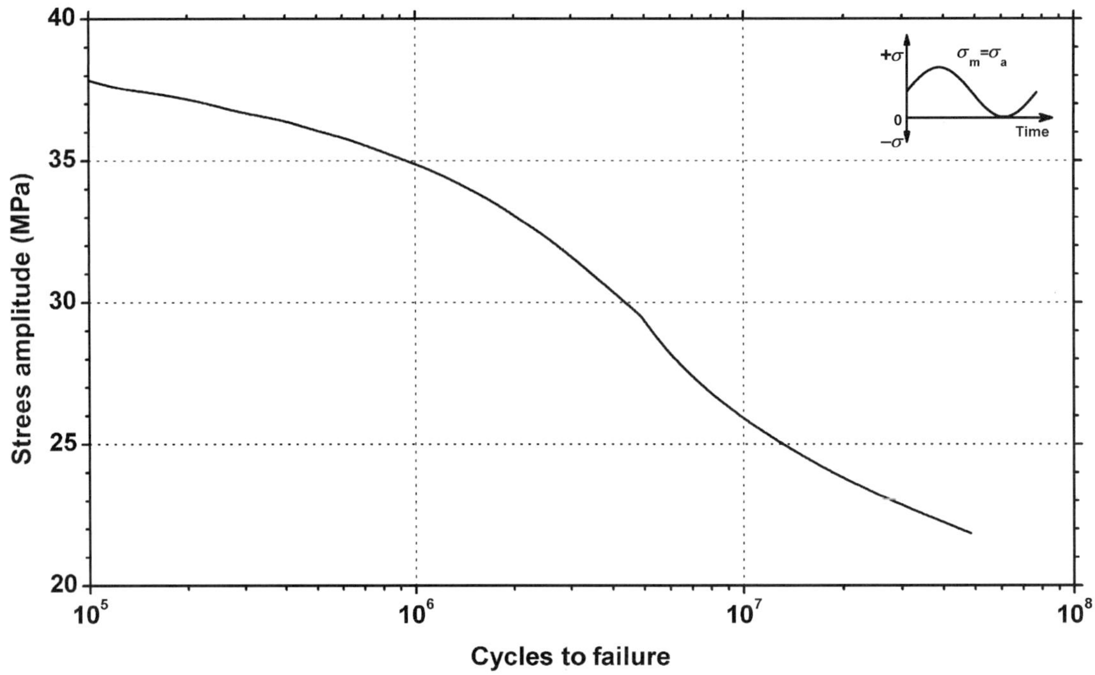

Figure 5.13 Tensile stress amplitude vs. cycles to failure at 23°C and 10 Hz for Ticona Hostaform® C 9021—Standard Injection Molding Grade POM-Co.

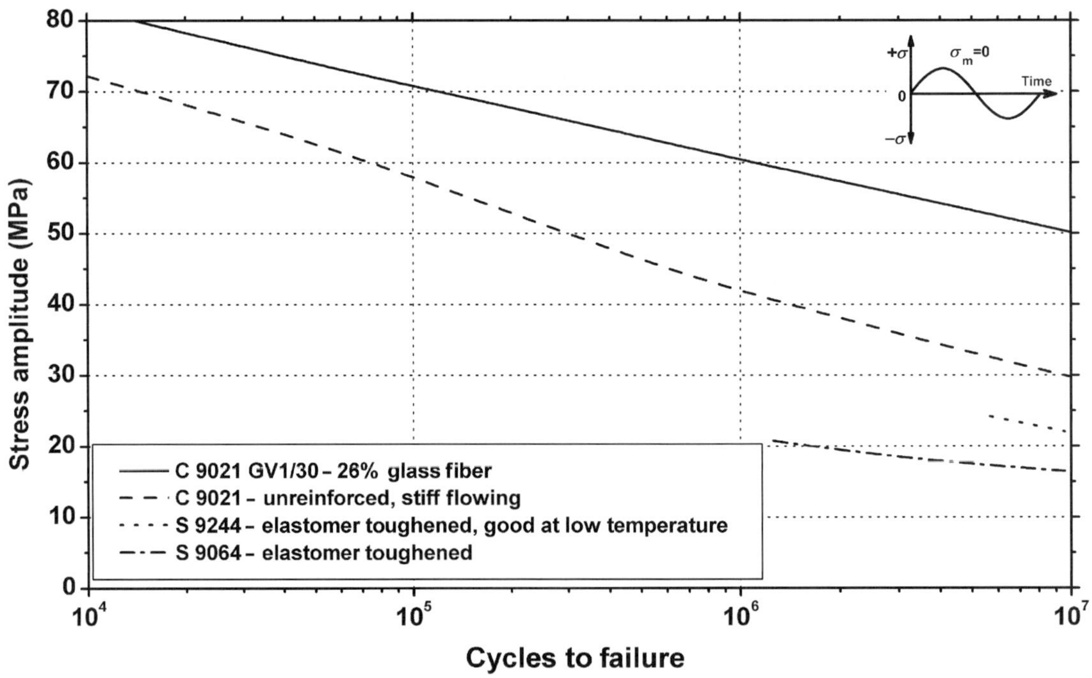

Figure 5.14 Flexural stress amplitude vs. cycles to failure at 23°C and 10 Hz for several Ticona Hostaform® POM-Co plastics.

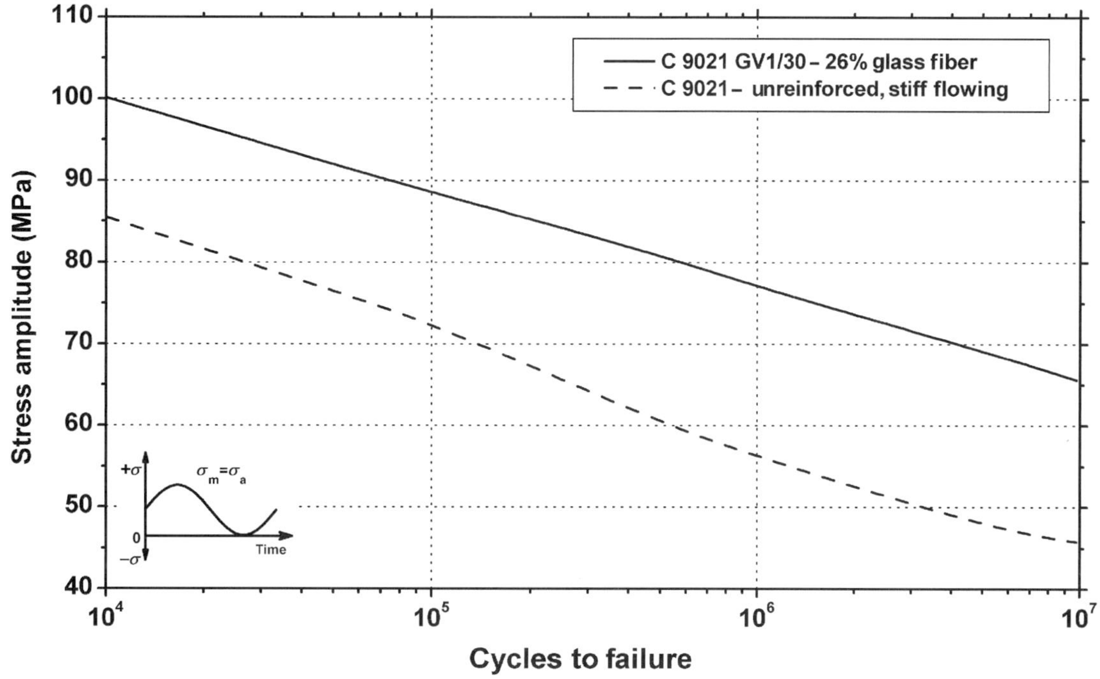

Figure 5.15 Flexural stress amplitude vs. cycles to failure at 23°C and 10 Hz for two Ticona Hostaform® C 9021 POM-Co plastics.

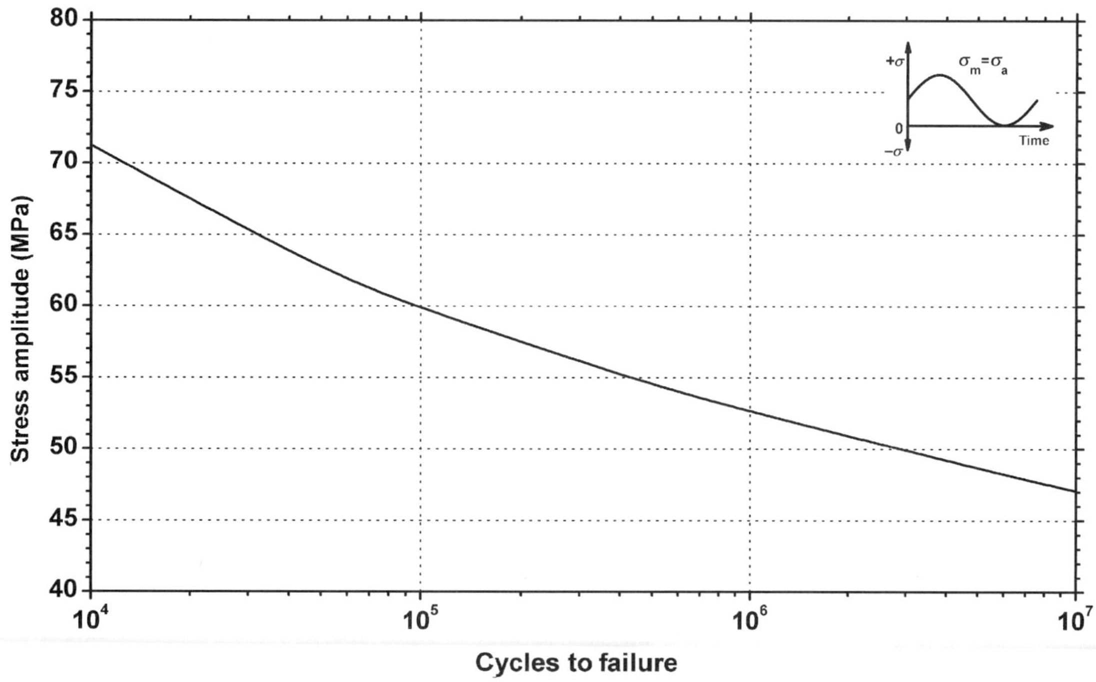

Figure 5.16 Torsional stress amplitude vs. cycles to failure at 23°C and 10 Hz for Ticona Hostaform® C 9021—Standard Injection Molding Grade POM-Co.

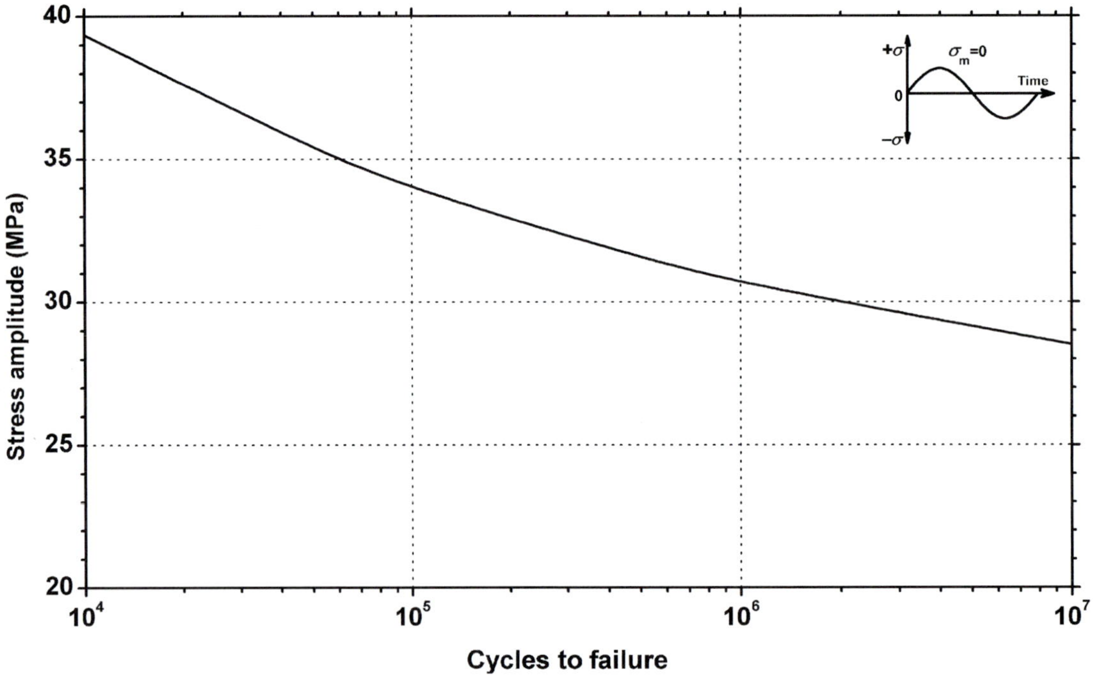

Figure 5.17 Torsional stress amplitude vs. cycles to failure at 23°C and 10 Hz for Ticona Hostaform® C 9021—Standard Injection Molding Grade POM-Co.

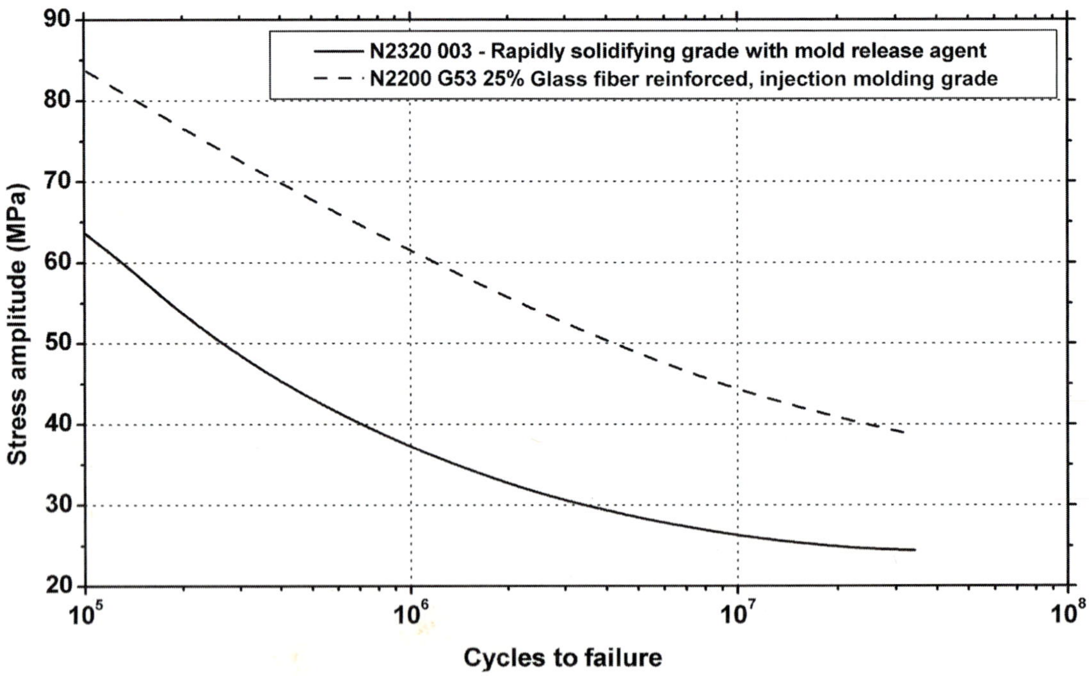

Figure 5.18 Flexural stress amplitude vs. cycles to failure at 23°C and 10 Hz for two BASF Ultraform® POM-Co plastics.

5.3.2 Tribology Data

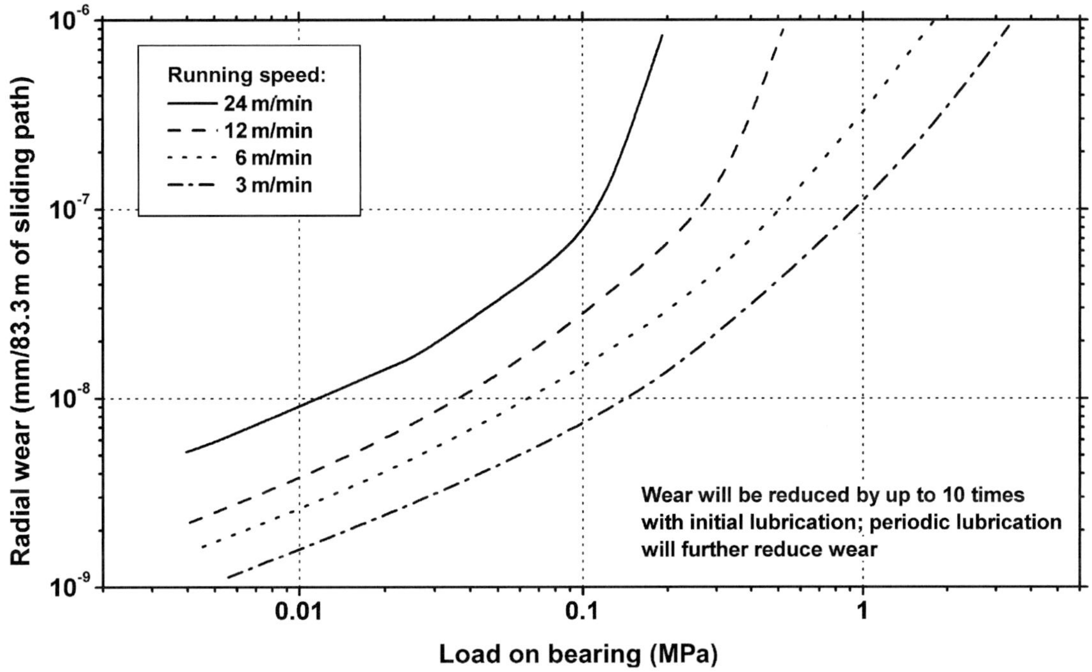

Figure 5.19 Radial wear vs. load on unlubricated journal bearing of Ticona Celcon® POM-Co at various running speeds at 23°C.

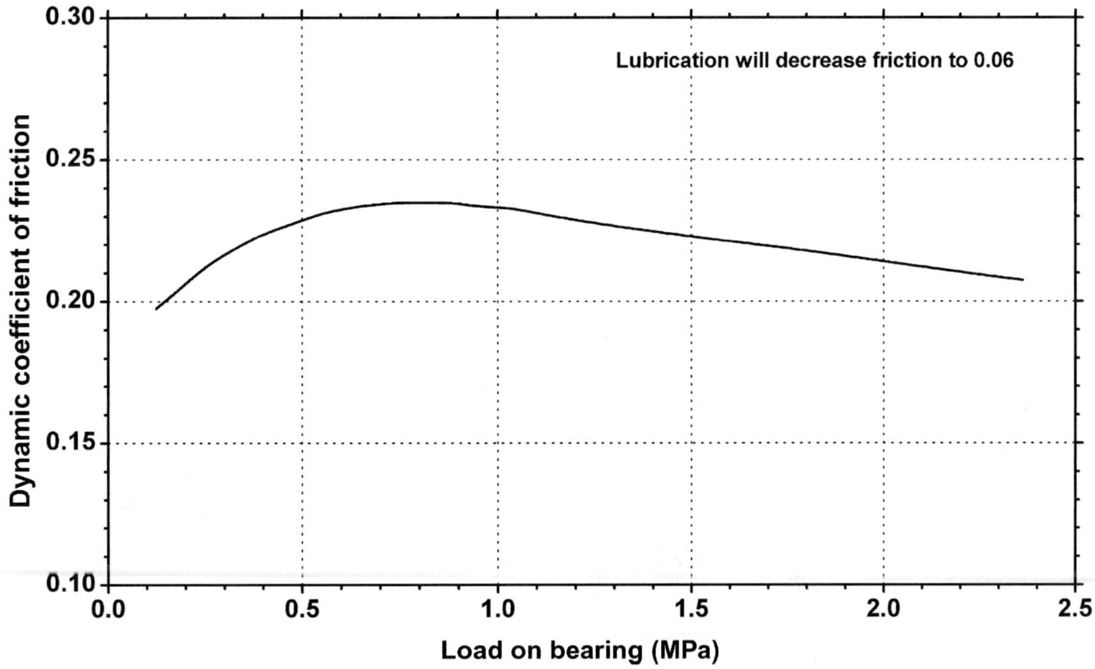

Figure 5.20 Dynamic coefficient of friction vs. bearing pressure on unlubricated hardened and polished steel shaft of Ticona Celcon® POM-Co at a running speed of 10 m/min.

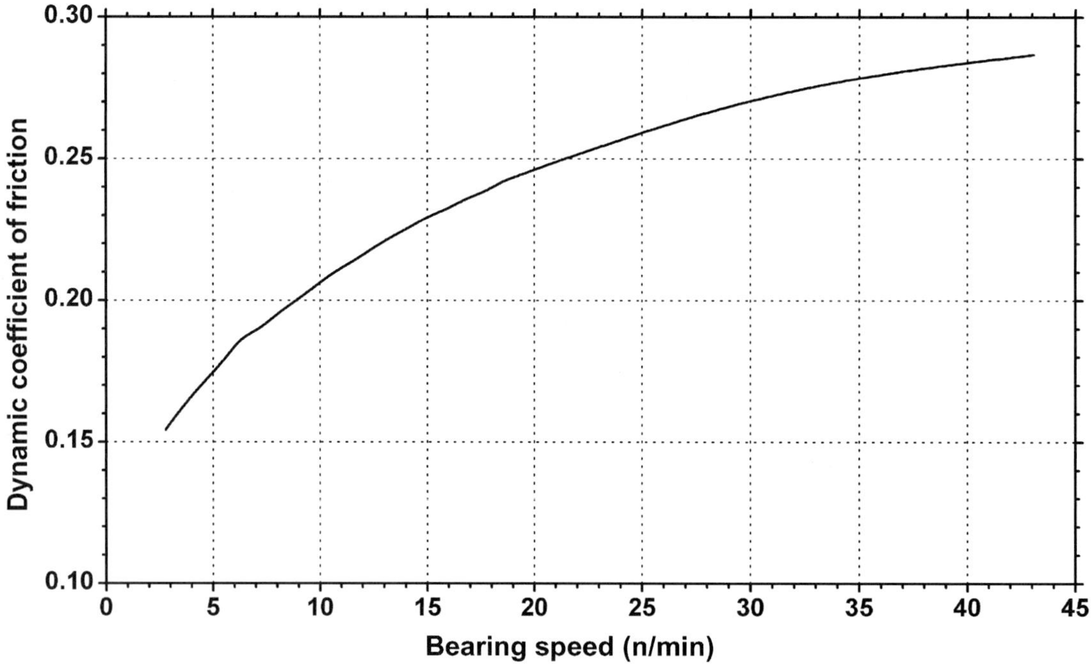

Figure 5.21 Dynamic coefficient of friction vs. bearing speed on unlubricated hardened and polished steel shaft of Ticona Celcon® POM-Co at a bearing pressure of 0.26 Mpa.

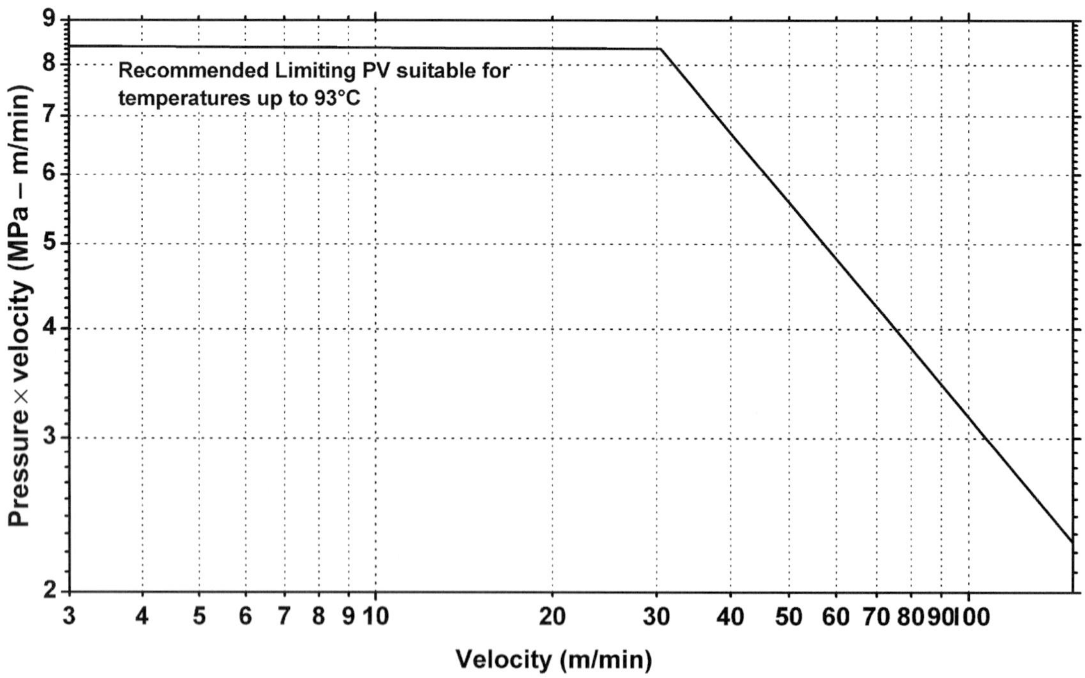

Figure 5.22 Limiting PV curve for unlubricated Ticona Celcon® POM-Co.

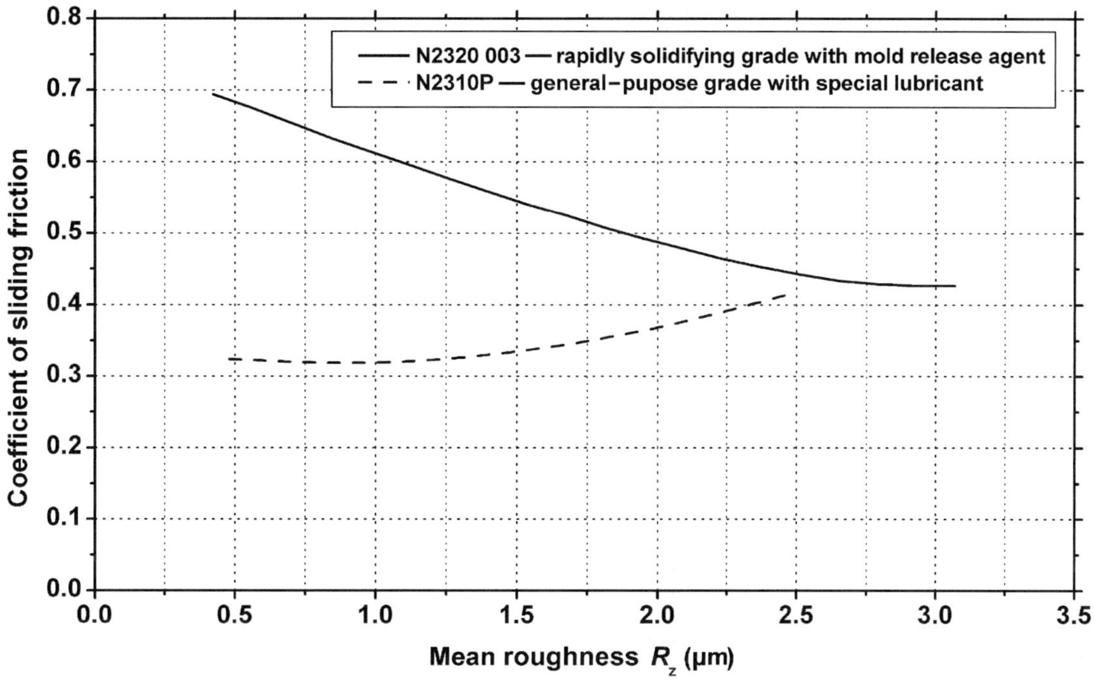

Figure 5.23 Coefficient of sliding friction vs. roughness of a sliding steel disk (HRC = 54–56 at 40°C, P = 1 MPa, V = 0.5 m/s) for two BASF Ultraform® POM-Co plastics.

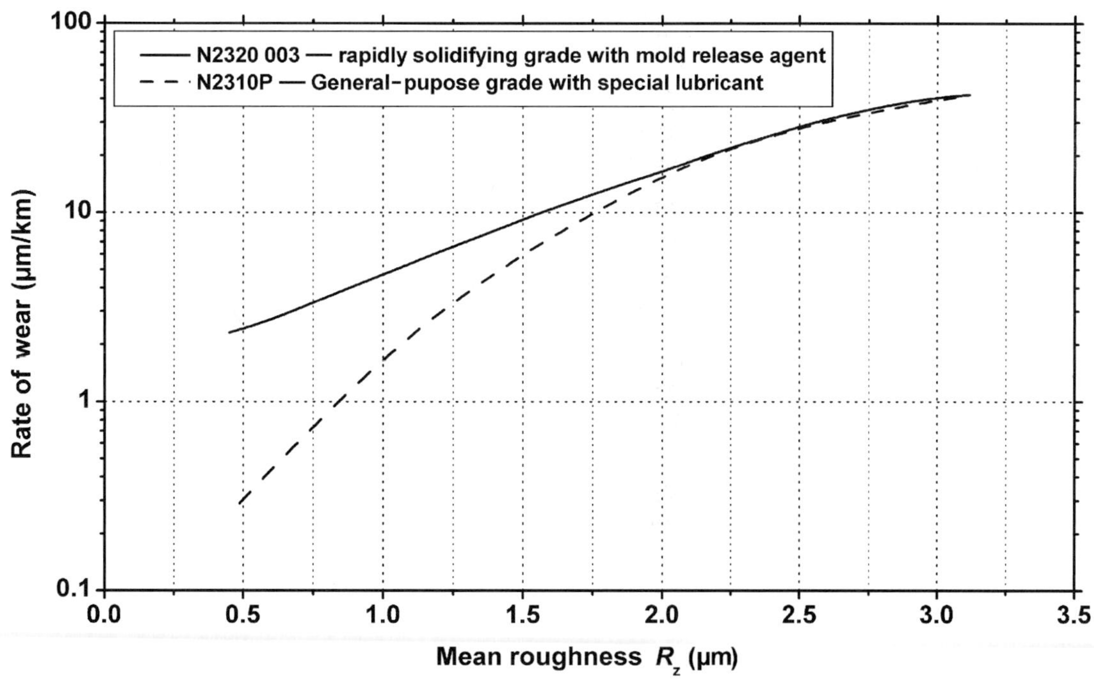

Figure 5.24 Wear rate vs. roughness of a sliding steel disk (HRC = 54–56 at 40°C, P = 1 MPa, V = 0.5 m/s) for two BASF Ultraform® POM-Co plastics.

Table 5.4 Tribological Properties of RTP Company ESD 800 Static Dissipative (Base Polymer POM-Co Ticona Celcon® M90) vs. 1018 C Steel (Data Obtained per ASTM 3702)

PV (KPa m/s)	Load (N)	Speed (m/s)	Wear Factor × 10^{-8} (mm^3/N m)	Dynamic Coefficient of Friction
70	1.80	0.25	87	0.60
175	2.25	0.50	39	0.57
350	2.25	1.00	6	0.44

Table 5.5 Tribological Properties of RTP Company RTP 800 Base Resin (Base Polymer POM-Co Ticona Celcon® M90) vs. 1018 C Steel (Data Obtained per ASTM 3702)

PV (KPa m/s)	Load (N)	Speed (m/s)	Wear Factor × 10^{-8} (mm^3/N m)	Dynamic Coefficient of Friction
70	1.80	0.25	187	0.40
175	2.25	0.50	206	0.42
350	2.25	1.00	302	0.44

Table 5.6 Tribological Properties of RTP Company RTP 800 SI 2 (Base Polymer POM-Co Ticona Celcon® M90 with Silicone 2%) vs. 1018 C Steel (Data Obtained per ASTM 3702)

PV (KPa m/s)	Load (N)	Speed (m/s)	Wear Factor × 10^{-8} (mm^3/N m)	Dynamic Coefficient of Friction
70	1.80	0.25	85	0.43
70	0.45	1.00	313	0.39
175	2.25	0.50	92	0.41
350	8.99	0.25	121	0.51
350	2.25	1.00	151	0.51

Table 5.7 Tribological Properties of RTP Company RTP 800 TFE 5 (Base Polymer POM-Co Ticona Celcon® M90 with PTFE 5%) vs. 1018 C Steel (Data Obtained per ASTM 3702)

PV (KPa m/s)	Load (N)	Speed (m/s)	Wear Factor × 10^{-8} (mm^3/N m)	Dynamic Coefficient of Friction
70	1.80	0.25	36	0.37
175	2.25	0.50	85	0.35
350	2.25	1.00	32	0.30

Table 5.8 Tribological Properties of RTP Company RTP 800 TFE 10 (Base Polymer POM-Co Ticona Celcon® M90 with PTFE 10%) vs. 1018 C Steel (Data Obtained per ASTM 3702)

PV (KPa m/s)	Load (N)	Speed (m/s)	Wear Factor × 10^{-8} (mm^3/N m)	Dynamic Coefficient of Friction
70	1.80	0.25	170	0.68
70	0.45	1.00	380	0.92
175	2.25	0.50	26	0.29
350	8.99	0.25	14	0.18
350	2.25	1.00	1245	0.10

Table 5.9 Tribological Properties of RTP Company RTP 800 TFE 10 SI 2 (Base Polymer POM-Co Ticona Celcon® M90 with PTFE 10% Silicone 2%) vs. 1018 C Steel (Data Obtained per ASTM 3702)

PV (KPa m/s)	Load (N)	Speed (m/s)	Wear Factor × 10^{-8} (mm^3/N m)	Dynamic Coefficient of Friction
70	1.80	0.25	19	0.29
70	0.45	1.00	114	0.63
175	2.25	0.50	12	0.24
350	8.99	0.25	18	0.14
350	2.25	1.00	219	0.27

Table 5.10 Wear and Dynamic Coefficient of Friction of Various Ticona Hostform® Grades in Dry Sliding Contact with a Rotating Polished Steel Shaft (Wear Conditions: Roughness Height 0.8 μm, Peripheral Speed of Shaft $v = 136$ m/min, Load = 3.1 N)

Grade	Wear Volume (mm^2)	Dynamic COF[a]
C 9021 AW	0.62	
C 9021 K	0.73	0.39
C 9021 TF + 3% Si oil	1.23	0.27
C 9021 G	1.96	0.29
C 9021 TF	3.64	0.23
C 9021 + 3% Si oil	6.56	0.39
C 9021	7.96	0.34

[a]Speed = 20 m/min, pressure = 1.25 N/mm^2, test duration 30 min.

5.4 Modified Polyphenylene Ether/ Polyphenylene Oxide

5.4.1 Fatigue Data

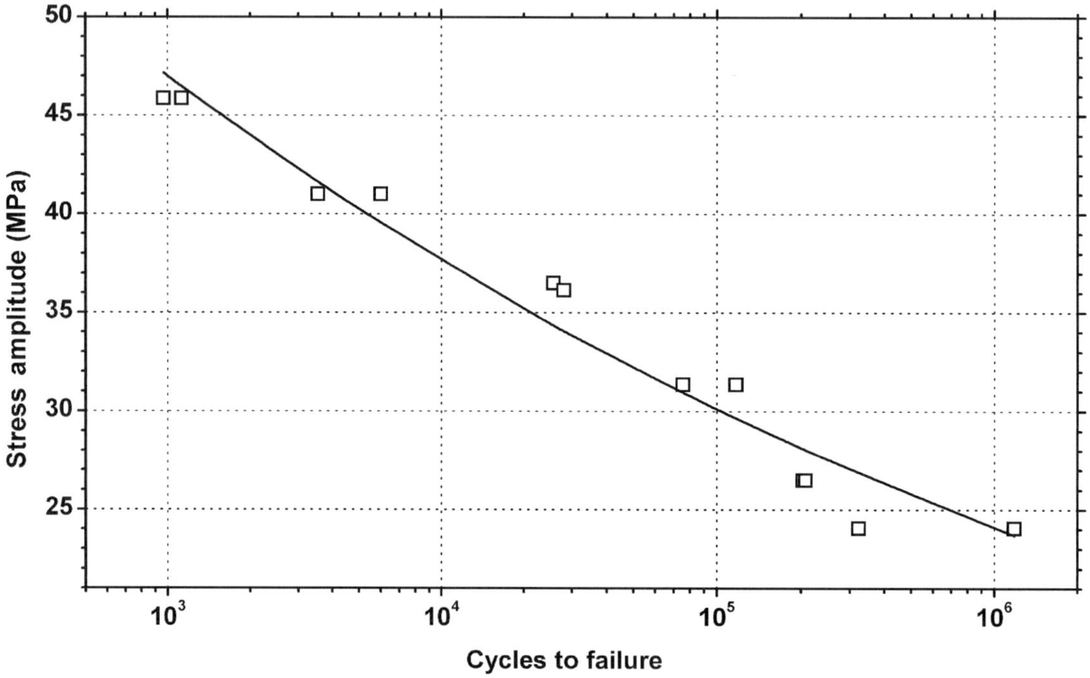

Figure 5.25 Tensile stress amplitude vs. cycles to failure at 23°C of SABIC Innovative Plastics Noryl® GTX954—unfilled impact modified PPE/PS/PA plastic.

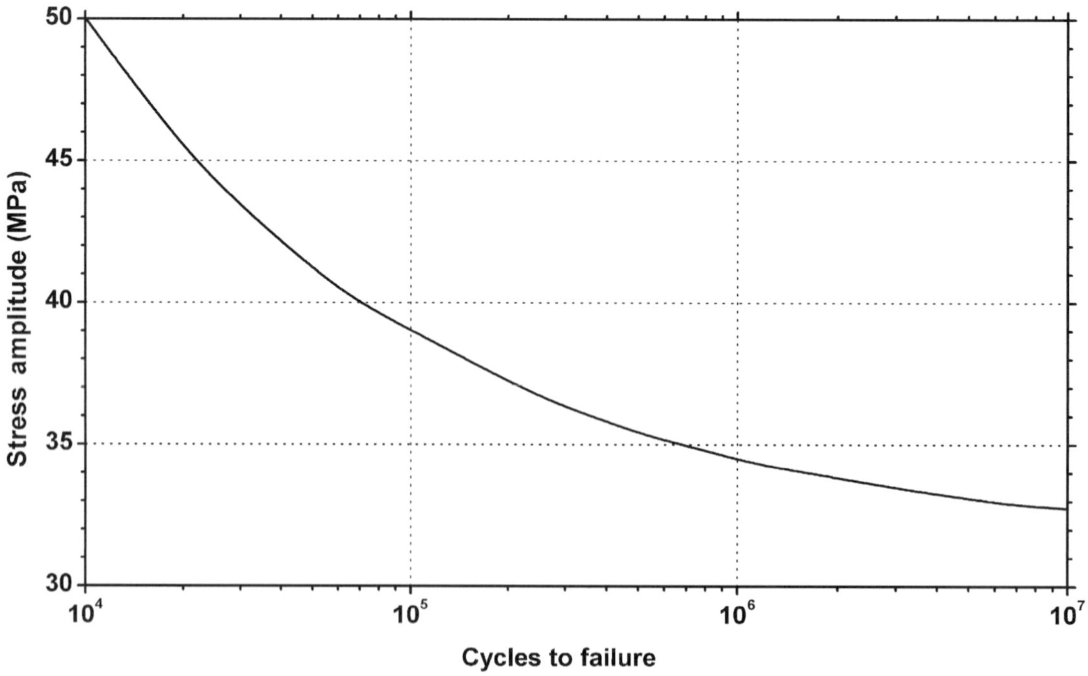

Figure 5.26 Flexural stress amplitude vs. cycles to failure at 23°C of SABIC Innovative Plastics Thermocomp® ZF-1006—30% glass fiber PPO.

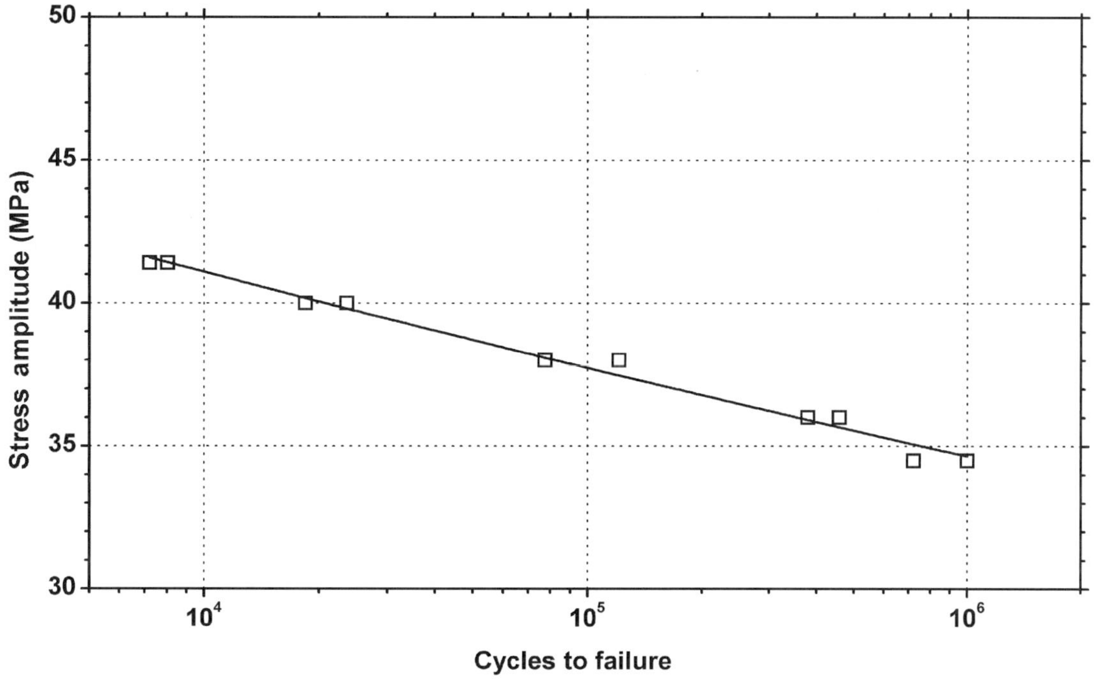

Figure 5.27 Tensile stress amplitude vs. cycles to failure at 23°C of SABIC Innovative Plastics Noryl® PPX615—15% glass fiber reinforced, impact modified PPE/PP plastic.

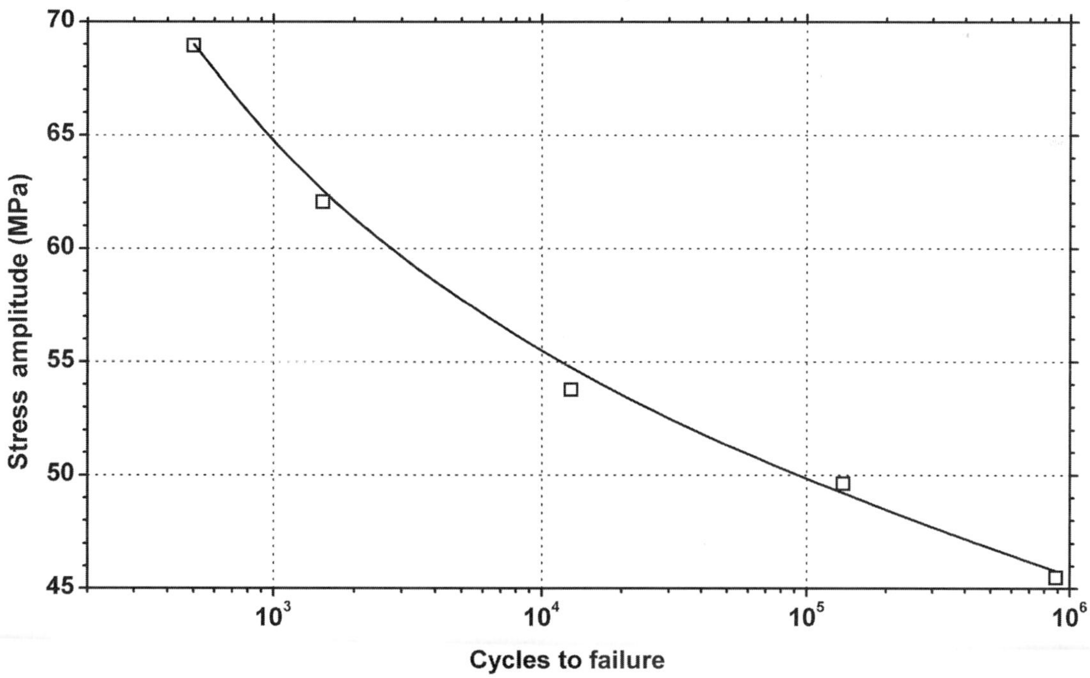

Figure 5.28 Tensile stress amplitude vs. cycles to failure at 23°C of SABIC Innovative Plastics Noryl® PPX630—30% glass fiber reinforced PPE/PP plastic.

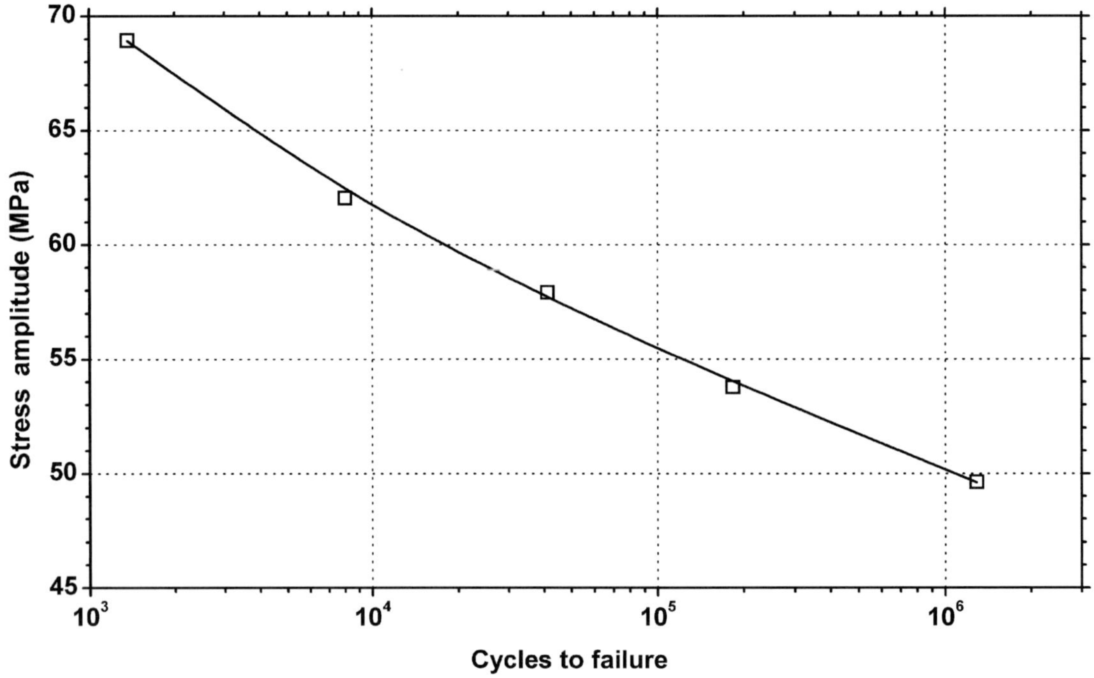

Figure 5.29 Tensile stress amplitude vs. cycles to failure at 23°C of SABIC Innovative Plastics Noryl® PPX640—40% glass fiber reinforced PPE/PP plastic.

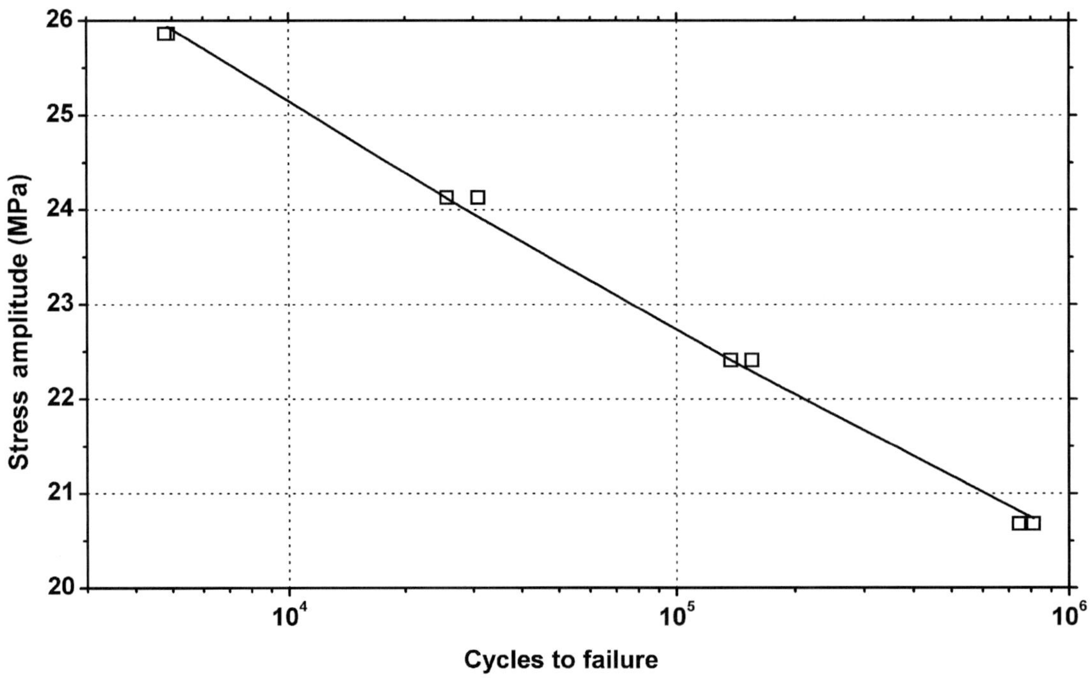

Figure 5.30 Tensile stress amplitude vs. cycles to failure at 23°C of SABIC Innovative Plastics Noryl® PPX7110—high-impact, good heat resistant PPE/PP plastic.

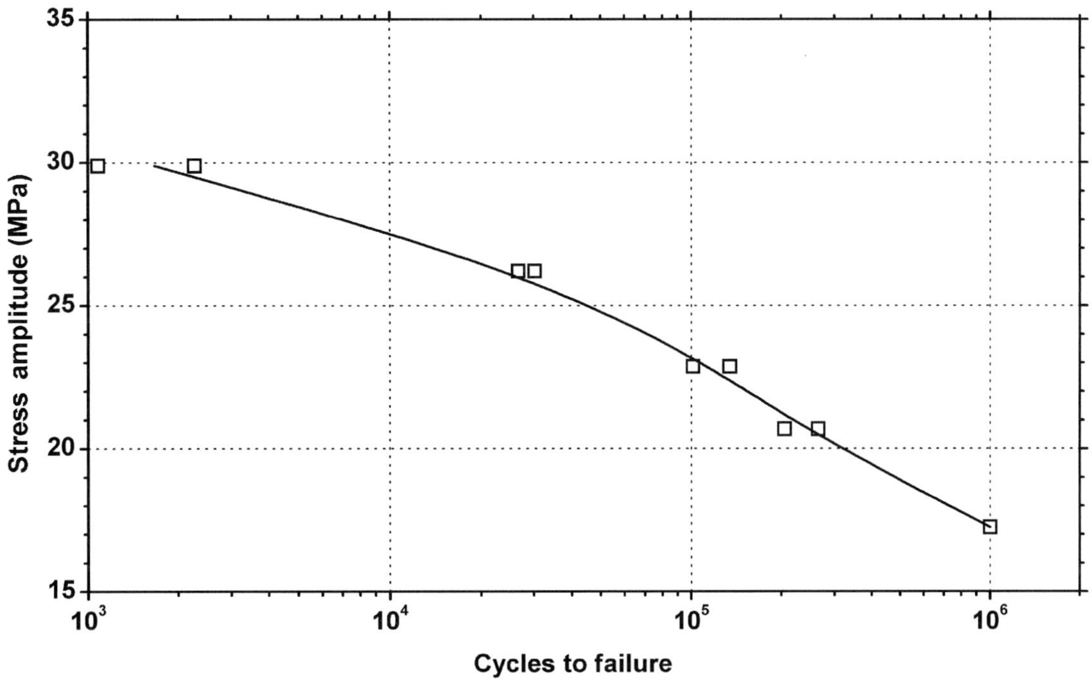

Figure 5.31 Tensile stress amplitude vs. cycles to failure at 23°C of SABIC Innovative Plastics Noryl® PPX7112—paintable, exterior automotive PPE/PP plastic.

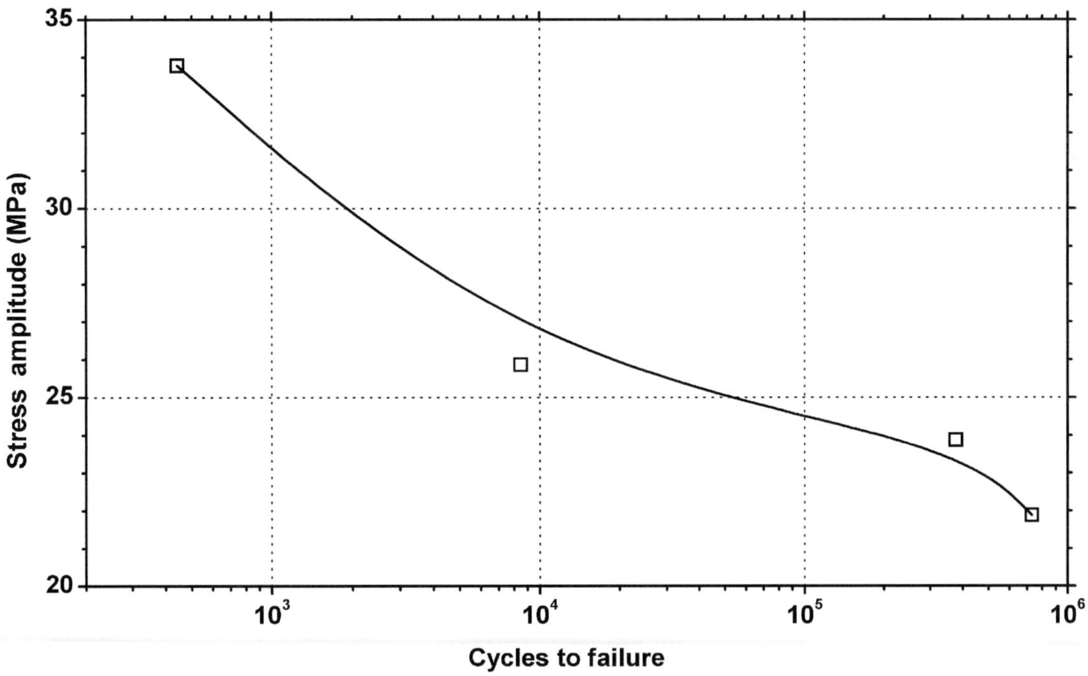

Figure 5.32 Tensile stress amplitude vs. cycles to failure at 23°C of SABIC Innovative Plastics Noryl® PPX7115—high-modulus/impact/heat PPE/PP plastic.

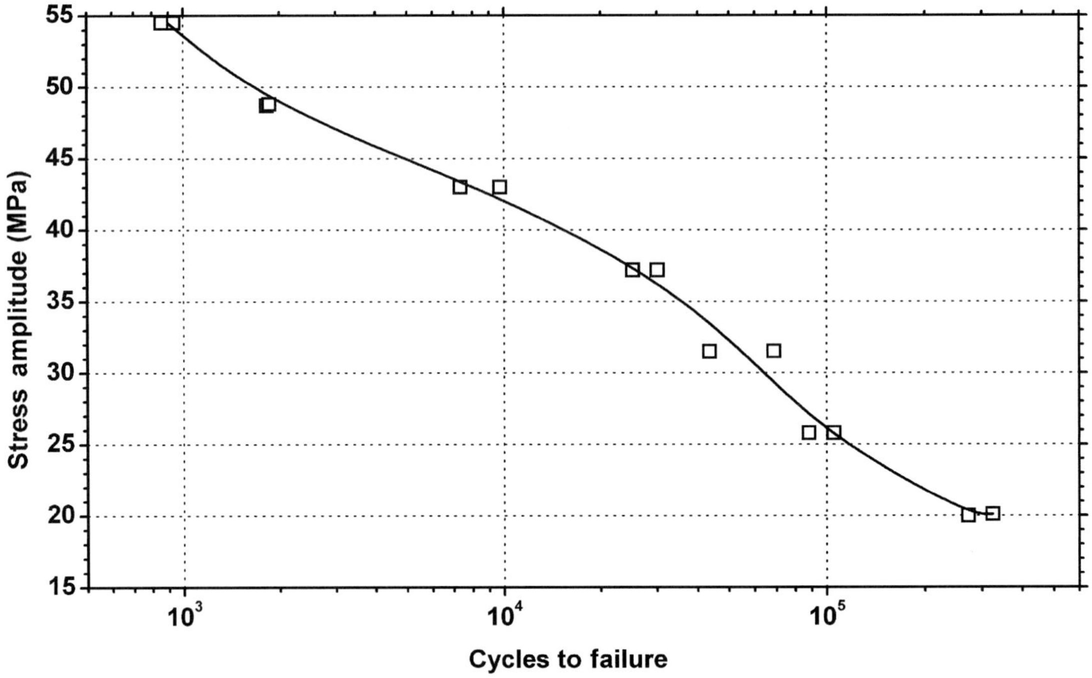

Figure 5.33 Tensile stress amplitude vs. cycles to failure at 23°C of SABIC Innovative Plastics Noryl® 731 general-purpose, UL94 HB rated PPE/PS plastic.

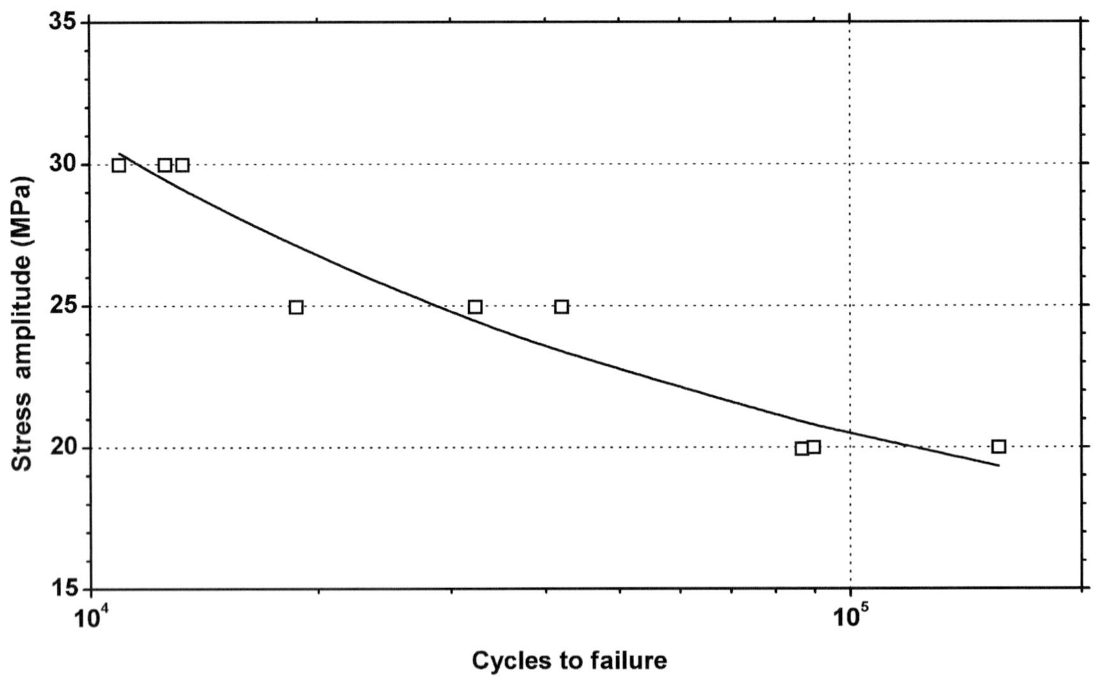

Figure 5.34 Tensile stress amplitude vs. cycles to failure at 23°C of SABIC Innovative Plastics Noryl® HH195—PPE/PS plastic.

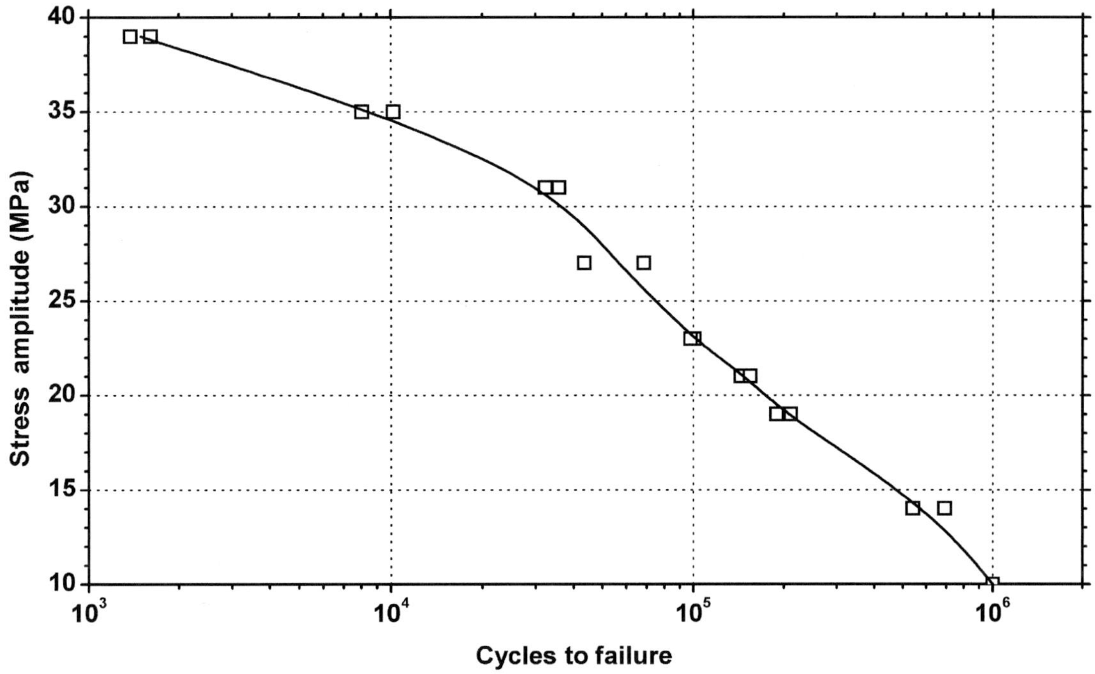

Figure 5.35 Tensile stress amplitude vs. cycles to failure at 23°C of SABIC Innovative Plastics Noryl® EM6100F—automotive interiors. 240°F HDT, high-impact, PPE/PS plastic.

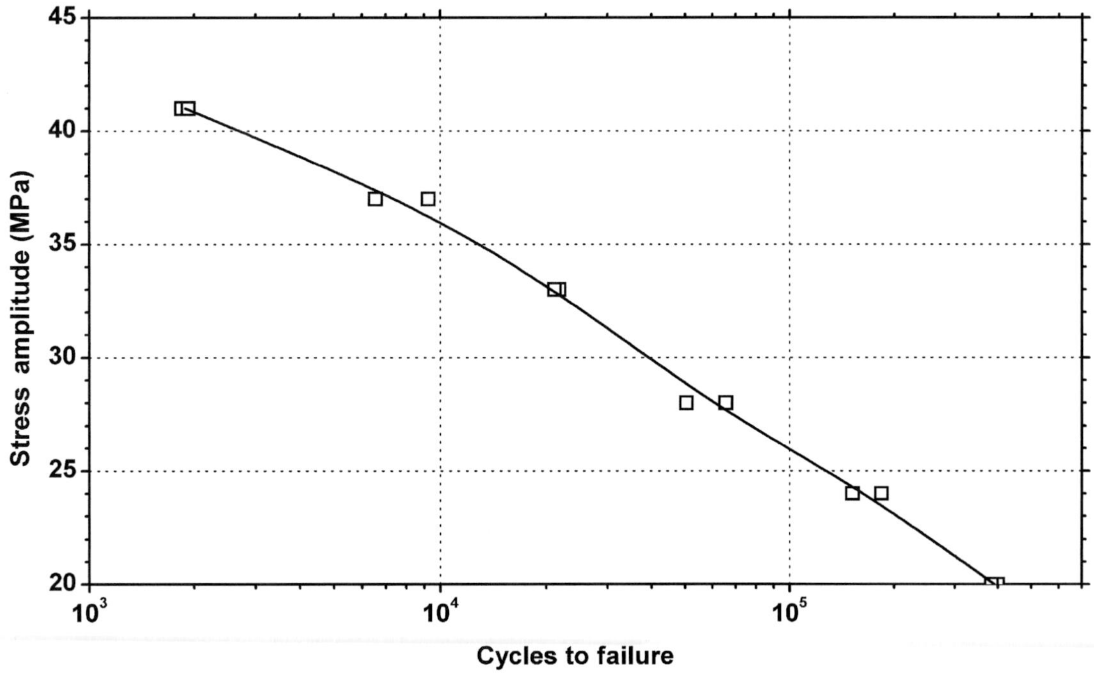

Figure 5.36 Tensile stress amplitude vs. cycles to failure at 23°C of SABIC Innovative Plastics Noryl® EM6101—automotive interiors, improved flow and processing 250°F DTUL PPE/PS plastic.

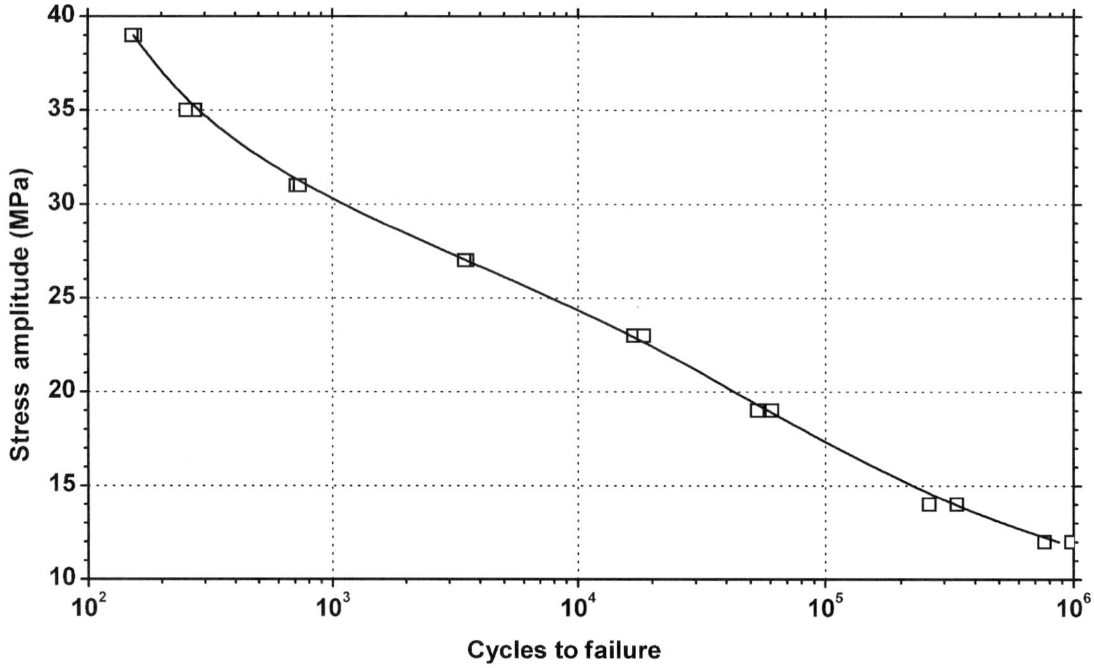

Figure 5.37 Tensile stress amplitude vs. cycles to failure at 23°C of SABIC Innovative Plastics Noryl® EM7100—automotive interiors. 200F DTUL. Excellent processability/economy. PPE/PS plastic.

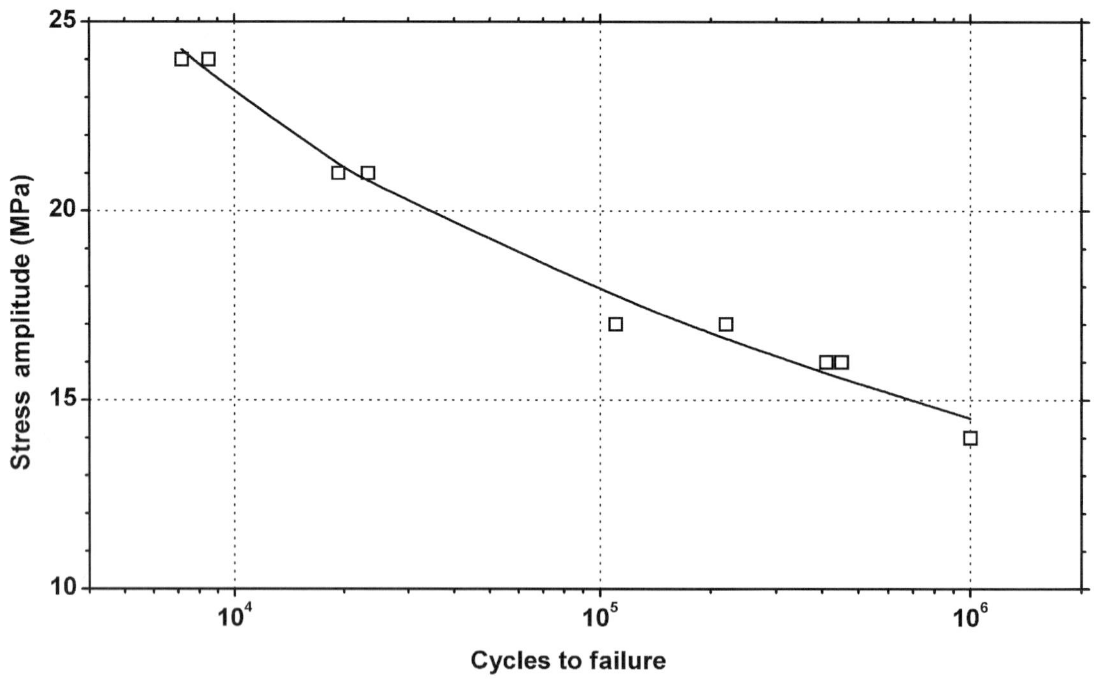

Figure 5.38 Tensile stress amplitude vs. cycles to failure at 23°C of SABIC Innovative Plastics Noryl® EM7304F—15% glass fiber reinforced, high flow, for automotive instrument panel retainers PPE/PS plastic.

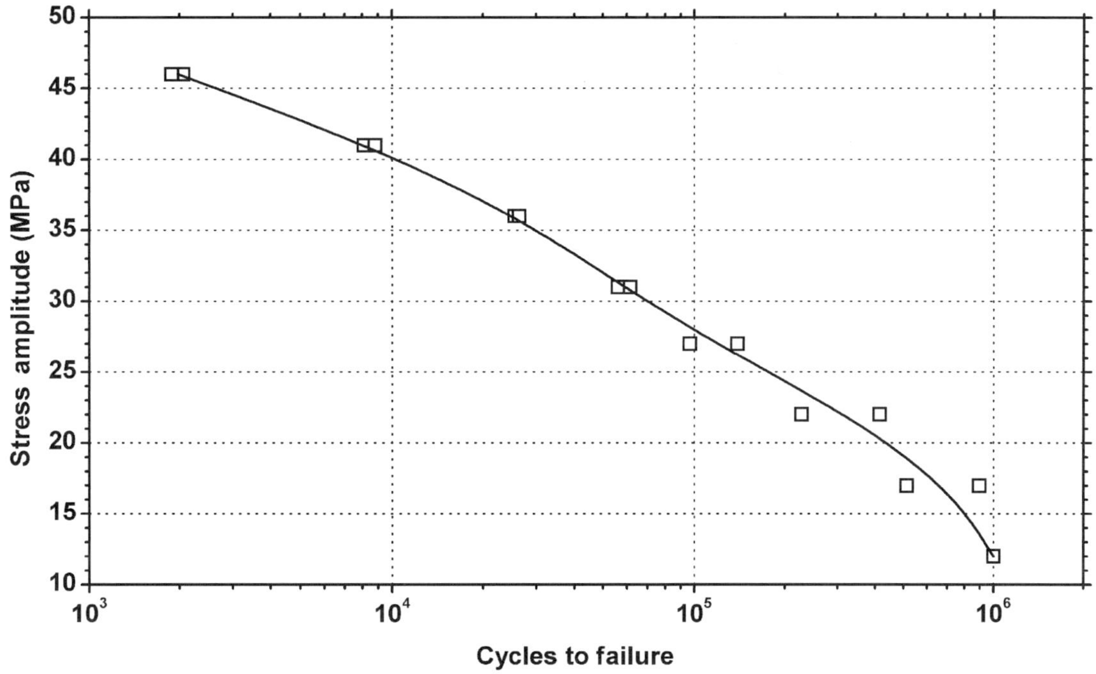

Figure 5.39 Tensile stress amplitude vs. cycles to failure at 23°C of SABIC Innovative Plastics Noryl® FN150X—improved productivity, thin wall capable, UL94 V-0/5VA rated PPE/PS plastic.

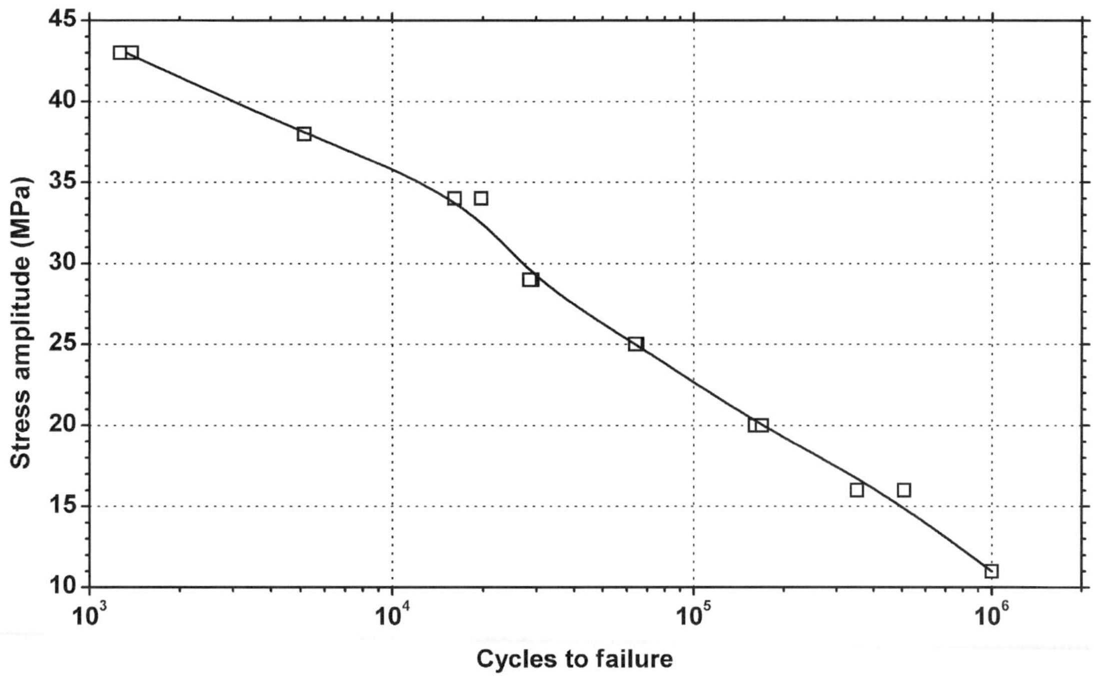

Figure 5.40 Tensile stress amplitude vs. cycles to failure at 23°C of SABIC Innovative Plastics Noryl® FN215X—structural foam PPE/PS plastic.

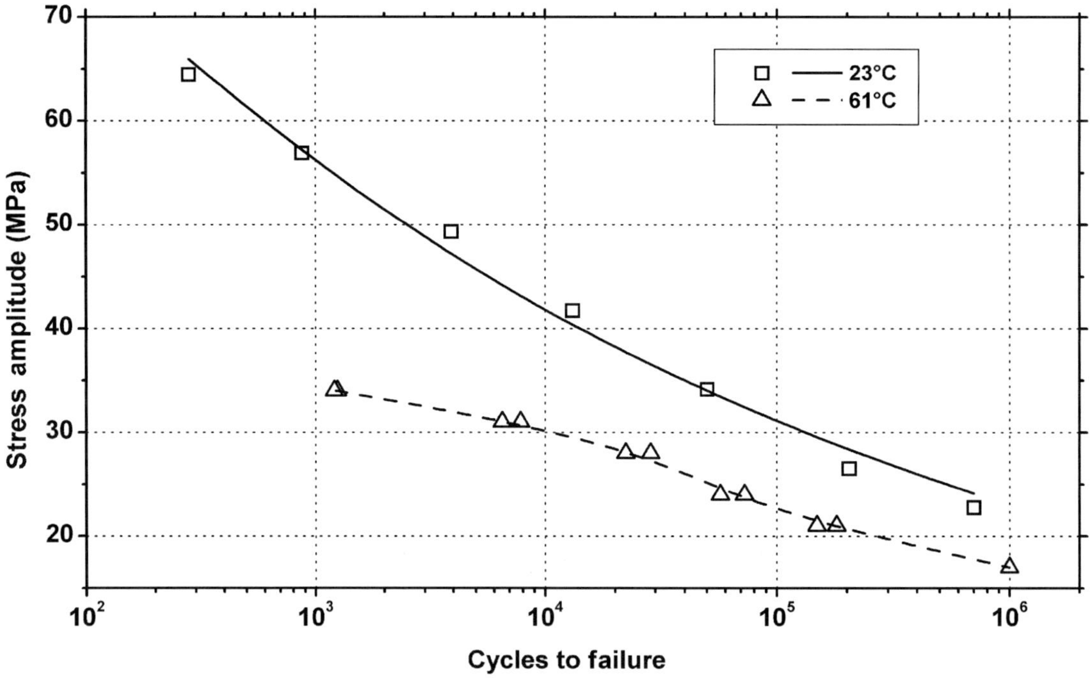

Figure 5.41 Tensile stress amplitude vs. cycles to failure at 23°C of SABIC Innovative Plastics Noryl® GFN1—10% glass fiber reinforced, UL94 HB rated PPE/PS plastic.

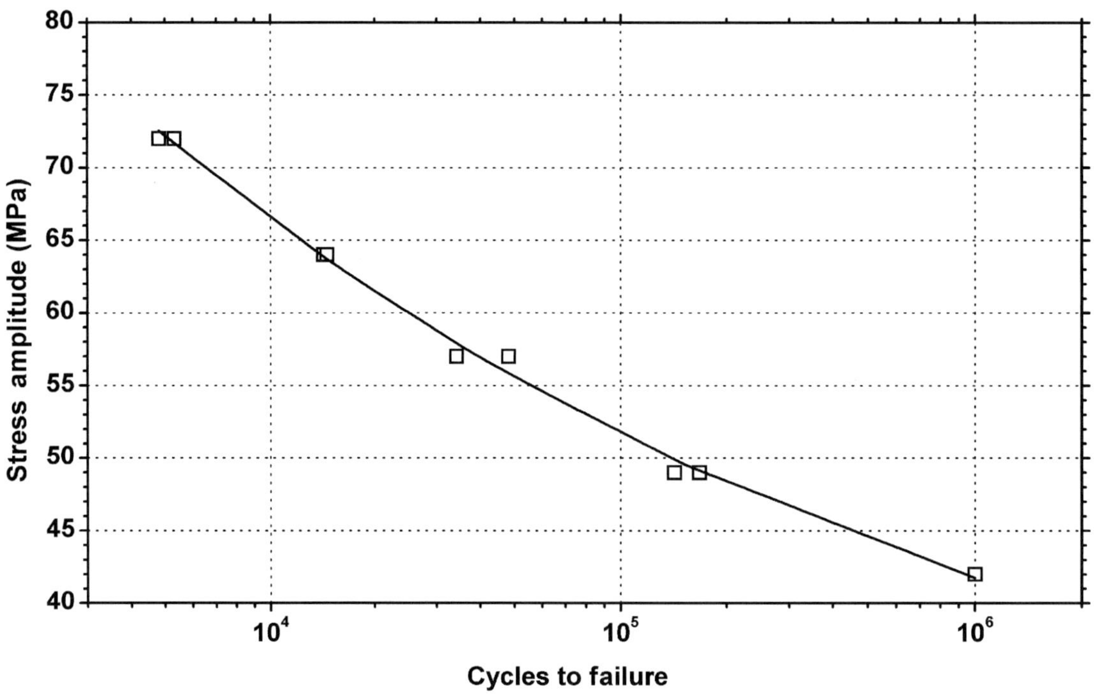

Figure 5.42 Tensile stress amplitude vs. cycles to failure at 23°C of SABIC Innovative Plastics Noryl® GFN2—20% glass fiber reinforced, UL94 HB rated PPE/PS plastic.

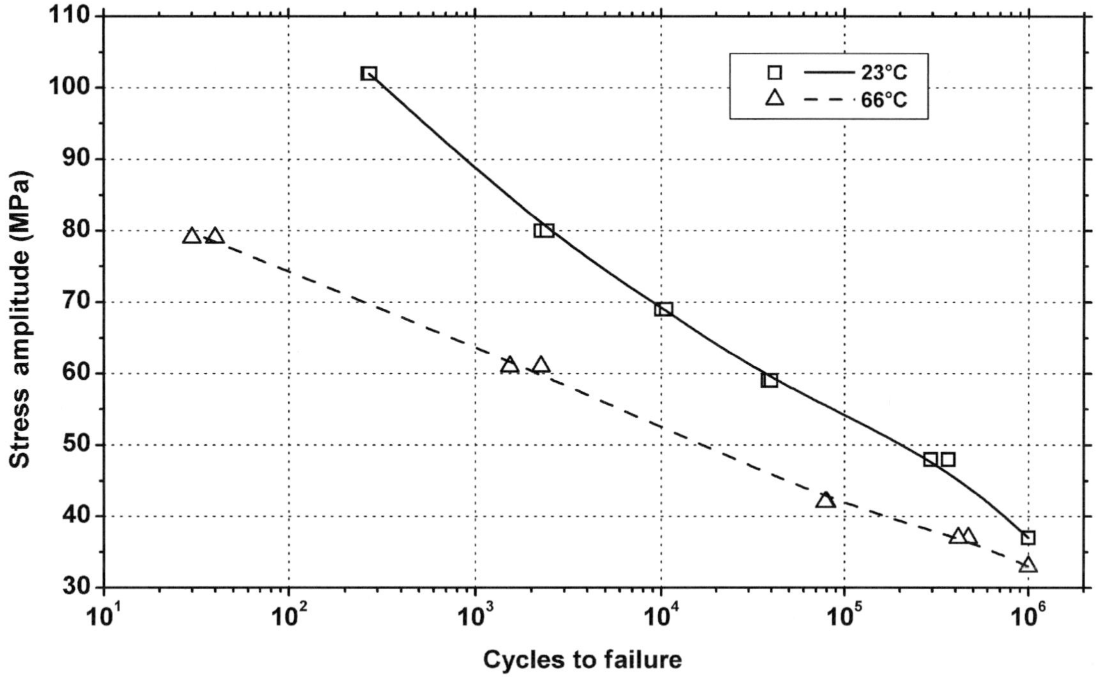

Figure 5.43 Tensile stress amplitude vs. cycles to failure of SABIC Innovative Plastics Noryl® GFN3—30% glass fiber reinforced, UL94 HB rated PPE/PS plastic.

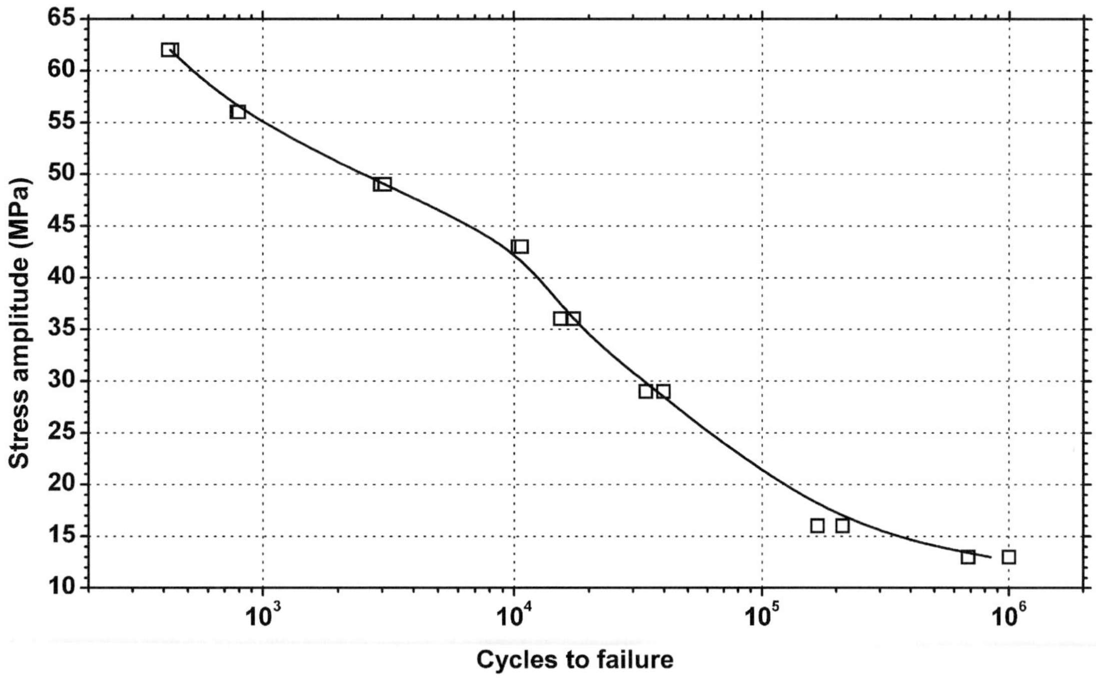

Figure 5.44 Tensile stress amplitude vs. cycles to failure at 23°C of SABIC Innovative Plastics Noryl® HS1000X—13% mineral reinforced. Nonbrominated, nonchlorinated fire resistant, UL94 V0 and 5VA listed PPE/PS plastic.

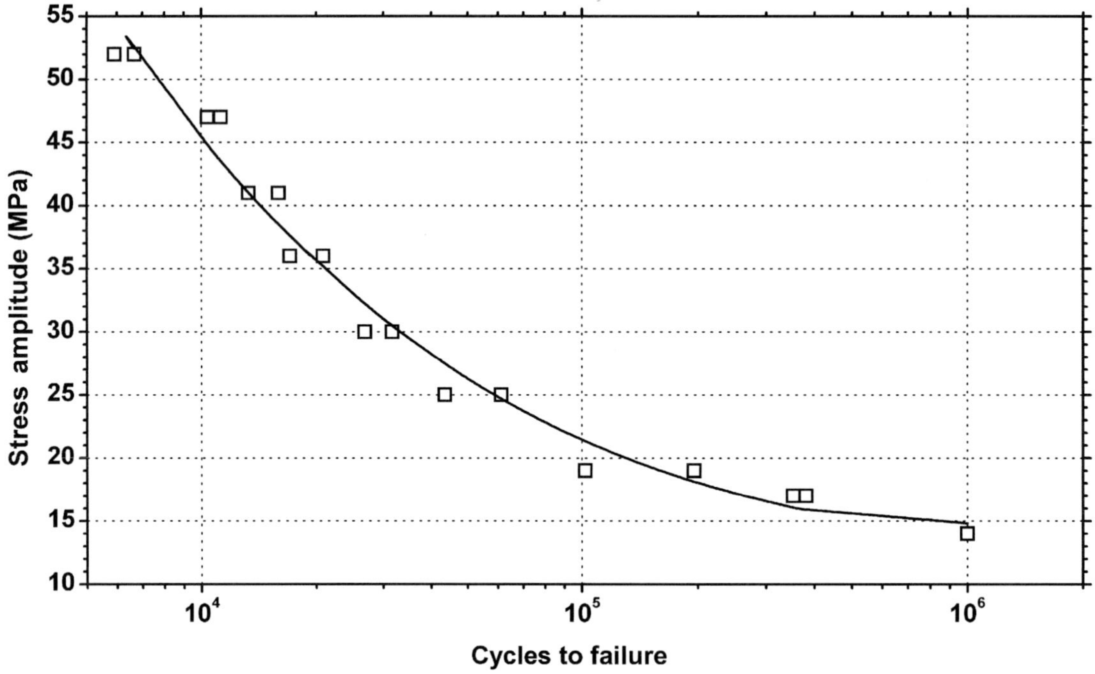

Figure 5.45 Tensile stress amplitude vs. cycles to failure at 23°C of SABIC Innovative Plastics Noryl® HS2000X—17% mineral reinforced, nonbrominated, nonchlorinated fire resistant, UL94 V0 and 5VA listed PPE/PS plastic.

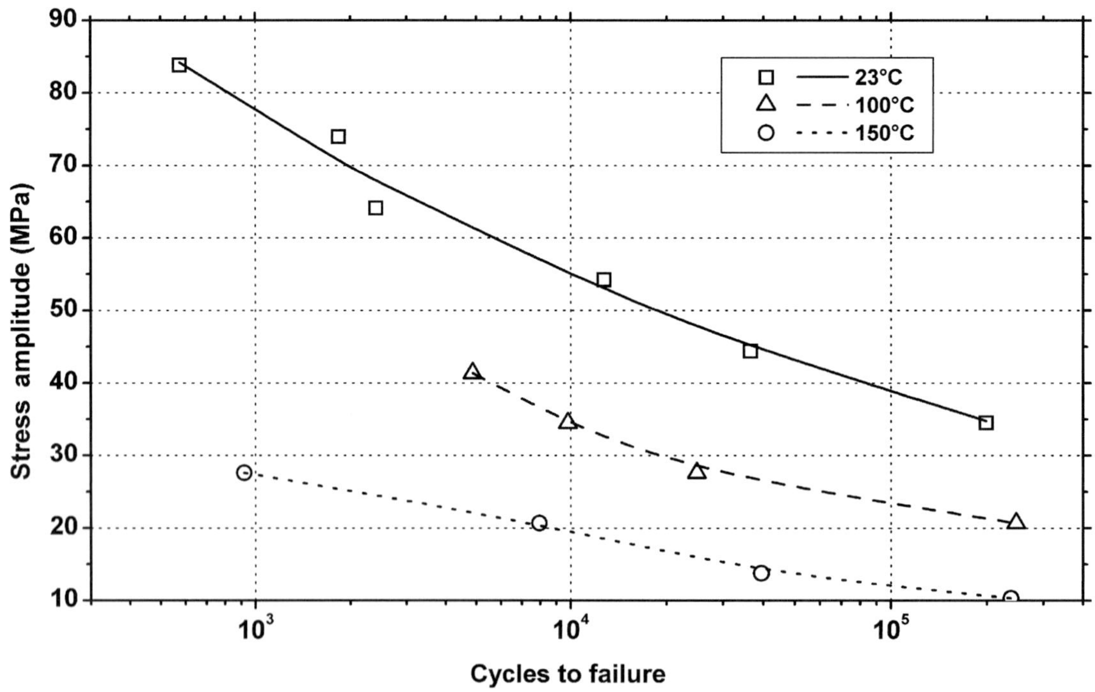

Figure 5.46 Tensile stress amplitude vs. cycles to failure at various temperatures of SABIC Innovative Plastics Noryl® IGN320—PPE/PS plastic.

6.1 Background

Polyesters are formed by a condensation reaction that is very similar to the reaction used to make polyamide or nylons. A diacid and dialcohol are reacted to form the polyester with the elimination of water as shown in Figure 6.1.

While the actual commercial route to making the polyesters may be more involved, the end result is the same polymeric structure. The diacid is usually aromatic. Polyester resins can be formulated to be brittle and hard, tough and resilient, or soft and flexible. In combination with reinforcements such as glass fibers, they offer outstanding strength, a high strength-to-weight ratio, chemical resistance, and other excellent mechanical properties. The three dominant materials in this plastics family are polycarbonate (PC), PET, and polybutylene terephthalate (PBT). Thermoplastic polyesters are similar in properties to Nylon 6 and Nylon 66, but have lower water absorption and higher dimensional stability than the nylons.

6.1.1 Polycarbonate

Theoretically, PC is formed from the reaction of bis-phenol A and carbonic acid. The structures of these two monomers are given in Figure 6.2.

Commercially, different routes are used, but the PC polymer of the structure shown in Figure 6.3 is the result.

Polycarbonate performance properties include:

- Very impact resistant and is virtually unbreakable and remains tough at low temperatures
- "Clear as glass" clarity
- High heat resistance
- Dimensional stability
- Resistant to UV light, allowing exterior use
- Flame retardant properties

Applications include glazing, safety shields, lenses, casings and housings, light fittings, kitchenware (microwaveable), medical apparatus (sterilizable), and CDs (the discs).

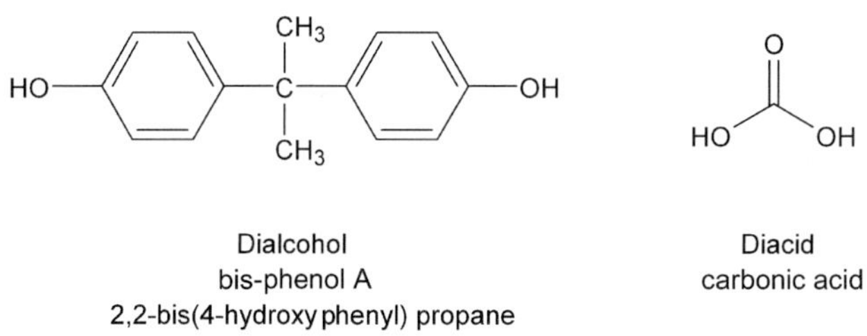

Figure 6.2 Chemical structures of monomers used to make PC polyester.

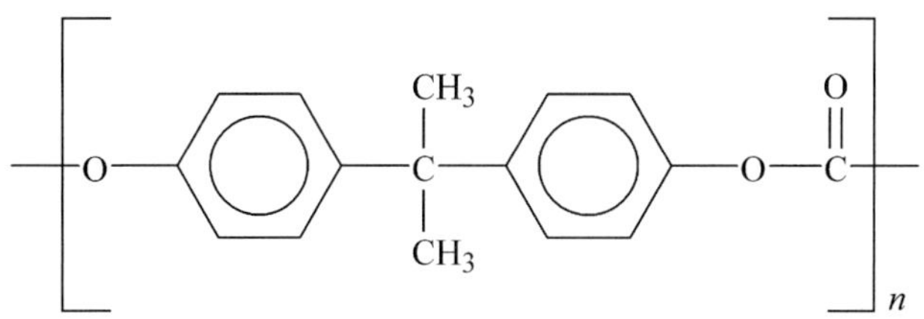

Figure 6.3 Chemical structure of PC polyester.

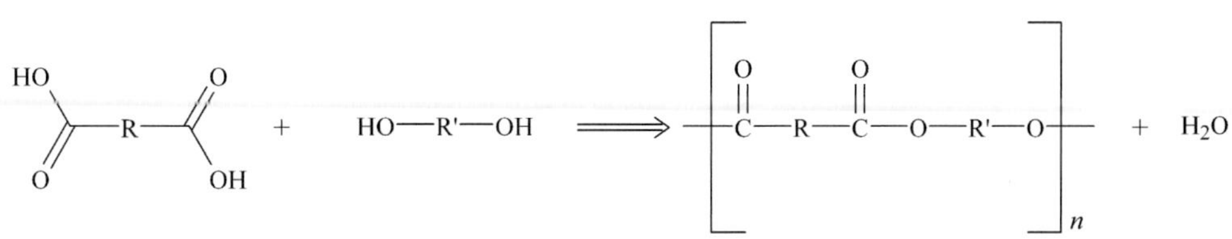

Figure 6.1 Chemical structure of PC polyester.

Copyright © 2010 Laurence W. McKeen. All rights reserved.

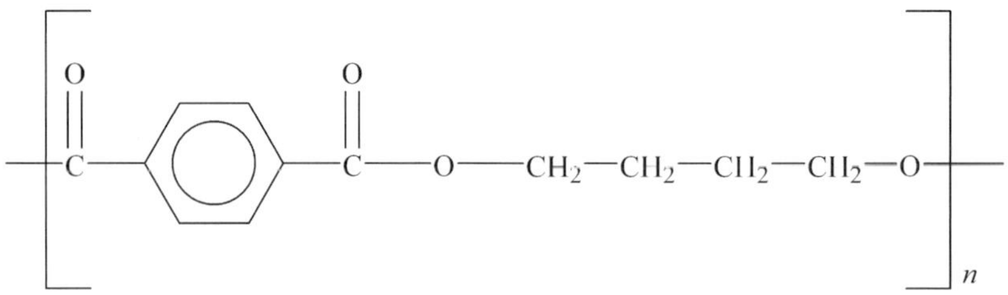

Figure 6.4 Chemical structure of PBT polyester.

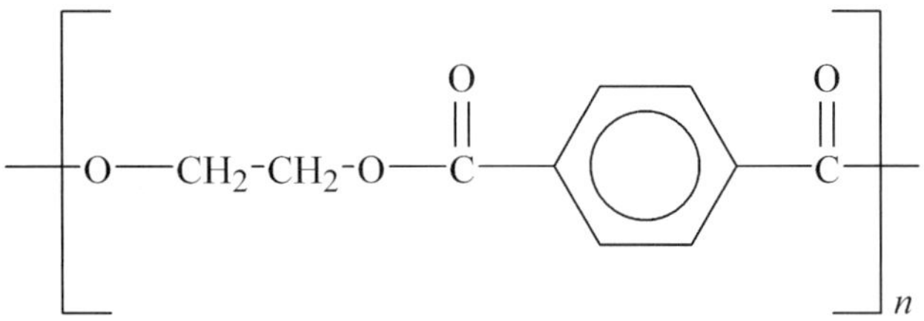

Figure 6.5 Chemical structure of PET polyester.

6.1.2 Polybutylene Terephthalate

PBT is a semi-crystalline, white or off-white polyester similar in both composition and properties to PET. It has somewhat lower strength and stiffness than PET, is a little softer but has higher impact strength and similar chemical resistance. As it crystallizes more rapidly than PET, it tends to be preferred for industrial scale molding. Its structure is shown in Figure 6.4.

PBT performance properties include:

- High mechanical properties
- High thermal properties
- Good electrical properties
- Dimensional stability
- Excellent chemical resistance
- Flame retardancy

6.1.3 Polyethylene Terephthalate

PET polyester is the most common thermoplastic polyester and is often called just "polyester". This often causes confusion with the other polyesters in this chapter. PET exists both as an amorphous (transparent) and as a semi-crystalline (opaque and white) thermoplastic material. The semi-crystalline PET has good strength, ductility, stiffness, and hardness. The amorphous PET has better ductility but less stiffness and hardness.

It absorbs very little water. Its structure is shown in Figure 6.5.

PET has good barrier properties against oxygen and carbon dioxide. Therefore, it is utilized in bottles for mineral water. Other applications include food trays for oven use, roasting bags, audio/video tapes as well as mechanical components.

6.1.4 Liquid Crystalline Polymers

Liquid crystalline polymers (LCP) are a relatively unique class of partially crystalline aromatic polyesters based on 4-hydroxybenzoic acid and related monomers shown in Figure 6.6. Liquid crystal polymers are capable of forming regions of highly ordered structure while in the liquid phase. However, the degree of order is somewhat less than that of a regular solid crystal. Typically, LCPs have outstanding mechanical properties at high temperatures, excellent chemical resistance, inherent flame retardancy and good weatherability. Liquid crystal polymers come in a variety of forms from sinterable high temperature to injection moldable compounds.

LCPs are exceptionally inert. They resist stress cracking in the presence of most chemicals at elevated temperatures, including aromatic or halogenated hydrocarbons, strong acids, bases, ketones, and other aggressive industrial substances. Hydrolytic stability in boiling water is excellent. Environments that deteriorate these polymers are high-temperature steam, concentrated sulfuric acid, and boiling caustic materials.

As an example, the structure of Ticona Vectra® A950 LCP is shown in Figure 6.7.

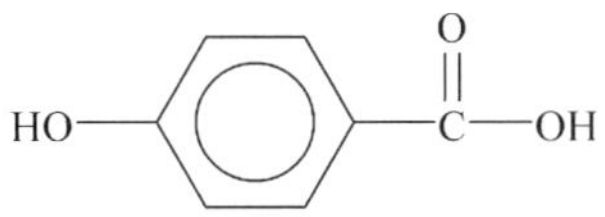

HBA
4-hydroxybenzoic acid

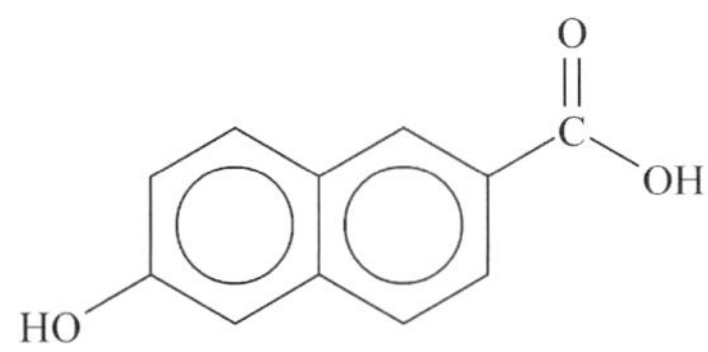

HNA
6-hydroxynaphthalene-2-carboxylic acid

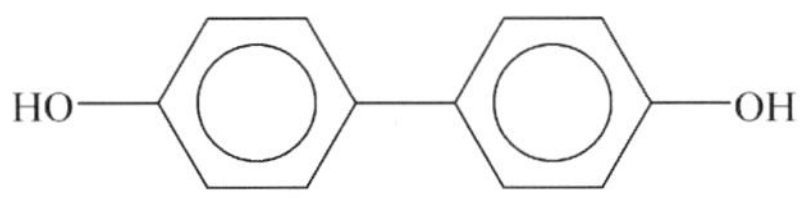

BP
4-(4-hydroxyphenyl)phenol

HQ
benzene-1,4-diol
(hydroquinone)

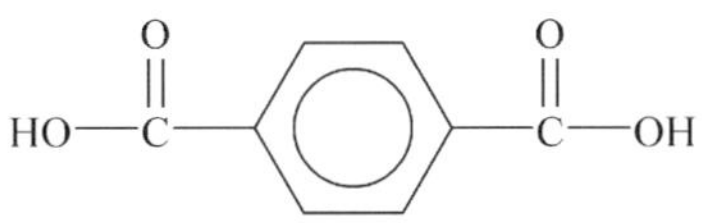

TA
benzene-1,4-dicarboxylic acid
(terephthalic acid)

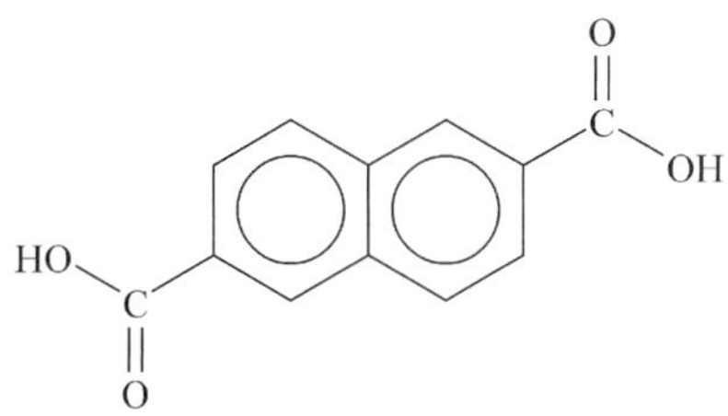

NDA
Naphthalene-2,6-dicarboxylic acid

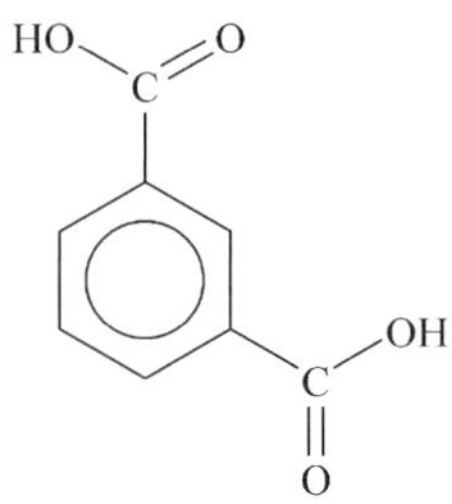

IA
benzene-1,3-dicarboxylic acid
(isophthalic acid)

Figure 6.6 Chemical structures of monomers used to make LCP polyesters.

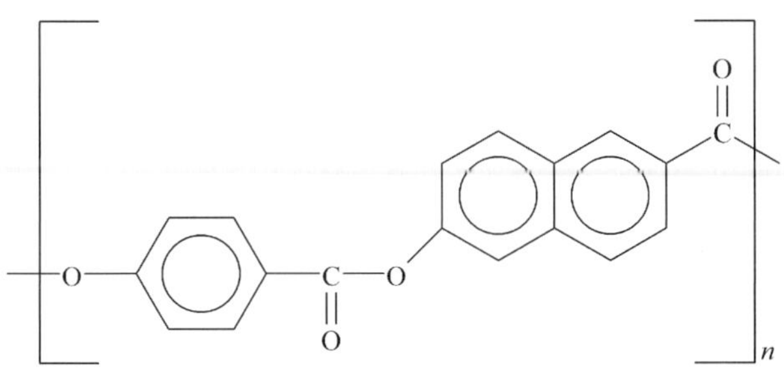

Figure 6.7 Chemical structure of Ticona Vectra® A950 LCP.

6.1.5 Polycyclohexylene-dimethylene Terephthalate

Polycyclohexylene-dimethylene terephthalate (PCT) is a high-temperature polyester that possesses the chemical resistance, processability, and dimensional stability of polyesters PET and PBT. However, the aliphatic cyclic ring shown in Figure 6.8 imparts added heat resistance. This puts it between the common polyesters and the LCP polyesters described

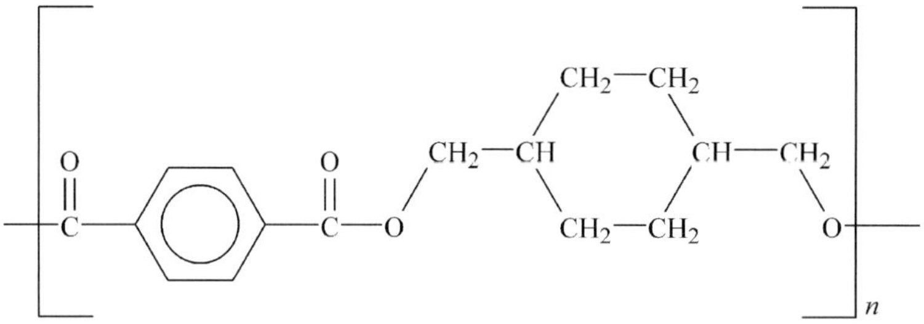

Figure 6.8 Chemical structure of PCT polyester.

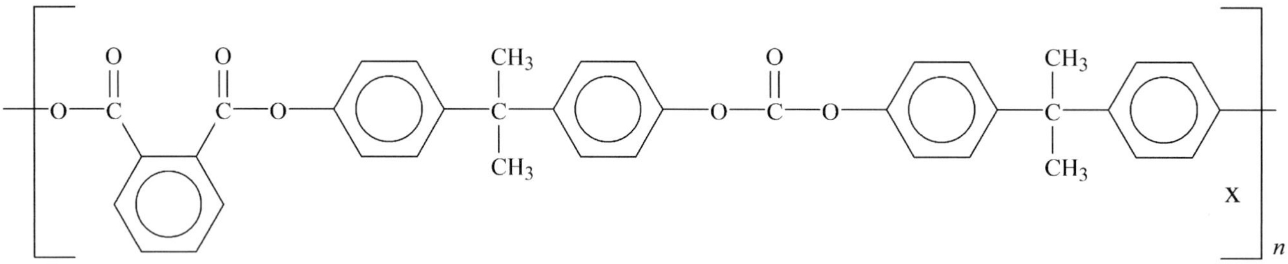

Figure 6.9 Chemical structure of PPC polyester.

in the previous section. At this time only DuPont makes this plastic under the trade name Thermx®.

This material has found use in automotive, electrical, and housewares applications.

6.1.6 Polyphthalate Carbonate

Amorphous polyphthalate carbonate copolymer (PPC) is another high-temperature PC. It provides excellent impact resistance, optical clarity, and abrasion resistance. The plastic offers UV protection as well. It is lightweight, impact-resistant, and can be reused after multiple exposures to sterilization. Its structure is shown in Figure 6.9.

6.1.7 Polytrimethylene Terephthalate

Polytrimethylene terephthalate (PTT) is a semi-crystalline polyester polymer that has many of the same property advantages as PBT and PET. However, compared to PBT, compounds composed of PTT exhibit better tensile strengths, flexural strengths, and stiffness. They also have excellent flow and surface finish. PTT can also be more cost-effective than PBT.

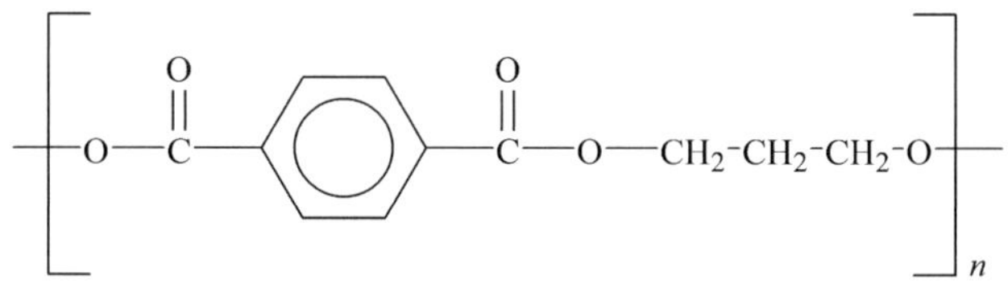

Figure 6.10 Chemical structure of PTT polyester.

PTT may have more uniform shrinkage and better dimensional stability in some applications. PTT, like PBT, has excellent resistance to a broad range of chemicals at room temperature, including aliphatic hydrocarbons, gasoline, carbon tetrachloride, perchloroethylene, oils, fats, alcohols, glycols, esters, ethers, and dilute acids and bases. Strong bases may attack PTT and many polyester resins.

The two monomer units used in producing this polymer are 1,3-propanediol and terephthalic acid and its structure is shown in Figure 6.10.

6.1.8 Polyester Blends and Alloys

There are numerous polyester blends and alloys. Often the different polyesters are blended.

6.2 Polycarbonate

6.2.1 Fatigue Data

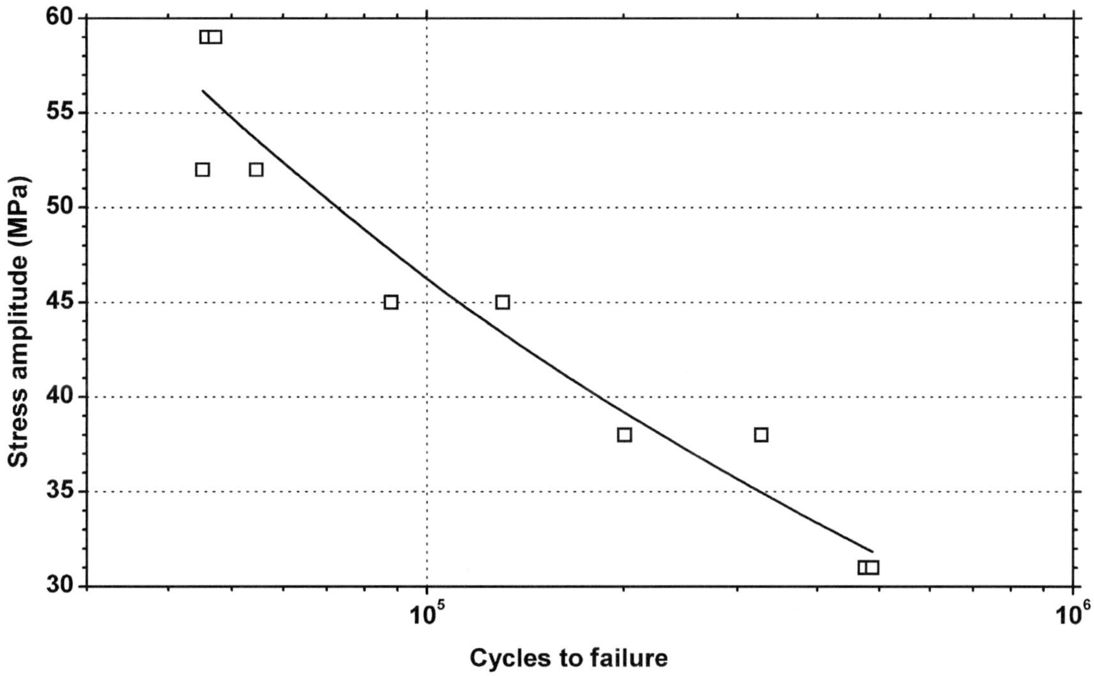

Figure 6.11 Tensile stress amplitude vs. cycles to failure of SABIC Innovative Plastics Lexan® 101—unreinforced, high viscosity, general-purpose extrusion PC.

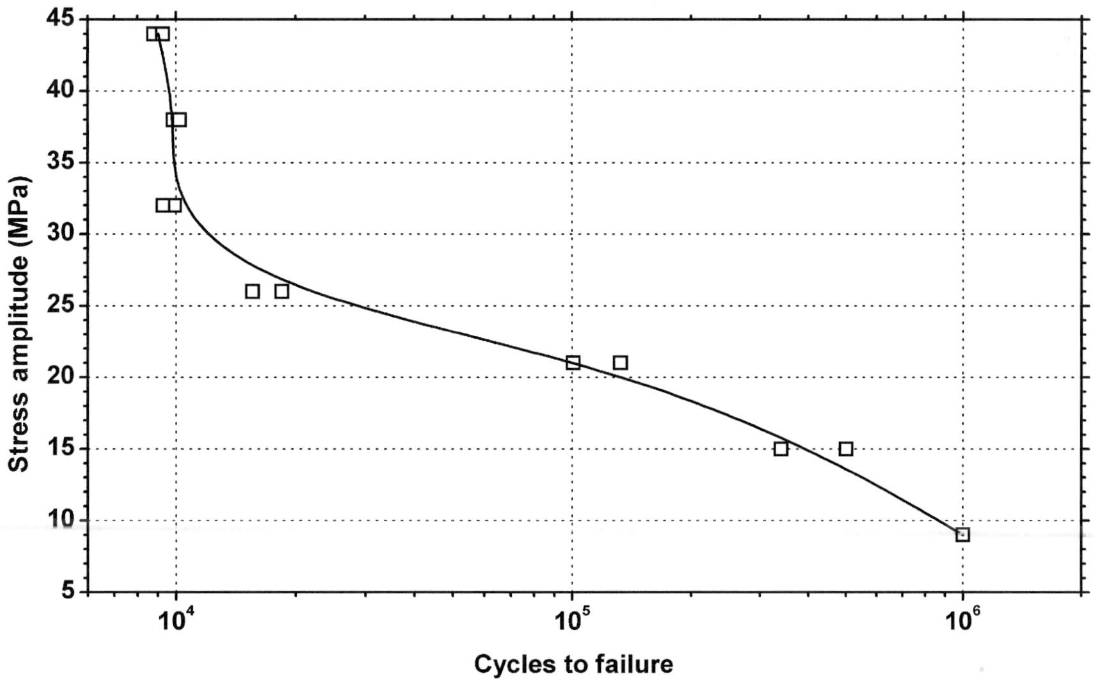

Figure 6.12 Tensile stress amplitude vs. cycles to failure of SABIC Innovative Plastics Lexan® 121—unreinforced, low viscosity, general-purpose PC.

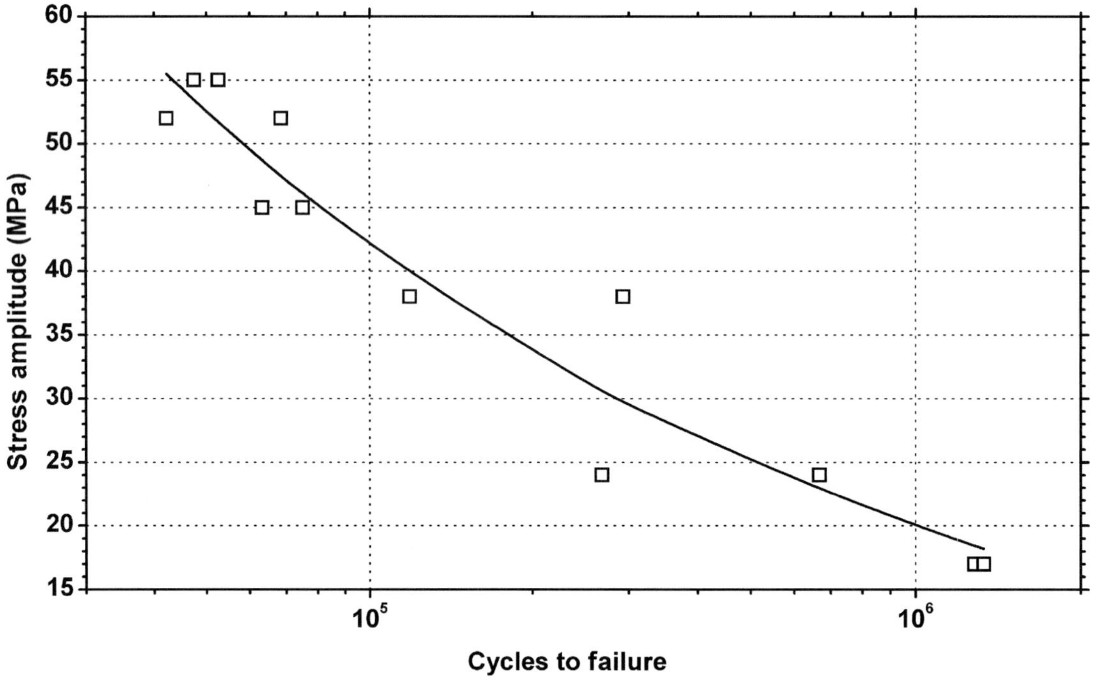

Figure 6.13 Tensile stress amplitude vs. cycles to failure of SABIC Innovative Plastics Lexan® 141—unreinforced, low–medium viscosity, general-purpose PC.

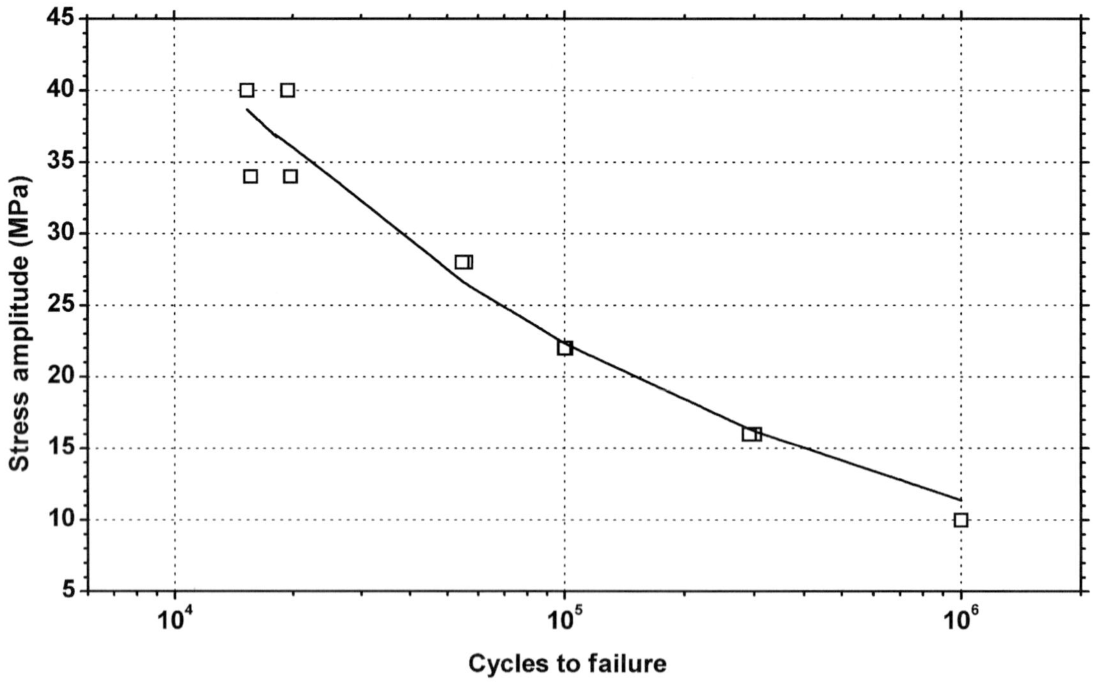

Figure 6.14 Tensile stress amplitude vs. cycles to failure of SABIC Innovative Plastics Lexan® 143R—unreinforced, low–medium viscosity, UV stabilized general-purpose PC.

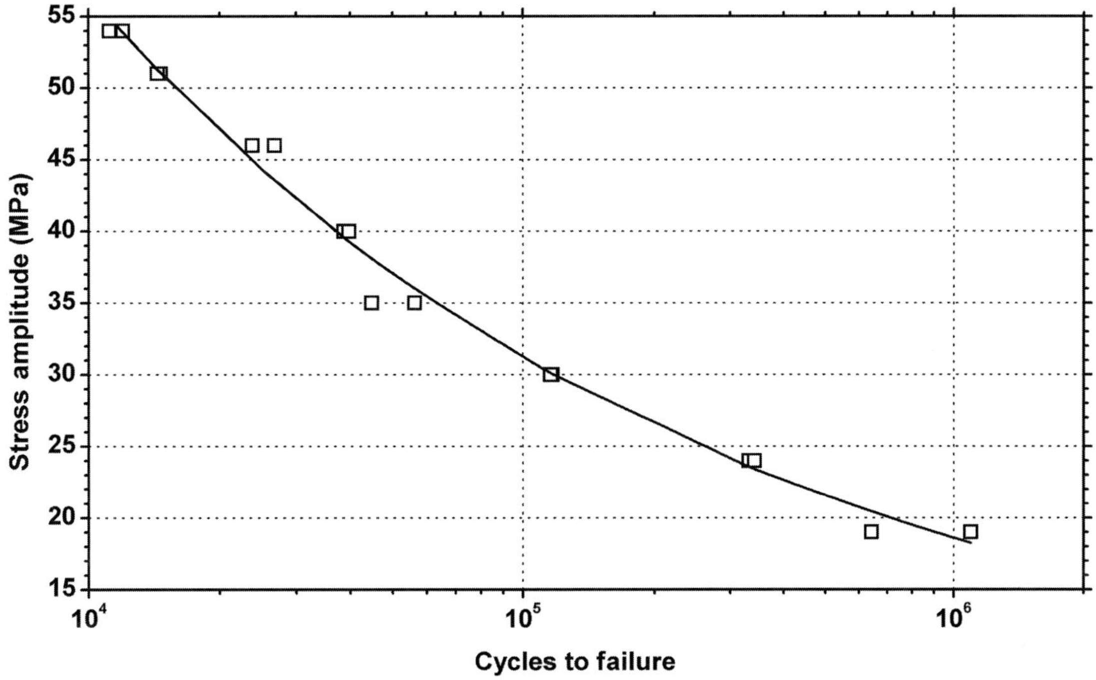

Figure 6.15 Tensile stress amplitude vs. cycles to failure of SABIC Innovative Plastics Lexan® 191—high impact PC.

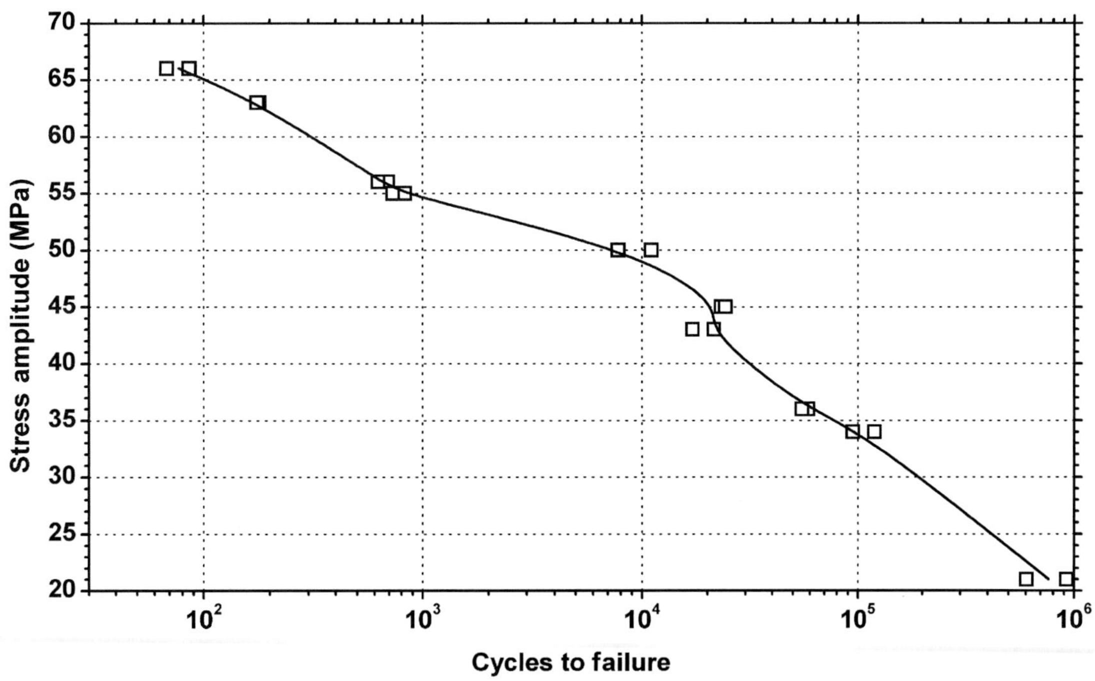

Figure 6.16 Tensile stress amplitude vs. cycles to failure of SABIC Innovative Plastics Lexan® 500—10% glass fiber reinforced PC.

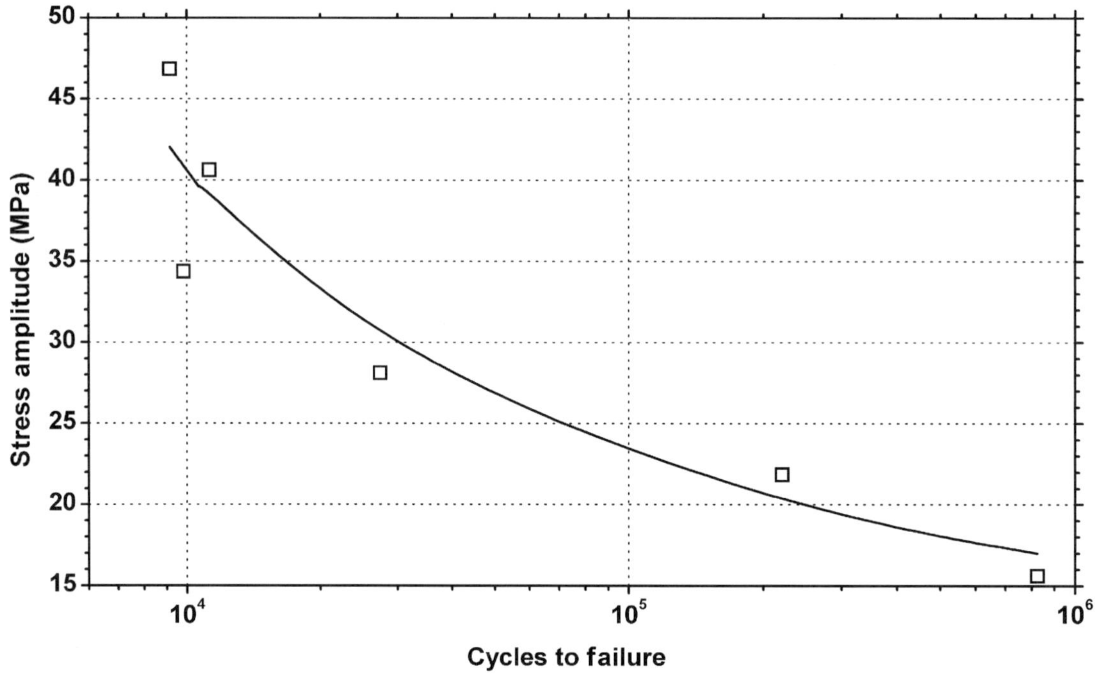

Figure 6.17 Tensile stress amplitude vs. cycles to failure of SABIC Innovative Plastics Lexan® 915R—unreinforced, flame retardant, easy release reinforced PC.

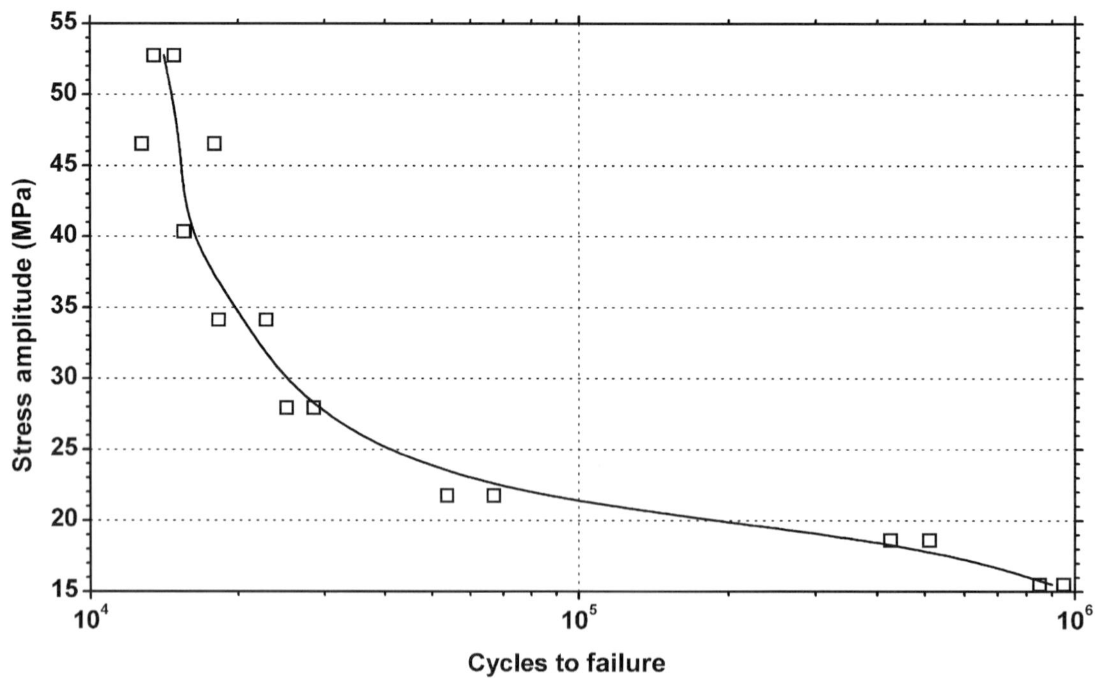

Figure 6.18 Tensile stress amplitude vs. cycles to failure of SABIC Innovative Plastics Lexan® 920—low viscosity, unreinforced, flame retardant PC.

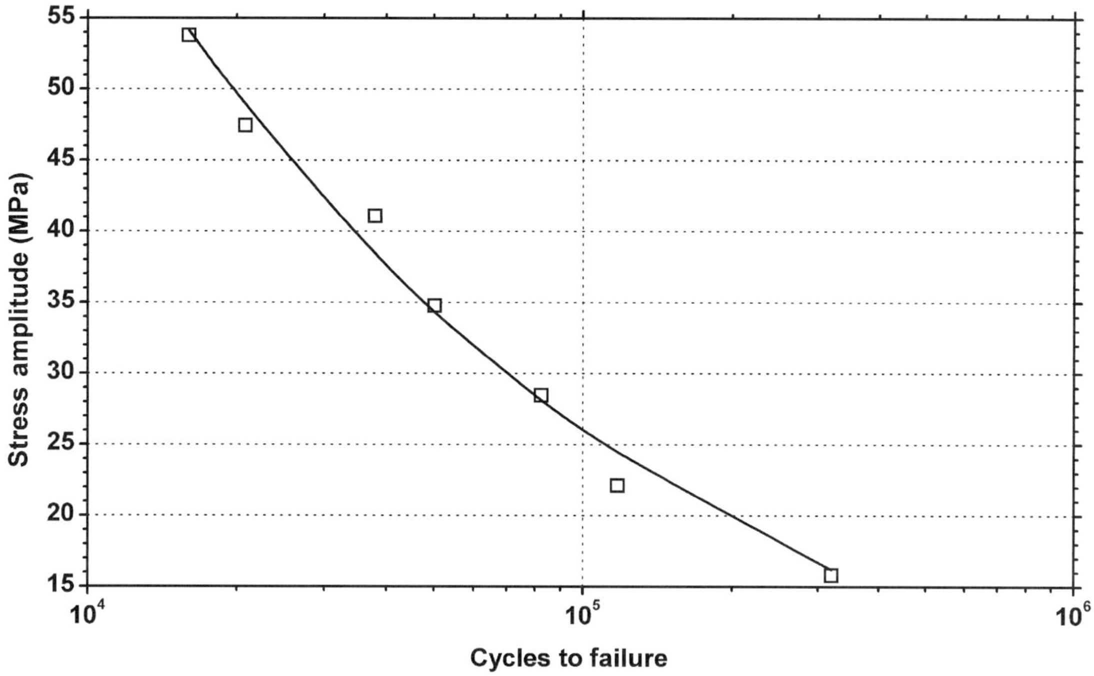

Figure 6.19 Tensile stress amplitude vs. cycles to failure of SABIC Innovative Plastics Lexan® 925—low viscosity, unreinforced, flame retardant, ECO conforming label grade PC.

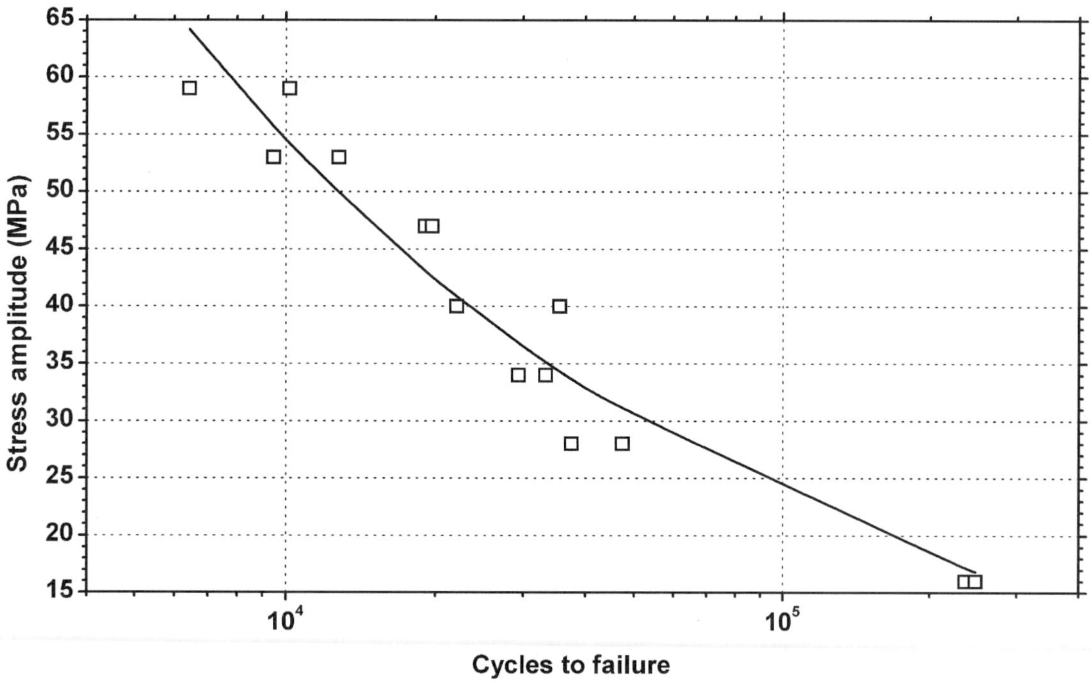

Figure 6.20 Tensile stress amplitude vs. cycles to failure of SABIC Innovative Plastics Lexan® 940—medium viscosity, unreinforced, flame retardant, ECO conforming label grade PC.

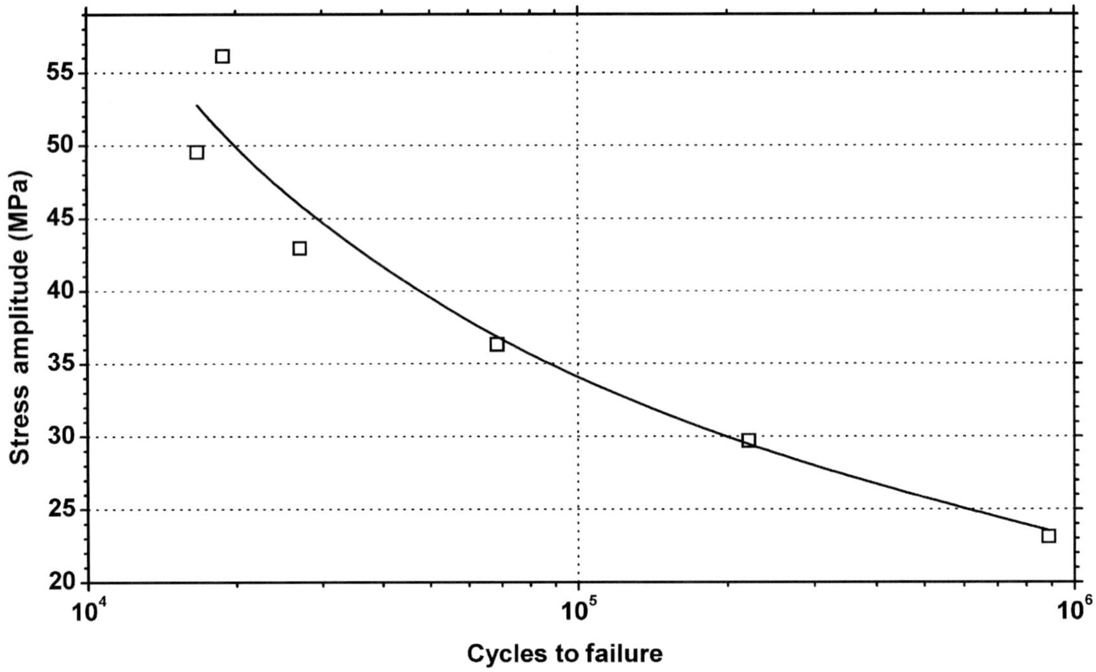

Figure 6.21 Tensile stress amplitude vs. cycles to failure of SABIC Innovative Plastics Lexan® 945—low–medium viscosity, unreinforced, flame retardant, ECO conforming label grade PC.

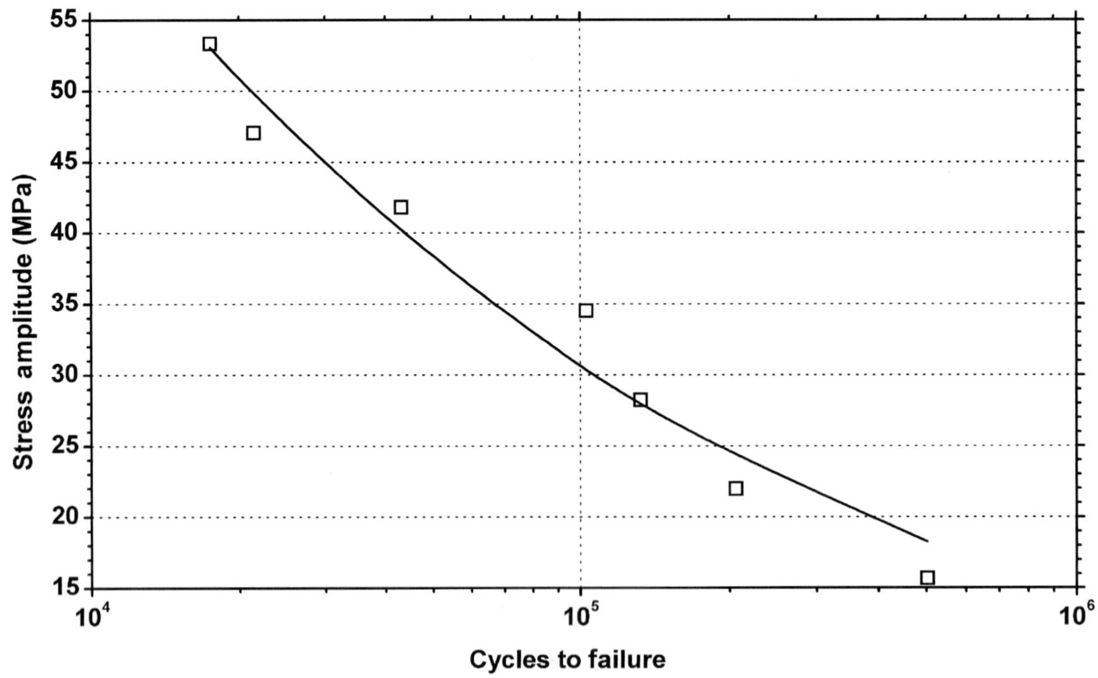

Figure 6.22 Tensile stress amplitude vs. cycles to failure of SABIC Innovative Plastics Lexan® 955—medium viscosity, unreinforced, flame retardant, ECO conforming label grade PC.

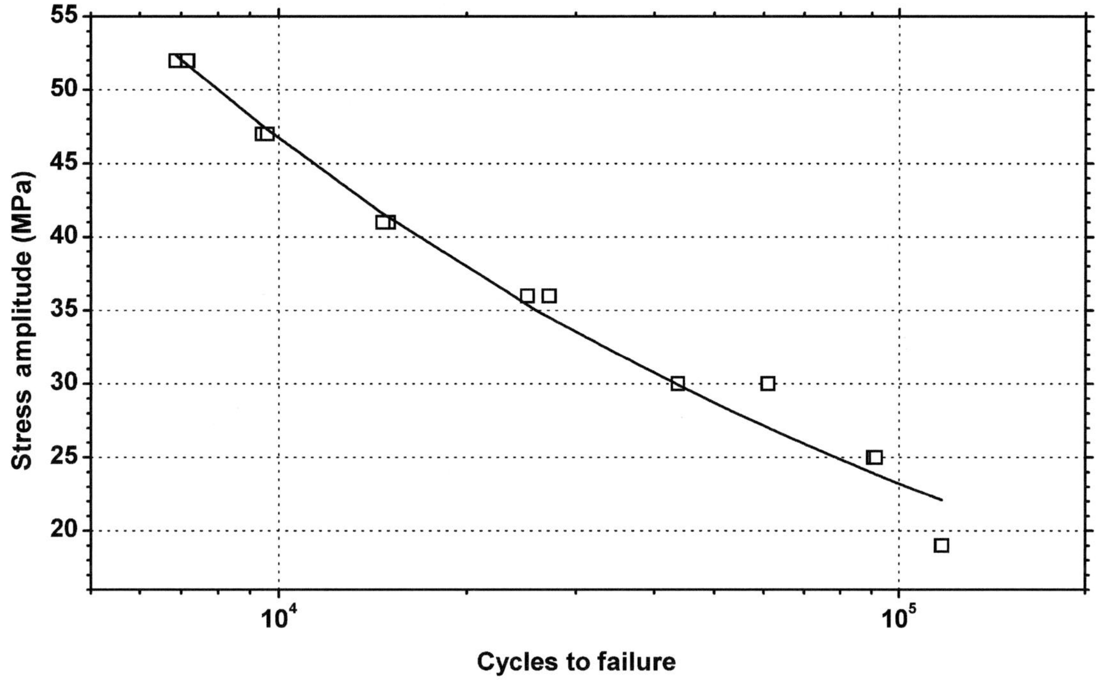

Figure 6.23 Tensile stress amplitude vs. cycles to failure of SABIC Innovative Plastics Lexan® EM1210—automotive interiors, heat and impact-resistant PC.

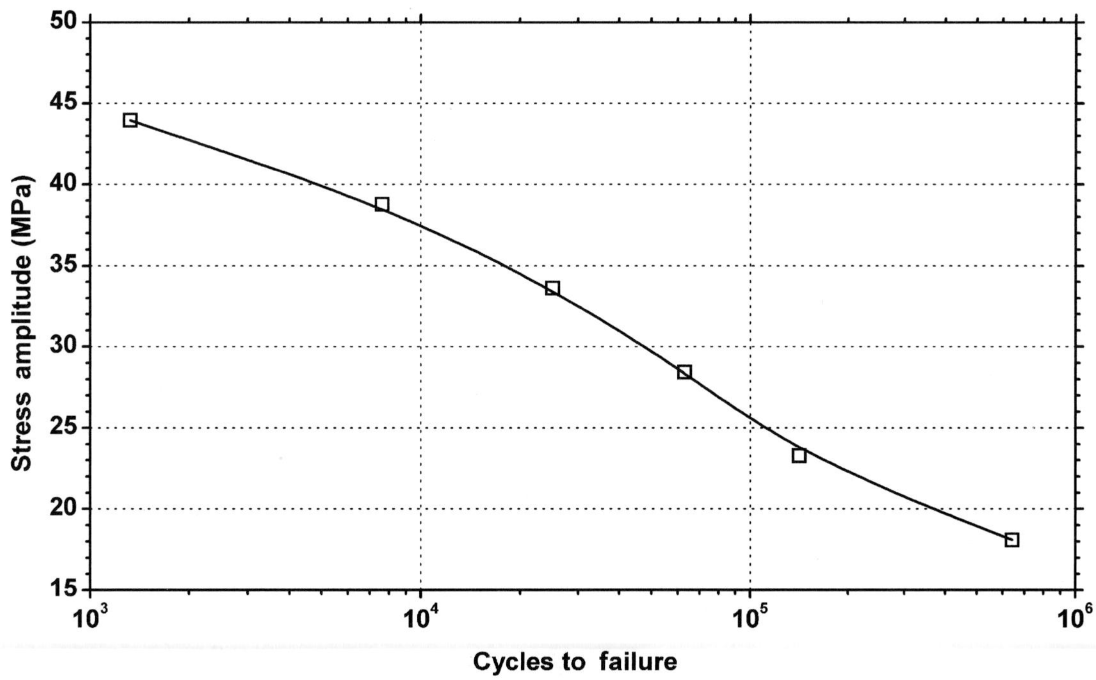

Figure 6.24 Tensile stress amplitude vs. cycles to failure of SABIC Innovative Plastics Lexan® EM2212—automotive interiors, 10% glass-reinforced PC.

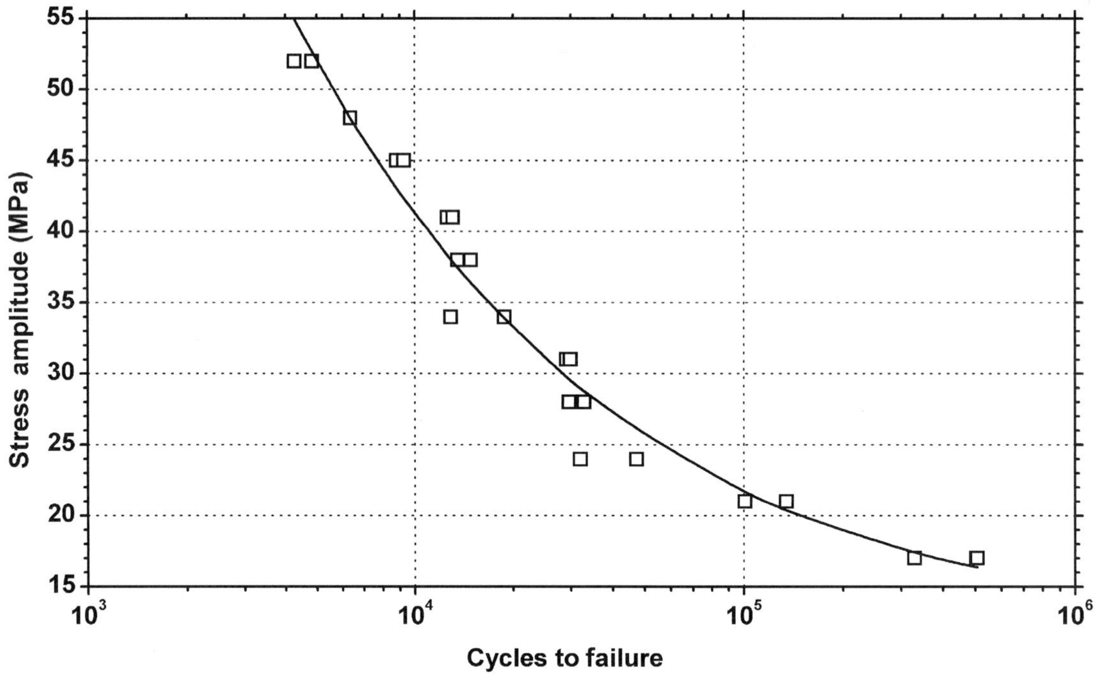

Figure 6.25 Tensile stress amplitude vs. cycles to failure of SABIC Innovative Plastics Lexan® EM3110—automotive interiors, optimized flow and processability for thinner wall uses PC.

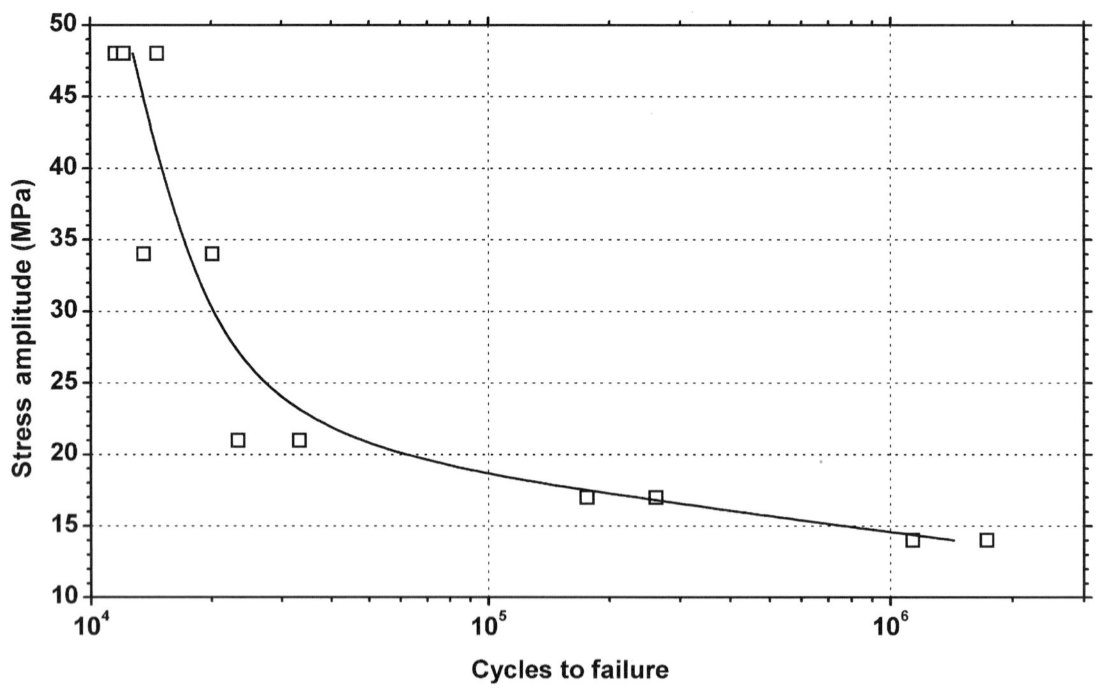

Figure 6.26 Tensile stress amplitude vs. cycles to failure of SABIC Innovative Plastics Lexan® HF1110—high flow, heat-resistant PC.

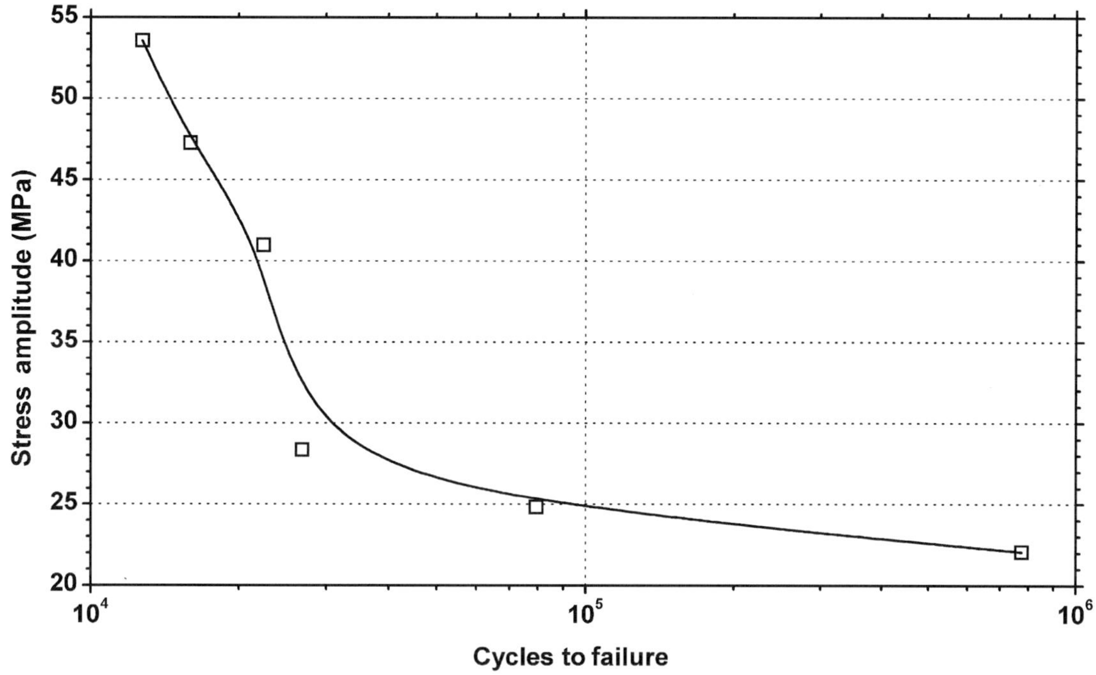

Figure 6.27 Tensile stress amplitude vs. cycles to failure of SABIC Innovative Plastics Lexan® HF1130—high flow, UV stabilized, heat resistance PC.

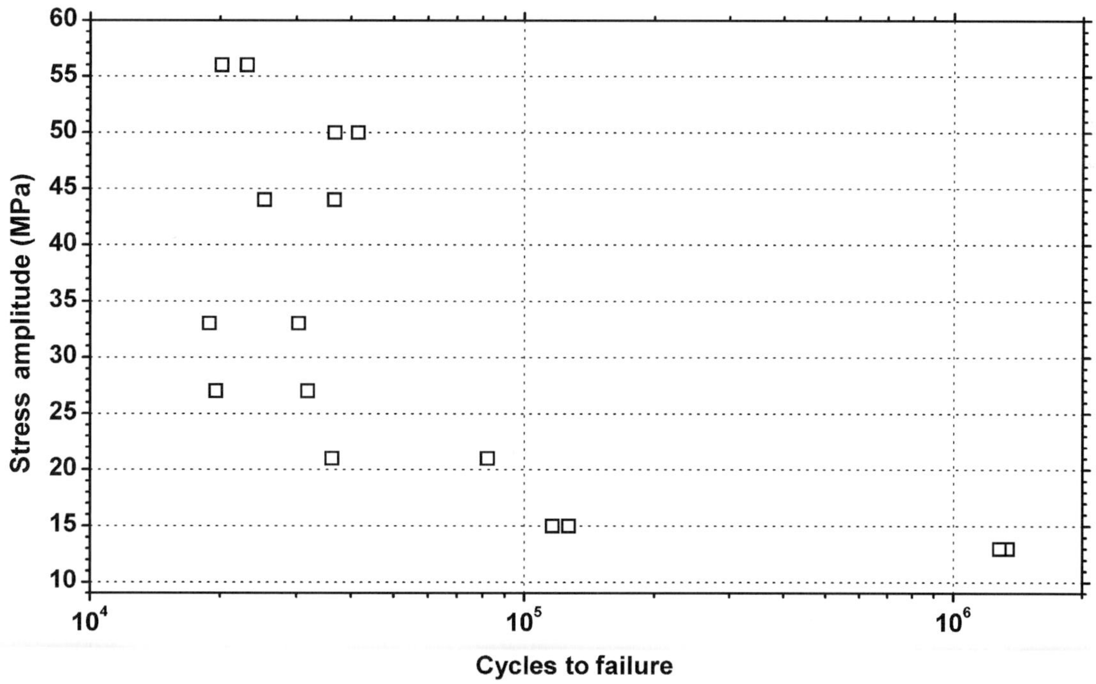

Figure 6.28 Tensile stress amplitude vs. cycles to failure of SABIC Innovative Plastics Lexan® HF1140—high flow, FDA food compliant for disposable end-uses PC.

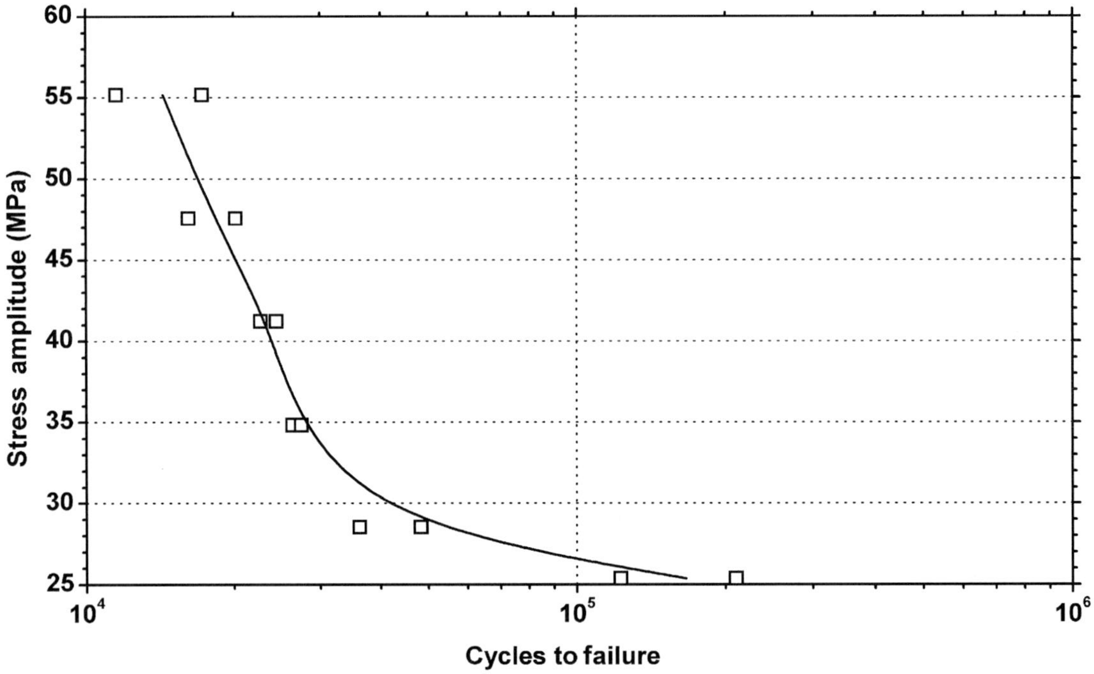

Figure 6.29 Tensile stress amplitude vs. cycles to failure of SABIC Innovative Plastics Lexan® LS1—automotive lens system, low-viscosity PC.

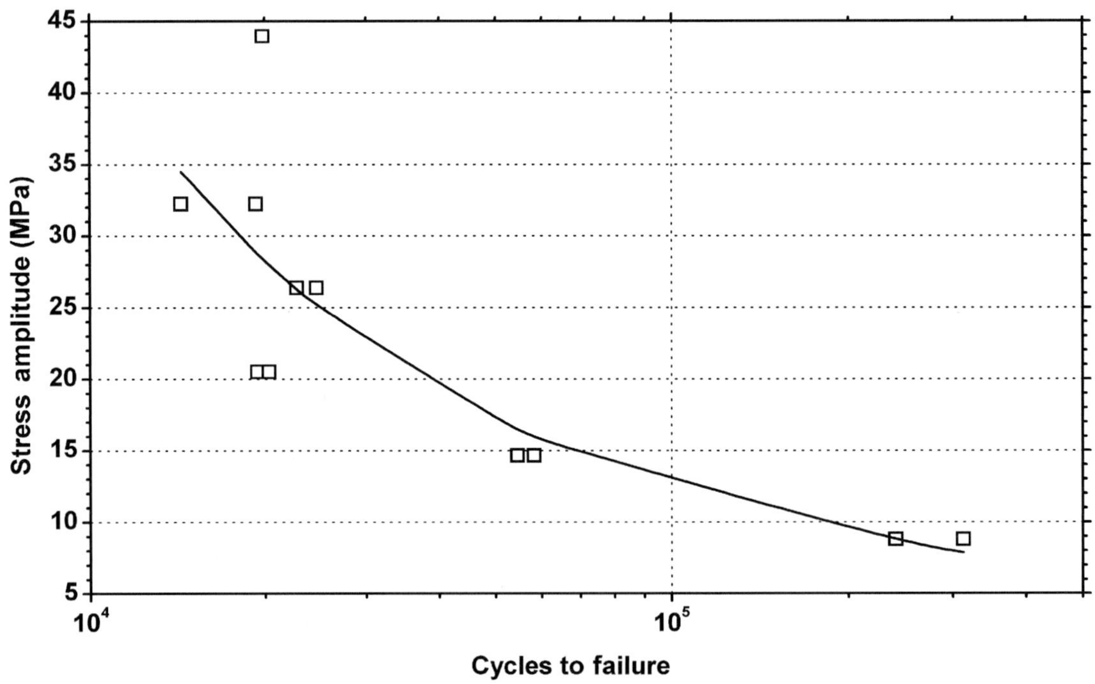

Figure 6.30 Tensile stress amplitude vs. cycles to failure of SABIC Innovative Plastics Lexan® OQ1030—optical quality for CD/DVD PC.

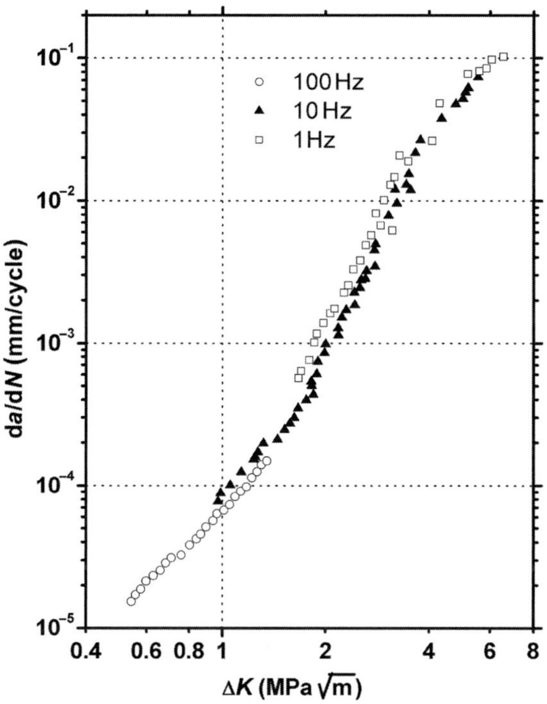

Figure 6.31 Fatigue crack propagation rate dependence on cyclic frequency and stress intensity factor range for generic PC.

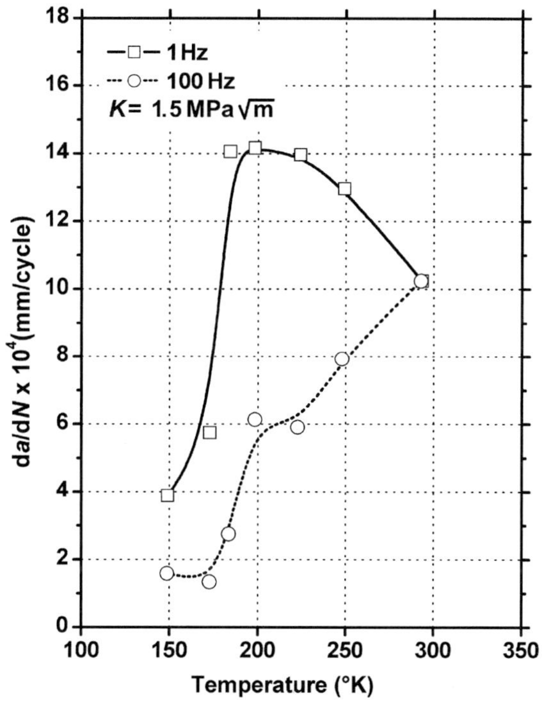

Figure 6.32 Fatigue crack propagation rate dependence on cyclic frequency and temperature for generic PC.

6.2.2 Tribology Data

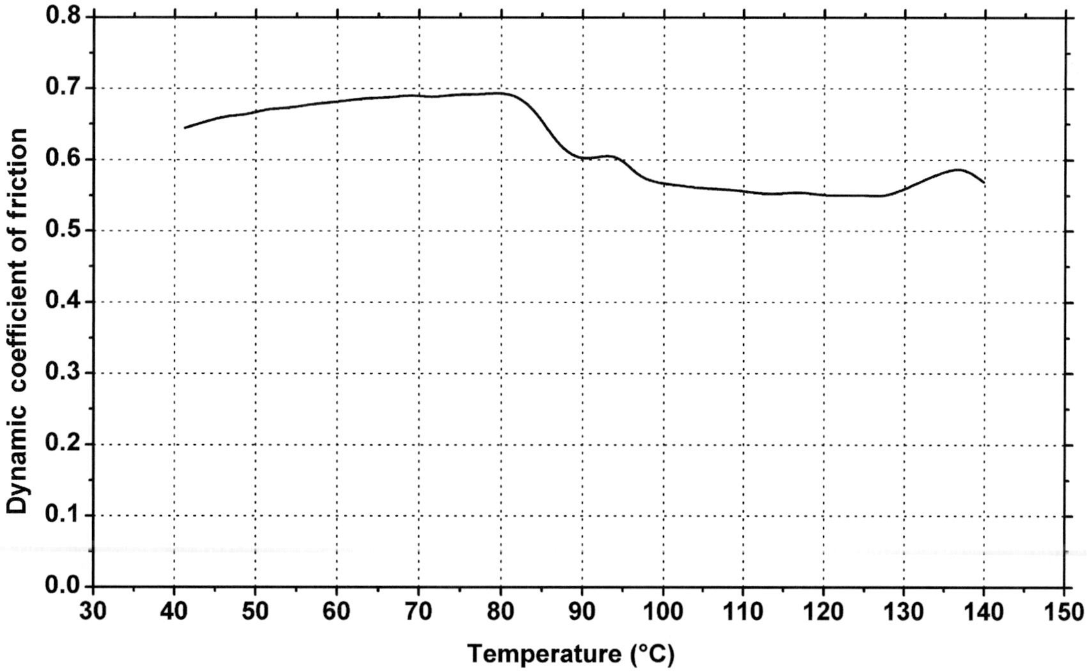

Figure 6.33 Coefficient of friction vs. temperature for SABIC Innovative Plastics Lexan® 101R—unreinforced, high viscosity, release agent, general-purpose extrusion PC (against 100Cr6 stainless steel, Ra = 0.1 μm, sliding speed = 0.1 m/s, pressure = 1.5 MPa).

Table 6.1 Tribological Properties of RTP Company RTP 300 TFE 5 (PC with 5% PTFE) vs. 1018 C Steel (Data obtained per ASTM 3702)

PV (KPa m/s)	Load (N)	Speed (m/s)	Wear Factor × 10^{-8} (mm^3/N m)	Dynamic Coefficient of Friction
70	0.45	1.00	5628	0.29
70	1.80	0.25	1373	0.33
175	2.25	0.50	939	0.42
350	9.00	0.25	386	0.26

Table 6.2 Tribological Properties of RTP Company RTP 300 TFE 5 (PC with 5% PTFE) vs. RTP 300 TFE 5 (Data obtained per ASTM 3702)

PV (KPa m/s)	Load (N)	Speed (m/s)	Wear Factor × 10^{-8} (mm^3/N m)	Dynamic Coefficient of Friction
17.5	0.90	0.15	26	0.17
35	0.90	0.25	639	0.24
70	1.80	0.25	320	0.24
70	0.90	0.50	152	0.18

Table 6.3 Tribological Properties of RTP Company RTP 300 TFE 10 (PC with 10% PTFE) vs. 1018 C Steel (Data obtained per ASTM 3702)

PV (KPa m/s)	Load (N)	Speed (m/s)	Wear Factor × 10^{-8} (mm^3/N m)	Dynamic Coefficient of Friction
70	0.45	1.00	1373	0.16
70	1.80	0.25	655	0.24
175	2.25	0.50	830	0.22
350	9.00	0.25	253	0.13

Table 6.4 Tribological Properties of RTP Company RTP 300 TFE 10 (PC with 10% PTFE) vs. RTP 300 TFE 10 (Data obtained per ASTM 3702)

PV (KPa m/s)	Load (N)	Speed (m/s)	Wear Factor × 10^{-8} (mm^3/N m)	Dynamic Coefficient of Friction
17.5	0.90	0.15	21	0.19
35	0.90	0.25	72	0.17
70	1.80	0.25	291	0.18
70	0.90	0.50	187	0.11

Table 6.5 Tribological Properties of RTP Company RTP 300 TFE 15 (PC with 15% PTFE) vs. 1018 C Steel (Data obtained per ASTM 3702)

PV (KPa m/s)	Load (N)	Speed (m/s)	Wear Factor × 10^{-8} (mm^3/N m)	Dynamic Coefficient of Friction
70	0.45	1.00	778	0.56
70	1.80	0.25	263	0.33
175	2.25	0.50	190	0.27
350	9.00	0.25	69	0.21

Table 6.6 Tribological Properties of RTP Company RTP 300 TFE 15 (PC with 15% PTFE) vs. RTP 300 TFE 10 (Data obtained per ASTM 3702)

PV (KPa m/s)	Load (N)	Speed (m/s)	Wear Factor × 10^{-8} (mm^3/N m)	Dynamic Coefficient of Friction
17.5	0.90	0.13	16	0.30
35	0.90	0.25	49	0.19
70	1.80	0.25	167	0.12
70	0.90	0.50	129	0.16

Table 6.7 Tribological Properties of RTP Company RTP 300 TFE 20 (PC with 20% PTFE) vs. 1018 C Steel (Data obtained per ASTM 3702)

PV (KPa m/s)	Load (N)	Speed (m/s)	Wear Factor × 10^{-8} (mm^3/N m)	Dynamic Coefficient of Friction
70	0.45	1.00	555	0.33
70	1.80	0.25	354	0.31
175	2.25	0.50	253	0.24
350	9.00	0.25	116	0.18

Table 6.8 Tribological Properties of RTP Company RTP 300 TFE 10 SI 2 (PC with 10% PTFE and 2% Silicone) vs. 1018 C Steel (Data obtained per ASTM 3702)

PV (KPa m/s)	Load (N)	Speed (m/s)	Wear Factor × 10^{-8} (mm^3/N m)	Dynamic Coefficient of Friction
70	0.45	1.00	1284	0.31
70	1.80	0.25	995	0.35
175	2.25	0.50	326	0.26
350	9.00	0.25	117	0.18

Table 6.9 Tribological Properties of RTP Company RTP 300 AR 10 (PC with 10% Aramid Fiber) vs. 1018 C Steel (Data obtained per ASTM 3702)

PV (KPa m/s)	Load (N)	Speed (m/s)	Wear Factor × 10^{-8} (mm^3/N m)	Dynamic Coefficient of Friction
70	0.45	1.00	58	0.22
70	1.80	0.25	985	0.38
175	2.25	0.50	12100	0.29
350	9.00	0.25	14733	0.31

Table 6.10 Tribological Properties of RTP Company RTP 300 AR 10 TFE 10 (PC with 10% Aramid Fiber and 10% PTFE) vs. 1018 C Steel (Data obtained per ASTM 3702)

PV (KPa m/s)	Load (N)	Speed (m/s)	Wear Factor × 10^{-8} (mm^3/N m)	Dynamic Coefficient of Friction
70	0.45	1.00	663	0.03
70	1.80	0.25	113	0.10
175	2.25	0.50	177	0.14
350	9.00	0.25	277	0.14

Table 6.11 Tribological Properties of RTP Company RTP 302 TFE 15 (PC with 15% Glass Fiber and 15% PTFE) vs. 1018 C Steel (Data obtained per ASTM 3702)

PV (KPa m/s)	Load (N)	Speed (m/s)	Wear Factor × 10^{-8} (mm^3/N m)	Dynamic Coefficient of Friction
70	0.45	1.00	94	0.22
70	1.80	0.25	111	0.35
175	2.25	0.50	1938	0.34
350	9.00	0.25	635	0.34

Table 6.12 Tribological Properties of RTP Company RTP 305 TFE 15 (PC with 30% Glass Fiber and 15% PTFE) vs. 1018 C Steel (Data obtained per ASTM 3702)

PV (KPa m/s)	Load (N)	Speed (m/s)	Wear Factor × 10^{-8} (mm^3/N m)	Dynamic Coefficient of Friction
70	0.45	1.00	198	0.31
70	1.80	0.25	176	0.35
175	2.25	0.50	93	0.53
350	9.00	0.25	1007	0.27

Table 6.13 Tribological Properties of RTP Company RTP 382 TFE 15 (PC with 15% Carbon Fiber and 15% PTFE) vs. 1018 C Steel (Data obtained per ASTM 3702)

PV (KPa m/s)	Load (N)	Speed (m/s)	Wear Factor × 10^{-8} (mm^3/N m)	Dynamic Coefficient of Friction
70	1.80	0.25	72	0.57
175	2.25	0.50	45	0.54

Table 6.14 Tribological Properties of RTP Company RTP 382 TFE 15 (PC with 15% Carbon Fiber and 15% PTFE) vs. RTP 382 TFE 15 (Data obtained per ASTM 3702)

PV (KPa m/s)	Load (N)	Speed (m/s)	Wear Factor × 10^{-8} (mm^3/N m)	Dynamic Coefficient of Friction
17.5	0.90	0.15	196	0.31
35	0.90	0.25	169	0.22
70	1.80	0.25	91	0.12
70	0.90	0.50	112	0.13

Table 6.15 Tribological Properties of RTP Company RTP 385 TFE 15 (PC with 30% Carbon Fiber and 15% PTFE) vs. 1018 C Steel (Data obtained per ASTM 3702)

PV (KPa m/s)	Load (N)	Speed (m/s)	Wear Factor × 10^{-8} (mm^3/N m)	Dynamic Coefficient of Friction
70	0.45	1.00	40	0.24
70	1.80	0.25	285	0.30
175	2.25	0.50	205	0.39
350	9.00	0.25	72	0.36

Table 6.16 Taber Abrasion Performance of SABIC Innovative Plastics Lexan® (Data obtained per ASTM D 1044, CS-17 wheels, 1 kg load)

Lexan® product	Taber abrasion mg/1000 cycles
101—Unreinforced, high viscosity, general purpose	10
121—Unreinforced, low viscosity, general purpose	10
141—Unreinforced, low–medium viscosity, general purpose	10
143R—Unreinforced, low–medium viscosity, UV stabilized general purpose	10
191—High impact	20
500—10% Glass fiber reinforced	11
920—Low viscosity, unreinforced, flame retardant	10
940—Medium viscosity, unreinforced, flame retardant, ECO conforming label grade	10

6.3 Polybutylene Terephthalate

6.3.1 Fatigue Data

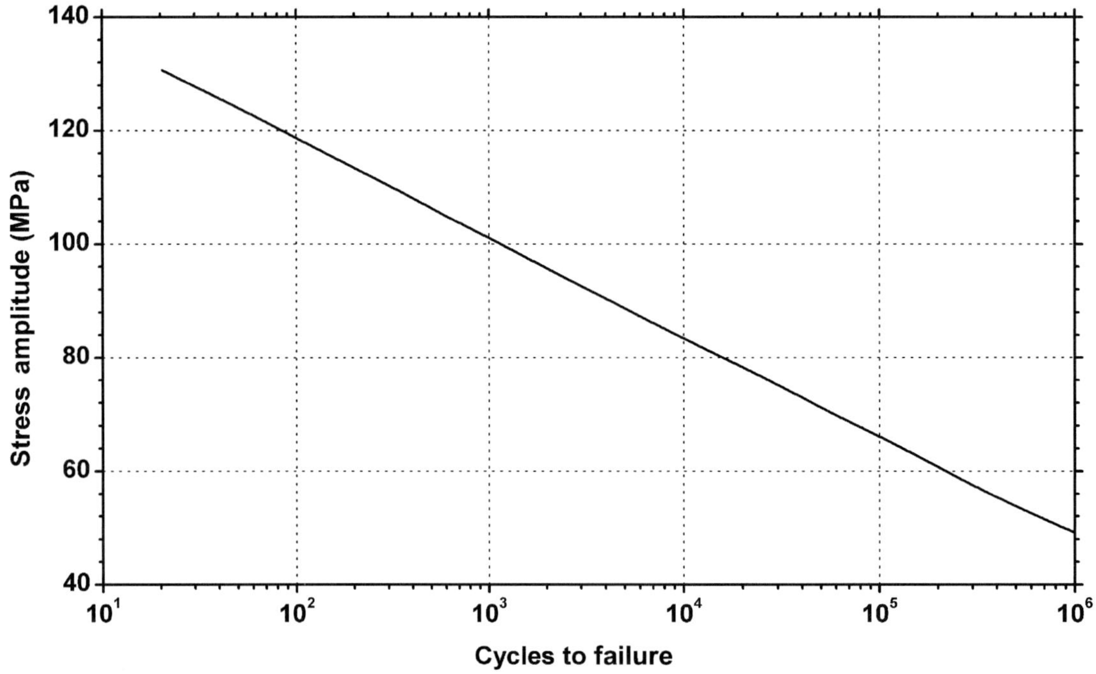

Figure 6.34 Stress amplitude vs. cycles to failure at 20°C and 50% relative humidity of DSM Arnite®—unreinforced PBT.

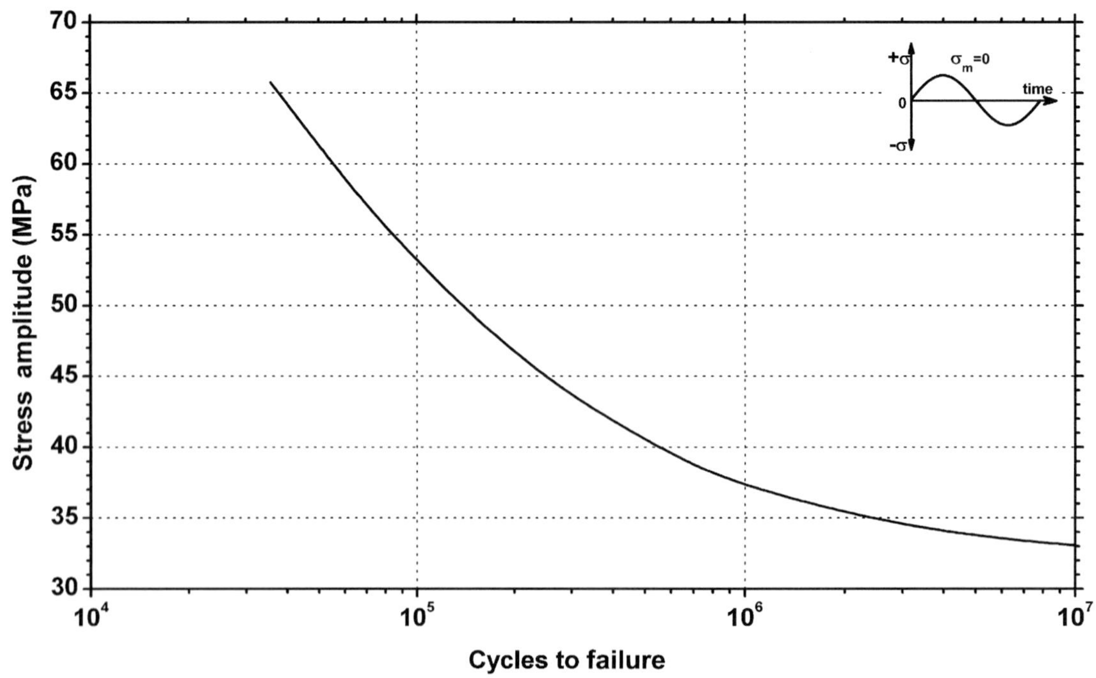

Figure 6.35 Flexural stress amplitude vs. cycles to failure of Ticona Celanex® 2300 GV/30—general-purpose, 30% glass fiber reinforced PBT.

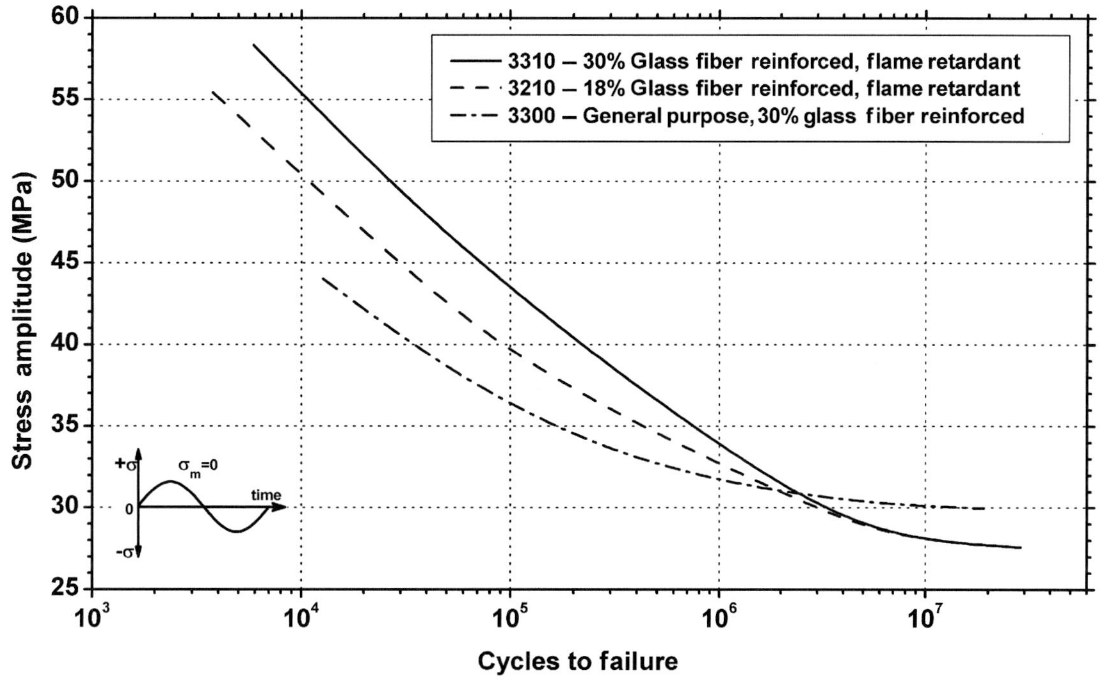

Figure 6.36 Flexural stress amplitude vs. cycles to failure of several Ticona Celanex® PBT plastics.

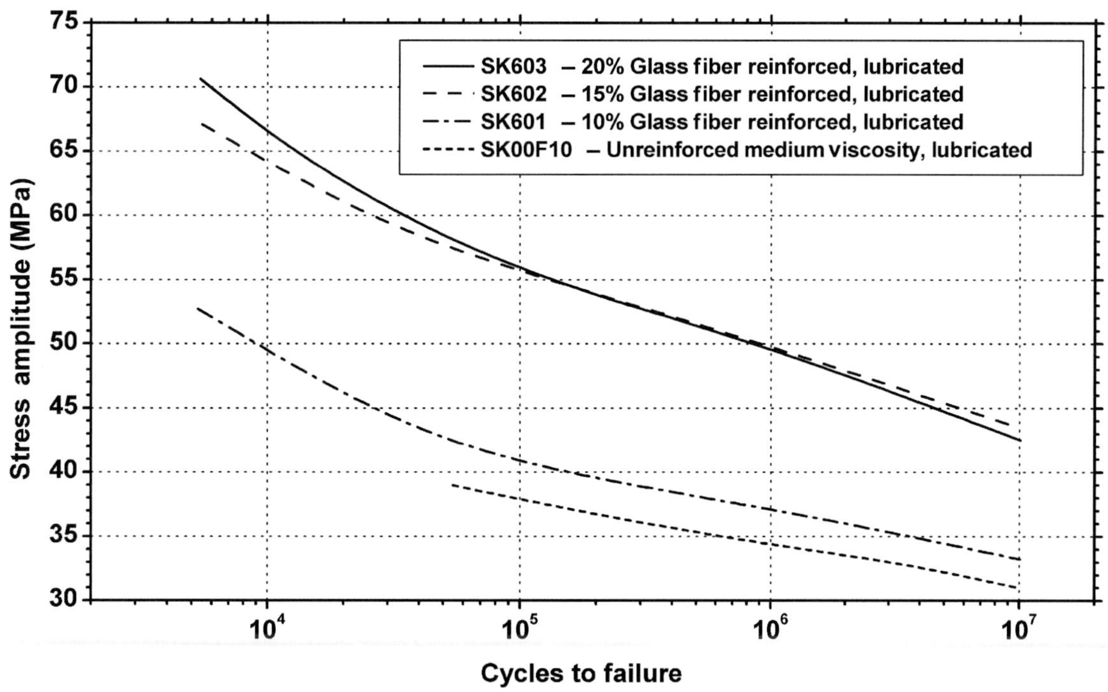

Figure 6.37 Flexural stress amplitude vs. cycles to failure at 23°C of several DuPont Engineering Polymers Crastin® PBT plastics.

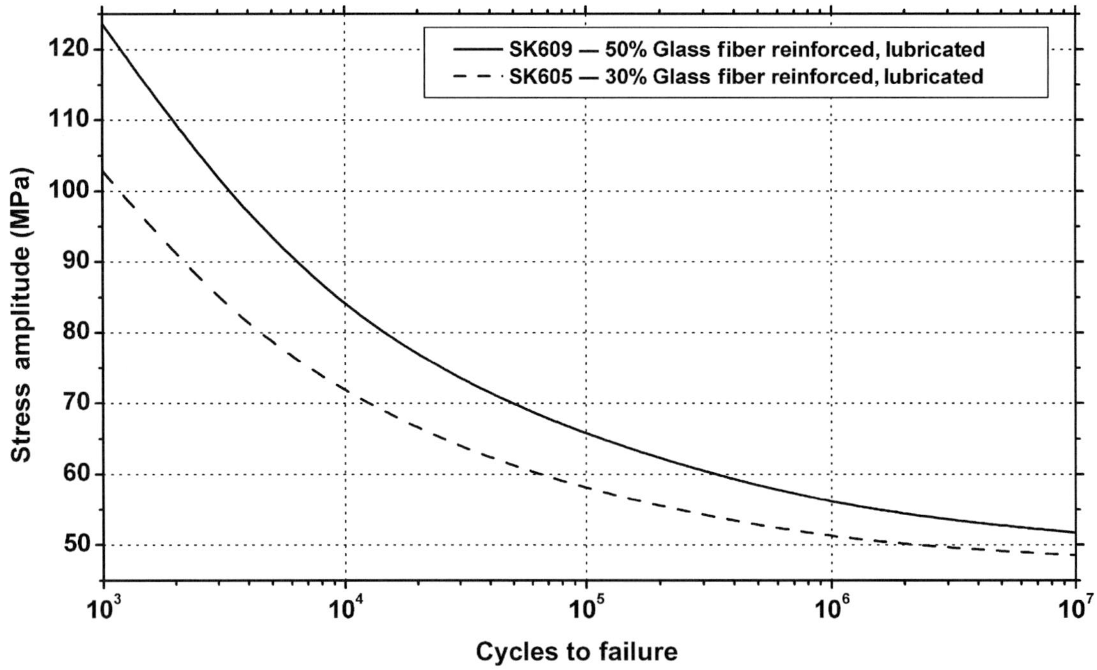

Figure 6.38 Flexural stress amplitude vs. cycles to failure at 23°C of several other DuPont Engineering Polymers Crastin® PBT plastics.

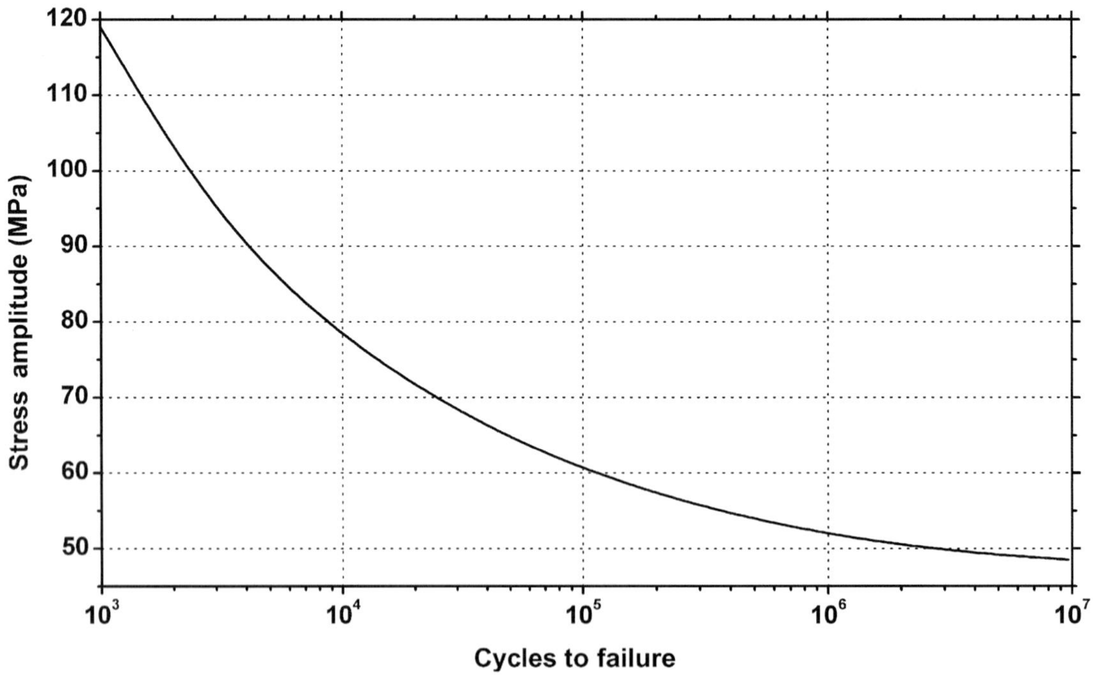

Figure 6.39 Flexural stress amplitude vs. cycles to failure at 23°C of DuPont Engineering Polymers Crastin® SK645FR—30% glass fiber reinforced, UL94 V-0 flame retardant PBT.

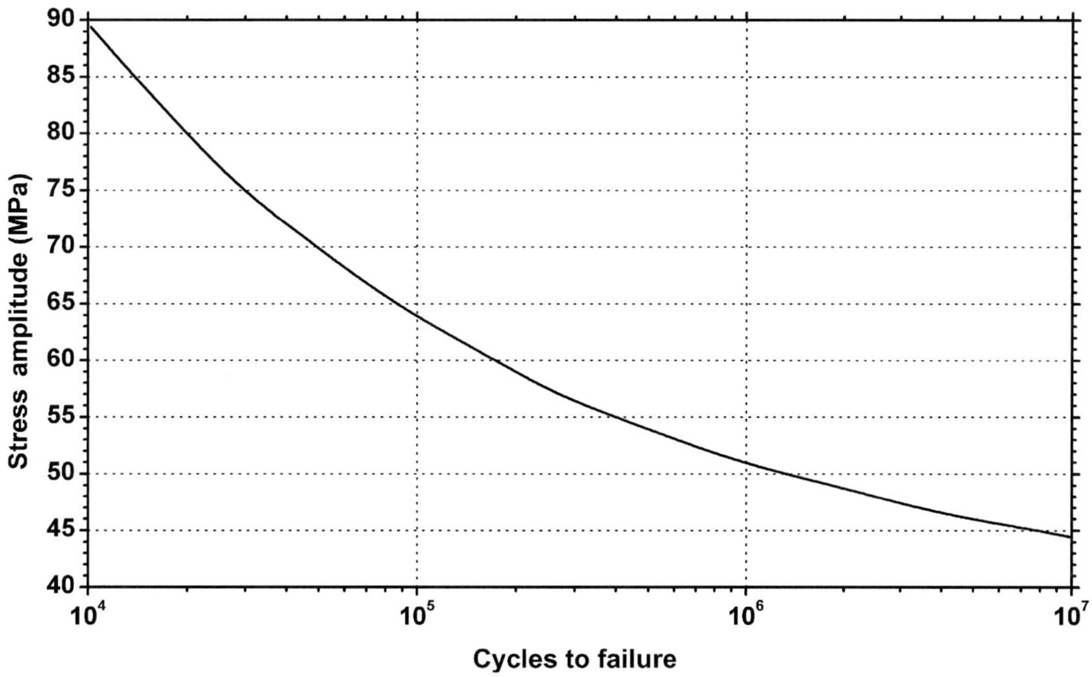

Figure 6.40 Flexural stress amplitude vs. cycles to failure at 23°C of SABIC Innovative Plastics LNP Stat-Kon® WC-4036—30% glass fiber reinforced PBT.

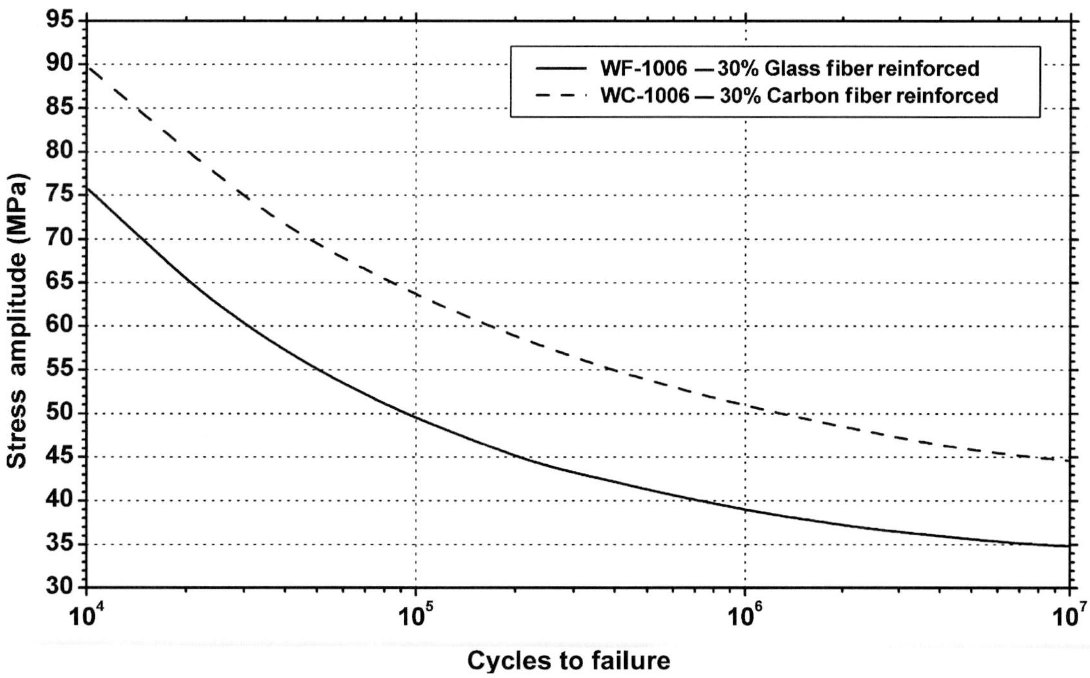

Figure 6.41 Flexural stress amplitude vs. cycles to failure at 23°C of two SABIC Innovative Plastics LNP Thermocomp® fiber reinforced PBT plastics.

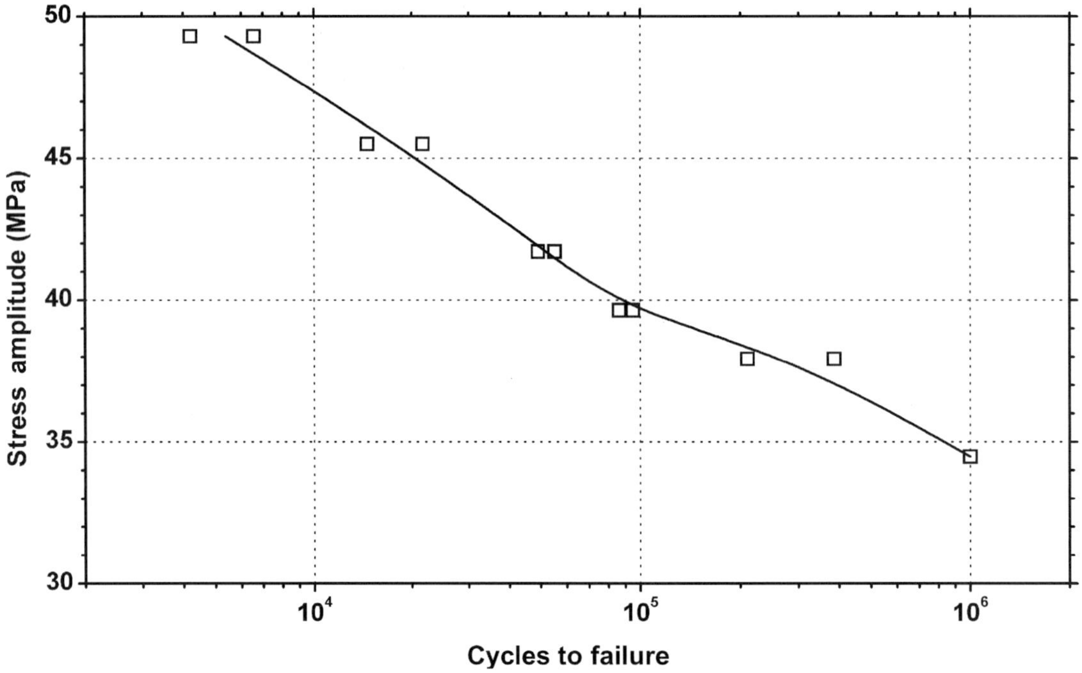

Figure 6.42 Tensile stress amplitude vs. cycles to failure at 23°C of SABIC Innovative Plastics Valox® 310—unreinforced, general-purpose PBT.

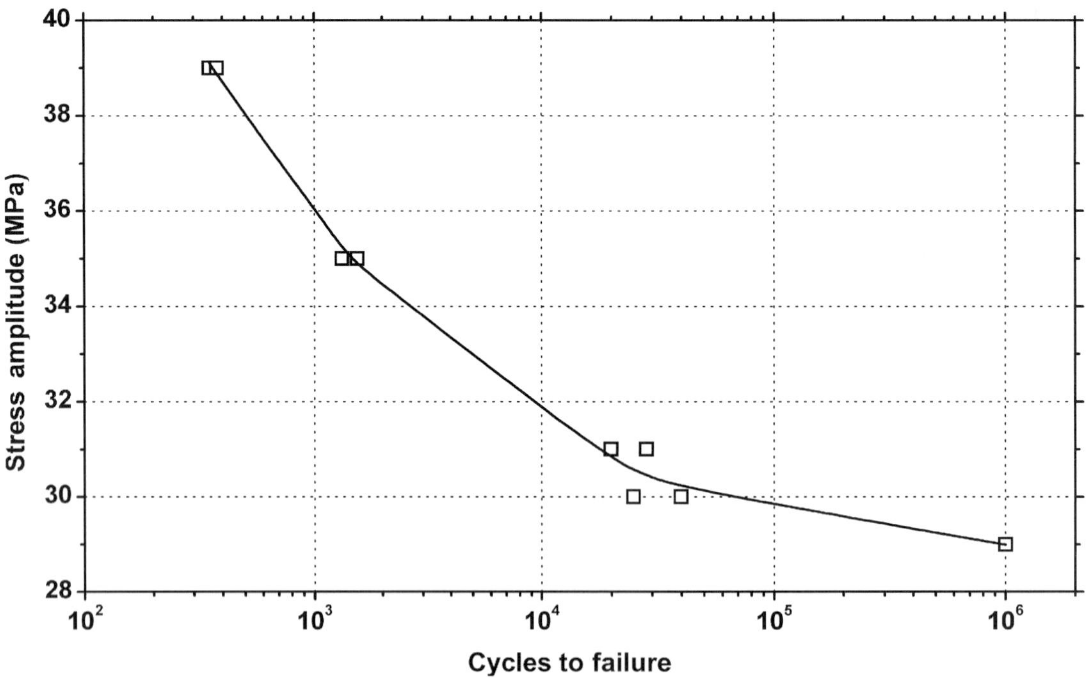

Figure 6.43 Tensile stress amplitude vs. cycles to failure at 23°C of SABIC Innovative Plastics Valox® 337—unfilled, impact modified grade for low-temperature PBT.

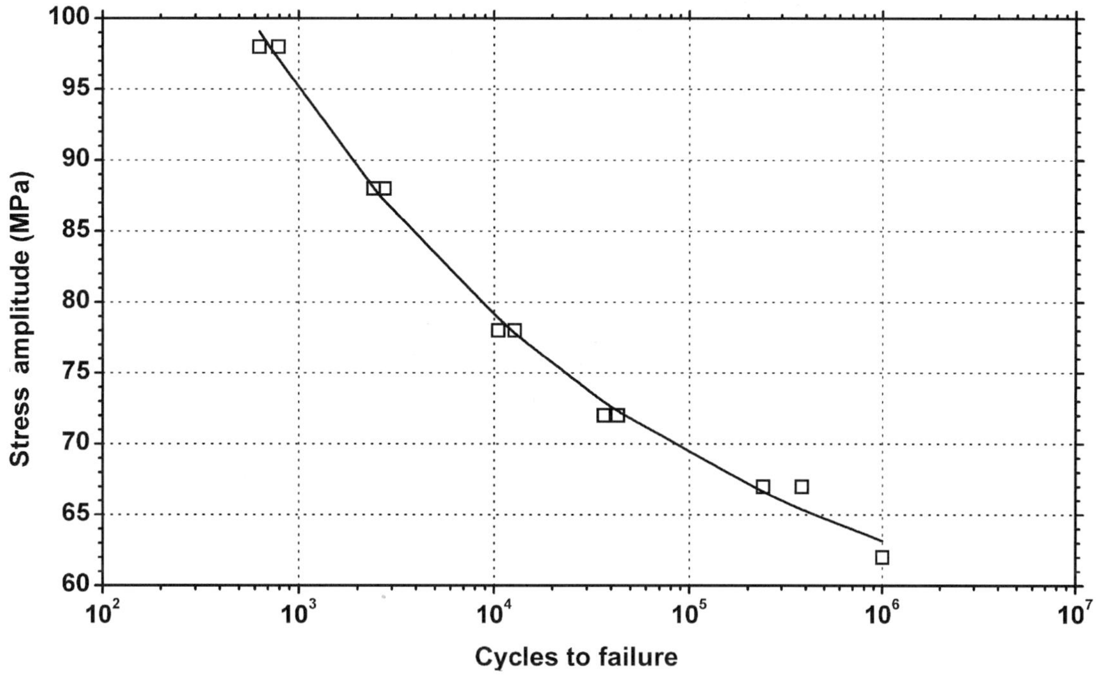

Figure 6.44 Tensile stress amplitude vs. cycles to failure at 23°C of SABIC Innovative Plastics Valox® 412E—20% glass fiber reinforced PBT.

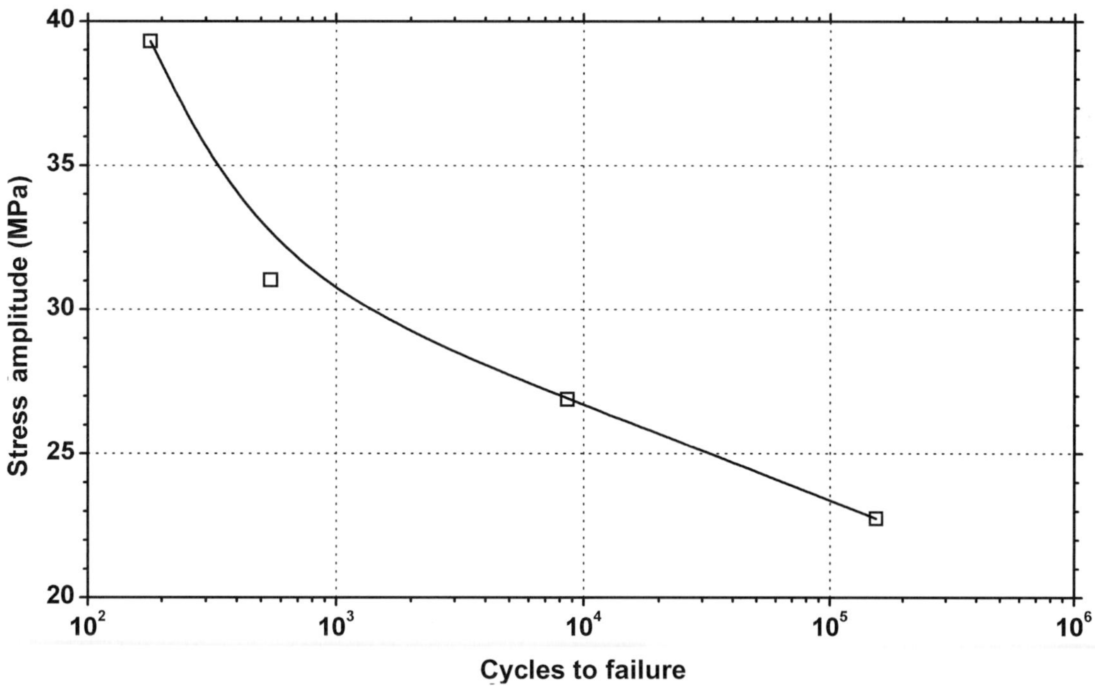

Figure 6.45 Tensile stress amplitude vs. cycles to failure at 23°C of SABIC Innovative Plastics Valox® 420—30% glass fiber reinforced, high heat PBT.

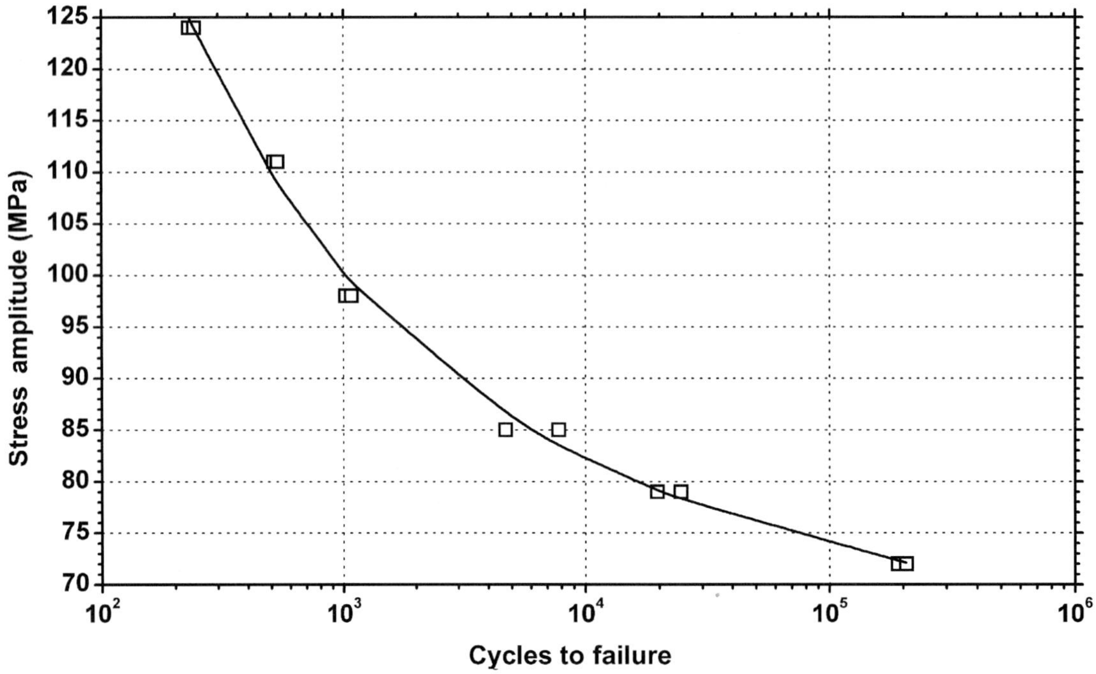

Figure 6.46 Tensile stress amplitude vs. cycles to failure at 23°C of SABIC Innovative Plastics Valox® 430—33% glass fiber reinforced, impact modified PBT.

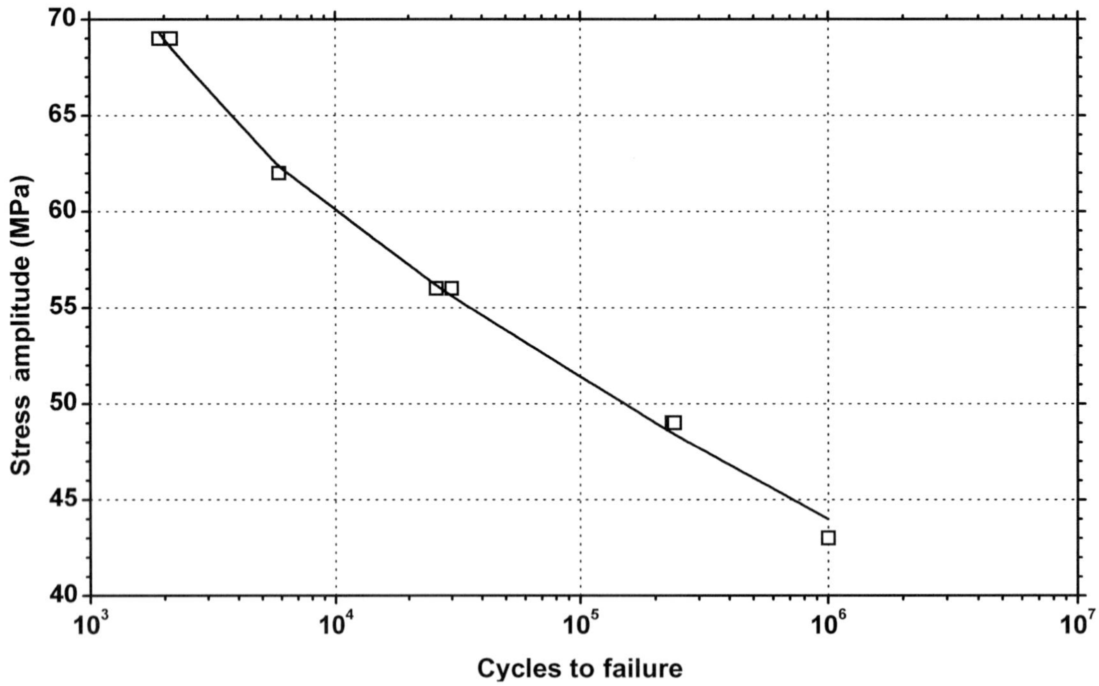

Figure 6.47 Tensile stress amplitude vs. cycles to failure at 23°C of SABIC Innovative Plastics Valox® 732E—30% glass/mineral filled, thermal stabilized, low warpage, enhanced flow PBT.

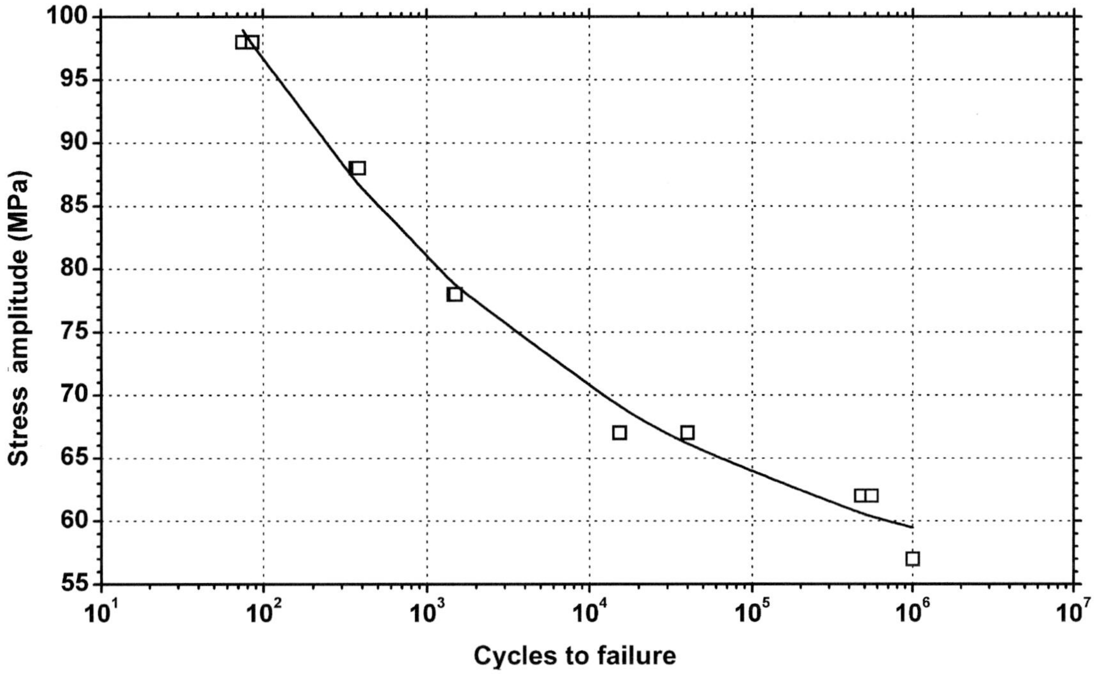

Figure 6.48 Tensile stress amplitude vs. cycles to failure at 23°C of SABIC Innovative Plastics Valox® 736—45% glass/mineral PBT.

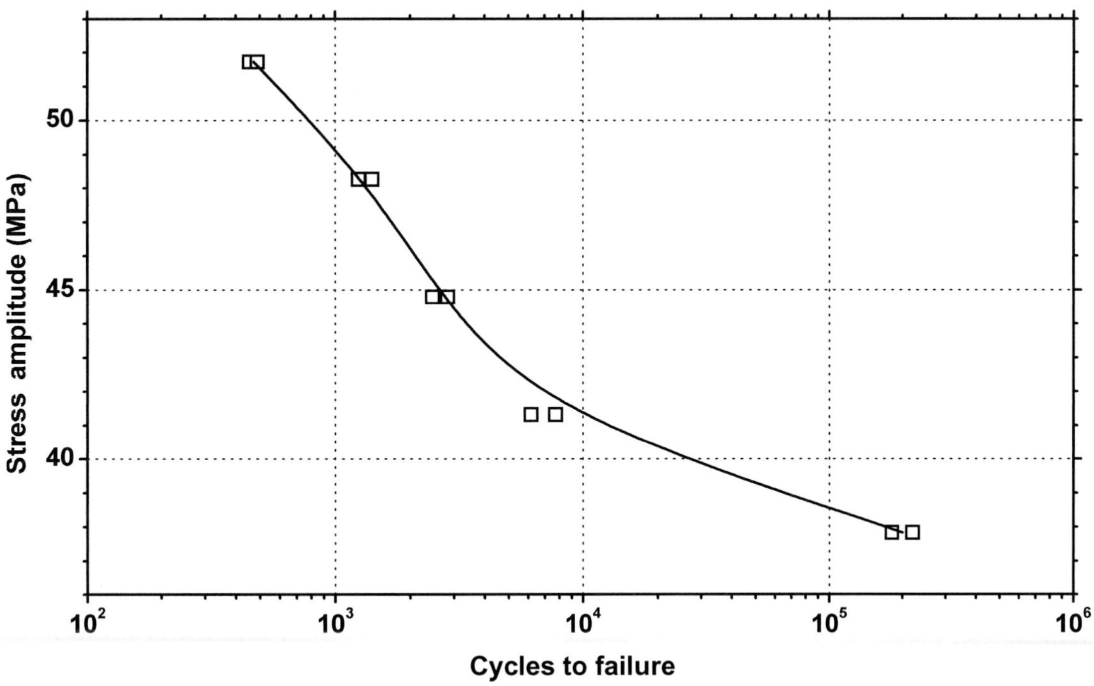

Figure 6.49 Tensile stress amplitude vs. cycles to failure at 23°C of SABIC Innovative Plastics Valox® 325—unreinforced, improved processing PBT.

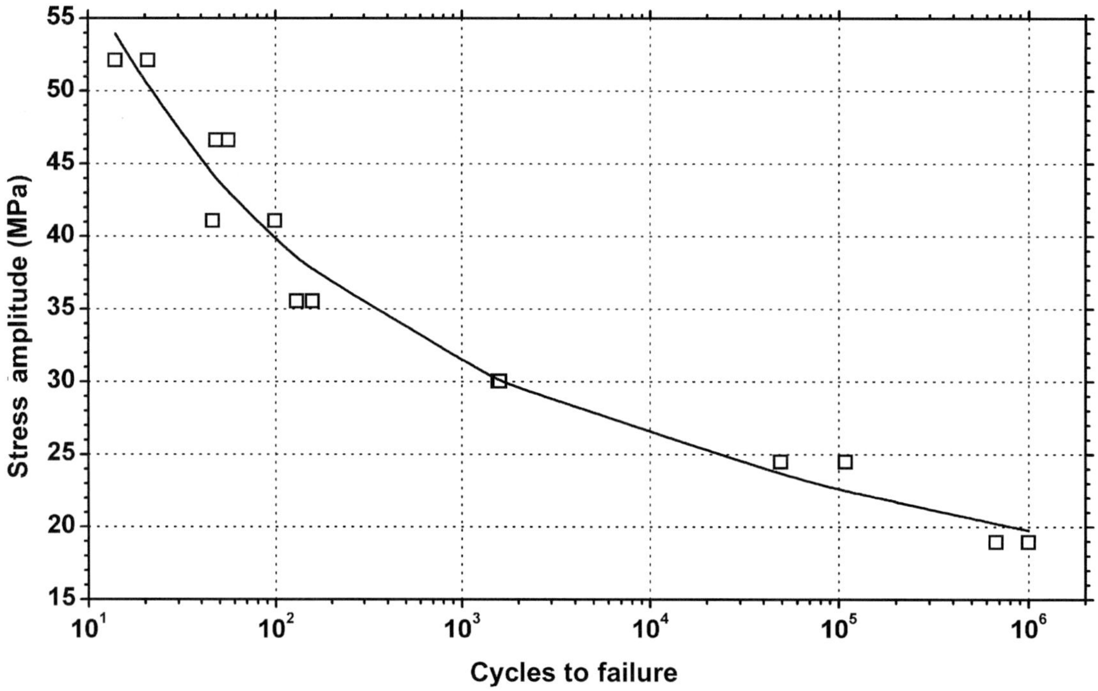

Figure 6.50 Tensile stress amplitude vs. cycles to failure at 23°C of SABIC Innovative Plastics Valox® HV7075 PBT.

6.3.2 Tribology Data

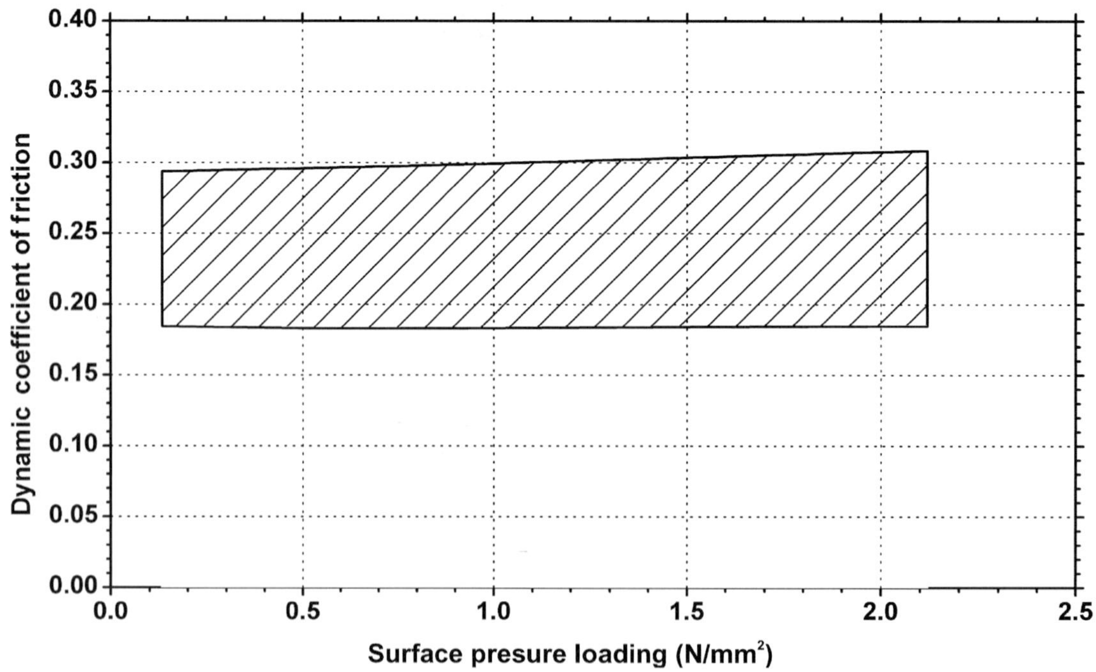

Figure 6.51 Dynamic coefficient of friction vs. pressure loading of Ticona Celanex® 2500—general purpose, nucleated, easy flow PBT (v = 10 m/min, against steel with Rz = 2 μm).

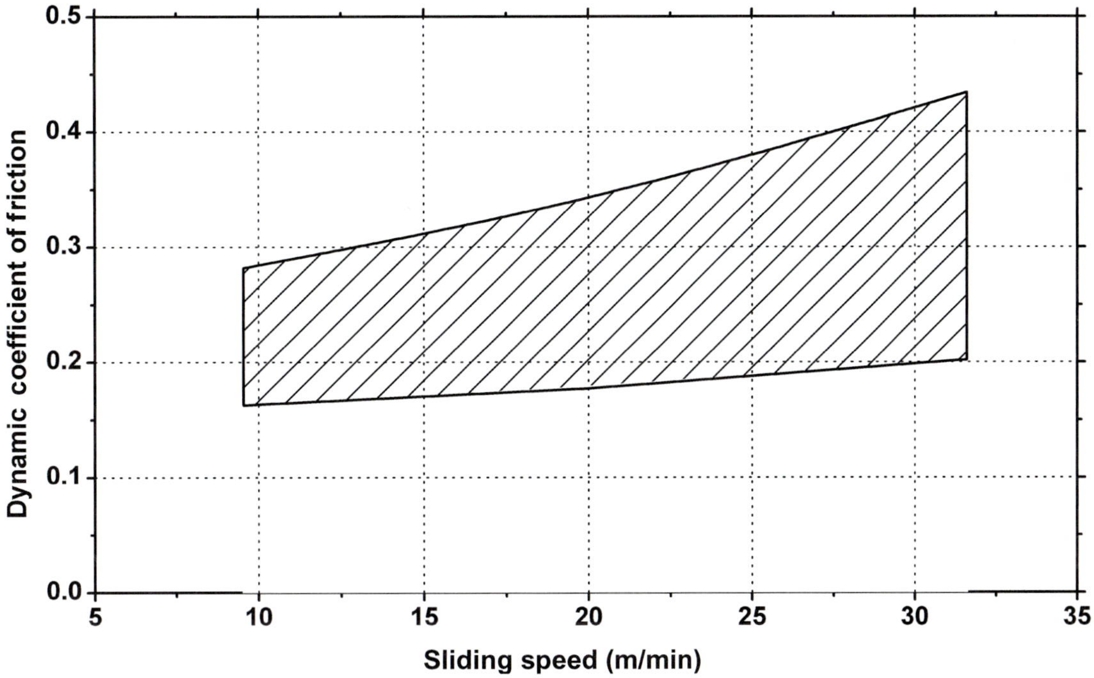

Figure 6.52 Dynamic coefficient of friction vs. sliding speed of Ticona Celanex® 2500—general purpose, nucleated, easy flow PBT (p = 1.25 N/mm^2, against steel with Rz = 2 μm).

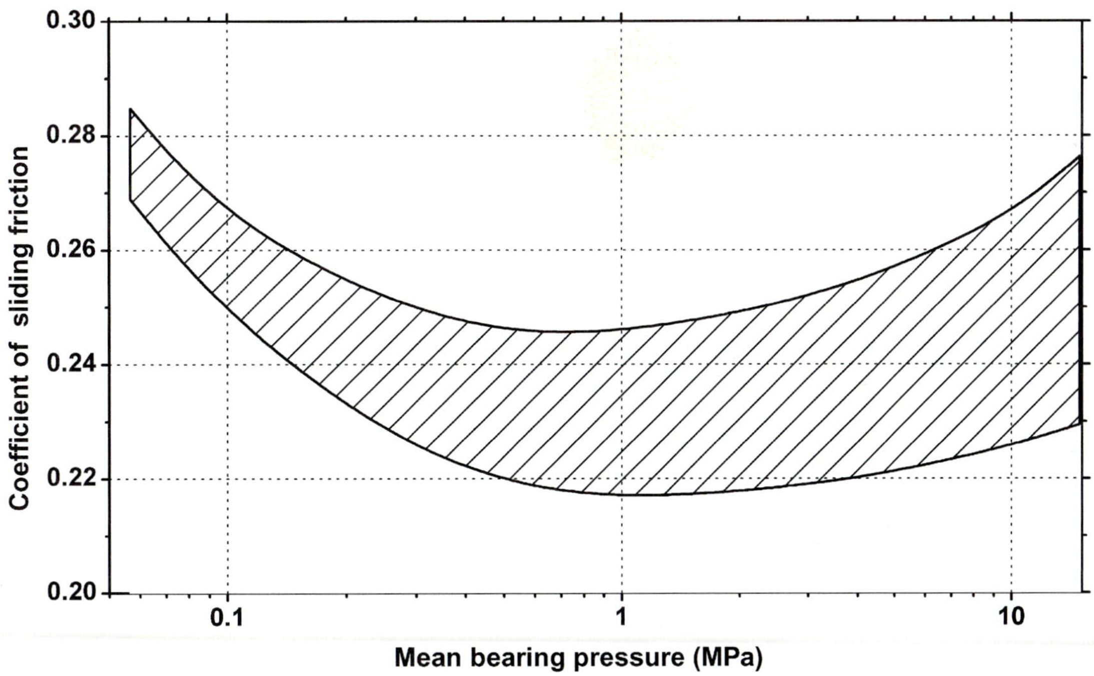

Figure 6.53 Range of sliding coefficient of friction vs. pressure of Evonik Industries Vestodur® 2000—unreinforced, medium viscosity PBT (v = 0.5 m/s).

Table 6.17 Taber Abrasion and Coefficient of Friction of Ticona Celanex® PBT Plastics

Property	Taber Abrasion (mg/1000 cycles)	Coefficient of Friction Dynamic	Coefficient of Friction Static
Celanex® 2000—Unfilled	13	0.13	0.13
Celanex® 2002—General purpose, unreinforced	14	0.13	0.13
Celanex® 2012—Flame retardant, unreinforced	–	0.10–0.13	0.10–0.13
Celanex® 3200—General purpose, 15% glass fiber reinforced	24	0.10–0.21	0.15–0.19
Celanex® 3210—18% Glass fiber reinforced, flame retardant	14	0.10–0.13	–
Celanex® 3211—18% Glass fiber reinforced, flame retardant	–	0.12–0.16	0.18–0.23
Celanex® 3300—General purpose, 30% glass fiber reinforced	40	0.12	0.16–0.34
Celanex® 3310—30% Glass fiber reinforced, flame retardant	3310	0.10–0.13	–
Celanex® 3311—30% Glass fiber reinforced, flame retardant	3311	0.12–0.16	0.17–0.26
Celanex® 3400—General purpose, 40% glass fiber reinforced	3400	0.12–0.16	0.17–0.19
Celanex® 4300—Improved impact, 30% glass fiber reinforced	4300	0.13–0.15	0.17–0.18
Celanex® 5300—Improved surface smoothness, 30% glass fiber reinforced	17	0.13	–
Celanex® 6400—Warp resistant, 40% glass fiber/mineral reinforced, good surface smoothness	25	0.13–0.15	0.17–0.23
Celanex® 7700—Warp resistant, 40% glass fiber/mineral reinforced, flame retardant	–	0.01–0.20	0.14–0.24

6.4 Polyethylene Terephthalate

6.4.1 Fatigue Data

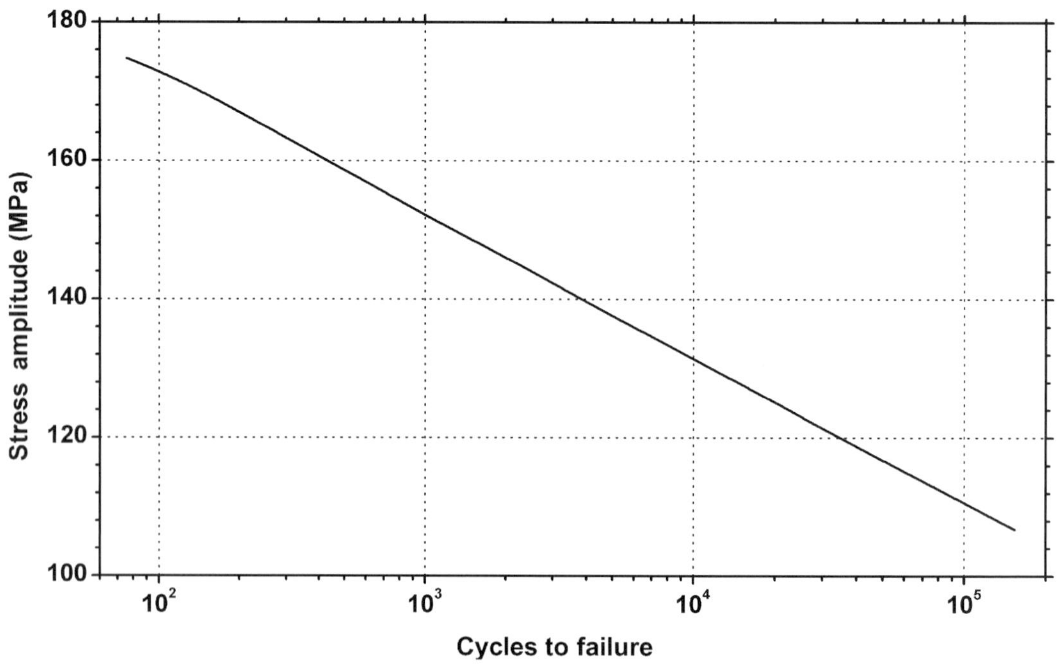

Figure 6.54 Stress amplitude vs. cycles to failure at 20°C and 50% relative humidity of DSM Arnite®—35% glass fiber reinforced PET.

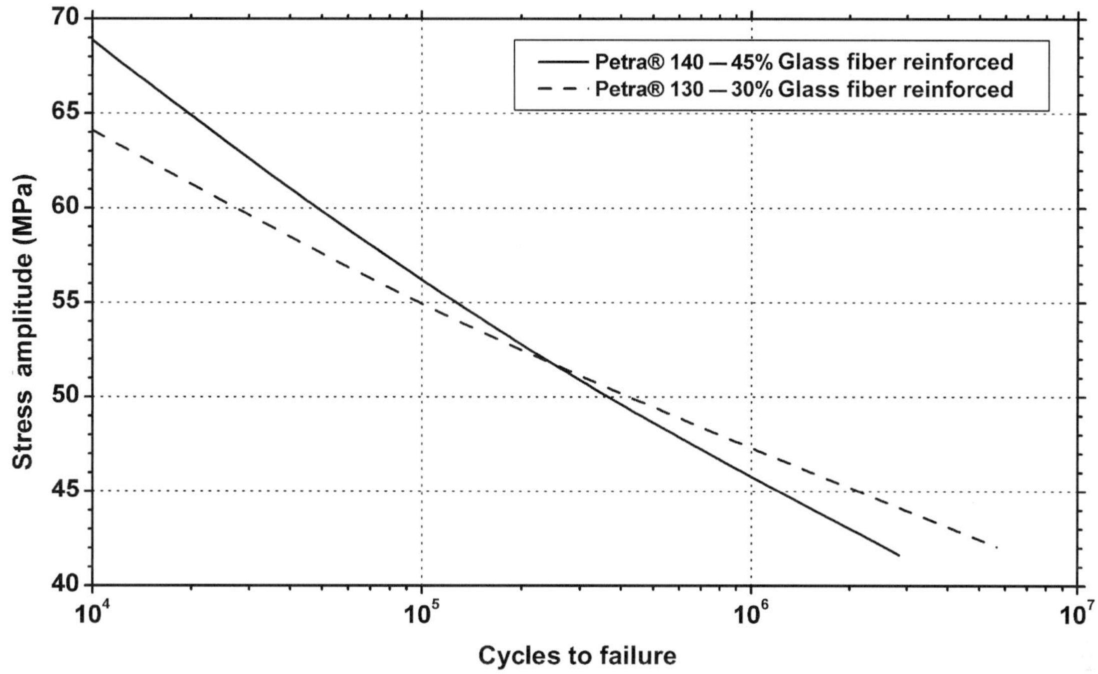

Figure 6.55 Flexural stress amplitude vs. cycles to failure of two BASF Petra®—glass fiber reinforced PET plastics.

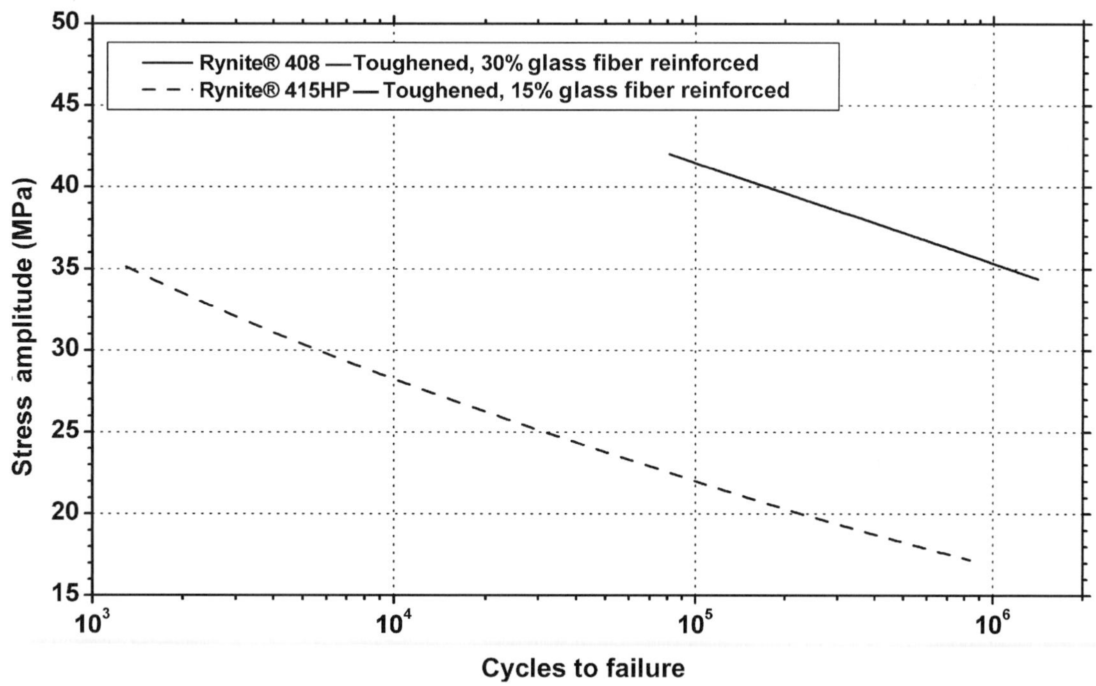

Figure 6.56 Flexural stress amplitude vs. cycles to failure at 23°C of two DuPont Engineering Polymers Rynite® PET plastics.

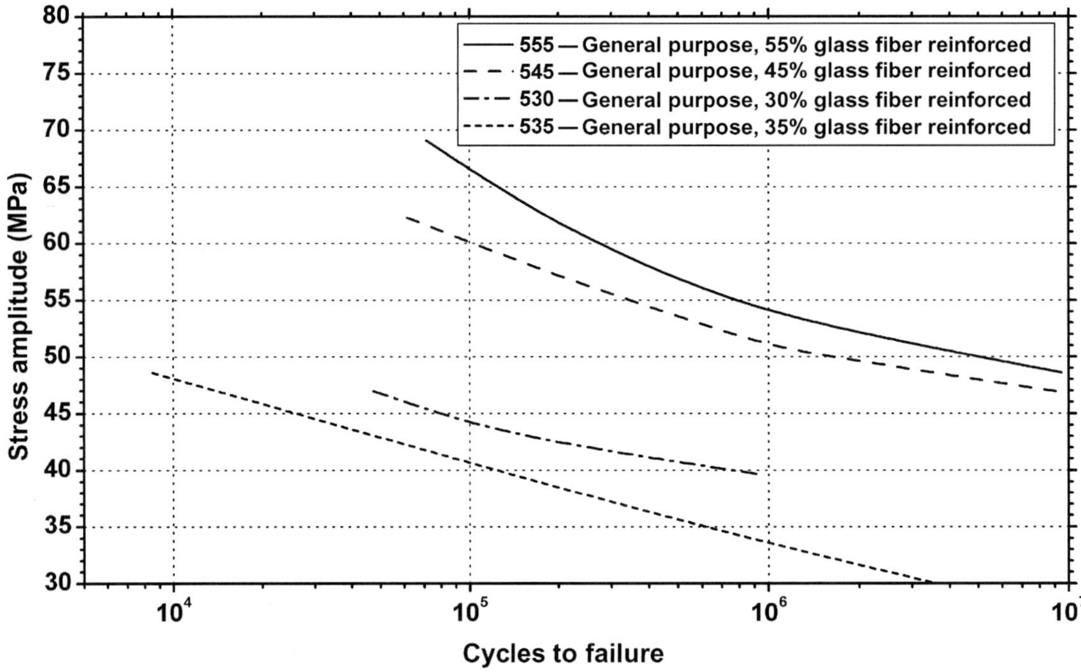

Figure 6.57 Flexural stress amplitude vs. cycles to failure of several DuPont Engineering Polymers Rynite® 500 Series—general purpose, glass fiber reinforced PET plastics.

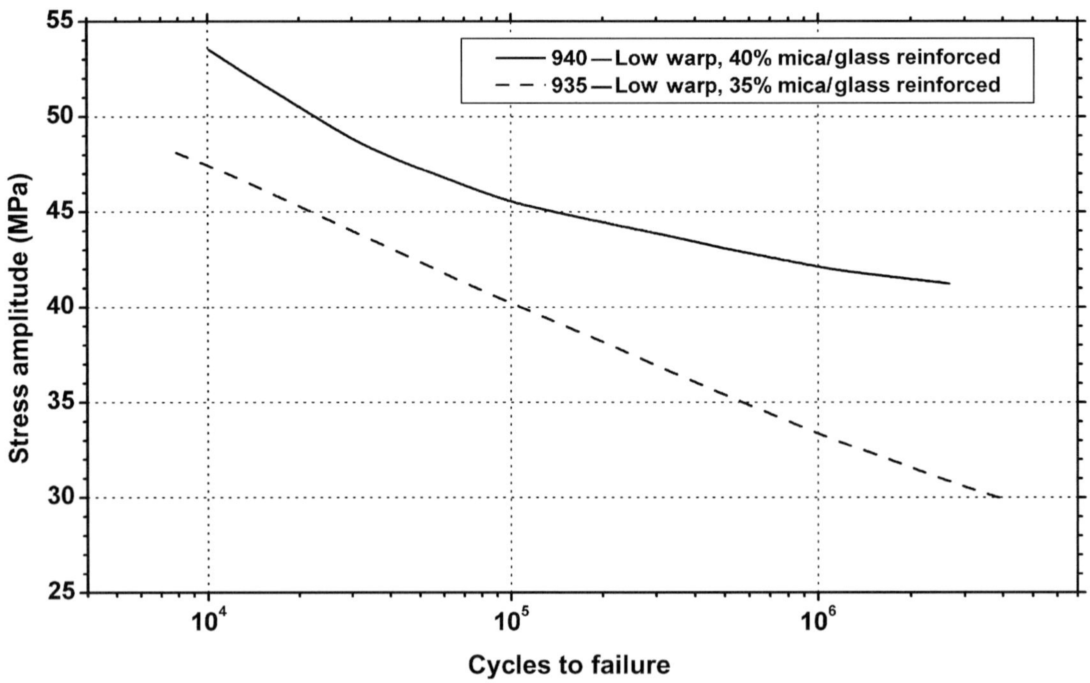

Figure 6.58 Flexural stress amplitude vs. cycles to failure of two DuPont Engineering Polymers Rynite® 900 Series—low warp, mica/glass fiber reinforced PET plastics.

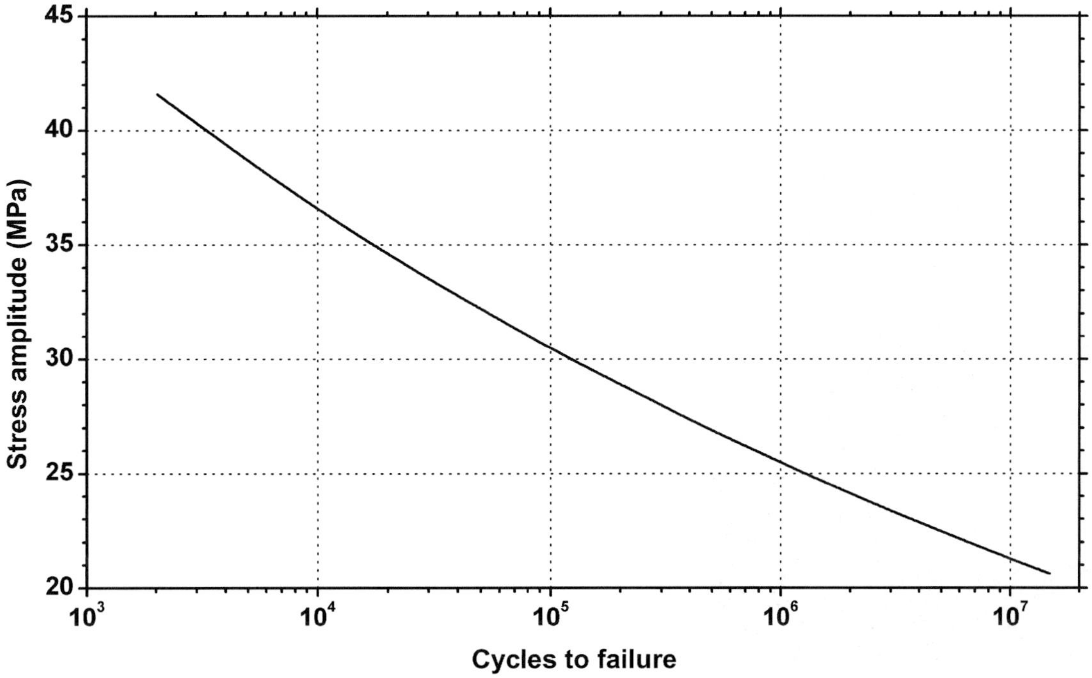

Figure 6.59 Flexural stress amplitude vs. cycles to failure at 23°C of DuPont Engineering Polymers Rynite® SST35—super tough, 35% glass fiber reinforced PET.

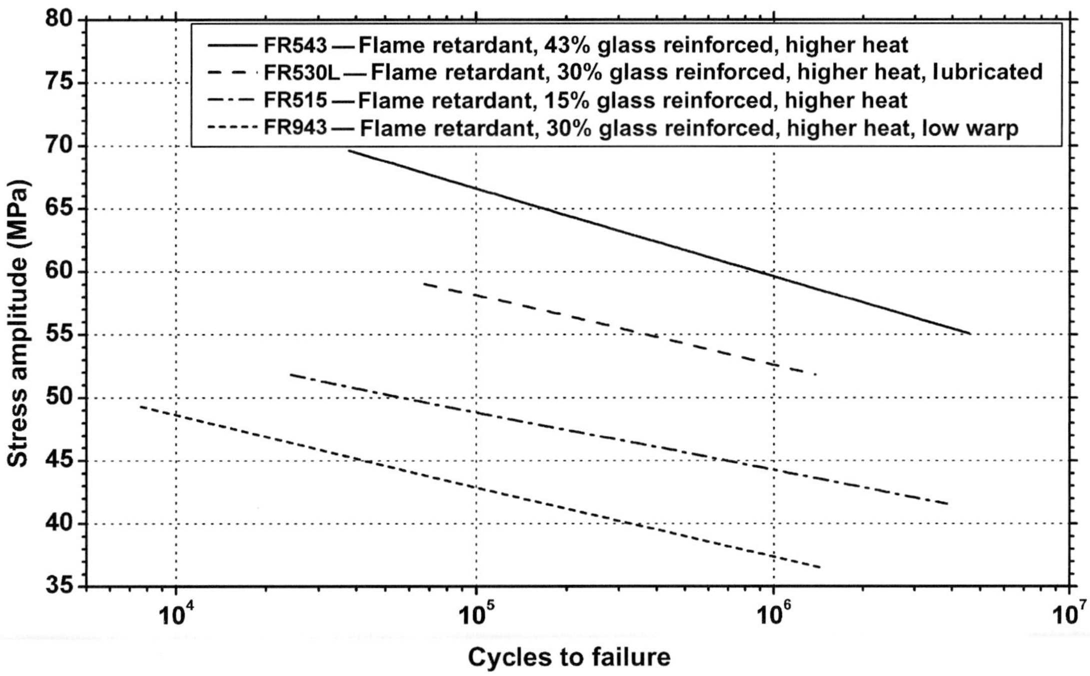

Figure 6.60 Flexural stress amplitude vs. cycles to failure at 23°C of several DuPont Engineering Polymers Rynite® FR400 Series—flame retardant reinforced PET plastics.

6.4.2 Tribology Data

Table 6.18 Coefficient of Friction and Taber Abrasion of DuPont Engineering Polymers Rynite® PET Plastics

Property	Coefficient of Friction (Against Self)	Coefficient of Friction (Against Steel)	Taber Abrasion (CS-17 Wheel, 1000 g)
Test Method	**ASTM D1894**	**ASTM D1894**	**–**
Units	**–**	**–**	**mg/1000 cycles**
Rynite® 530—General purpose, 30% glass fiber reinforced	0.18	0.17	30
Rynite® 545—General purpose, 45% glass fiber reinforced	0.17	0.20	44
Rynite® 555—General purpose, 55% glass fiber reinforced	0.27	0.18	–
Rynite® 935—Low warp, 35% mica/ glass fiber reinforced	0.21	0.19	–
Rynite® 940—Low warp, 40% mica/ glass fiber reinforced	–	–	81
Rynite® 415HP—Toughened, 15% glass fiber reinforced	0.42	0.27	35
Rynite® SST 35 Super Tough, 35% glass fiber reinforced	–	–	82
Rynite® FR330—Flame retardant, 30% glass fiber reinforced	0.24	0.18	88
Rynite® FR515—Flame retardant, 15% glass fiber reinforced, higher heat	0.21	0.18	88
Rynite® FR530—Flame retardant, 30% glass fiber reinforced, higher heat	0.18	0.19	38
Rynite® FR543—Flame retardant, 43% glass fiber reinforced, higher heat	0.18	0.16	69
Rynite® FR943—Flame retardant, 43% mica/glass fiber reinforced, higher heat, low warp	0.29	0.18	82
Rynite® FR945—Flame retardant, 45% mica/glass fiber reinforced, higher heat, low warp	0.20	0.20	81
Rynite® FR946—Flame retardant, 46% mica/glass fiber reinforced, higher heat, low warp	0.27	0.18	74

6.5 Liquid Crystal Polymer

6.5.1 Fatigue Data

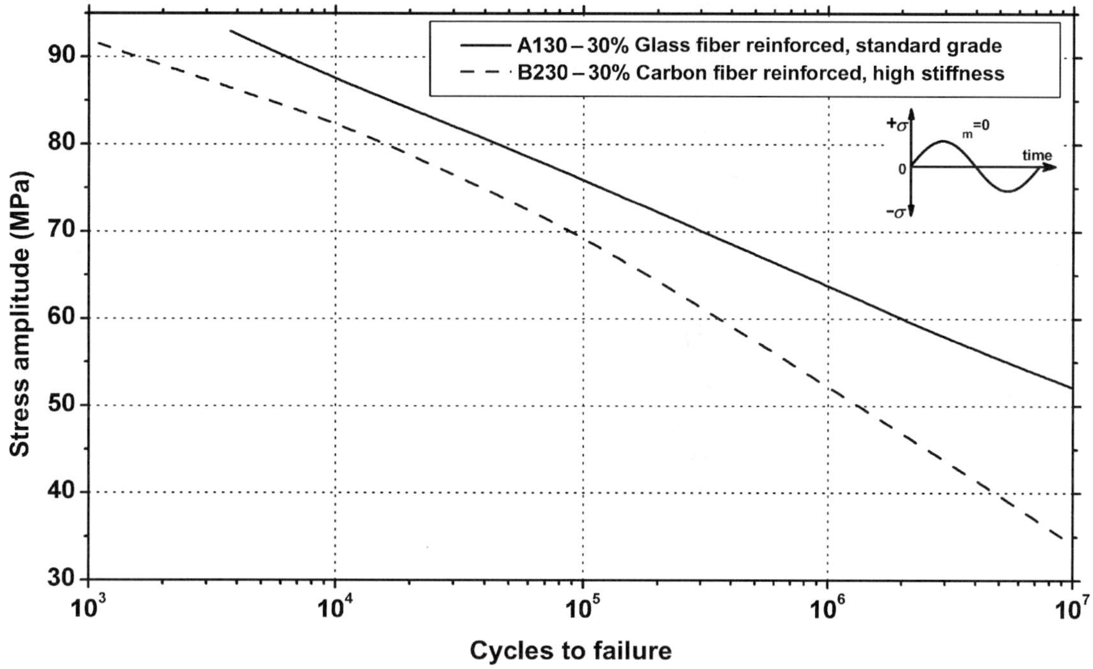

Figure 6.61 Flexural stress amplitude vs. cycles to failure at 23°C of two Ticona Vectra®—fiber reinforced LCP plastics (10 Hz).

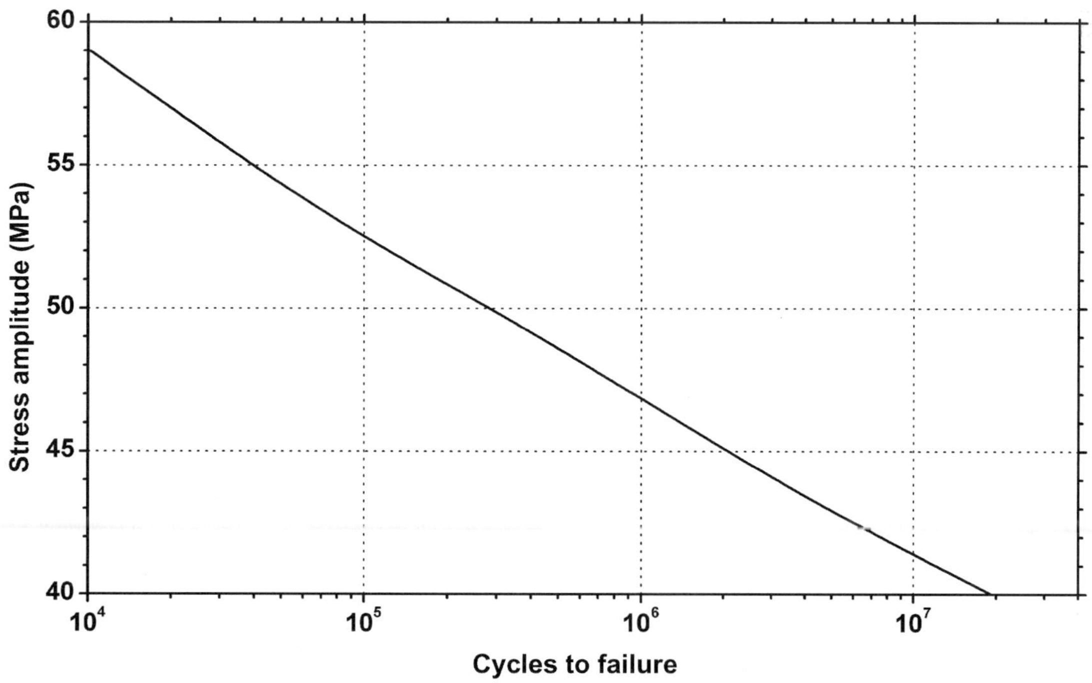

Figure 6.62 Flexural stress amplitude vs. cycles to failure at 23°C of DuPont Engineering Polymers Zenite® 6130 BK010—30% glass fiber reinforced LCP.

6.5.2 Tribology Data

Ticona Vectra® A130—30% Glass fiber reinforced, standard grade LCP
Ticona Vectra® A230—30% Carbon fiber reinforced, high-stiffness LCP
Ticona Vectra® A530—30% Mineral-filled LCP
Ticona Vectra® A430—PTFE modified, standard grade LCP
Ticona Vectra® A435—Glass/PTFE-filled LCP
Ticona Vectra® A625—25% Graphite-filled LCP
Ticona Vectra® B130—30% Glass fiber reinforced, high-stiffness LCP
Ticona Vectra® B230—30% Carbon fiber reinforced, high-stiffness LCP
Ticona Vectra® C130—30% Glass fiber reinforced, heat resistant LCP
Ticona Vectra® L130—30% Glass fiber reinforced, high flow LCP

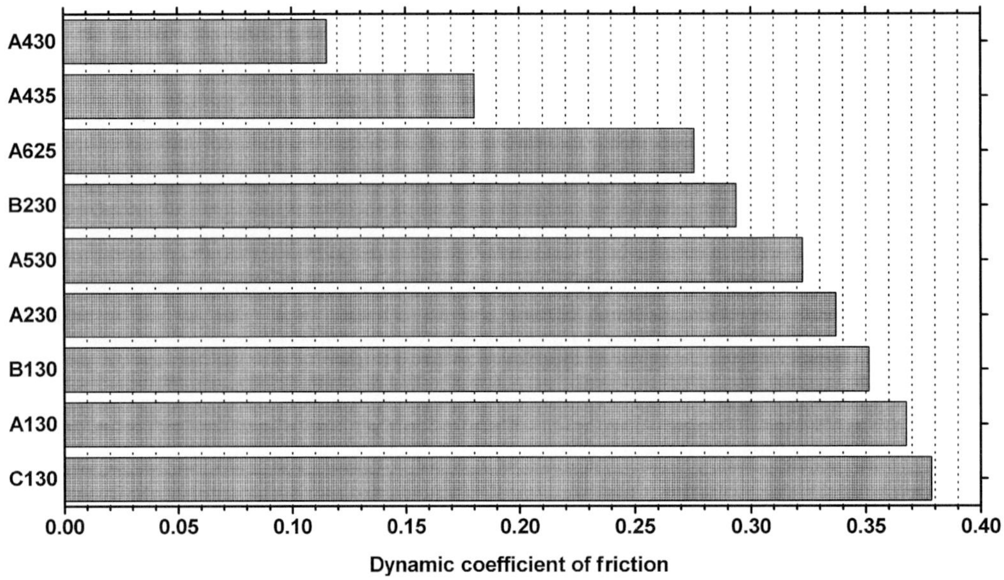

Figure 6.63 Dynamic coefficient of friction for various Ticona Vectra® LCP resins ($P = 6$ N, $v = 60$ cm/min).

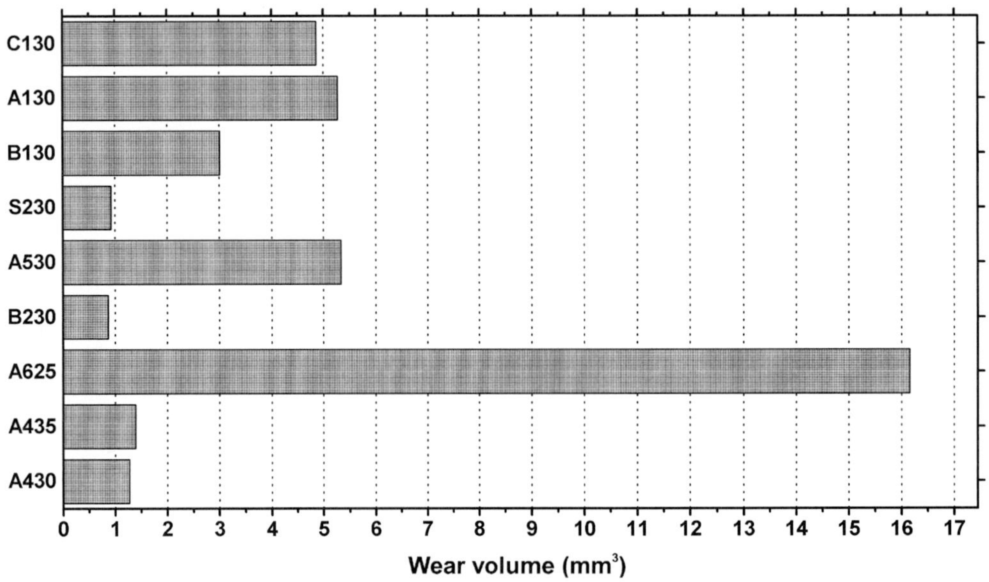

Figure 6.64 Wear volumes after 60 hours of testing for various Ticona Vectra® LCP resins ($P = 3$ N, $v = 136$ m/min).

Table 6.19 Coefficients of Friction for Various Vectra® LCP Grades

Vectra® LCP Grade	Coefficient of Friction—Flow Direction	
	Static	Dynamic
A115—15% Glass fiber reinforced, standard grade	0.11	0.11
A130—30% Glass fiber reinforced, standard grade	0.14	0.14
A150—50% Glass fiber reinforced, standard grade	0.16	0.19
A230—30% Carbon fiber reinforced, high stiffness	0.19	0.12
A410—10% Mineral/glass fiber filled	0.21	0.21
A430—PTFE modified, standard grade	0.11	0.11
A435—Glass/PTFE filled	0.16	0.18
A515—15% Mineral filled	0.20	0.19
A625—Graphite reinforced	0.21	0.15
B230—30% Carbon fiber reinforced, high stiffness	0.14	0.14
L130—30% Glass fiber reinforced, high flow	0.15	0.16

6.6 Polyphthalate Carbonate

6.6.1 Fatigue Data

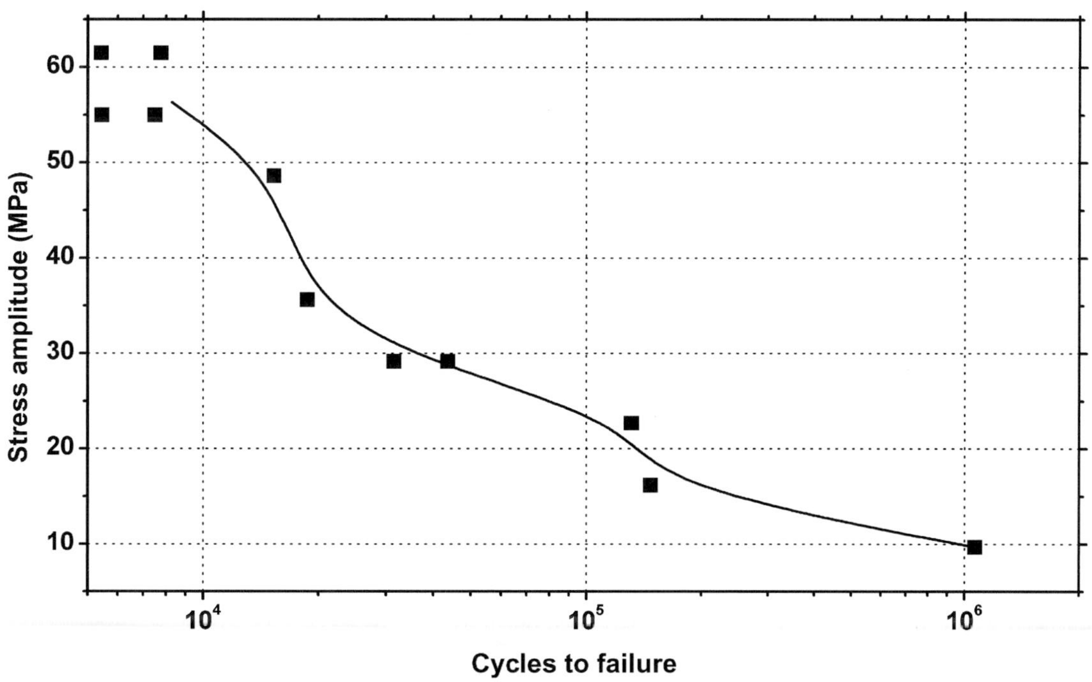

Figure 6.65 Tensile stress amplitude vs. cycles to failure at 23°C of SABIC Innovative Plastics Lexan® 4501—high heat-resistant PPC.

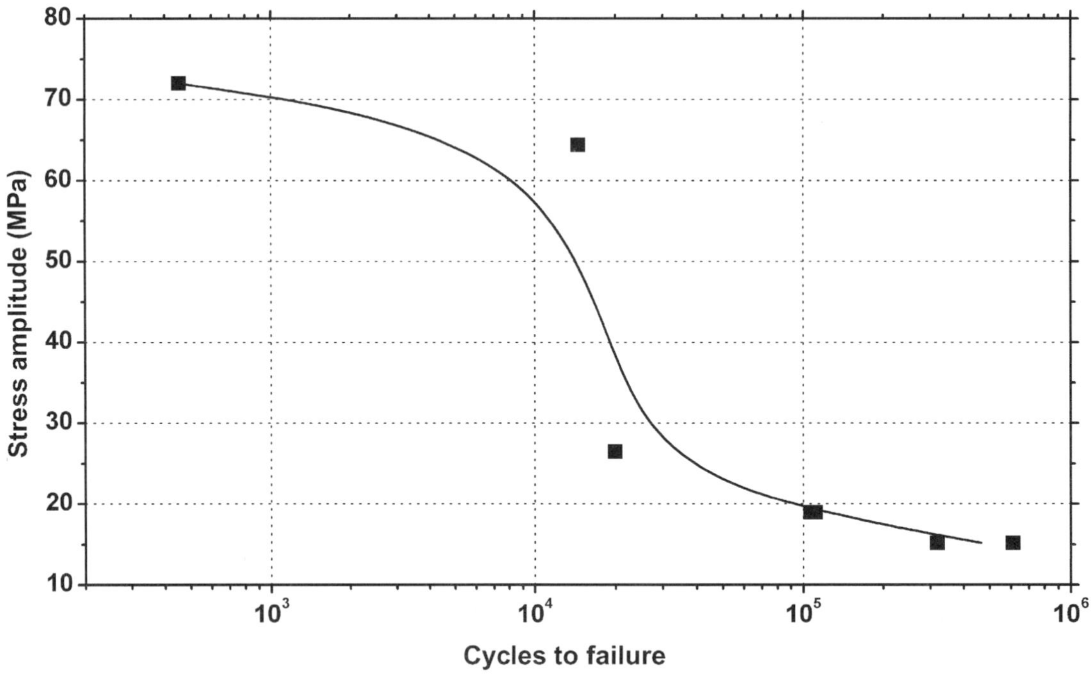

Figure 6.66 Tensile stress amplitude vs. cycles to failure at 23°C of SABIC Innovative Plastics Lexan® 4701R—high heat-resistant PPC.

6.7 Polycyclohexylene-dimethylene Terephthalate

6.7.1 Fatigue Data

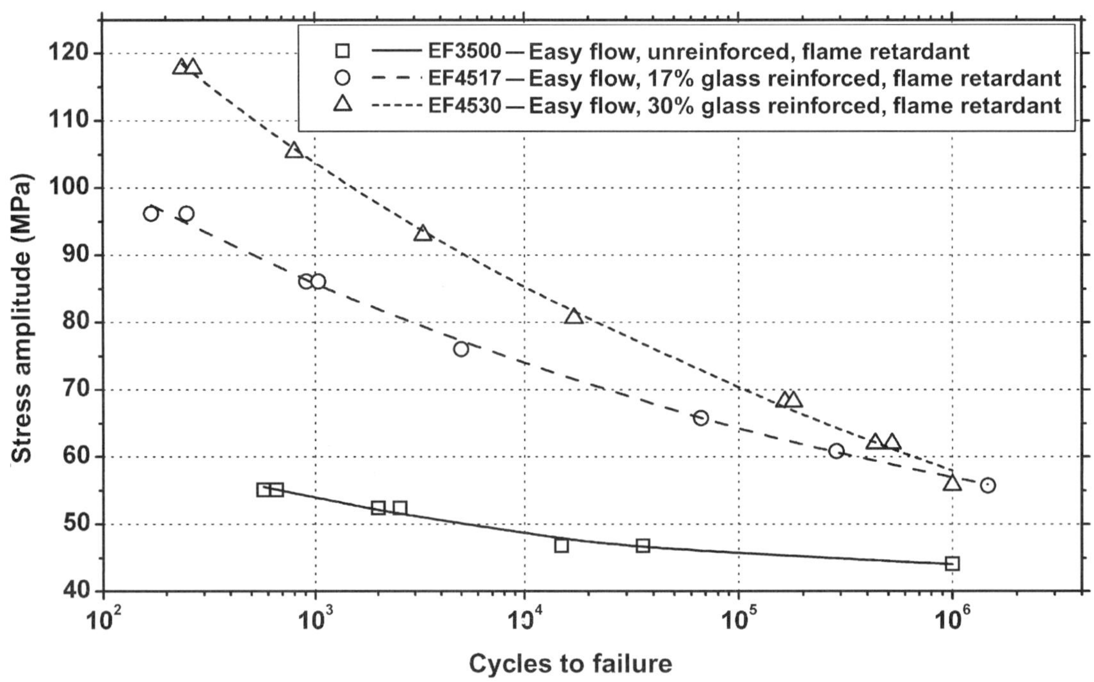

Figure 6.67 Tensile stress amplitude vs. cycles to failure at 23°C of several SABIC Innovative Plastics Valox®—fire retardant PCT plastics.

6.8 Polyester Blends and Alloys

6.8.1 Fatigue Data

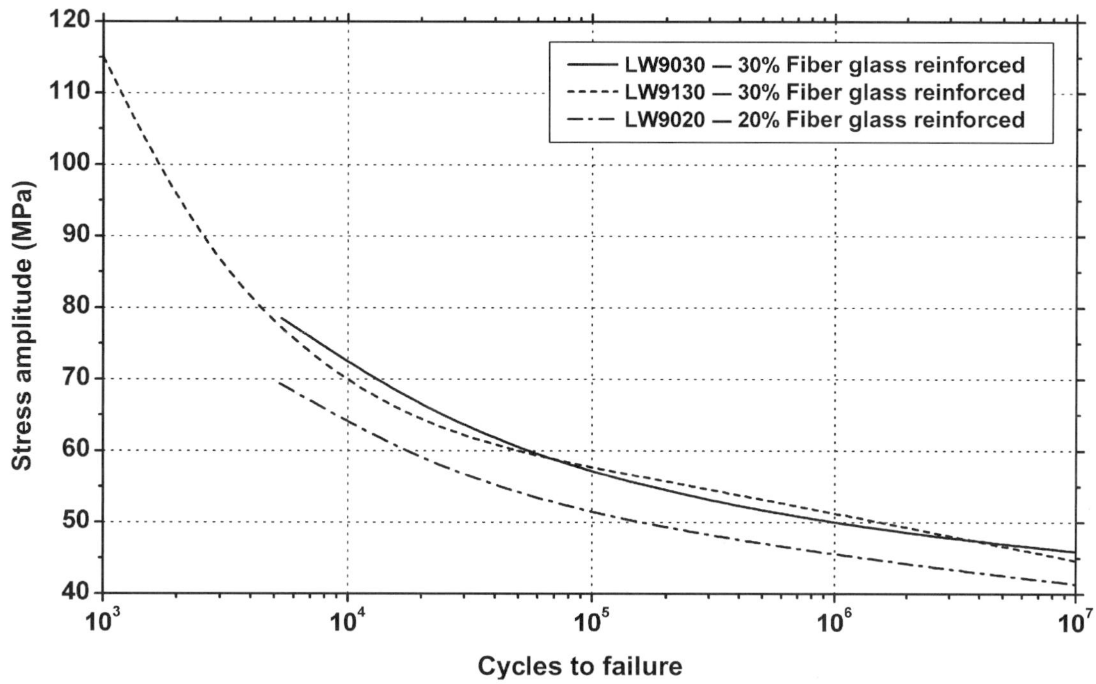

Figure 6.68 Flexural stress amplitude vs. cycles to failure at 23°C of several DuPont Engineering Polymers Crastin®—injection molding PBT/ASA Alloy plastics.

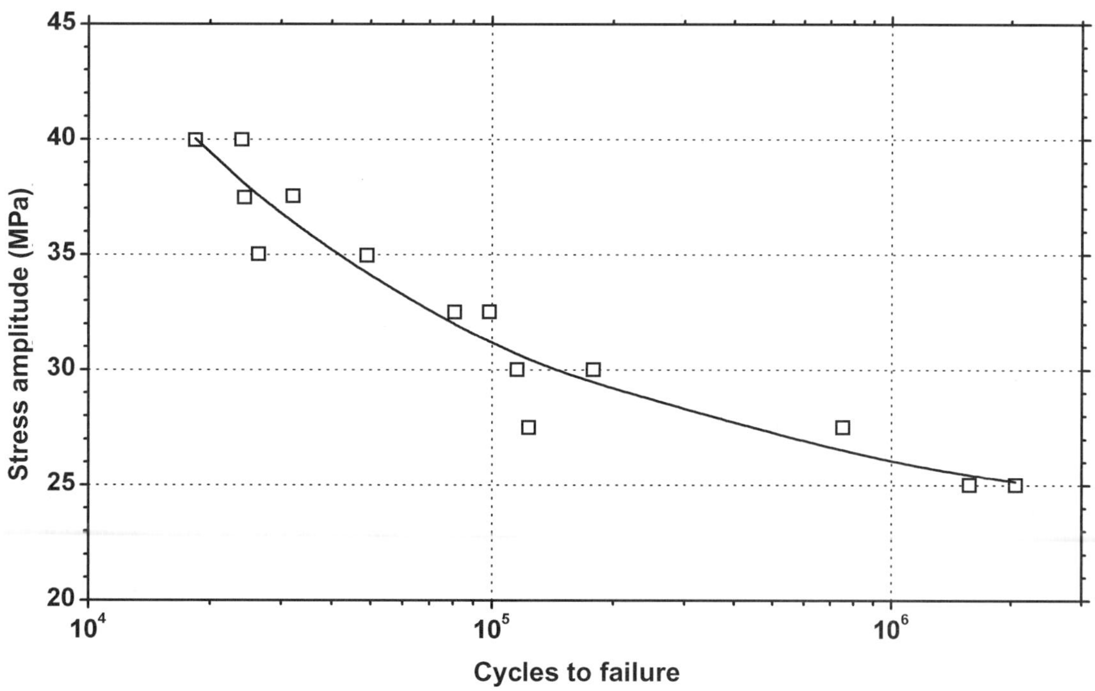

Figure 6.69 Tensile stress amplitude vs. cycles to failure at 23°C of SABIC Innovative Plastics Xenoy® CL101—automotive exterior PC/PBT Alloy.

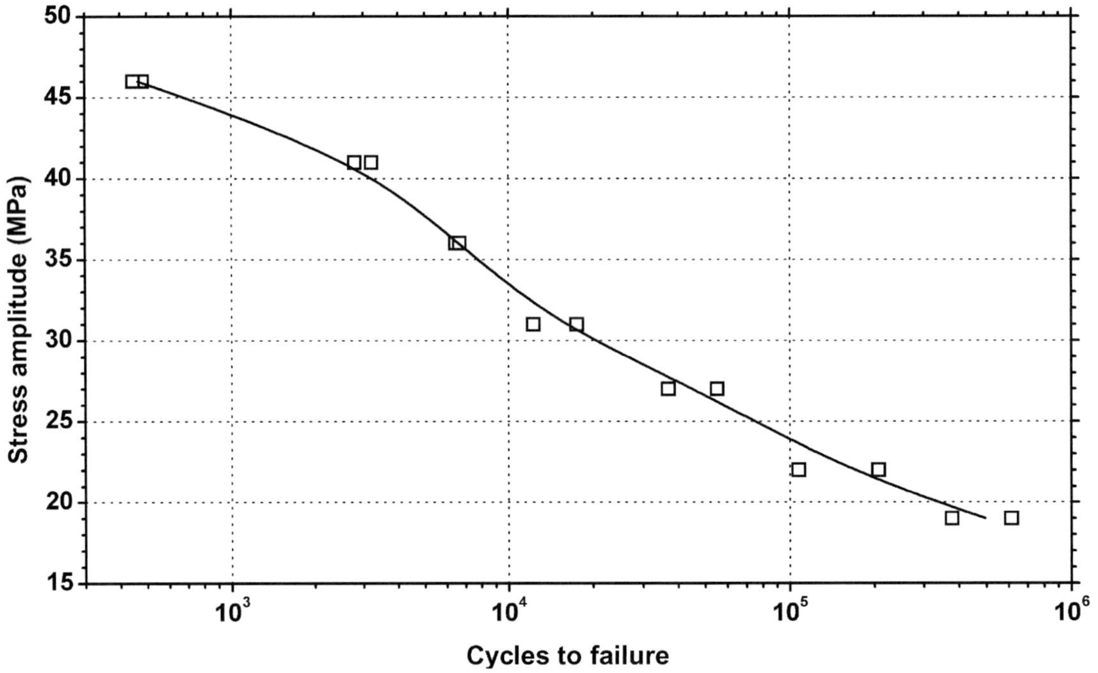

Figure 6.70 Tensile stress amplitude vs. cycles to failure at 23°C of SABIC Innovative Plastics Valox® 368—flame retardant, impact modified, mold release PBT/PC Alloy.

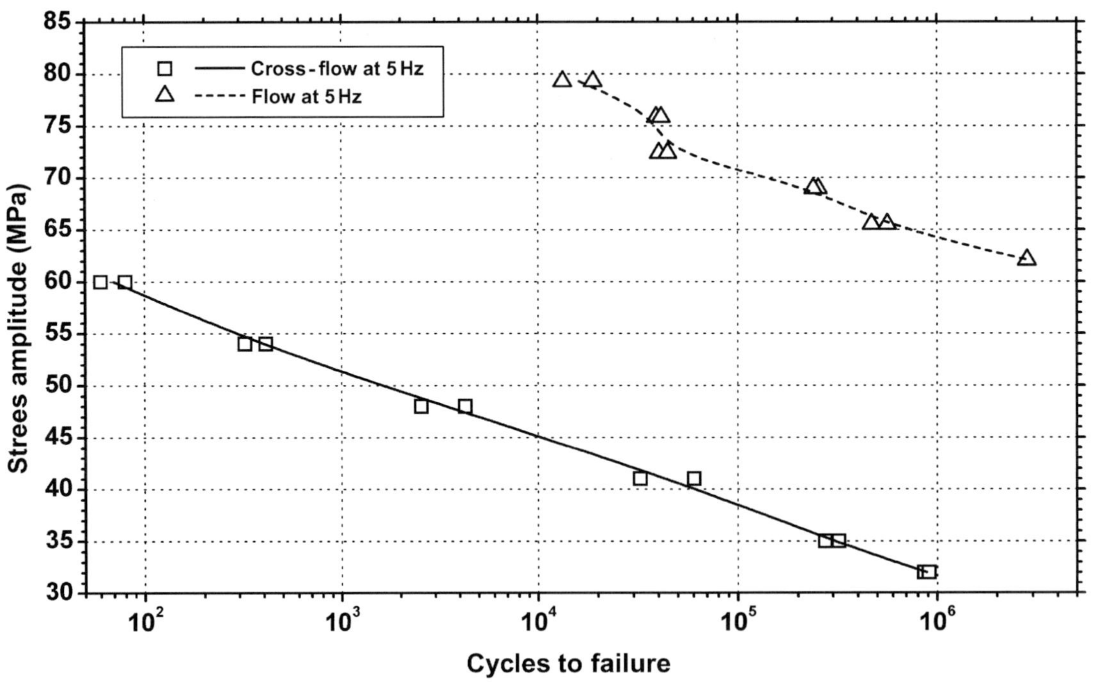

Figure 6.71 Tensile stress amplitude vs. cycles to failure at 23°C of SABIC Innovative Plastics Valox® 508—30% glass fiber reinforced PBT/PC Alloy.

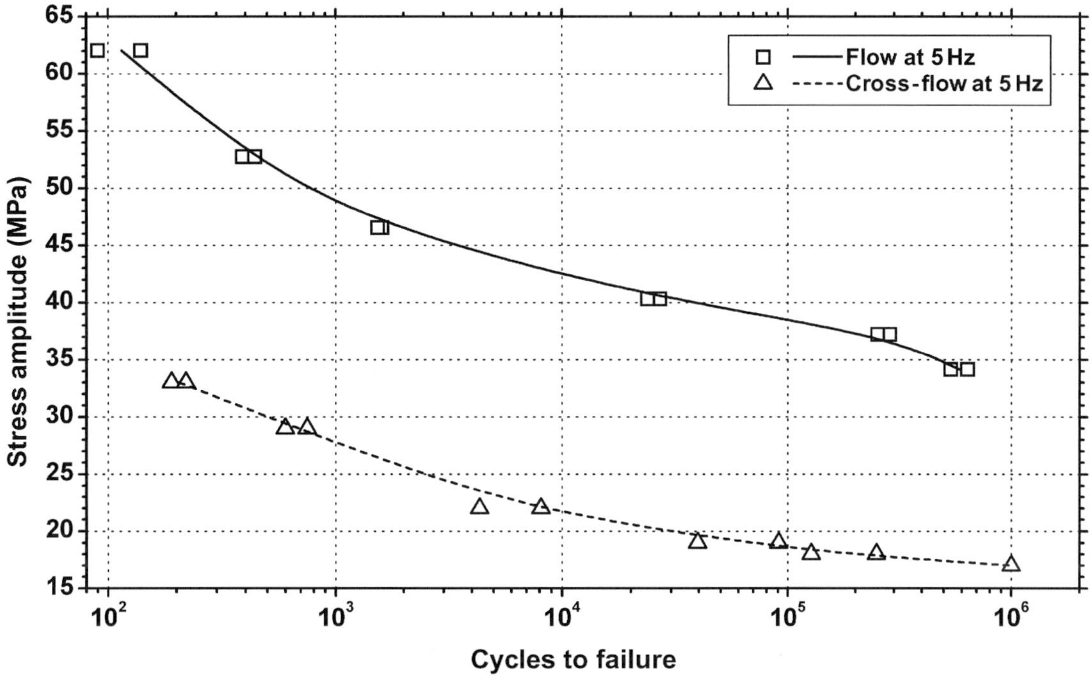

Figure 6.72 Tensile stress amplitude vs. cycles to failure at 82°C of SABIC Innovative Plastics Valox® 508—30% glass fiber reinforced PBT/PC Alloy.

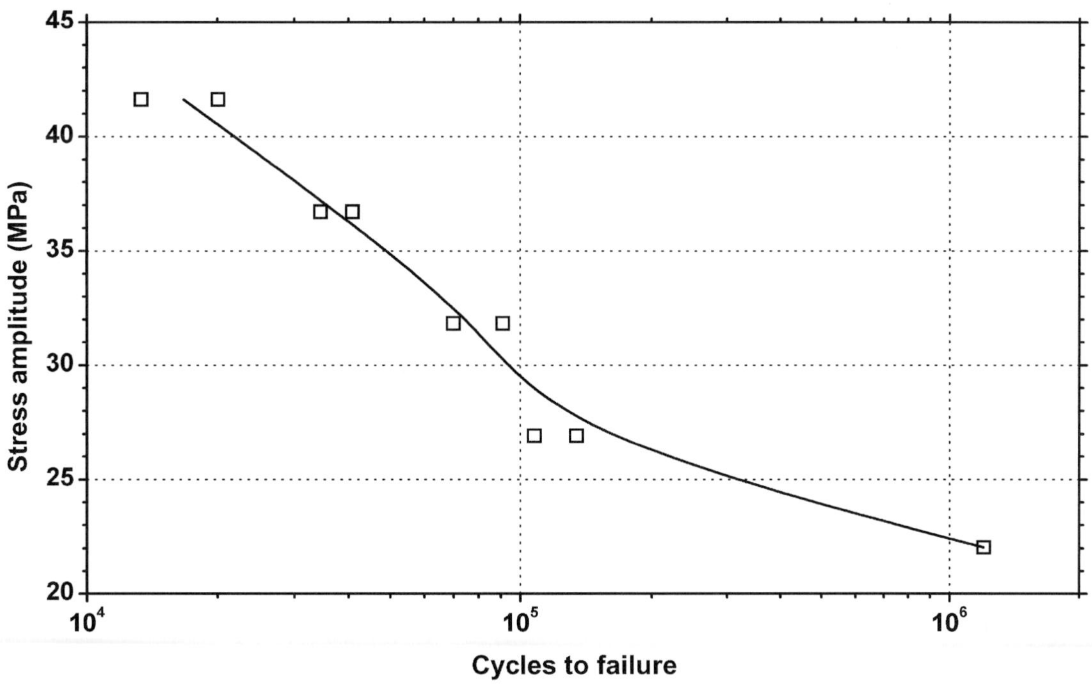

Figure 6.73 Tensile stress amplitude vs. cycles to failure at 23°C of SABIC Innovative Plastics Valox® 3706—impact modified PBT/PC Alloy.

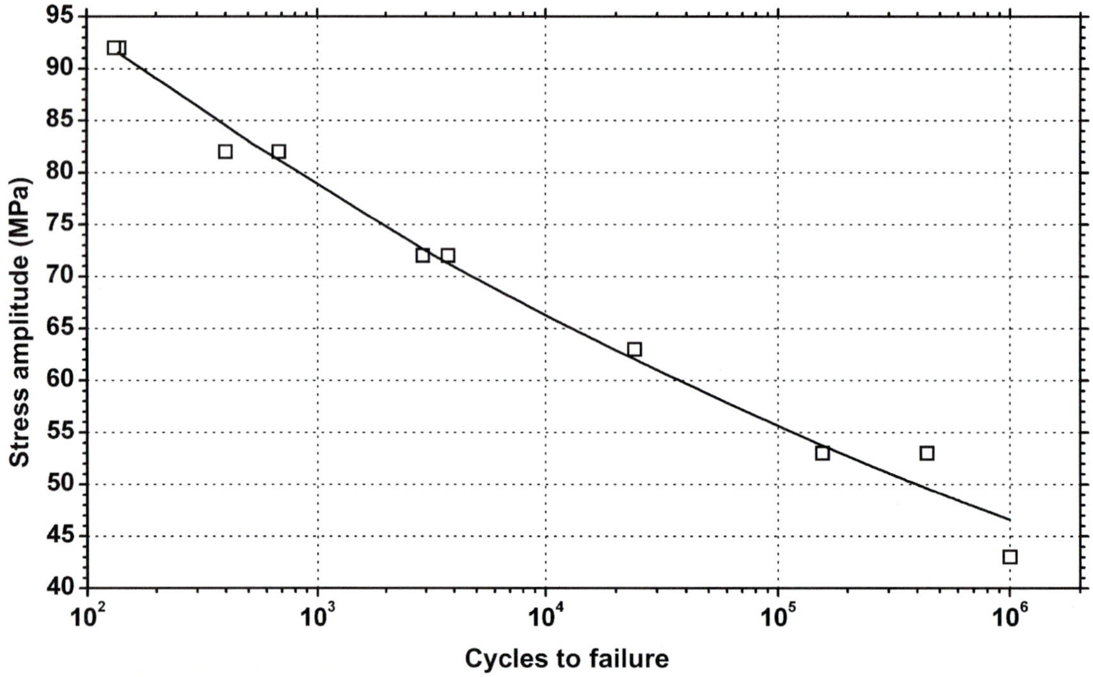

Figure 6.74 Tensile stress amplitude vs. cycles to failure at 23°C of SABIC Innovative Plastics Xenoy® K4630—17% glass fiber reinforced PC/PBT Alloy.

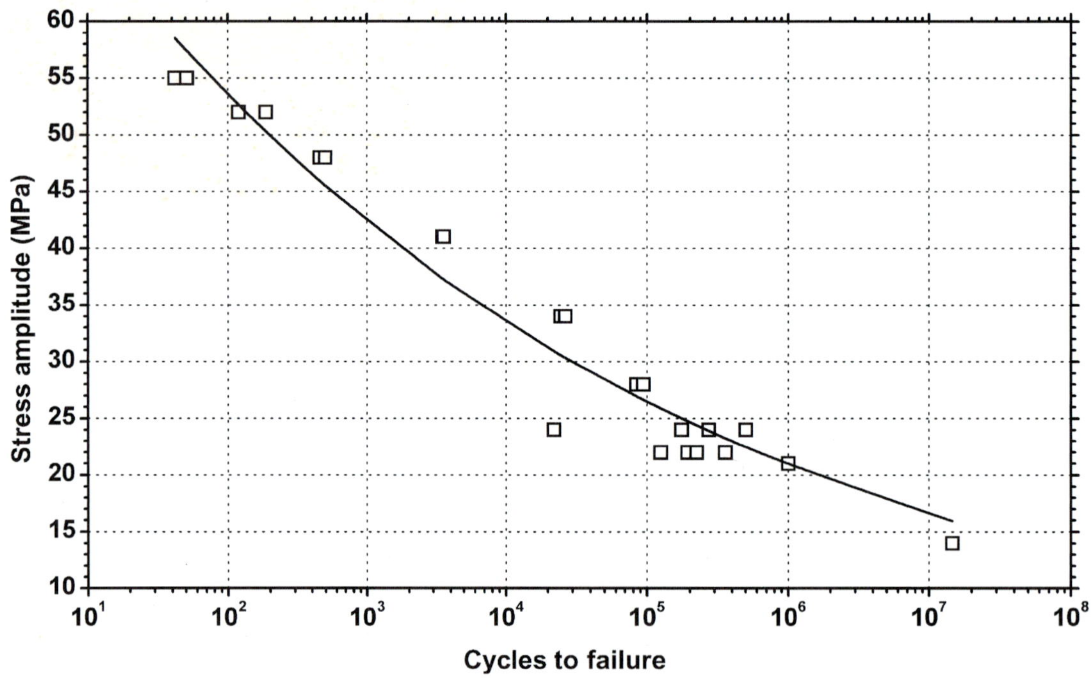

Figure 6.75 Tensile stress amplitude vs. cycles to failure at 23°C of SABIC Innovative Plastics Xenoy® 1102—unreinforced PBT/PC Alloy.

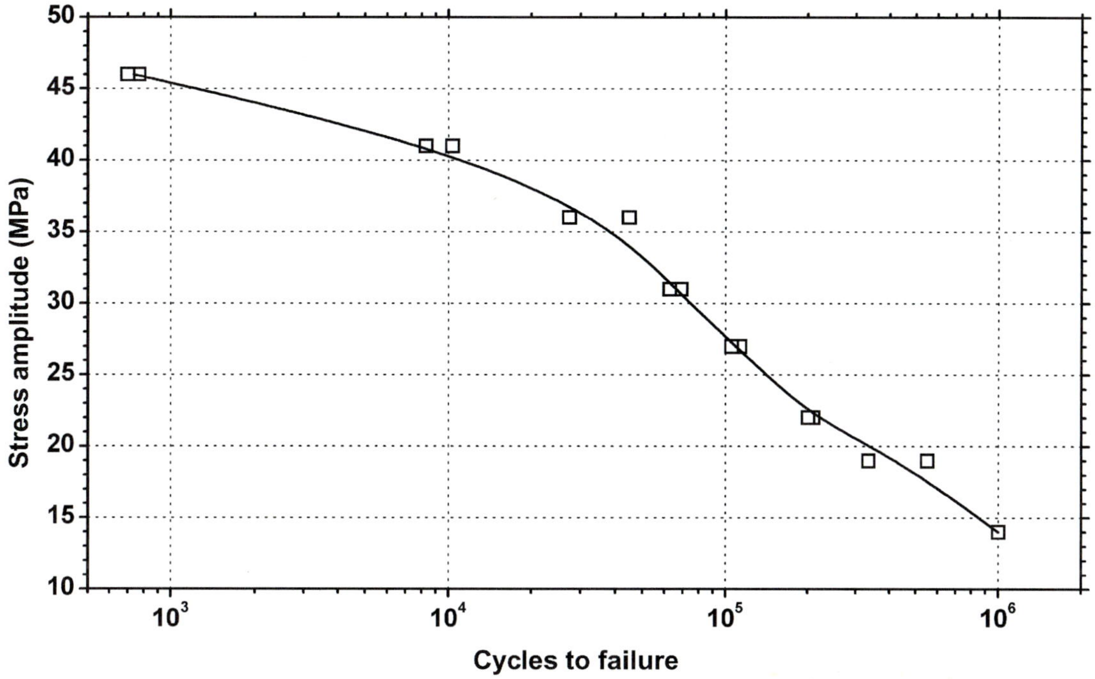

Figure 6.76 Tensile stress amplitude vs. cycles to failure at 23°C of SABIC Innovative Plastics Xenoy® 1103—unreinforced, impact modified PBT/PC Alloy.

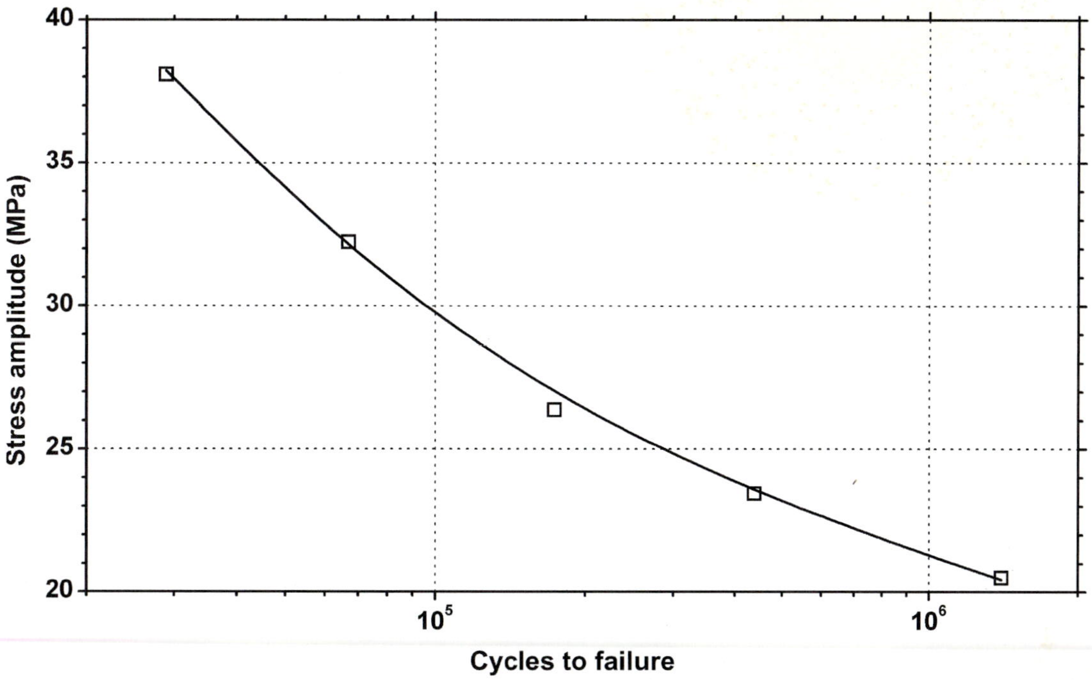

Figure 6.77 Tensile stress amplitude vs. cycles to failure at 23°C of SABIC Innovative Plastics Xenoy® 1402B—blowmoldable, unreinforced PBT/PC Alloy.

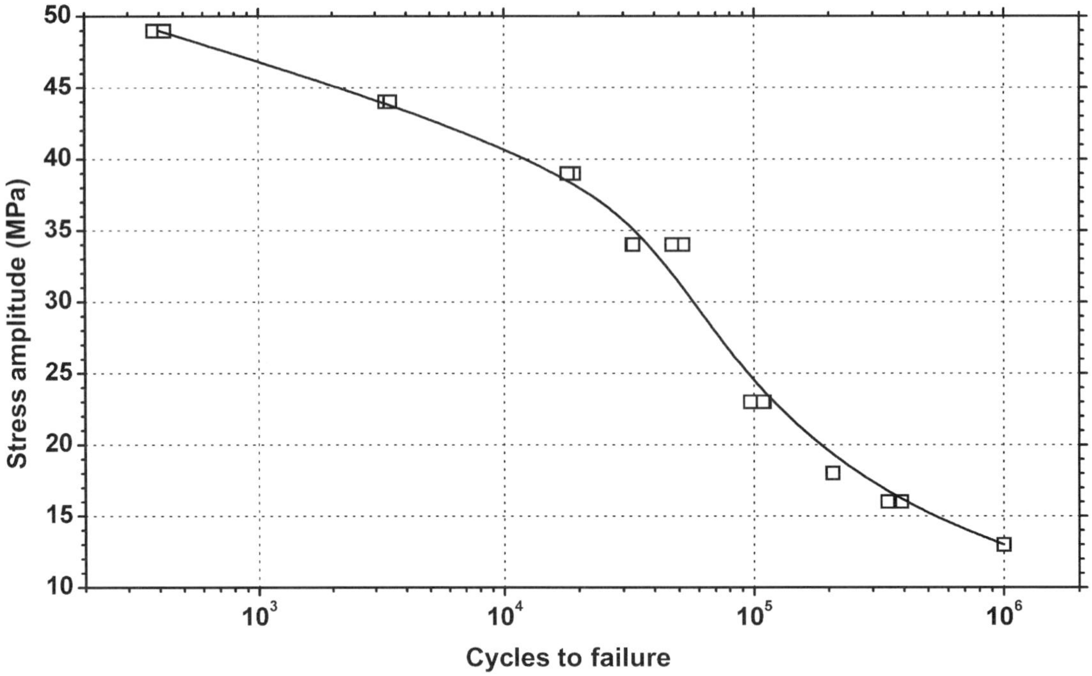

Figure 6.78 Tensile stress amplitude vs. cycles to failure at 23°C of SABIC Innovative Plastics Xenoy® 1403B—PBT/PC Alloy.

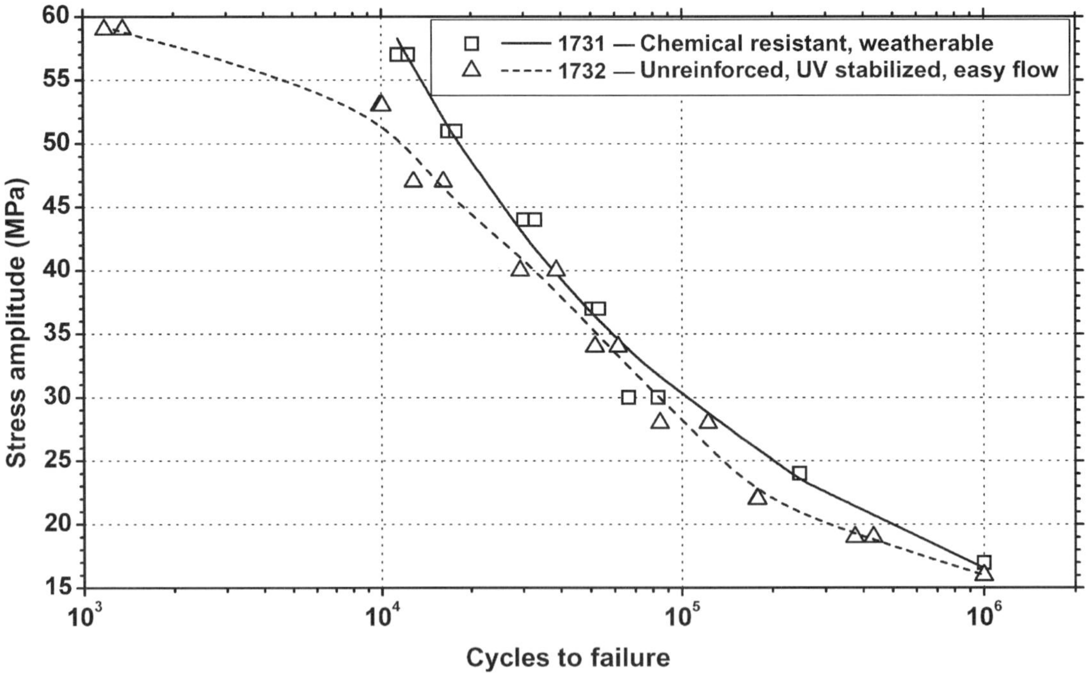

Figure 6.79 Tensile stress amplitude vs. cycles to failure at 23°C of two SABIC Innovative Plastics Xenoy®—PBT/PC Alloys.

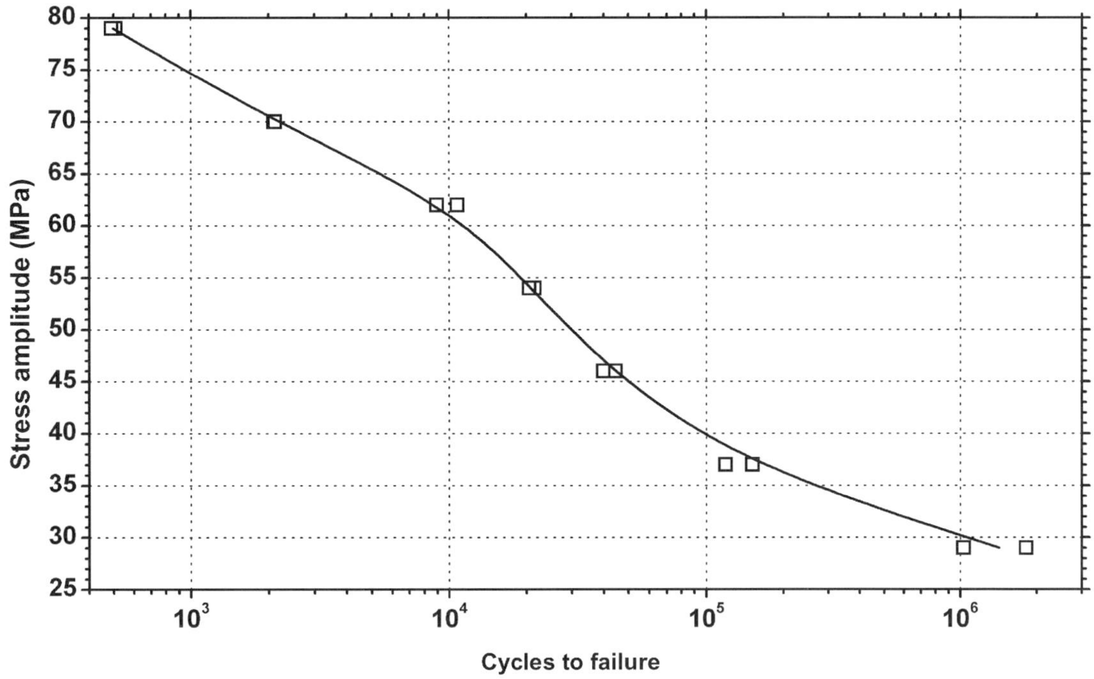

Figure 6.80 Tensile stress amplitude vs. cycles to failure at 23°C of SABIC Innovative Plastics Xenoy® 1760E—high flow, 11% glass-filled PBT/PC Alloy.

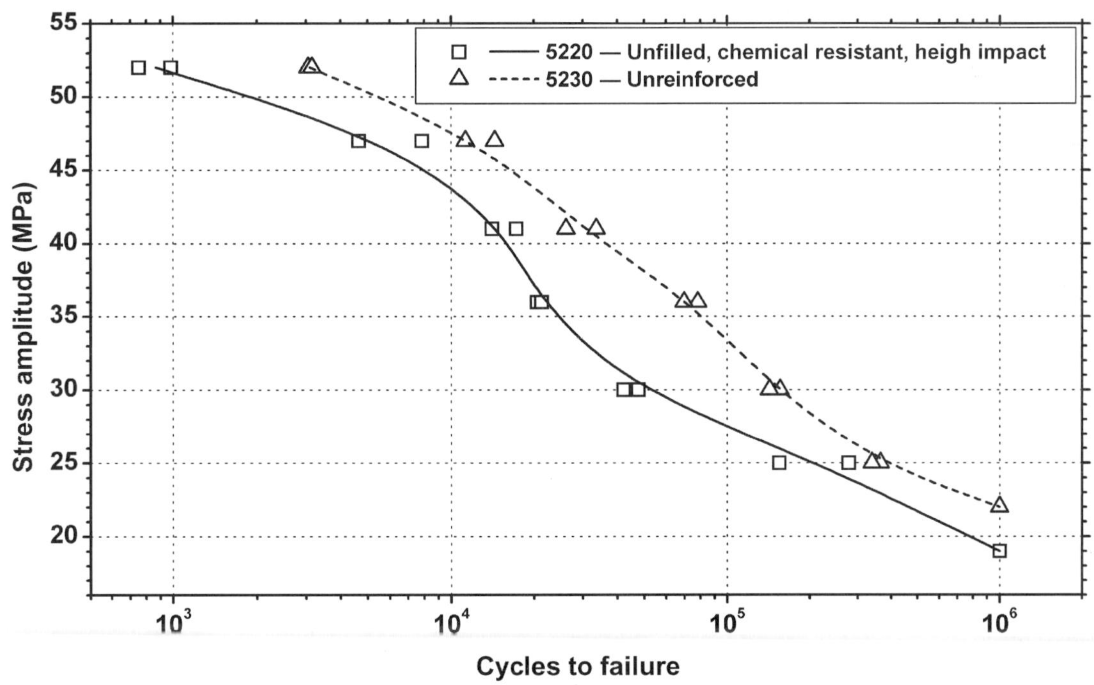

Figure 6.81 Tensile stress amplitude vs. cycles to failure at 23°C of two SABIC Innovative Plastics Xenoy® 52xx series PBT/PC Alloys.

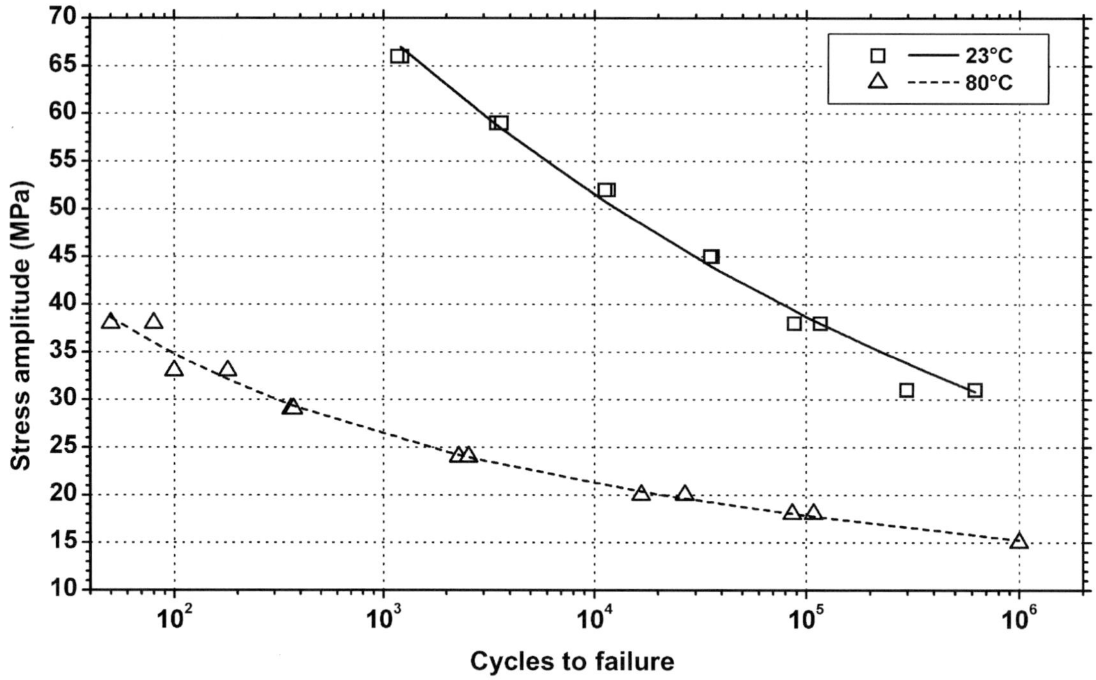

Figure 6.82 Tensile stress amplitude vs. cycles to failure at two temperatures of SABIC Innovative Plastics Xenoy® 5770—20% glass fiber/mineral filled, impact modified PBT/PC Alloy.

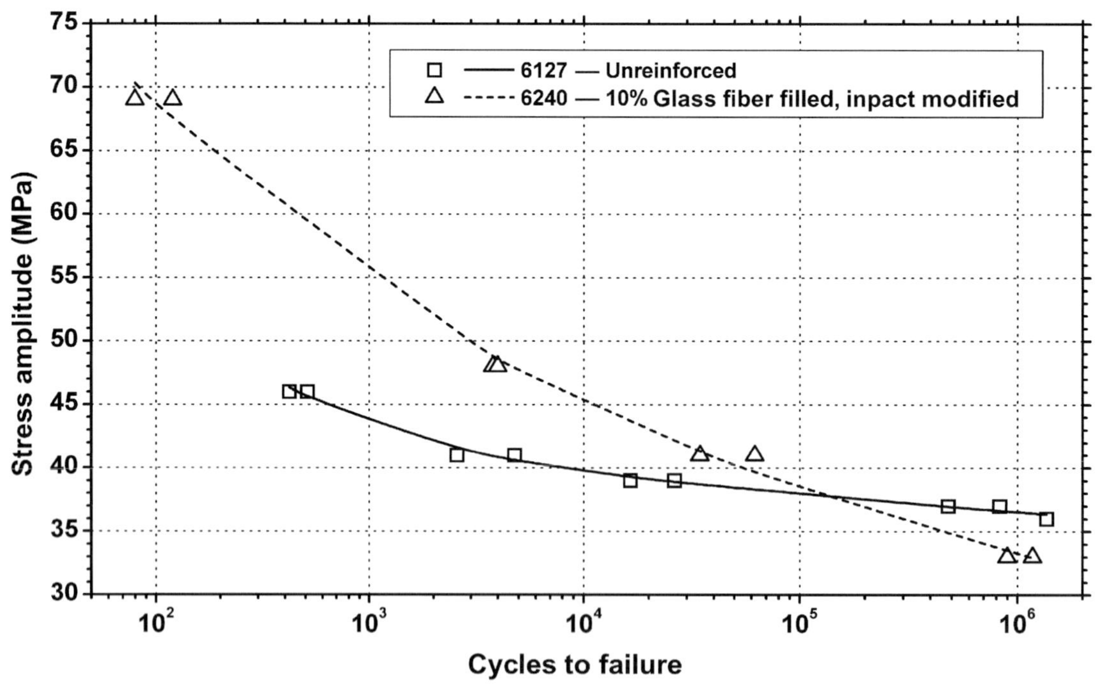

Figure 6.83 Tensile stress amplitude vs. cycles to failure at 23°C of two SABIC Innovative Plastics Xenoy® 5xxx series PBT/PC Alloys.

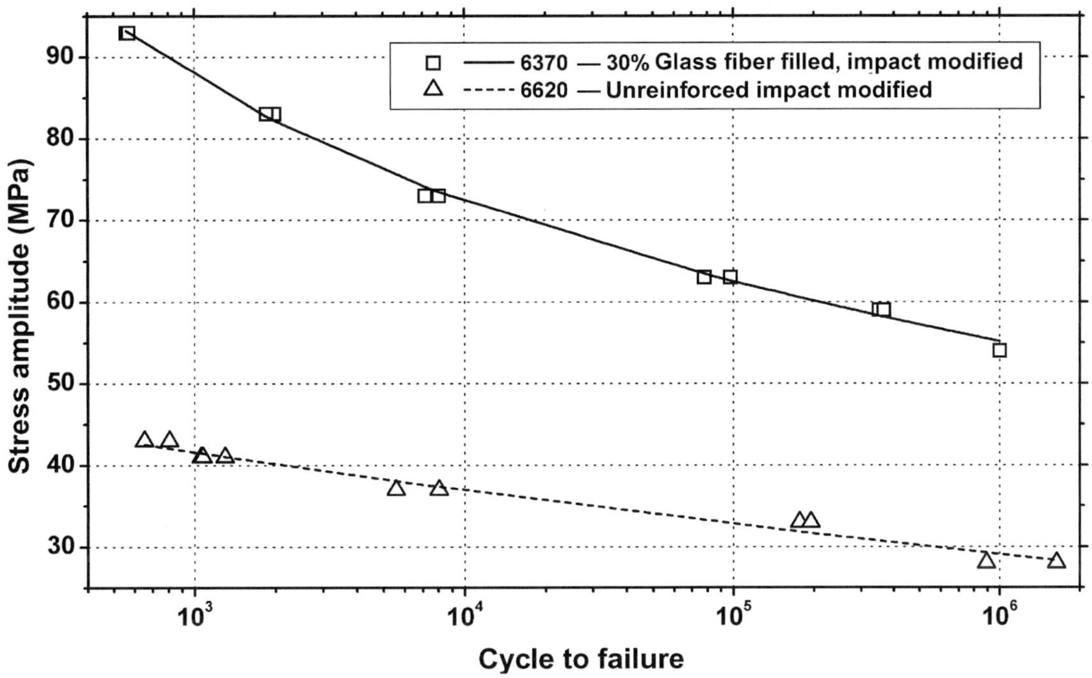

Figure 6.84 Tensile stress amplitude vs. cycles to failure at 23°C of two other SABIC Innovative Plastics Xenoy® 5xxx series PBT/PC Alloys.

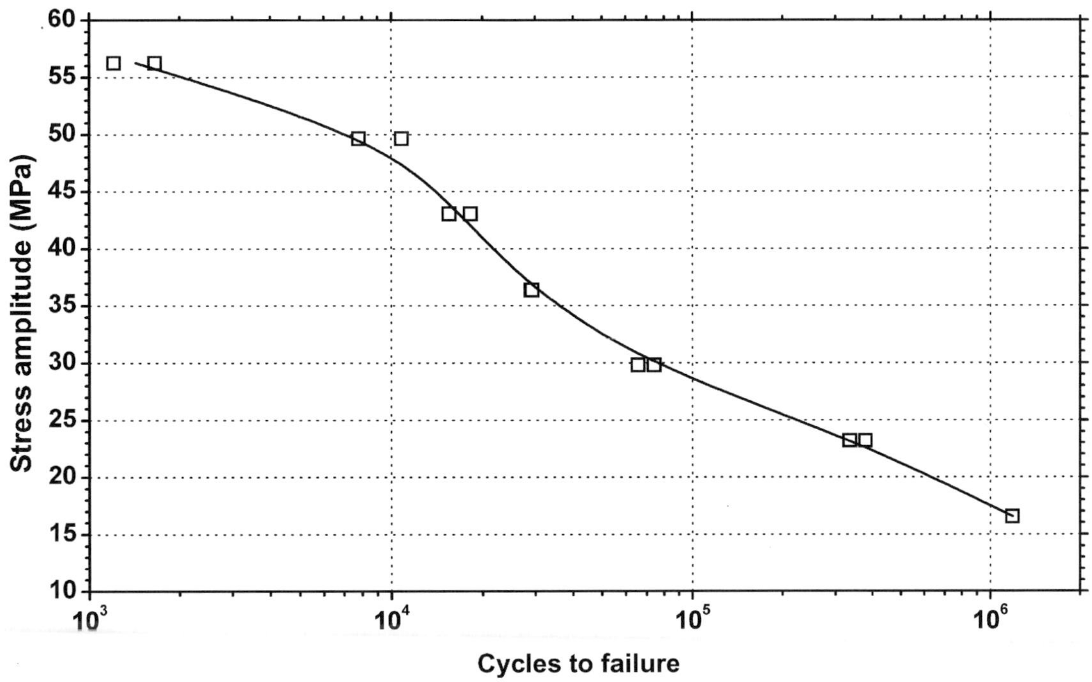

Figure 6.85 Tensile stress amplitude vs. cycles to failure at two temperatures of SABIC Innovative Plastics Xenoy® X5300WX—unreinforced, chemically resistant, UV stabilized PBT/PC Alloy.

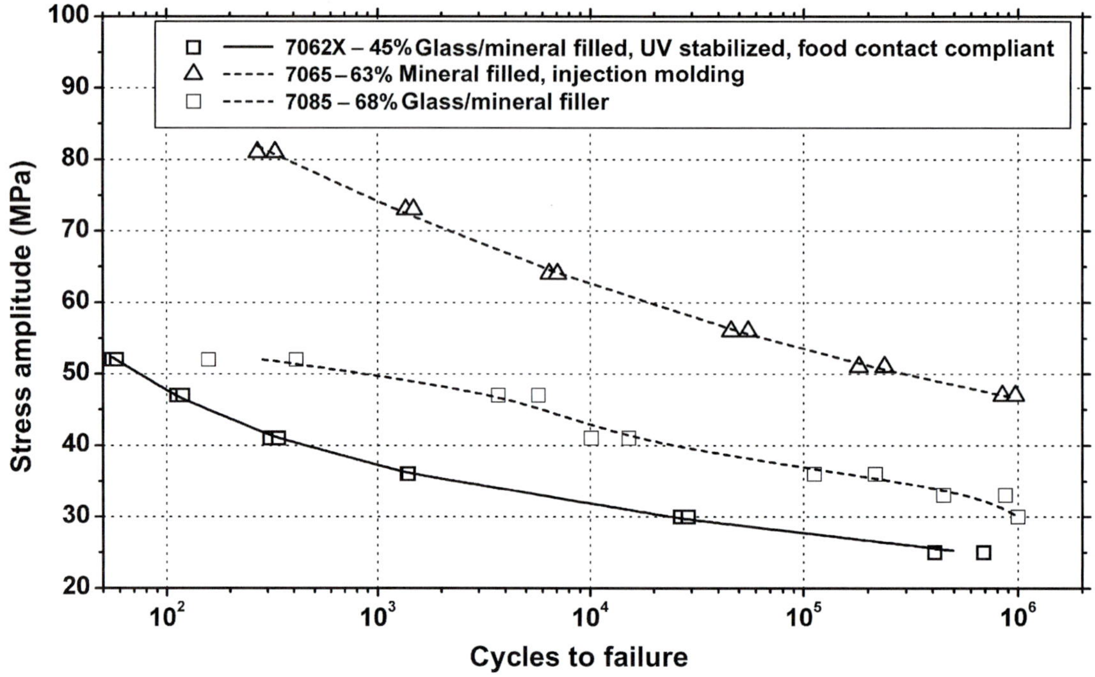

Figure 6.86 Tensile stress amplitude vs. cycles to failure at 23°C of several SABIC Innovative Plastics Enduran® PET/PBT Alloy plastics.

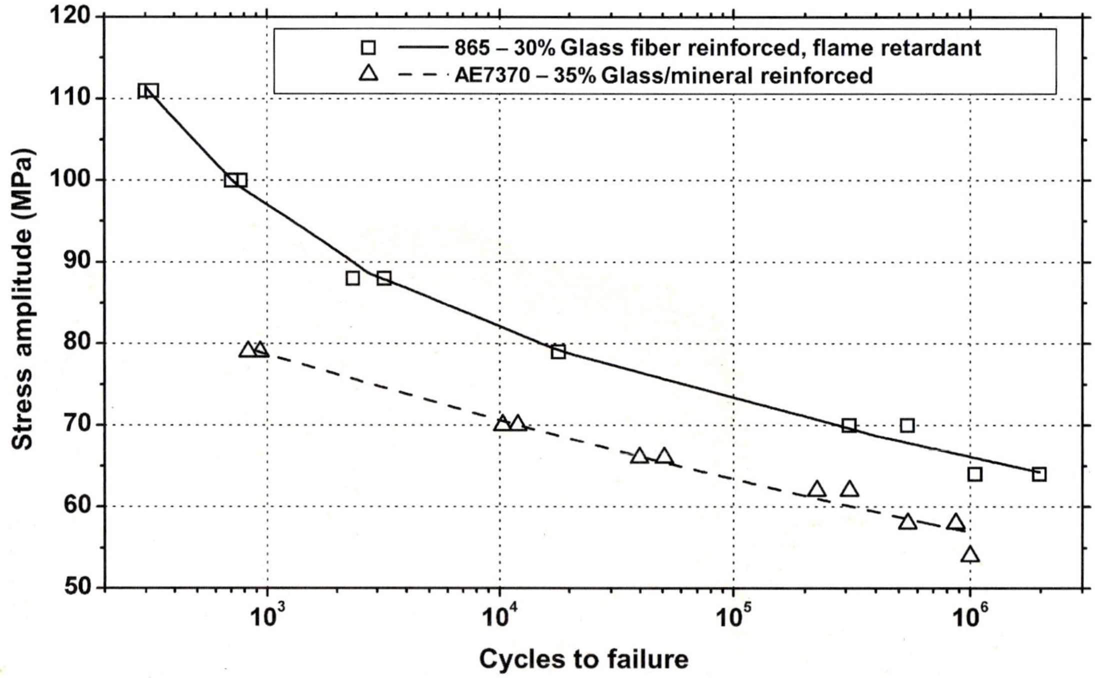

Figure 6.87 Tensile stress amplitude vs. cycles to failure at 23°C of two SABIC Innovative Plastics Valox® PET/PBT Alloy plastics.

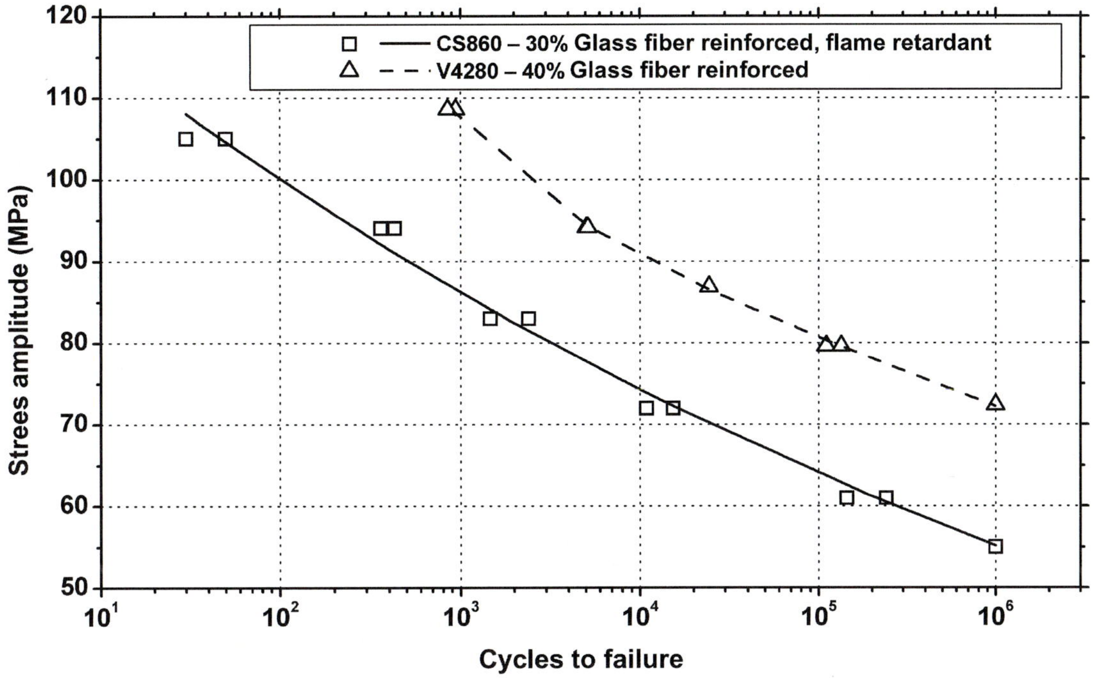

Figure 6.88 Tensile stress amplitude vs. cycles to failure at 23°C of two glass fiber reinforced SABIC Innovative Plastics Valox® PET/PBT Alloy plastics.

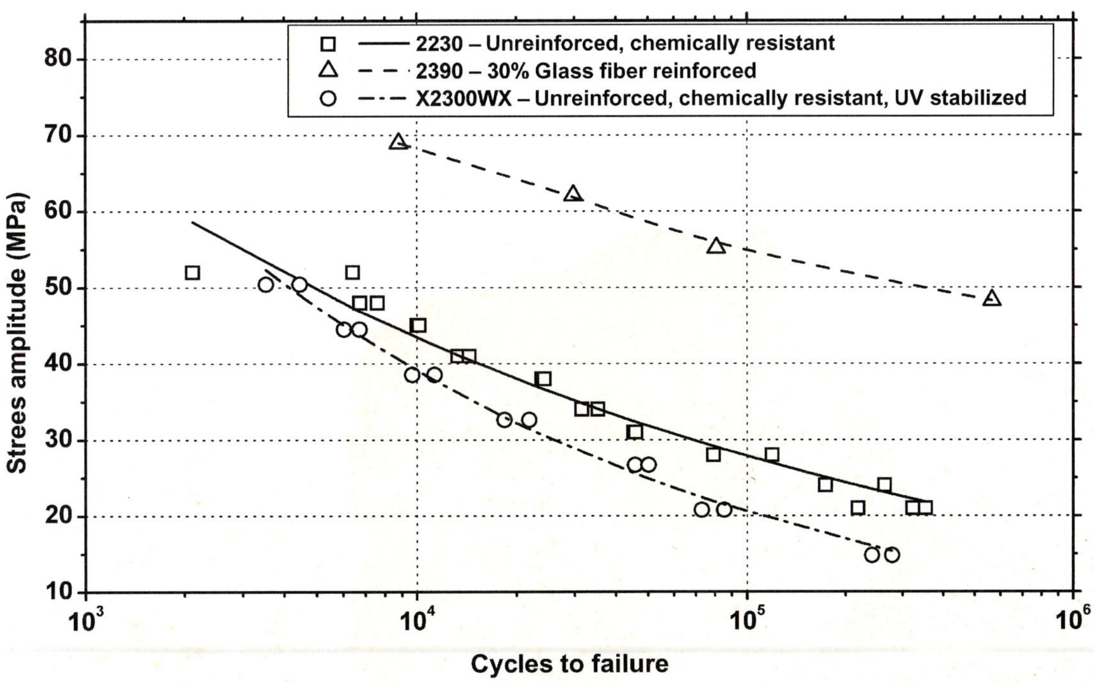

Figure 6.89 Tensile stress amplitude vs. cycles to failure at 23°C of several SABIC Innovative Plastics Xenoy® PET/PC Alloy plastics.

7 Polyimides

7.1 Background

This chapter covers a series of plastics of which the imide group is an important part of the molecule. The imide group is formed by a condensation reaction of an aromatic anhydride group with an aromatic amine as shown in Figure 7.1.

This group is very thermally stable. Aliphatic imides are possible, but the thermal stability is reduced, and thermal stability is one of the main reasons to use an imide type polymer.

7.1.1 Polyetherimide

Polyetherimide (PEI) is an amorphous engineering thermoplastic. Thermoplastic PEIs provide the strength, heat resistance, and flame retardancy of traditional polyimides (PIs) with the ease of simple melt processing seen in standard injection-molding resins like polycarbonate and ABS.

The key performance features of PEI resins include:

- excellent dimensional stability at high temperatures under load
- smooth as-molded surfaces
- transparency, though slightly yellow
- good optical properties
- very high strength and modulus
- high continuous-use temperature
- inherent ignition resistance without the use of additives
- good electrical properties with low ion content

There are several different polymers that are offered in various PEI plastics. The structures of these are shown in Figures 7.2–7.6 with references to one of the product lines that utilize that molecule.

The acid dianhydride used to make most of the PEIs is 4,4′-bisphenol A dianhydride (BPADA), the structure of which is shown in Figure 7.7.

Some of the other monomers used in these PEIs are shown in Figure 7.8.

Many products are called thermoplastic polyimide (TPI) by their manufacturer. These can usually be classified as PEIs.

7.1.2 Polyamide-Imide

Polyamide-imides (PAIs) are thermoplastic amorphous polymers that have useful properties:

- Exceptional chemical resistance
- Outstanding mechanical strength
- Excellent thermal stability
- Performs from cryogenic up to 260°C
- Excellent electrical properties

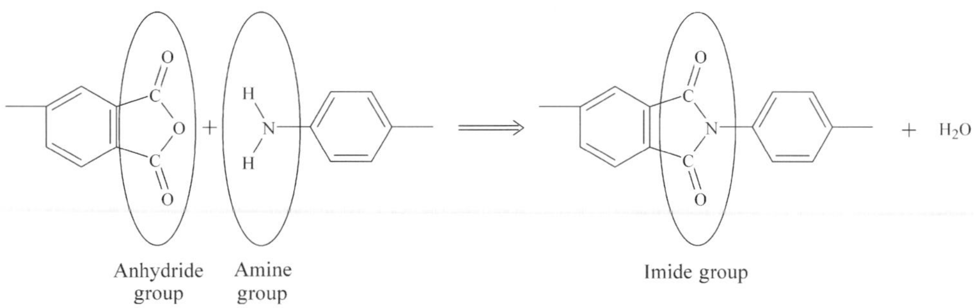

Figure 7.1 Reaction of amine with anhydride to form an imide.

Copyright © 2010 Laurence W. McKeen. All rights reserved.

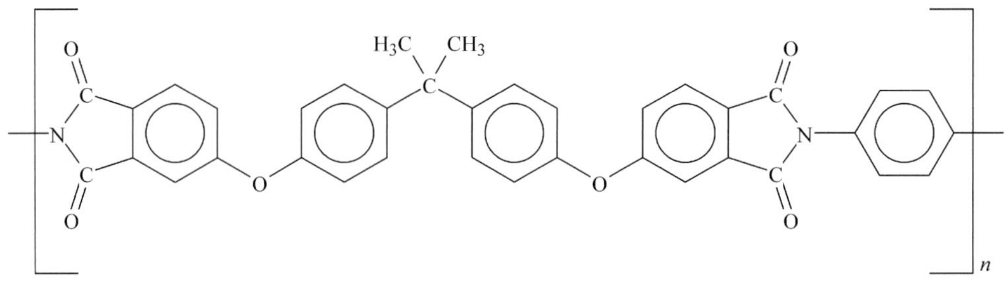

Figure 7.2 Chemical structure of BPADA–PPD PEI (Ultem® 5000 Series).

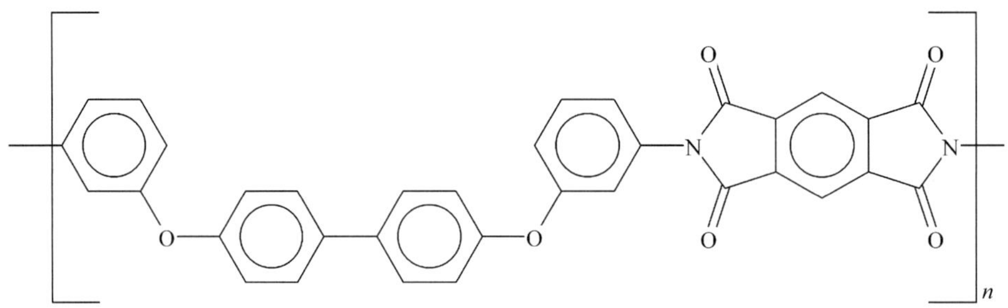

Figure 7.3 Chemical structure of biphenol diamine PMDA PEI (Aurum®, Vespel® TP-8000 Series).

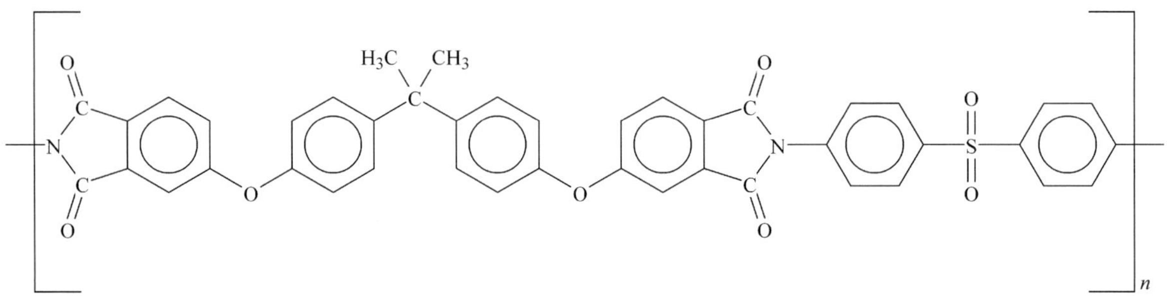

Figure 7.4 Chemical structure of BPADA–DDS PEI sulfone (Ultem® XH6050).

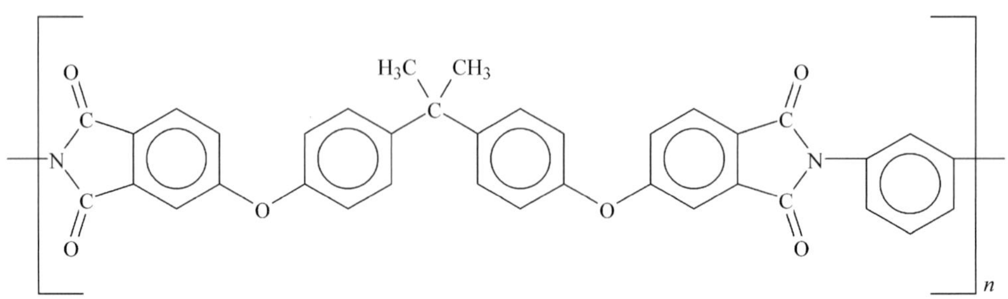

Figure 7.5 Chemical structure of BPADA–MPD PEI (Ultem® 1000 Series).

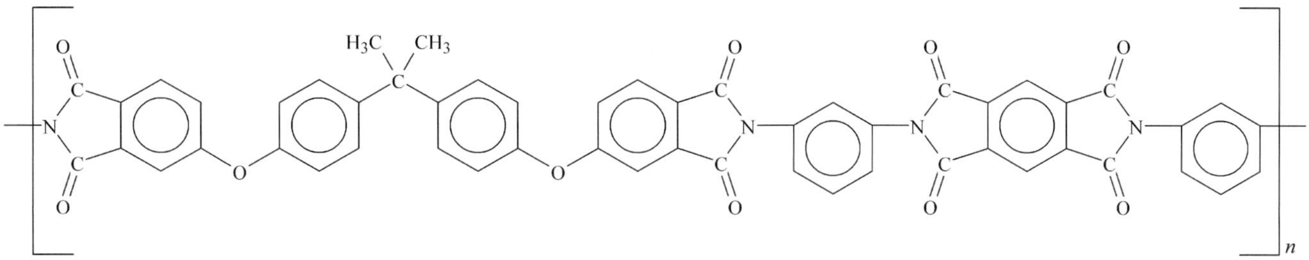

Figure 7.6 Chemical structure of BPADA–PMDA–MPD copolyetherimide (Ultem® 6000 Series).

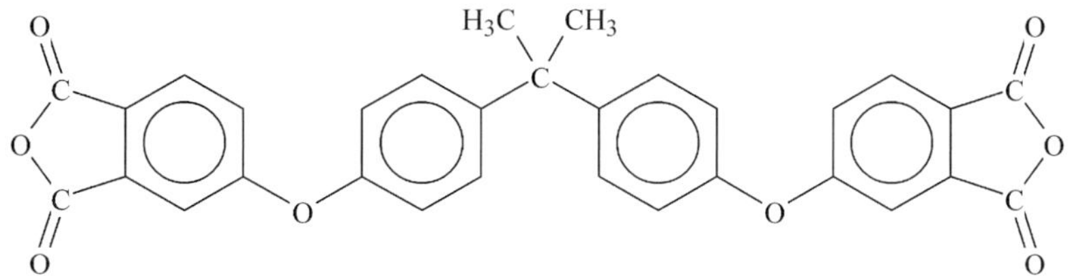

Figure 7.7 Chemical structure of BPADA monomer.

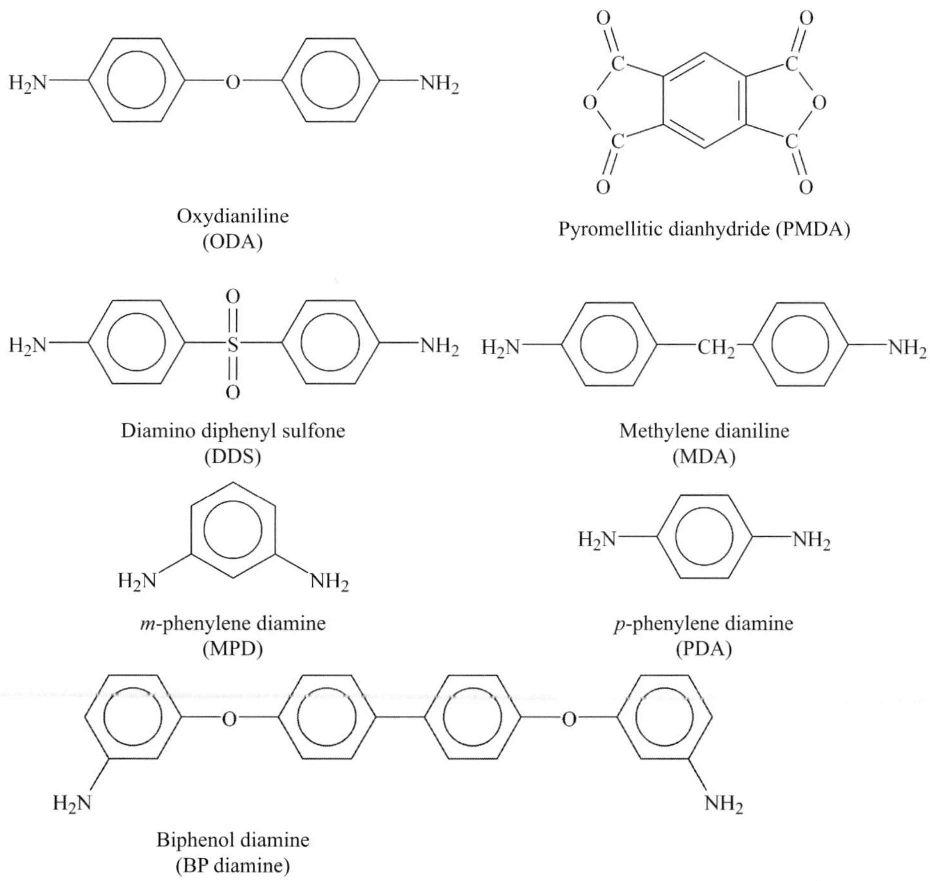

Figure 7.8 Chemical structures of other monomers used to make PIs.

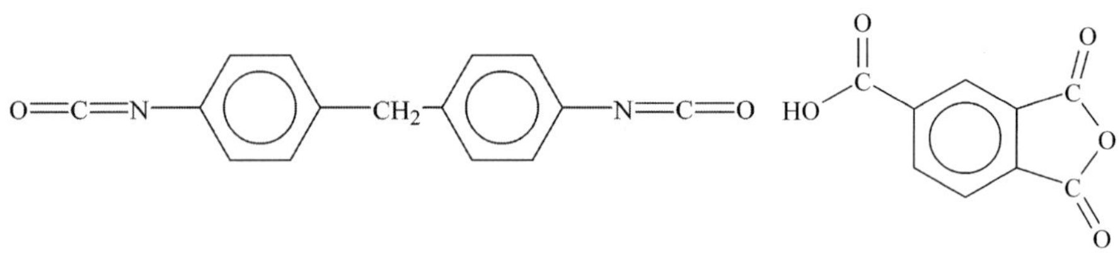

4,4′-Diphenyl methane diisocyanate (MDI) Trimellitic anhydride (TMA)

Figure 7.9 Chemical structures of monomer used to make PAIs.

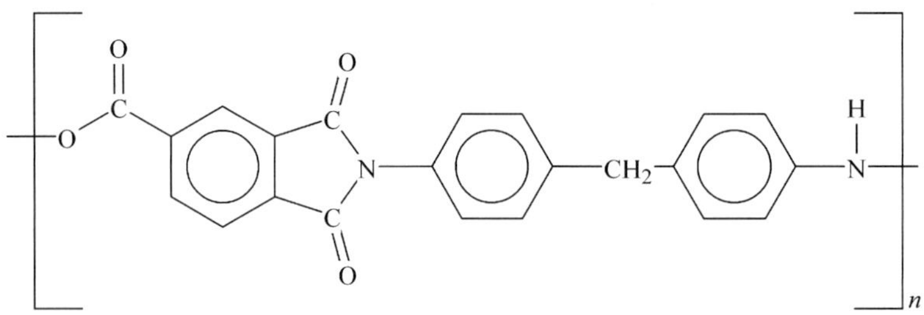

Figure 7.10 Chemical structure of a typical PAI.

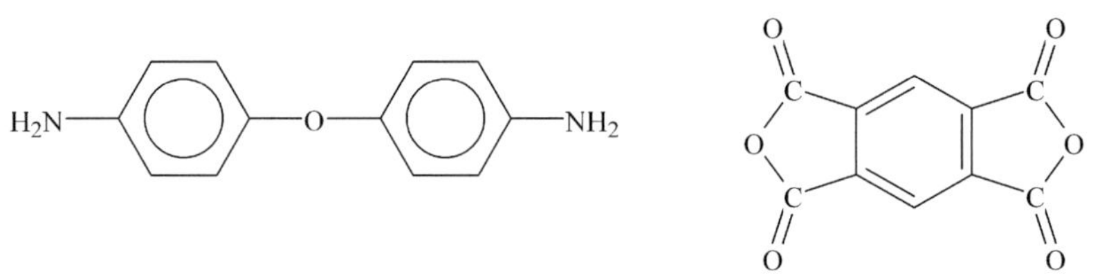

4,4′-Diaminodiphenyl ether oxydianiline (ODA) Pyromellitic dianhydride (PMDA)

Figure 7.11 Chemical structures of monomer used to make PIs.

The monomers used to make PAI resin are shown in Figure 7.9.

When these monomers are reacted carbon dioxide, rather than water, is generated. The closer the monomer ratio is to 1:1 the higher the molecular weight of the polymer shown in Figure 7.10.

7.1.3 Polyimide

PIs are high-temperature engineering polymers originally developed by the DuPont Company. PIs exhibit an exceptional combination of thermal stability (>500°C), mechanical toughness, and chemical resistance. They have excellent dielectric properties and inherently low coefficient of thermal expansion. They are formed from diamines and dianhydrides such as those shown in Figure 7.11.

Many other diamines and several other dianhydrides may be chosen to tailor the final properties of a polymer whose structure is like that shown in Figure 7.12.

7.1.4 Imide Polymer Blends

PI-based resins, especially PEI and PAI polymers, may also be combined with other polymers. The PEI resins have produced a surprising number of miscible

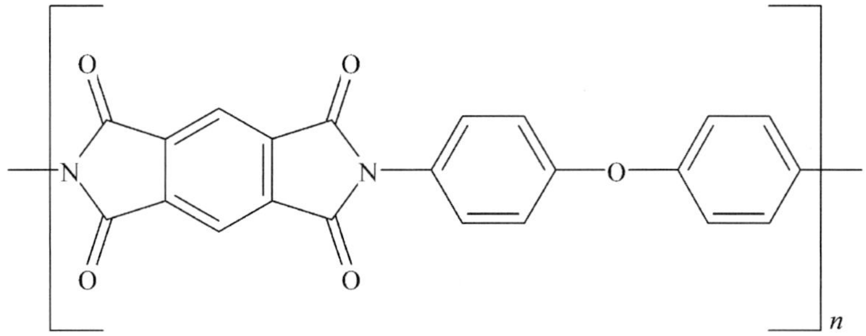

Figure 7.12 Chemical structure of a typical PI.

(one-phase) and compatible blends. Compatible blends are phase-separated mixtures having sufficient attraction between phases to provide some level of molecular adhesion, resulting in stable morphology and giving rise to good mechanical properties.

PEI forms miscible blends with polyesters such as PBT and PET. These blends have a single glass transition temperature between that of the PEI and polyester. However, few of these are commercial products yet.

Blends of BPADA-based PIs are also miscible with polyaryl ether ketones such as polyetheretherketone (PEEK). As injection molded, many PEEK–PEI blends are transparent.

7.2 Polyetherimides

7.2.1 Fatigue Data

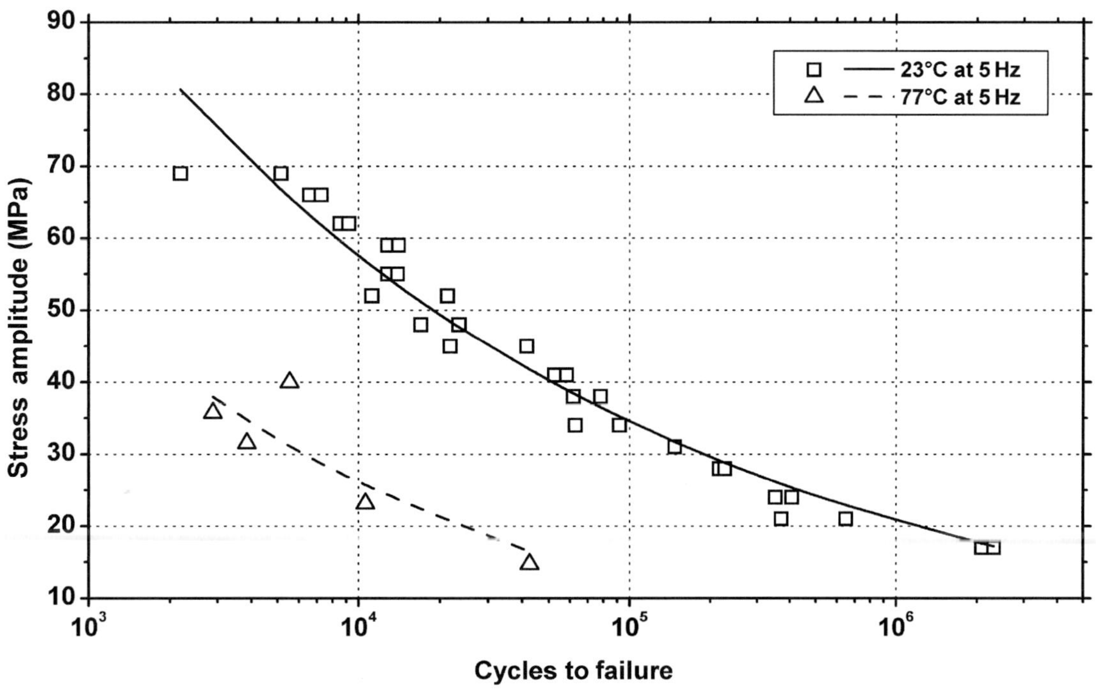

Figure 7.13 Tensile stress amplitude vs. cycles to failure of SABIC Innovative Plastics Ultem® 1000—transparent, standard flow, unreinforced general-purpose PEI.

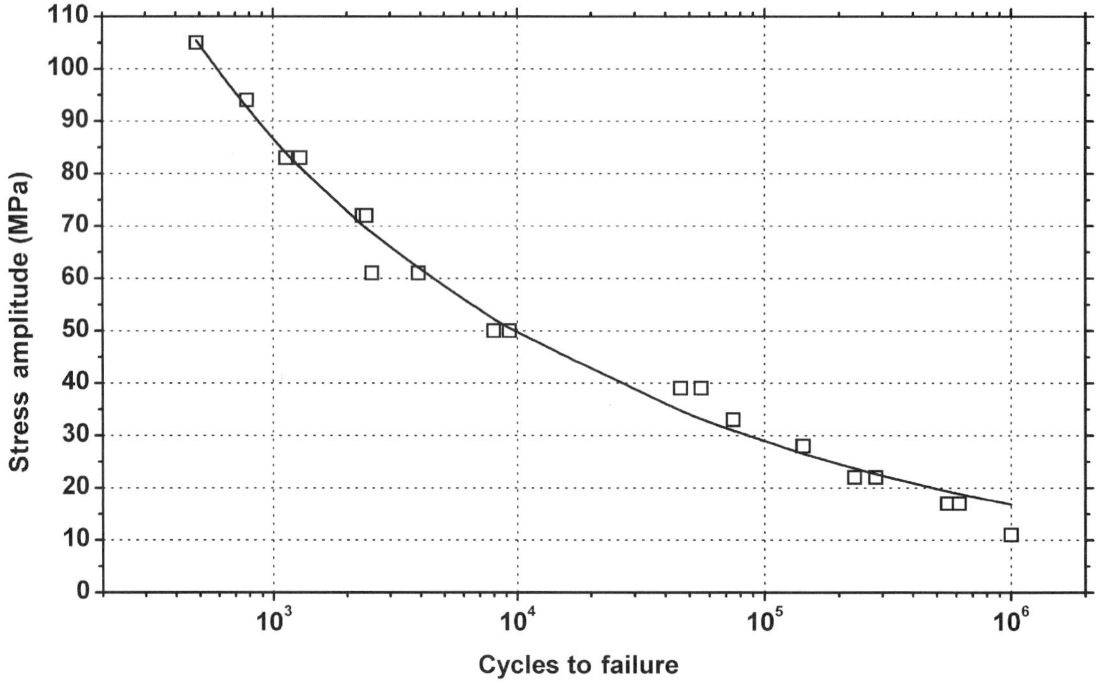

Figure 7.14 Tensile stress amplitude vs. cycles to failure at 23°C of SABIC Innovative Plastics Ultem® 1010—transparent, enhanced flow, unreinforced general-purpose PEI.

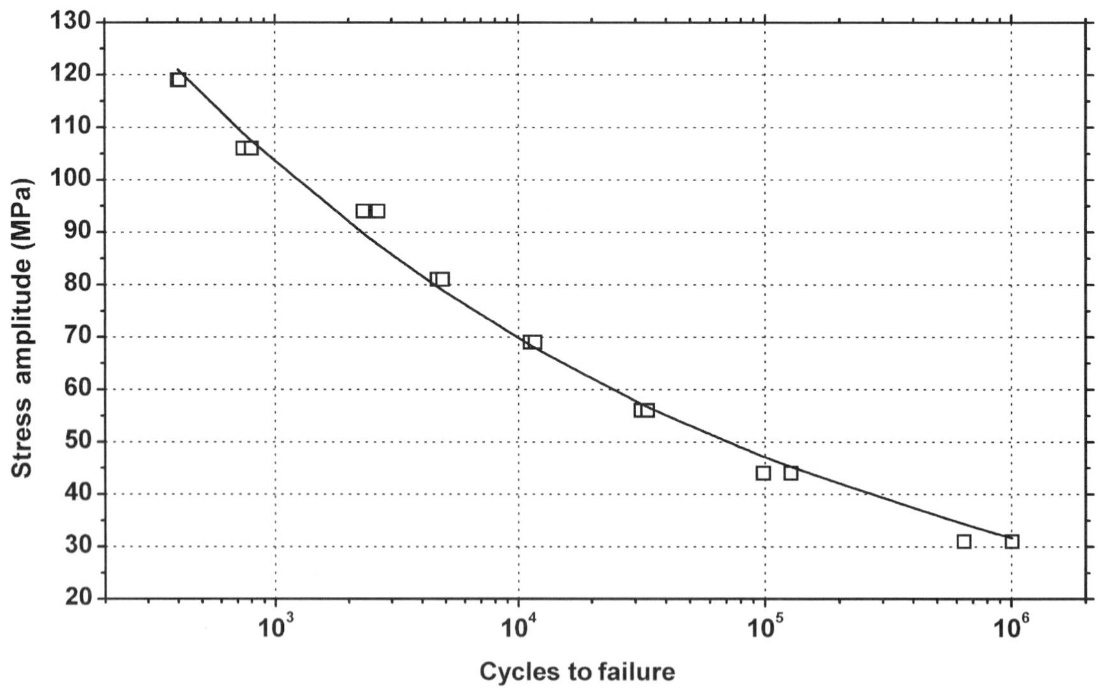

Figure 7.15 Tensile stress amplitude vs. cycles to failure at 23°C of SABIC Innovative Plastics Ultem® 2100—10% glass fiber reinforced, standard flow PEI.

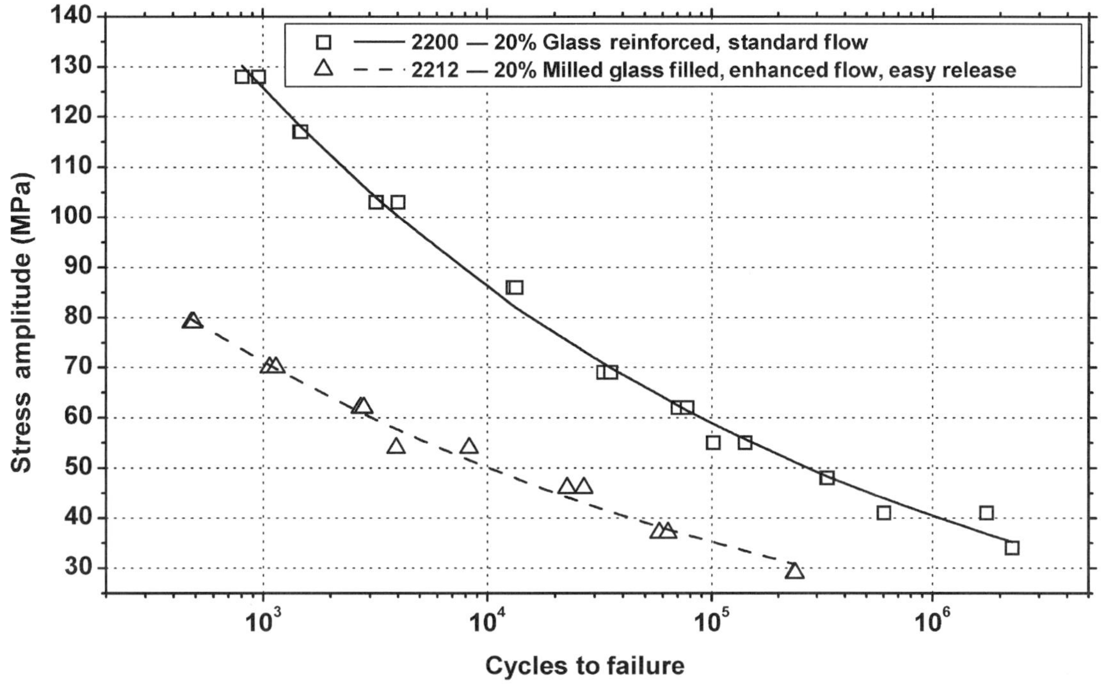

Figure 7.16 Tensile stress amplitude vs. cycles to failure at 23°C of two SABIC Innovative Plastics Ultem®—20% glass reinforced PEI plastics.

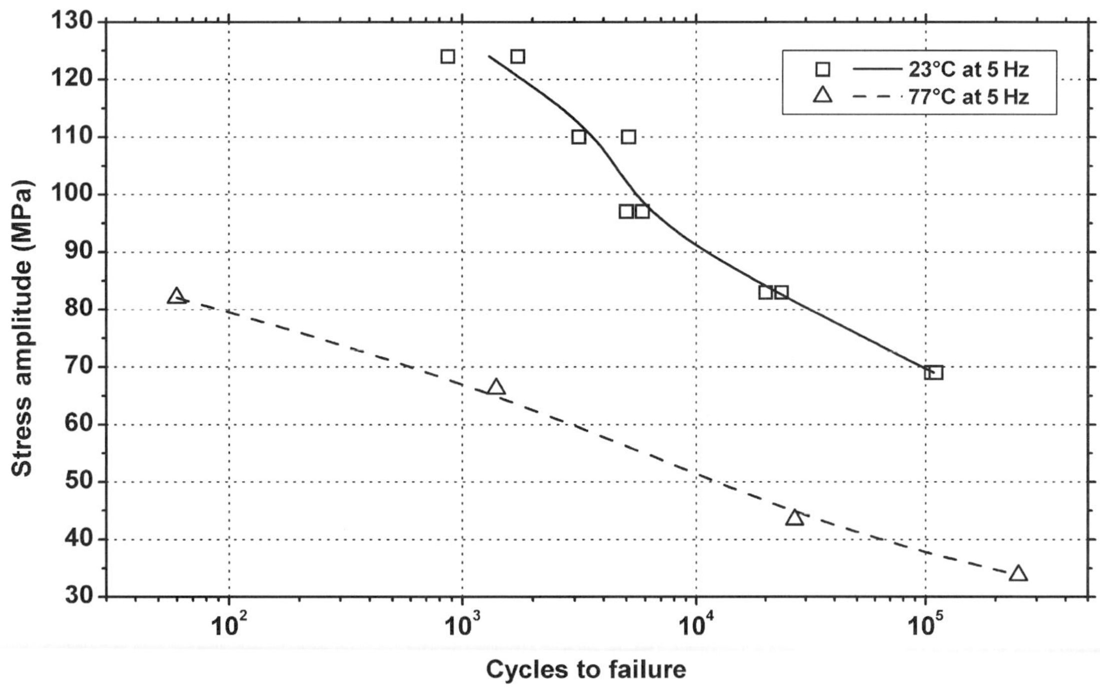

Figure 7.17 Tensile stress amplitude vs. cycles to failure of SABIC Innovative Plastics Ultem® 2300—30% glass fiber reinforced, standard flow PEI.

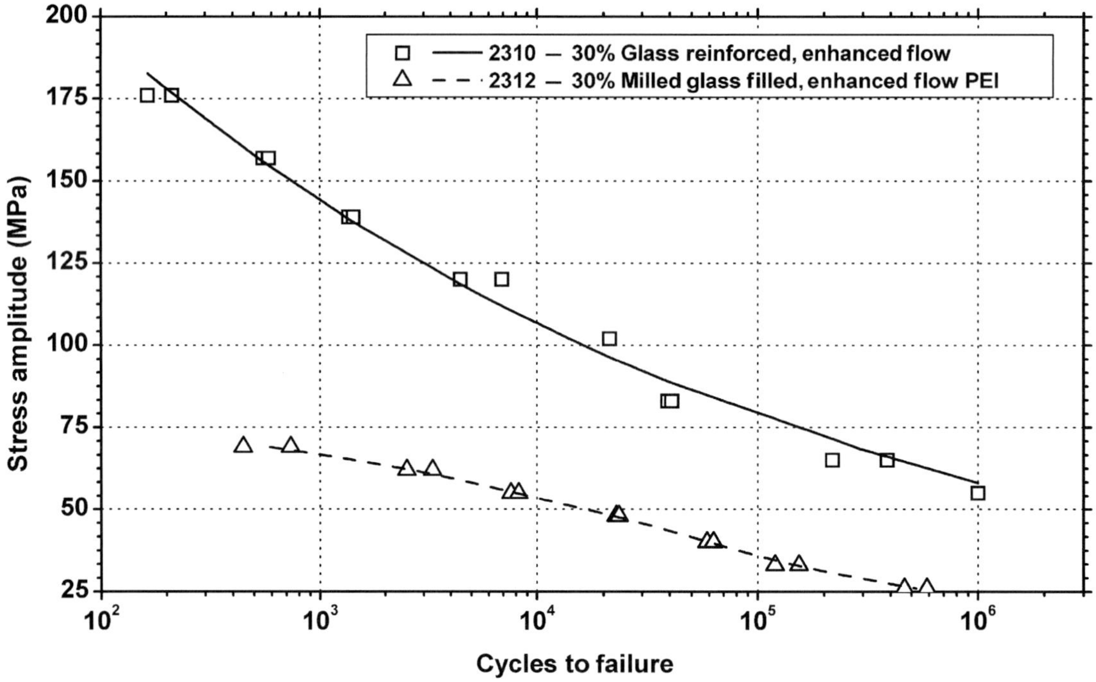

Figure 7.18 Tensile stress amplitude vs. cycles to failure at 23°C of two SABIC Innovative Plastics Ultem®—30% glass reinforced PEI plastics.

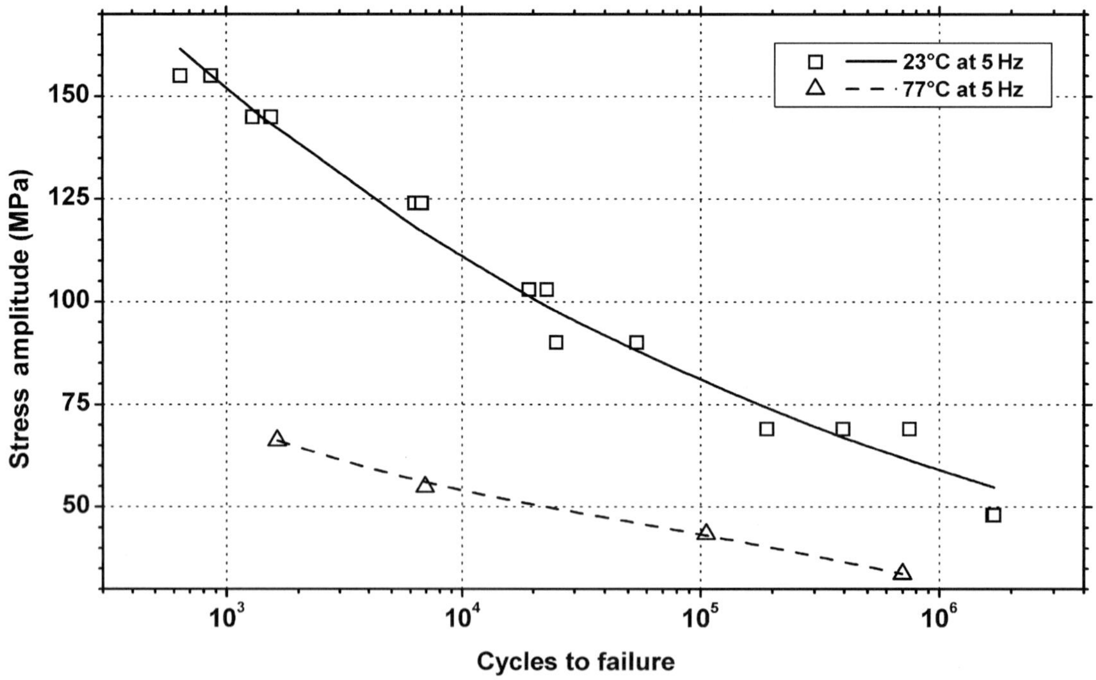

Figure 7.19 Tensile stress amplitude vs. cycles to failure of SABIC Innovative Plastics Ultem® 2400—40% glass fiber reinforced, standard flow PEI.

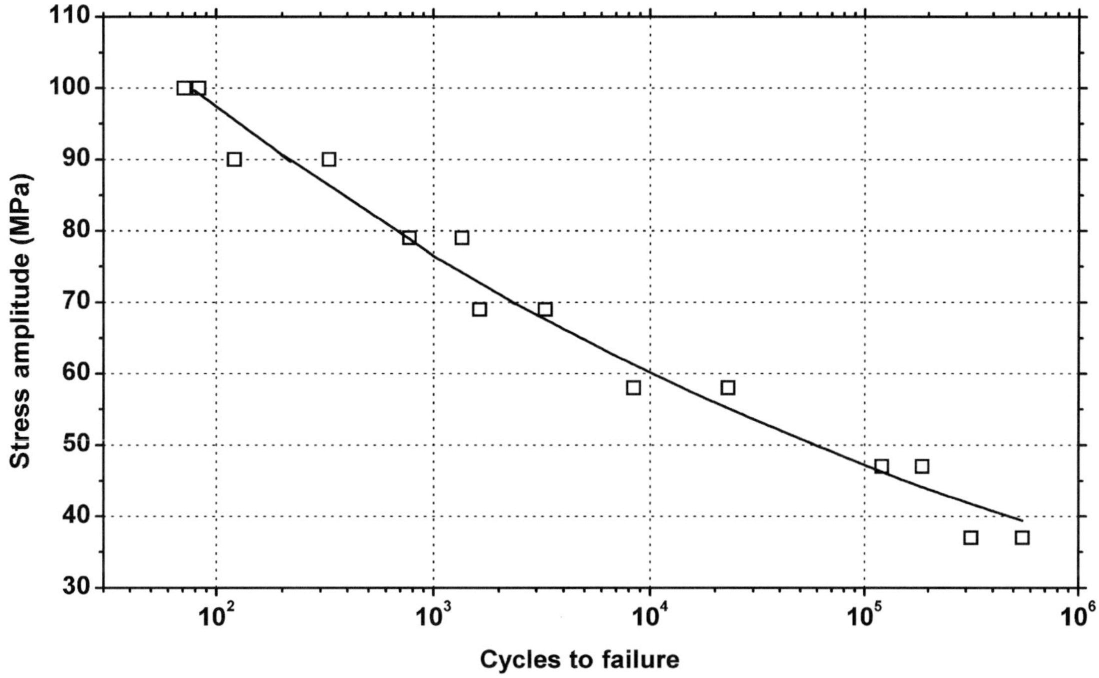

Figure 7.20 Tensile stress amplitude vs. cycles to failure of SABIC Innovative Plastics Ultem® 3452—45% glass/mineral reinforced, enhanced flow PEI.

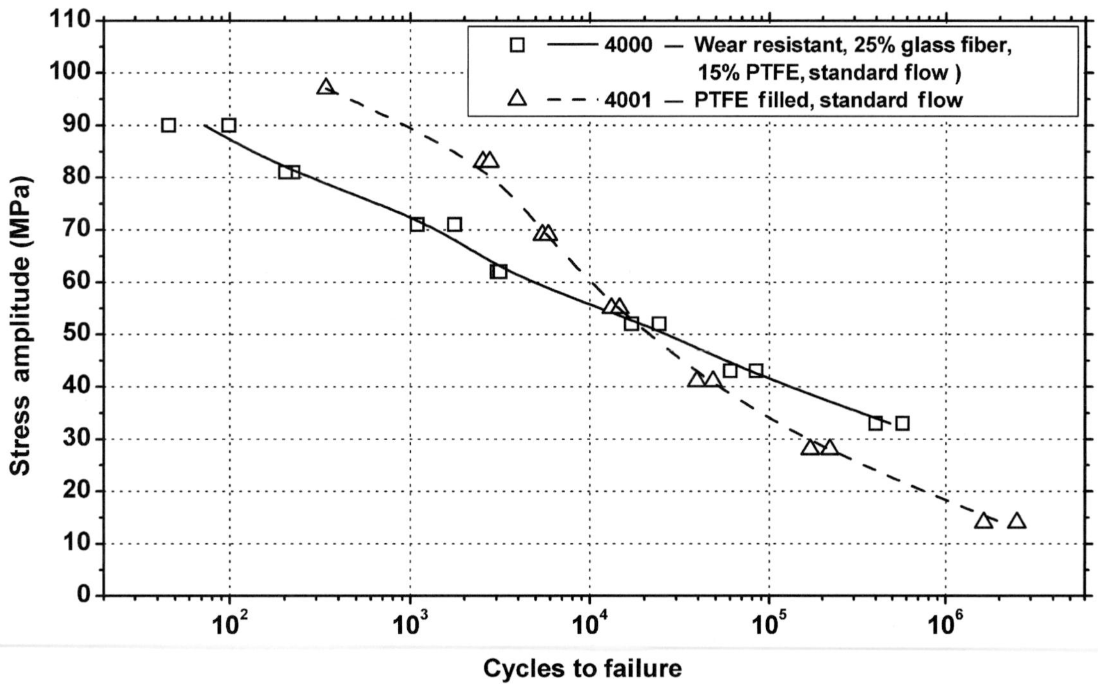

Figure 7.21 Tensile stress amplitude vs. cycles to failure of two SABIC Innovative Plastics Ultem® 4000 series PEI plastics.

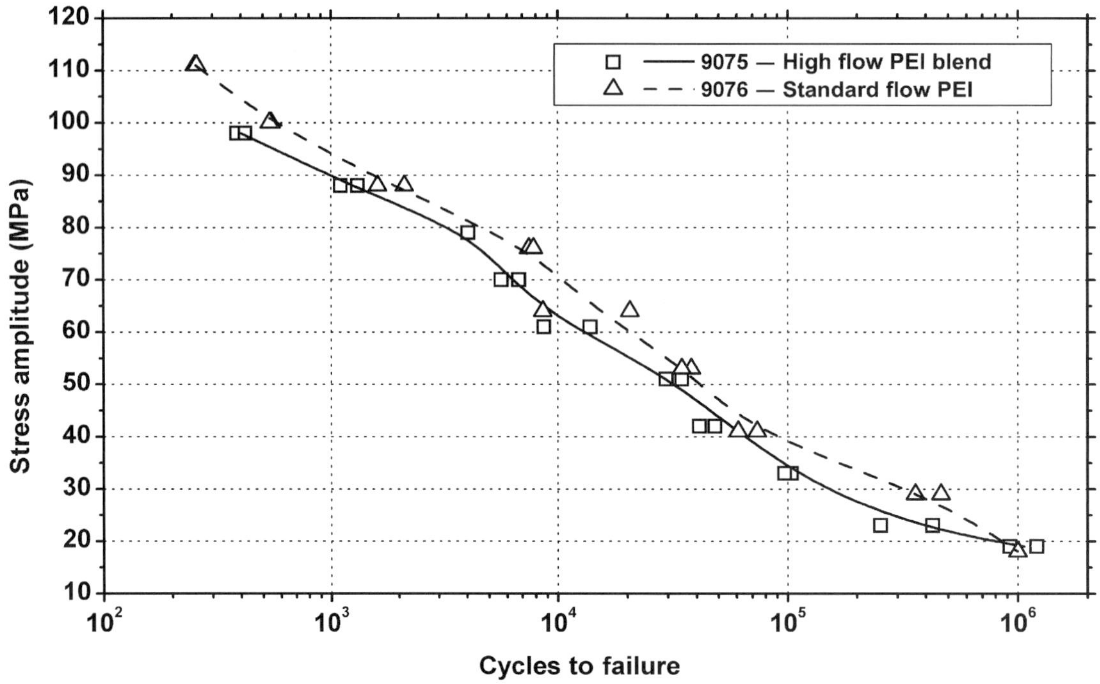

Figure 7.22 Tensile stress amplitude vs. cycles to failure of two SABIC Innovative Plastics Ultem® 9000 series PEI plastics.

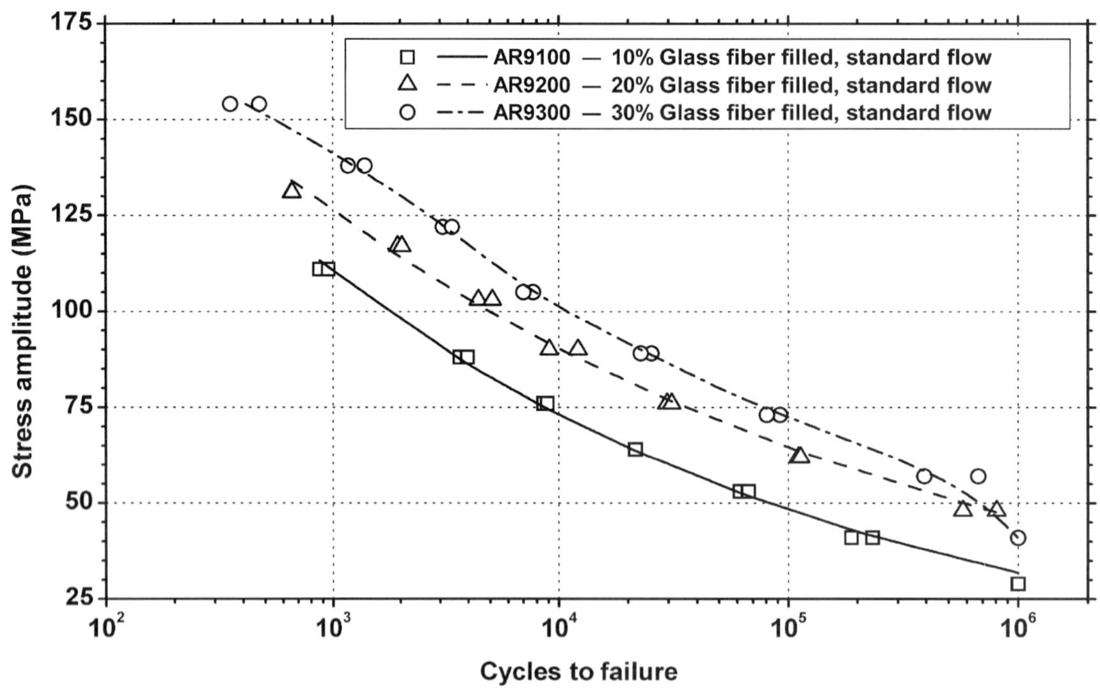

Figure 7.23 Tensile stress amplitude vs. cycles to failure of two SABIC Innovative Plastics Ultem® AR9000 series PEI plastics.

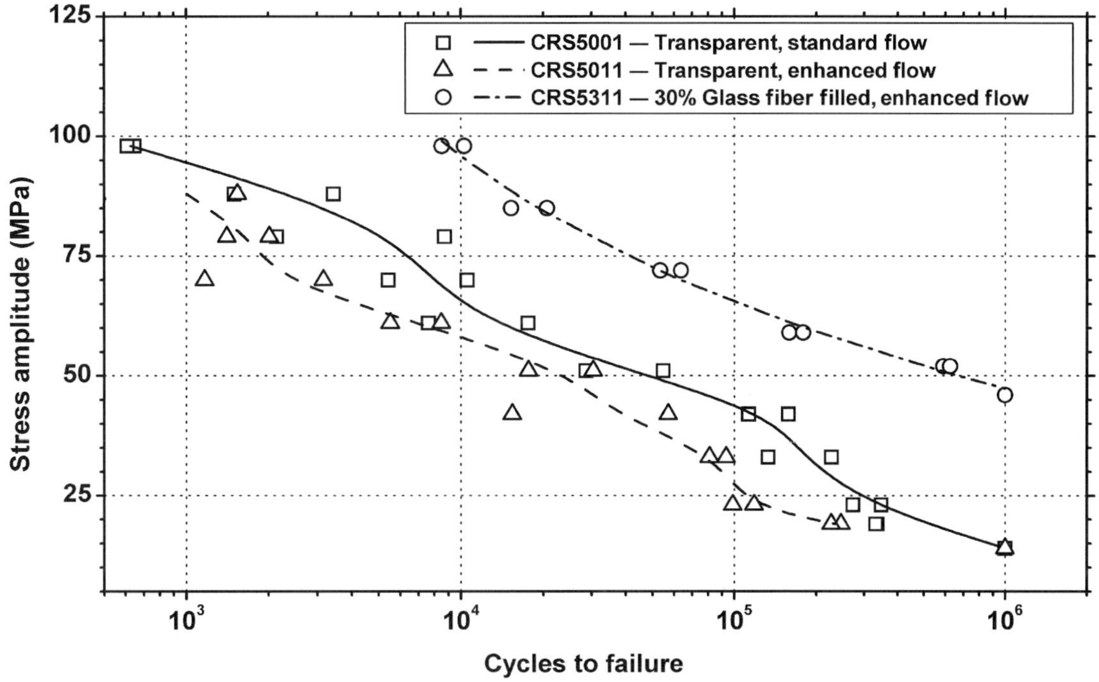

Figure 7.24 Tensile stress amplitude vs. cycles to failure of three SABIC Innovative Plastics Ultem® CRS5000 series PEI plastics.

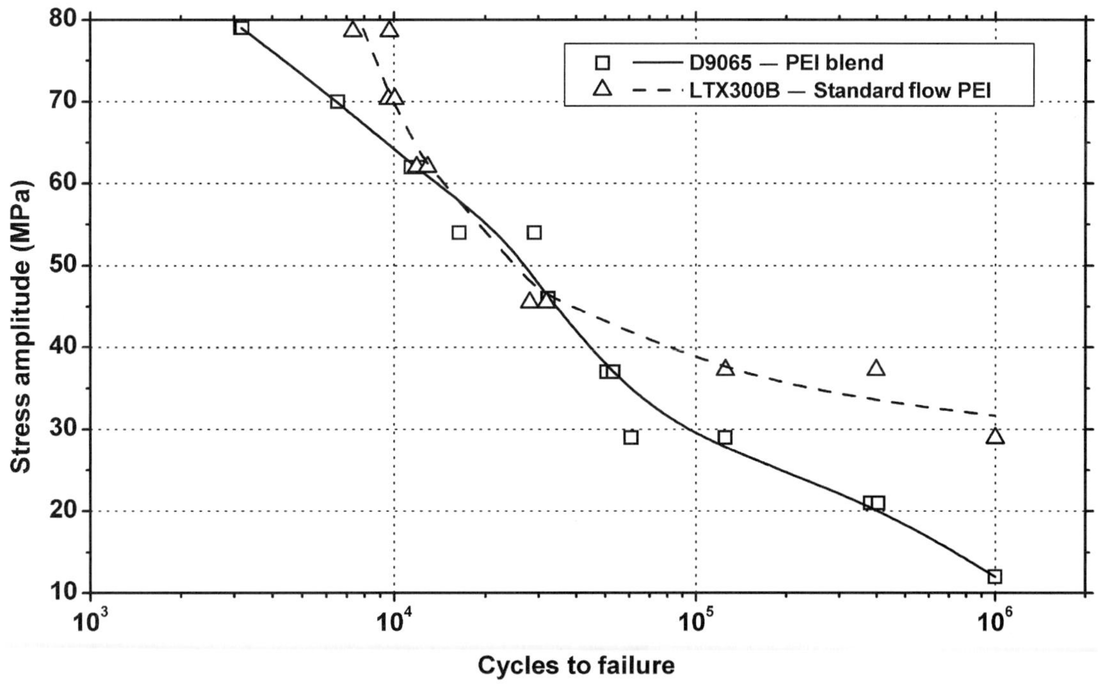

Figure 7.25 Tensile stress amplitude vs. cycles to failure of two SABIC Innovative Plastics Ultem® series PEI plastics.

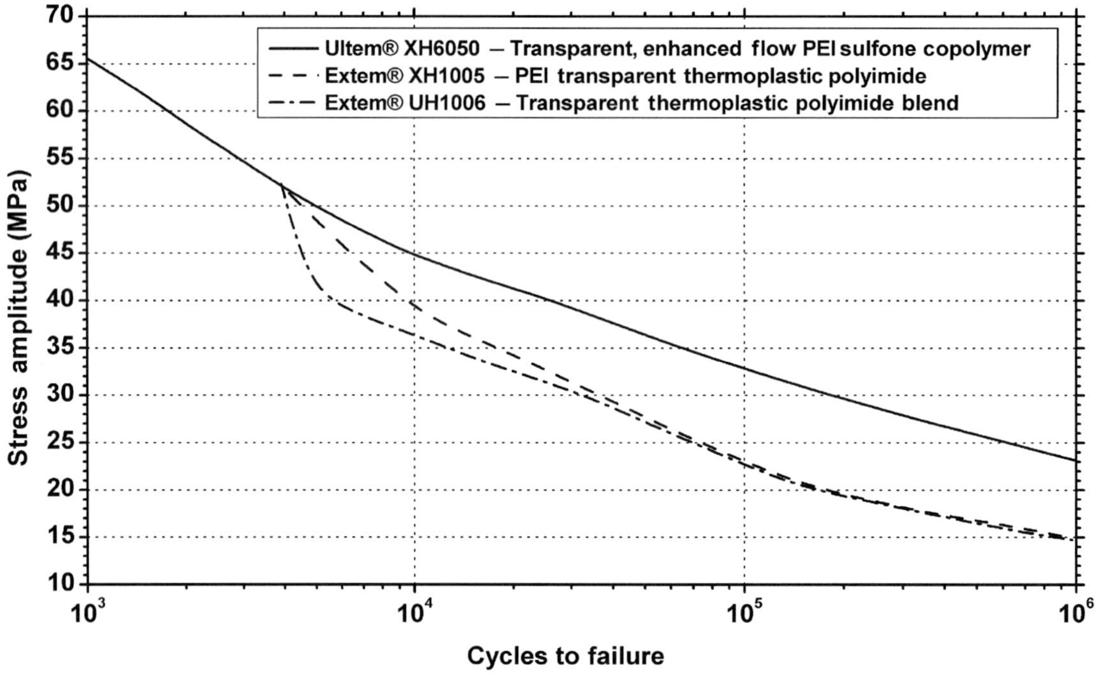

Figure 7.26 Tensile stress amplitude vs. cycles to failure of three SABIC Innovative Plastics PEI plastics.

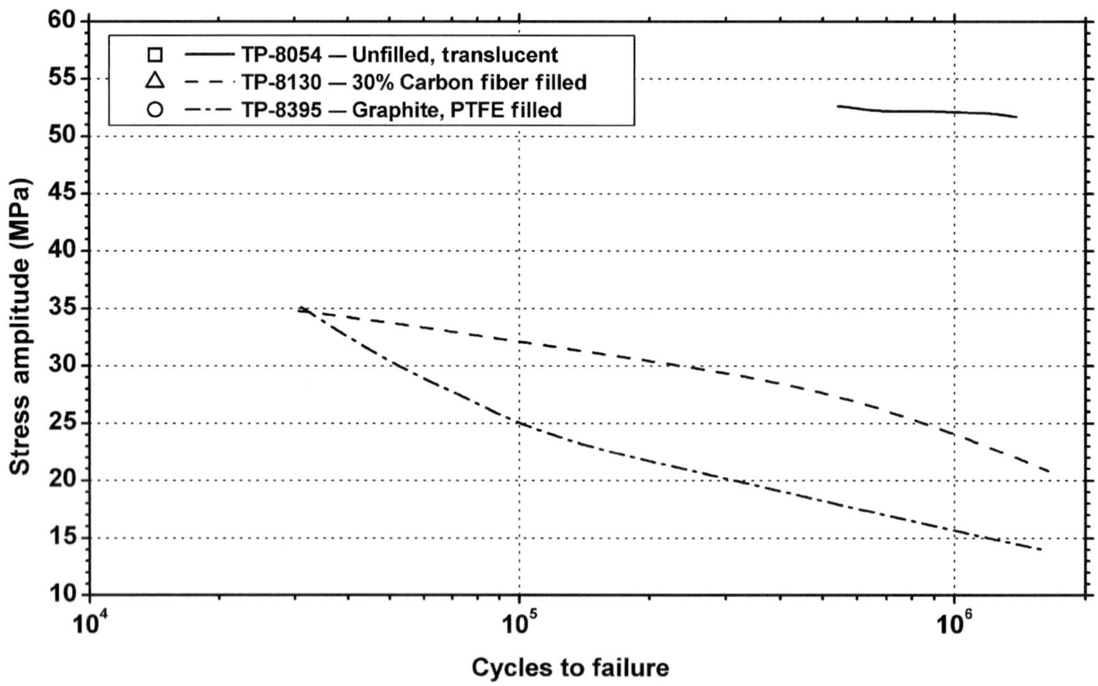

Figure 7.27 Flexural stress amplitude vs. cycles to failure of three DuPont Engineering Polymers Vespel® TP-8000 Series—semicrystalline PEI plastics.

7.2.2 Tribology Data

Table 7.1 Tribological Properties of RTP Company RTP 4205 TFE 15 (TPI with 30% Glass Fiber Reinforcement and 15% PTFE) vs. 1018 C Steel (Data Obtained per ASTM 3702)

PV (KPa m/s)	Load (N)	Speed (m/s)	Wear Factor × 10^{-8} (mm^3/N m)	Dynamic Coefficient of Friction
70	1.80	0.25	160	0.54
70	0.90	0.50	310	0.62
70	0.45	1.00	224	0.68
175	4.50	0.25	64	0.64
175	2.25	0.50	58	0.66
175	1.15	1.00	194	0.62
350	9.00	0.25	212	0.52
350	4.50	0.50	270	0.57
350	2.25	1.00	402	0.41

Table 7.2 Tribological Properties of RTP Company RTP 4285 TFE 15 (TPI with 30% Carbon Fiber Reinforcement and 15% PTFE) vs. 1018 C Steel (Data Obtained per ASTM 3702)

PV (KPa m/s)	Load (N)	Speed (m/s)	Wear Factor × 10^{-8} (mm^3/N m)	Dynamic Coefficient of Friction
70	1.80	0.25	66	0.31
70	0.90	0.50	66	0.28
70	0.45	1.00	92	0.20
175	4.50	0.25	130	0.31
175	2.25	0.50	78	0.31
175	1.15	1.00	92	0.30
350	9.00	0.25	68	0.55
350	4.50	0.50	164	0.77
350	2.25	1.00	96	0.92

Table 7.3 Tribological Properties of RTP Company RTP 4299 × 71927 (TPI with Proprietary Composition) vs. 1018 C Steel (Data Obtained per ASTM 3702)

PV (KPa m/s)	Load (N)	Speed (m/s)	Wear Factor × 10^{-8} (mm^3/N m)	Dynamic Coefficient of Friction
70	1.80	0.25	50	0.28
70	0.90	0.50	48	0.28
70	0.45	1.00	62	0.28
175	4.50	0.25	290	0.16
175	2.25	0.50	444	0.17
175	1.15	1.00	348	0.17
350	9.00	0.25	130	0.19
350	4.50	0.50	164	0.19
350	2.25	1.00	278	0.20

Table 7.4 Tribological Properties of RTP Company RTP 4299 × 64425 (TPI with Proprietary Composition) vs. 1018 C Steel (Data Obtained per ASTM 3702)

PV (KPa m/s)	Load (N)	Speed (m/s)	Wear Factor × 10^{-8} (mm^3/N m)	Dynamic Coefficient of Friction
70	1.80	0.25	34	0.24
70	0.90	0.50	20	0.29
70	0.45	1.00	50	0.27
175	4.50	0.25	68	0.53
175	2.25	0.50	22	0.59
175	1.15	1.00	34	0.58
350	9.00	0.25	66	0.48
350	4.50	0.50	84	0.57
350	2.25	1.00	94	0.58

Table 7.5 Tribological Properties of RTP Company RTP 2100 AR 15 TFE 15 (15% Aramid Fiber Reinforced and 15% PTFE) vs. 1018 C Steel (Data Obtained per ASTM 3702)

PV (KPa m/s)	Load (N)	Speed (m/s)	Wear Factor × 10^{-8} (mm^3/N m)	Dynamic Coefficient of Friction
70	1.80	0.25	18	0.19
70	0.45	1.00	10	0.21
175	2.25	0.50	64	0.23
350	9.00	0.25	50	0.25
350	2.25	1.00	45	0.33

Table 7.6 Tribological Properties of Several SABIC Innovative Plastics Ultem® Series PEI Plastics

Material and Test	Value	Units	Test Method
Ultem® 1000			
Taber Abrasion, CS-17, 1 kg	10	mg/1000 cycle	ASTM D 1044
Ultem® 1010			
Taber Abrasion, CS-17, 1 kg	10	mg/1000 cycle	ASTM D 1044
Ultem® 4000			
Taber Abrasion, CS-17, 1 kg	33	mg/1000 cycle	ASTM D 1044
PV Limit, 0.51 m/s	2.1	MPa m/s	SABIC Method
K-factor × E–10, *PV* = 2000 psi fpm vs. steel	62	–	SABIC Method
K-factor × E–10, *PV* = 2000 psi fpm vs. self	1900	–	SABIC Method
Coefficient of friction on steel, static	0.25	–	ASTM D 1894
Coefficient of friction on steel, kinetic	0.24	–	ASTM D 1894
Ultem® 4001			
Taber Abrasion, CS-17, 1 kg	2	mg/1000 cycle	ASTM D 1044
PV Limit, 0.51 m/s	1.9	MPa m/s	SABIC Method
K-factor × E–10, *PV* = 2000 psi fpm vs. steel	72	–	SABIC Method
K-factor × E–10, *PV* = 2000 psi fpm vs. self	27	–	SABIC Method
Coefficient of friction on steel, kinetic	0.25	–	ASTM D 1894
Ultem® CRS5001			
Taber Abrasion, CS-17, 1 kg	10	mg/1000 cycle	ASTM D 1044

Table 7.7 Tribological Properties of Several DuPont Engineering Plastics Vespel® TP Series TPI Plastics

Grade	PV (MPa m/s)	Coefficient of Friction	Wear Factor, *K* (10^{-10} cm³/kg fm)	Resin Wear (mg)	Metal Wear (mg)
TP-8130—30% carbon fiber filled	2.5	0.05	66	10	<1
TP-8130—30% carbon fiber filled	3.3	0.04	77	16	<1
TP-8549—30% carbon fiber filled, improved wear and chemical resistant	2.5	0.05	49	9	<1
TP-8549—30% carbon fiber filled, improved wear and chemical resistant	3.3	0.05	63	14	<1
TP-8311—10% carbon fiber filled	0.5	0.10	670	23	<1
TP-8311—10% carbon fiber filled	1.0	0.10	490	34	<1

Suzuki thrust wear test results—dry.

Table 7.8 Tribological Properties of Several DuPont Engineering Plastics Vespel® TP Series TPI Plastics

Grade	PV (MPa m/s)	Coefficient of Friction	Wear Factor, K (10^{-10} cm^3/kg fm)	Resin Wear (mg)	Metal Wear (mg)
TP-8130—30% carbon fiber filled	10.4	0.03	4	2	<1
TP-8130—30% carbon fiber filled	12.5	0.03	3	1	<1
TP-8549—30% carbon fiber filled, improved wear and chemical resistant	10.4	0.02	3	1	<1
TP-8549—30% carbon fiber filled, improved wear and chemical resistant	12.5	0.02	4	2	<1
TP-8311—10% carbon fiber filled	10.4	0.02	3	1	<1
TP-8311—10% carbon fiber filled	12.5	0.02	2	1	<1

Suzuki thrust wear test results—lubricated.

7.3 Polyamide-Imides

7.3.1 Fatigue Data

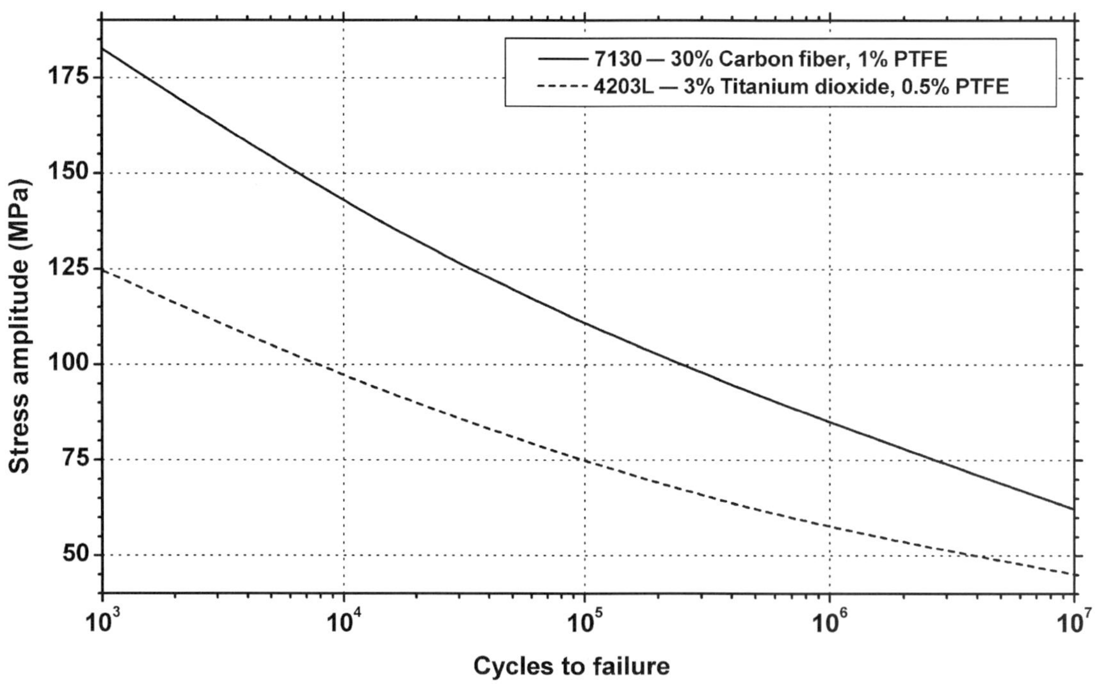

Figure 7.28 Tension/tension stress amplitude vs. cycles to failure at 30 Hz of two Solvay Torlon® PAI plastics.

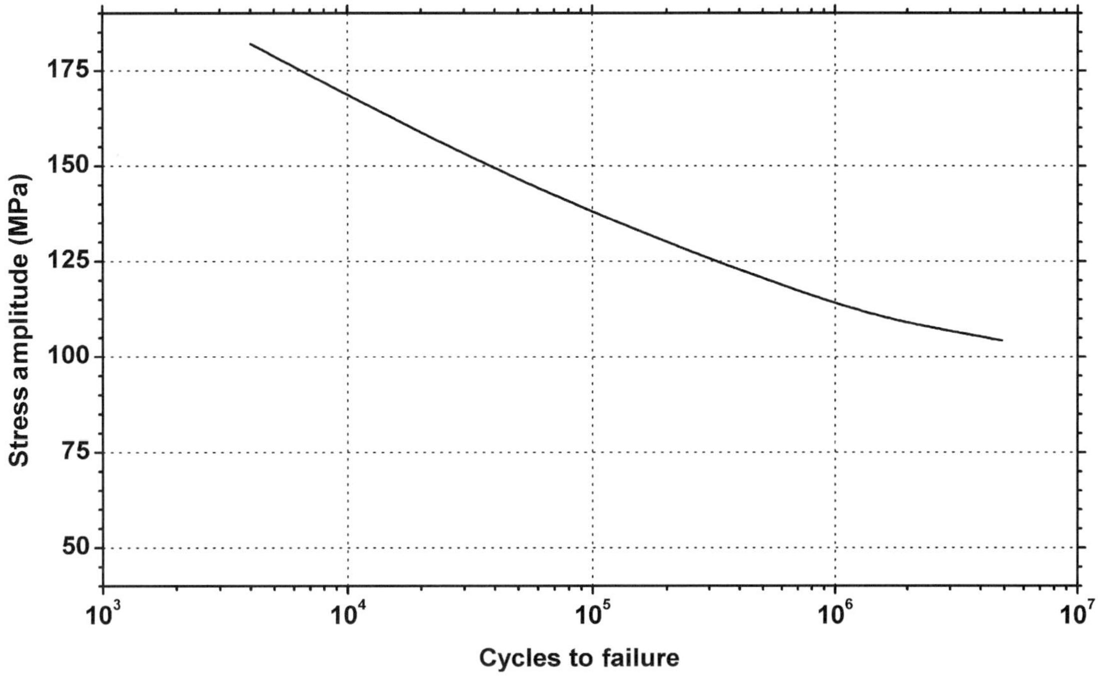

Figure 7.29 Tension/tension stress amplitude vs. cycles to failure at 2 Hz of Solvay Torlon® 7130—30% carbon fiber, 1% PTFE PAI.

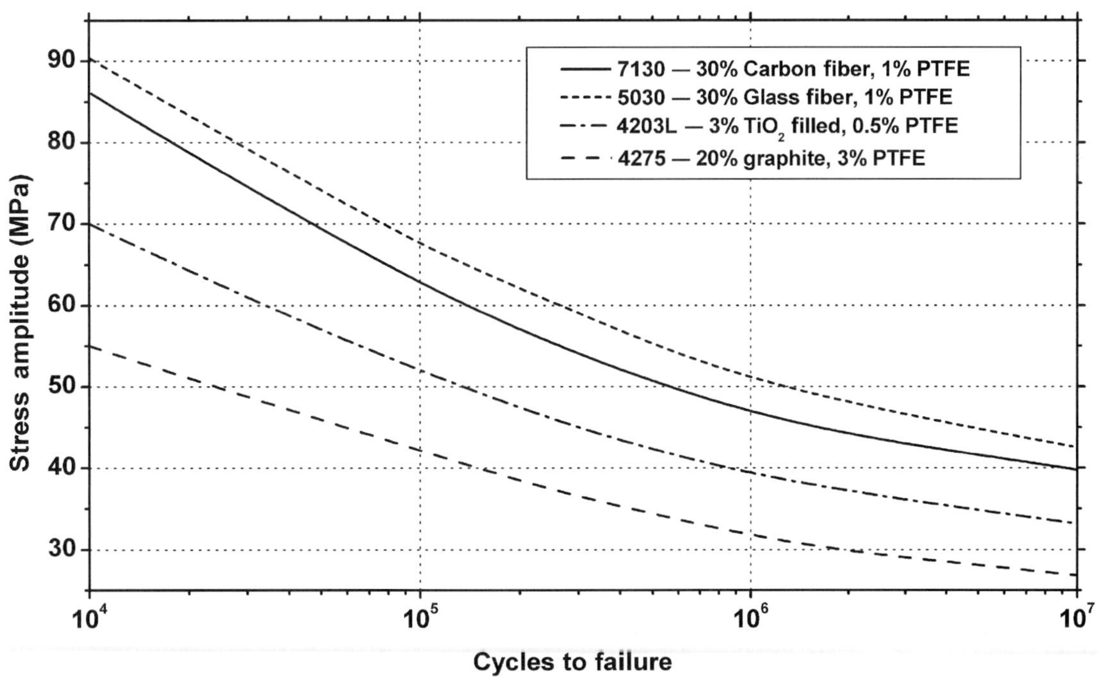

Figure 7.30 Flexural stress amplitude vs. cycles to failure at 30 Hz of several Solvay Torlon® PAI plastics.

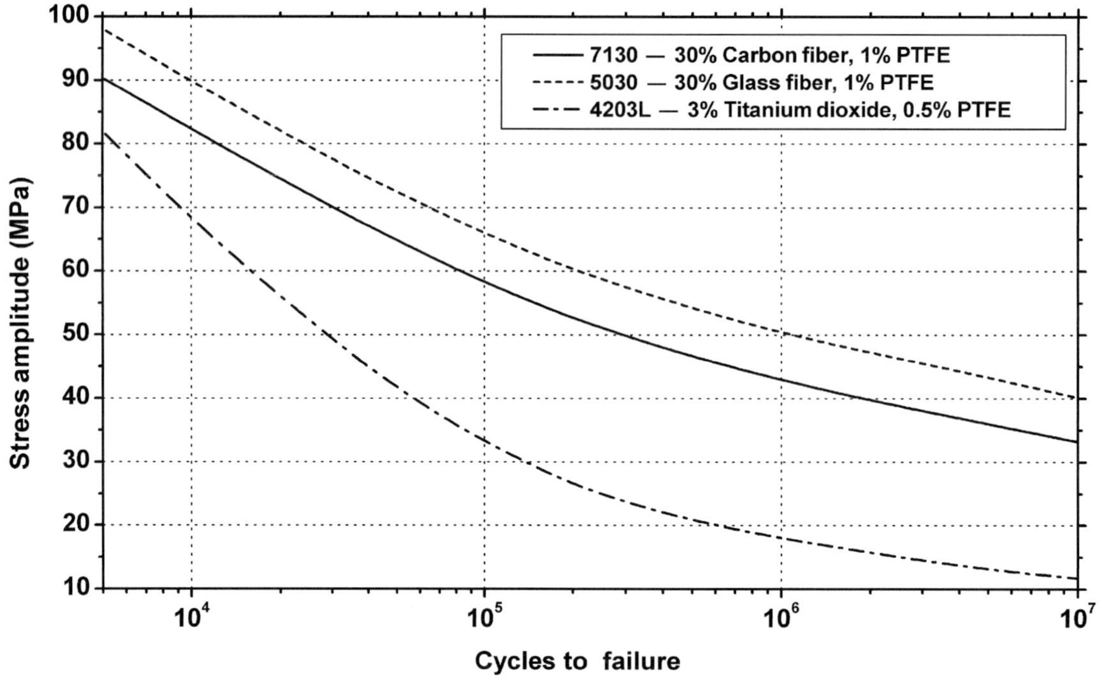

Figure 7.31 Flexural stress amplitude vs. cycles to failure at 30 Hz and 177°C of several Solvay Torlon® PAI plastics.

7.3.2 Tribology Data

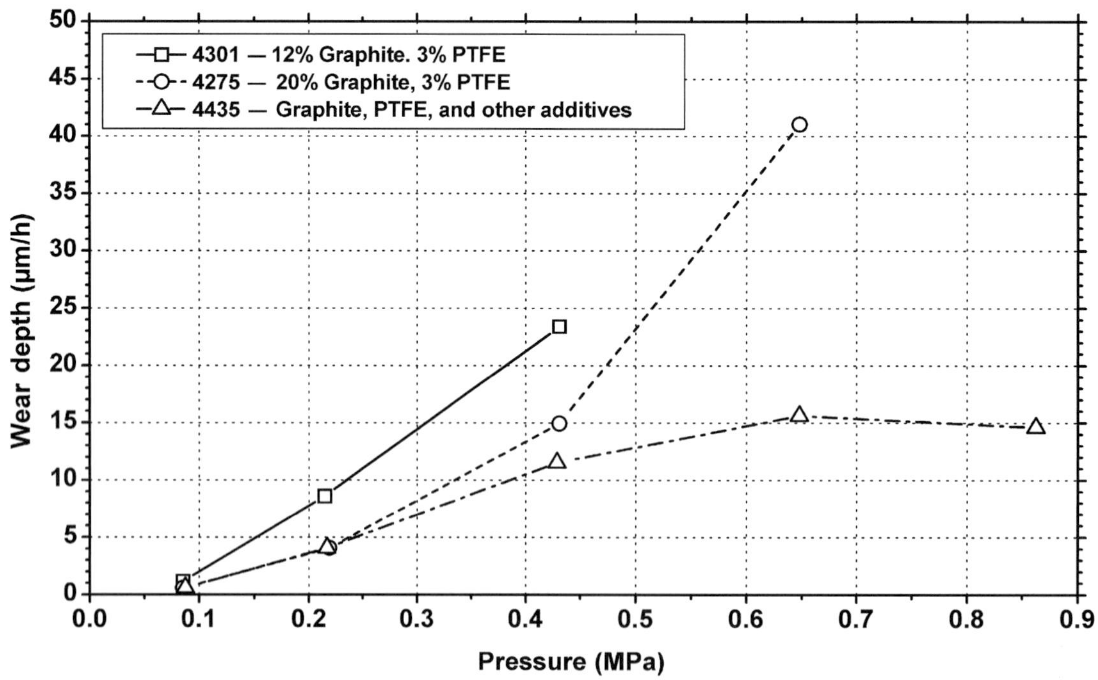

Figure 7.32 Wear resistance vs. pressure at high velocity (4.06 m/s) of several Solvay Torlon® PAI plastics.

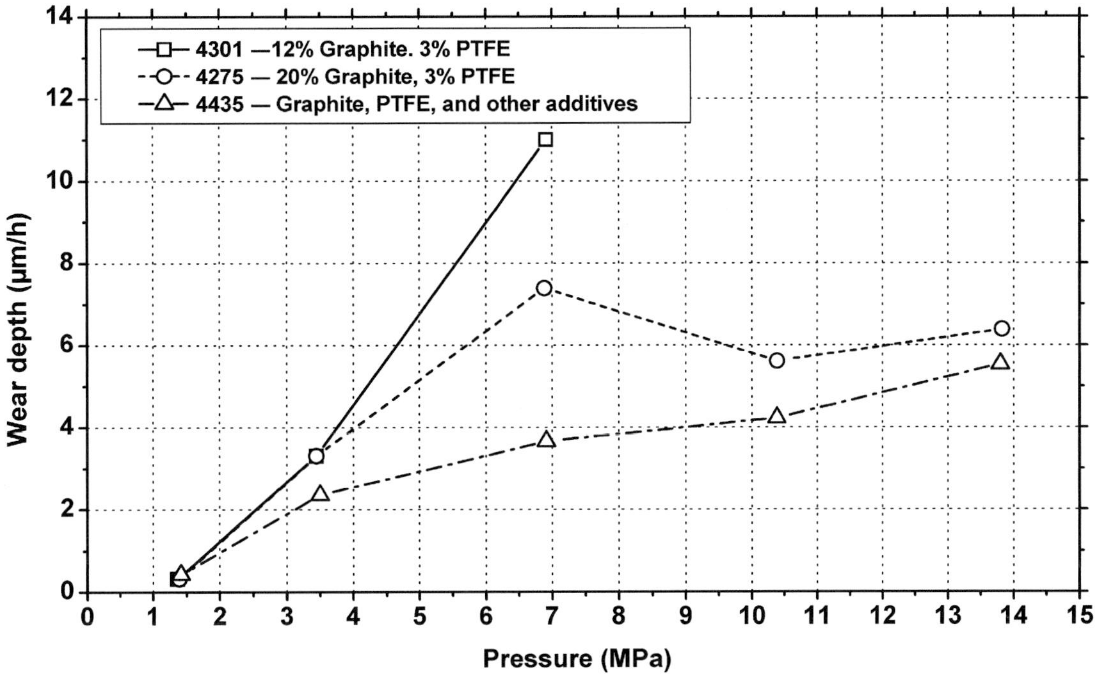

Figure 7.33 Wear resistance vs. pressure at low velocity (0.25 m/s) of several Solvay Torlon® PAI plastics.

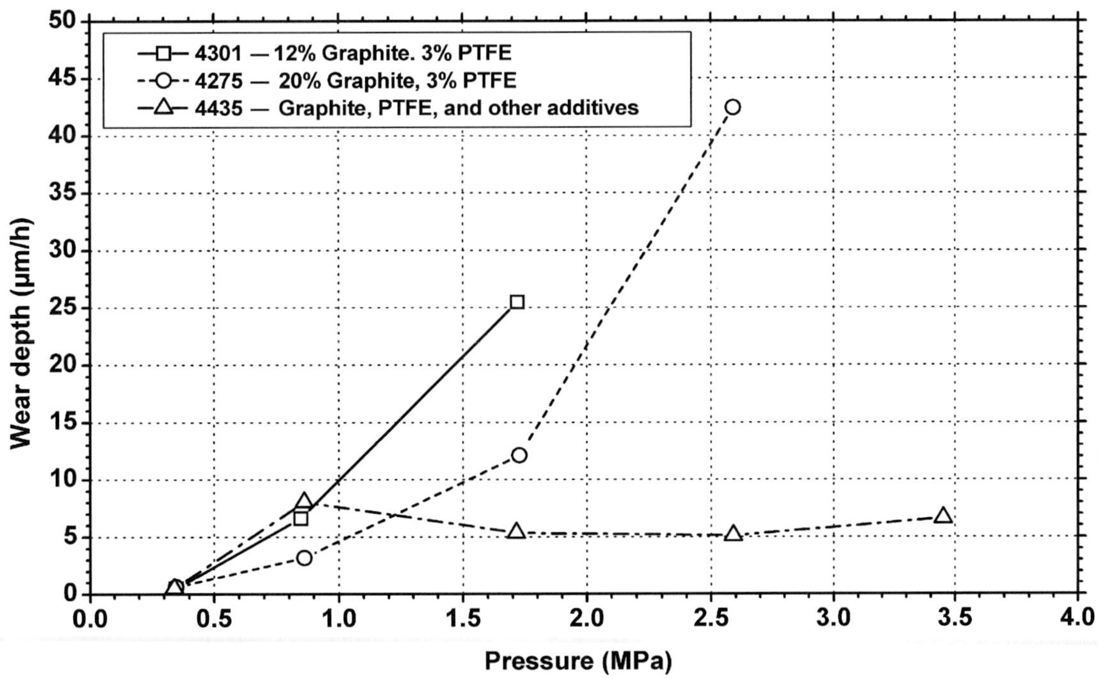

Figure 7.34 Wear resistance vs. pressure at moderate velocity (1.02 m/s) of several Solvay Torlon® PAI plastics.

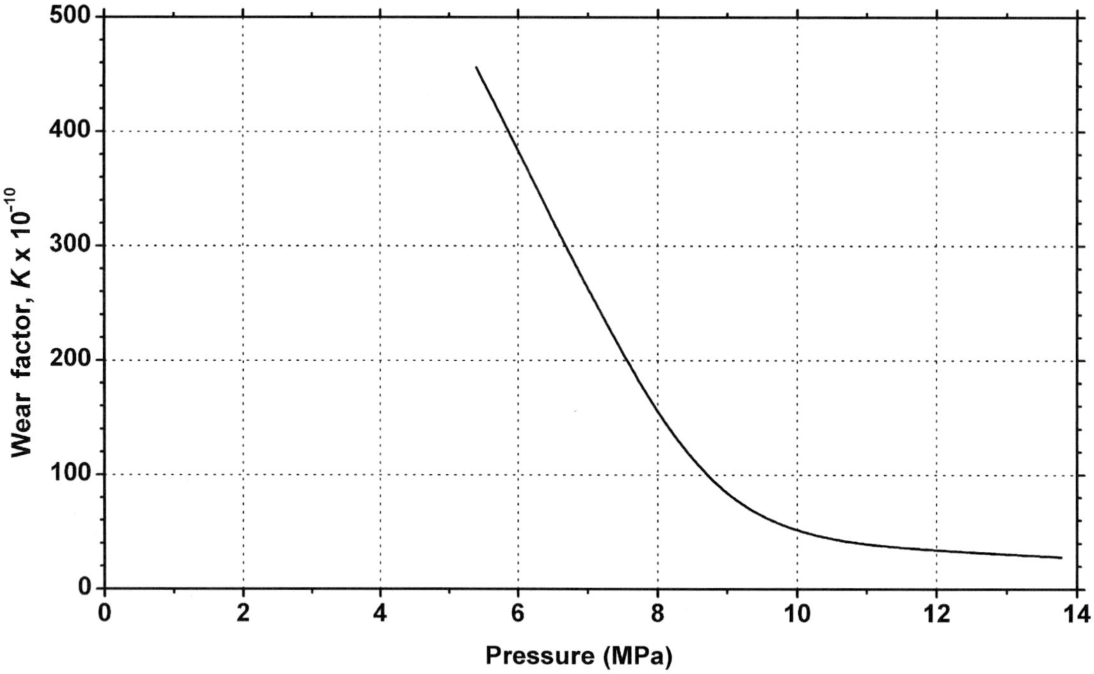

Figure 7.35 Wear factor vs. extended cure time at 260°C of Solvay Torlon® 4301—12% Graphite, 3% PTFE PAI.

Table 7.9 Wear Factor and Wear Rates of Several Solvay Torlon® PAI Plastics

Pressure (MPa)	PV	Wear Factor (10^{-10} mm s/mPa h)			Wear Rate (10^{-6} m/h)		
		4301	4275	4435	4301	4275	4435
Velocity—0.25 m/s							
1.379	0.350	8	6		0.3	0.2	
3.447	0.876	30	36		2.7	3.1	
6.895	1.751	59	40	20	10.4	7.0	3.4
10.342	2.627		20	15		5.3	3.8
13.790	3.503		17	15		6.1	5.1
Velocity—1.02 m/s							
0.345	0.350	12	13		0.4	0.5	
0.862	0.876	60	28	71	5.3	2.5	6.2
1.724	1.751	113	54	24	19.8	9.4	4.2
2.586	2.627		126	15		33.1	4.0
3.447	3.503		Melted	15		Melted	5.1
Velocity—4.06 m/s							
0.086	0.350	69	9		2.4	0.3	
0.215	0.876	102	50		8.9	4.4	
0.431	1.751	135	86	67	23.6	15.0	11.7
0.646	2.627		155	56		40.8	14.7
0.862	3.503		Melted	38		Melted	13.2

Torlon® 4301—12% graphite, 3% PTFE.
Torlon® 4275—20% graphite, 3% PTFE.
Torlon® 4435—graphite, PTFE, and other additives.

7.4 Polyimides

7.4.1 Fatigue Data

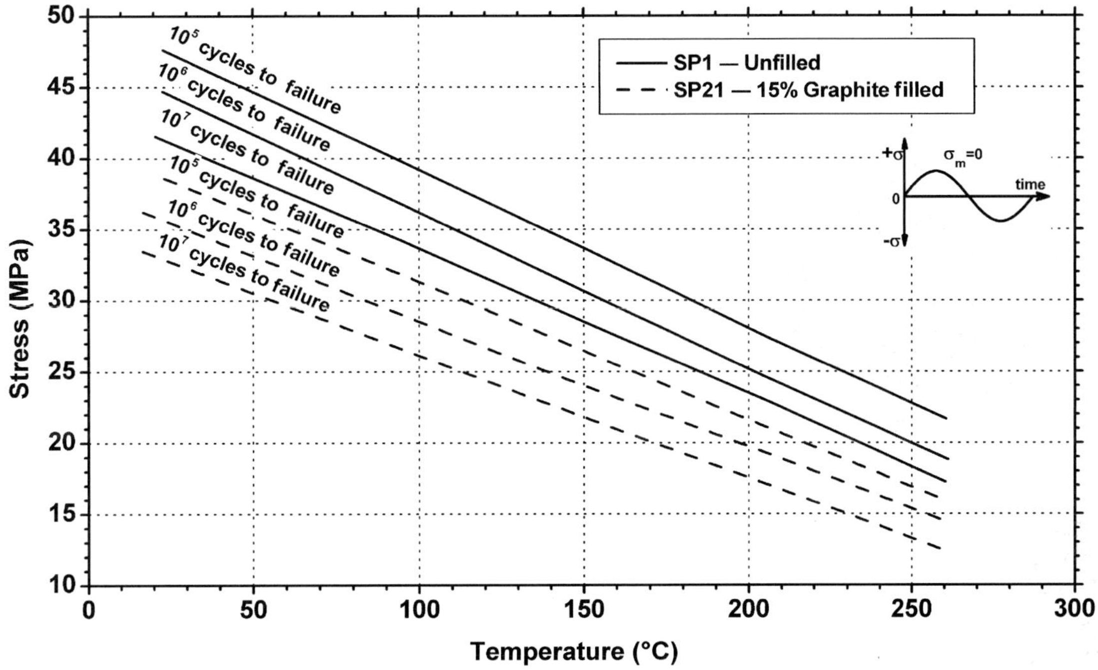

Figure 7.36 Fatigue resistance vs. temperature to failure at 30 Hz and various cycles of machined DuPont Engineering Polymers Vespel® SP PI plastics.

7.4.2 Tribology Data

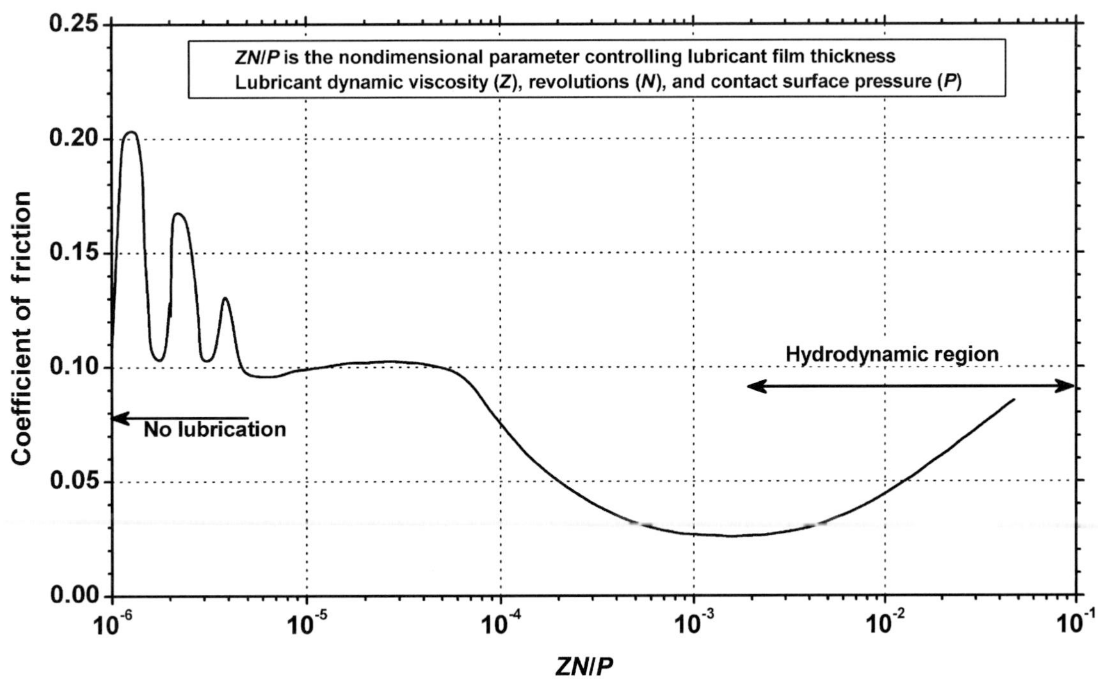

Figure 7.37 Lubricated friction test: dynamic coefficient of friction vs. *ZN/P* by thrust bearing test against steel with Sunvis 31 Oil lubricant of DuPont Engineering Polymers Vespel® SP21—15% graphite filled PI.

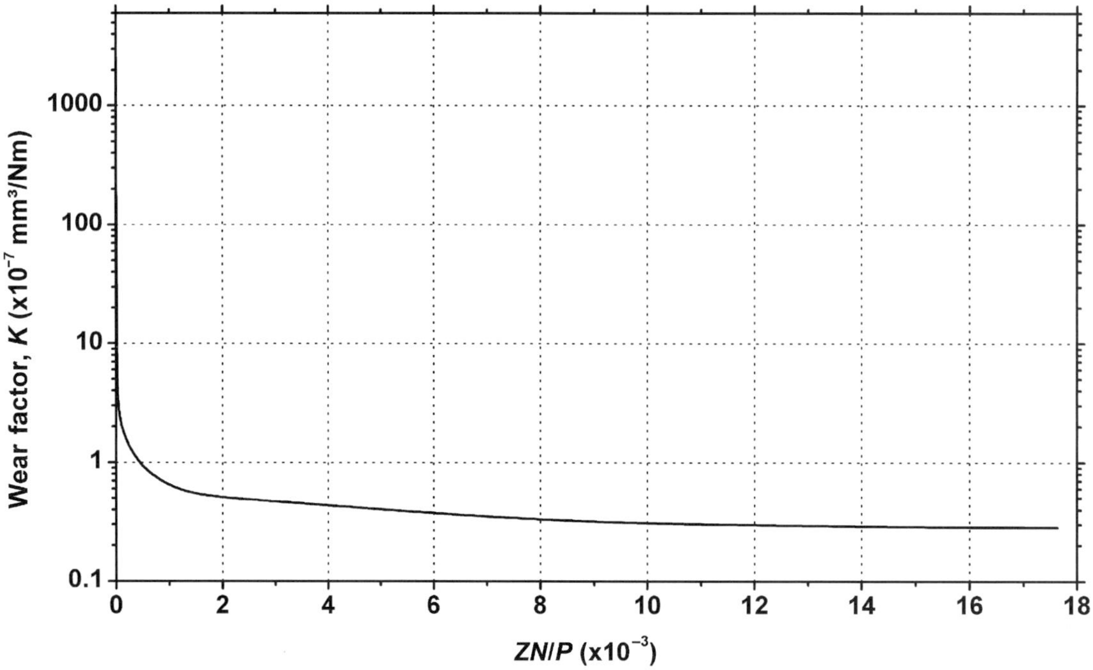

Figure 7.38 Lubricated friction test: wear factor vs. *ZN/P* by thrust bearing test against steel with Sunvis 31 Oil lubricant of DuPont Engineering Polymers Vespel® SP21—15% graphite filled PI.

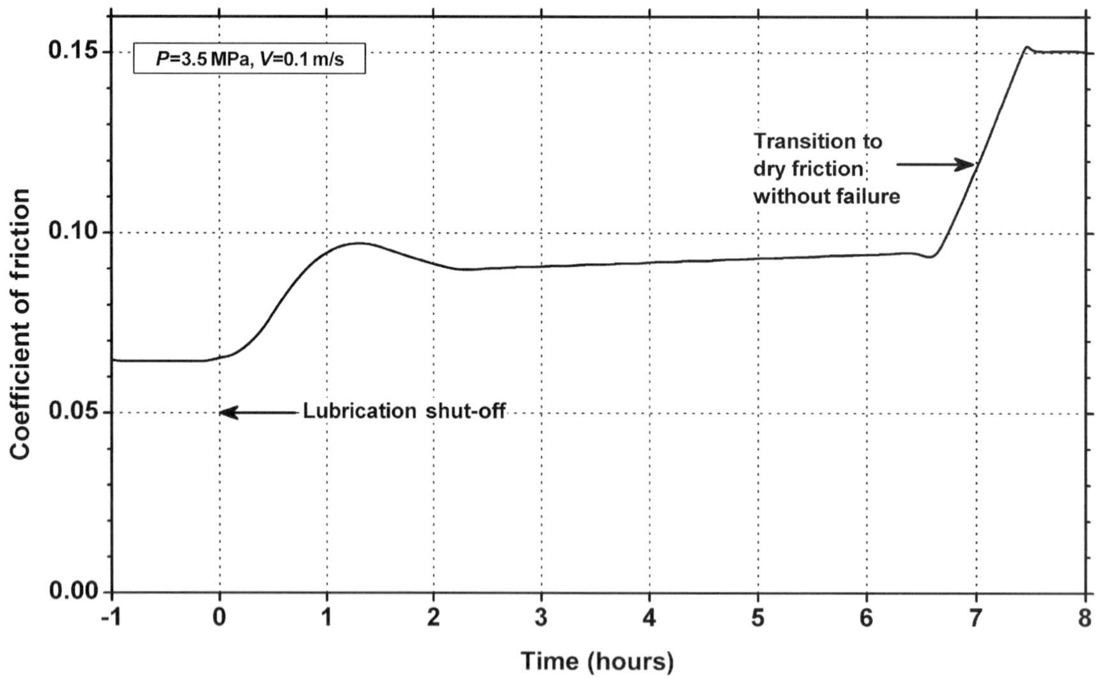

Figure 7.39 Lubricated starvation test: dynamic coefficient of friction vs. time in hours by thrust bearing test against steel with Sunvis 31 Oil lubricant of DuPont Engineering Polymers Vespel® SP21—15% graphite filled PI (redo this chart *X*-axis).

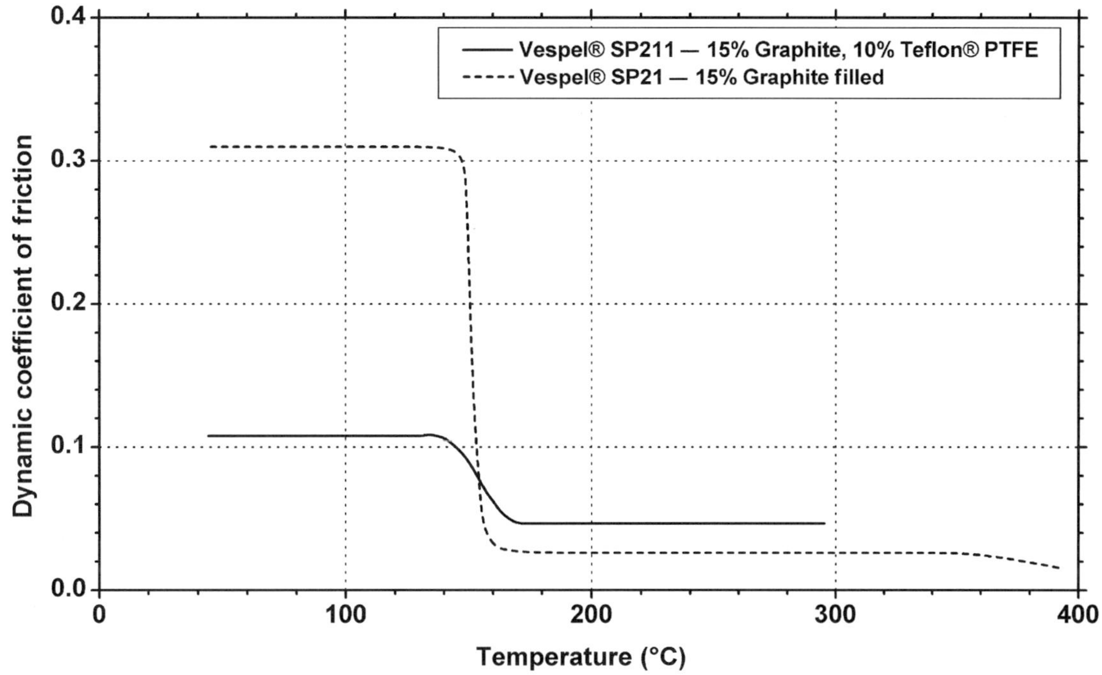

Figure 7.40 Dynamic coefficient of friction vs. temperature by thrust bearing test against unlubricated steel of two DuPont Engineering Polymers Vespel® SP PI plastics.

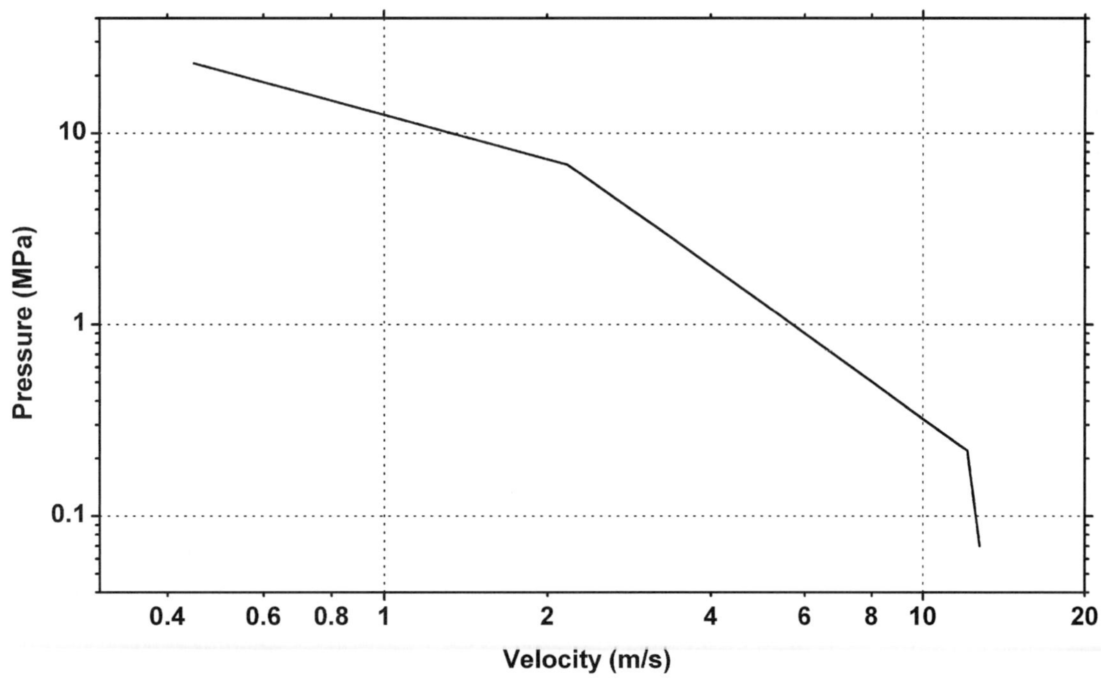

Figure 7.41 Pressure vs. velocity limit at 395°C against unlubricated carbon steel of DuPont Engineering Polymers Vespel® SP21—15% graphite filled PI.

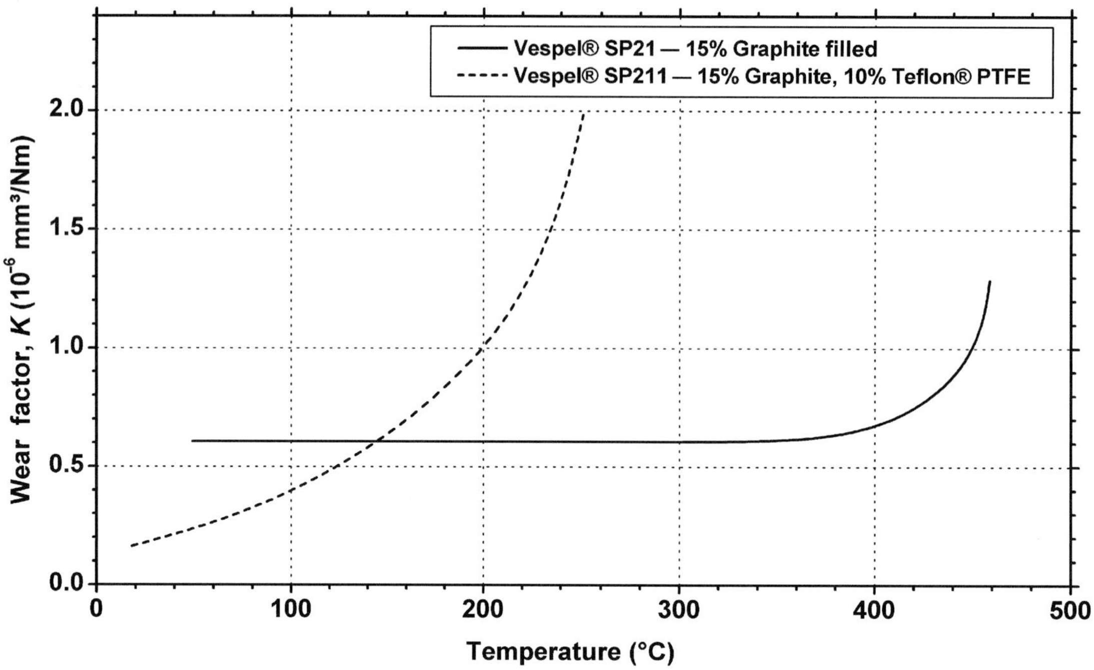

Figure 7.42 Wear factor vs. temperature against unlubricated mild carbon steel of two DuPont Engineering Polymers Vespel® SP PI plastics.

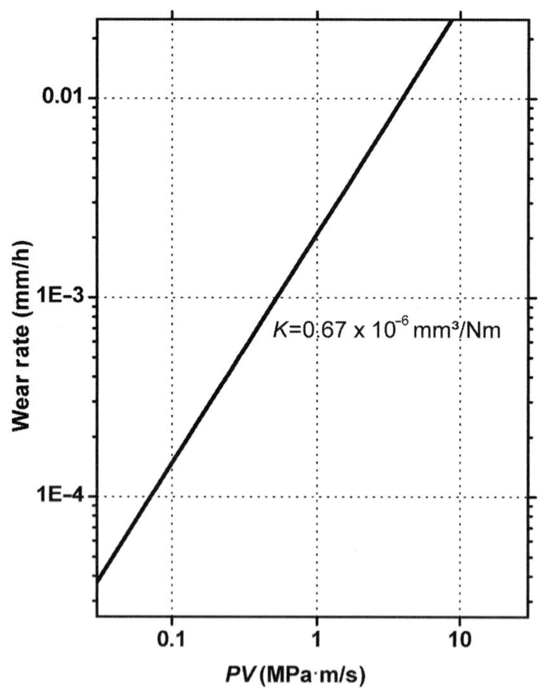

Figure 7.43 Wear rate vs. PV against unlubricated mild carbon steel in thrust bearing tester of DuPont Engineering Polymers Vespel® SP21—15% graphite filled PI.

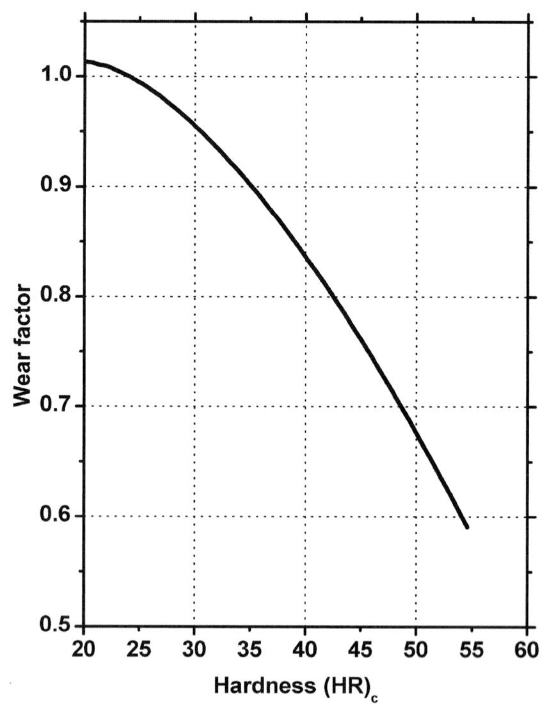

Figure 7.44 Wear factor vs. unlubricated counter material hardness in thrust bearing tester of DuPont Engineering Polymers Vespel® SP21—15% graphite filled PI.

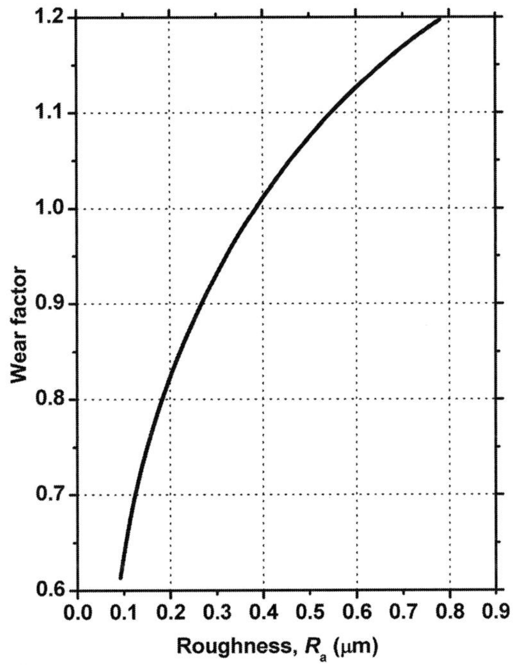

Figure 7.45 Wear factor vs. roughness of unlubricated counter material hardness in thrust bearing tester of

Table 7.10 Wear and Friction Properties of Several DuPont Engineering Polymers Vespel® PI Plastics

	SP1		SP21		SP22		SP211		SP3
	M[a]	DF[b]	M	DF	M	DF	M	DF	M
Wear Rate (m/s × 10^{-10})[c]	17–85	17–85	6.3	6.3	4.2	4.2	4.9	4.9	17–23
Coefficient of Friction:									
At *PV* = 0.875 MPa m/s	0.29	0.29	0.24	0.24	0.30	0.30	0.12	0.12	0.25
At *PV* = 3.5 MPa m/s	–	–	0.12	0.12	0.09	0.09	0.08	0.08	0.17
In vacuum	–	–	–	–	–	–	–	–	0.03
Static in air	0.35	–	0.30	–	0.27	–	0.20	–	–

Vespel® SP1—Unfilled.
Vespel® SP21—15% graphite filled.
Vespel® SP22—40% graphite filled.
Vespel® SP211—15% graphite, 10% Teflon® PTFE.
Vespel® SP3—15% molybdenum sulfide filled.
[a]M = machined part.
[b]DF = direct formed part.
[c]Unlubricated in air (*PV* = 0.875 MPa m/s).

Table 7.11 Maximum *PV* limits for unlubricated DuPont Engineering Polymers Vespel® PI Plastics

Material	Filler	PV (kg m/cm^3 s)	Maximum Contact Temperature (°C)
SP21	15% Graphite	107	393
SP22	40% Graphite	107	393
SP211	15% Graphite 10% PTFE	36	260

8 Polyamides (Nylons)

8.1 Background

High-molecular-weight polyamides are commonly known as nylon. Polyamides are crystalline polymers typically produced by the condensation of a diacid and a diamine. There are several types and each type is often described by a number, such as Nylon 66 or Polyamide 66 (PA66). The numeric suffixes refer to the number of carbon atoms present in the molecular structures of the amine and acid, respectively (or a single suffix if the amine and acid groups are part of the same molecule).

The polyamide plastic materials discussed in this book and the monomers used to make them are given in Table 8.1.

The general reaction is shown in Figure 8.1.

The –COOH acid group reacts with the $–NH_2$ amine group to form the amide. A molecule of water is given off as the nylon polymer is formed. The properties of the polymer are determined by the R and R′ groups in the monomers. In Nylon 6, 6, R′ = 6C and R = 4C alkanes, but one also has to include the two carboxyl carbons in the diacid to get the number it designates to the chain.

The structures of these diamine monomers are shown in Figure 8.2, the diacid monomers are shown in Figure 8.3. Figure 8.4 shows the amino acid monomers. These structures only show the functional groups, the CH_2 connecting groups are implied at the bond intersections.

All polyamides tend to absorb moisture which can affect their properties. Properties are often reported as DAM (dry as molded) or conditioned (usually at equilibrium in 50% relative humidity at 23°C). The absorbed water tends to act like a plasticizer and can have a significant effect on the plastics properties.

8.1.1 Nylon 6

Nylon 6 begins as pure caprolactam which is a ring structured molecule. This is unique in that the ring is opened and the molecule polymerizes with

Table 8.1 Monomers Used to Make Specific Polyamides/Nylons

Polyamide/Nylon Type	Monomers Used to Make
Nylon 6	Caprolactam
Nylon 11	Aminoundecanoic acid
Nylon 12	Aminolauric acid
Nylon 66	1,6-Hexamethylene diamine and adipic acid
Nylon 610	1,6-Hexamethylene diamine and sebacic acid
Nylon 612	1,6-Hexamethylene diamine and 1,12-dodecanedioic acid
Nylon 666	Copolymer based on Nylon 6 and Nylon 66
Nylon 46	1,4-Diaminobutane and adipic acid
Nylon amorphous	Trimethyl hexamethylene diamine and terephthalic acid
Polyphthalamide	Any diamine and isophthalic acid and/or terephthalic acid

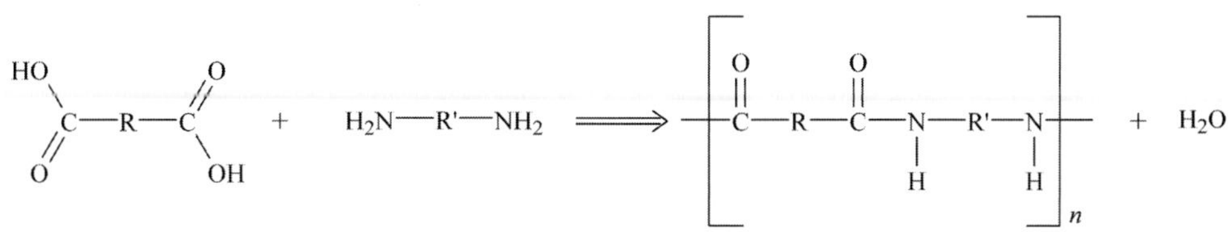

Figure 8.1 Generalized polyamide reaction.

Copyright © 2010 Laurence W. McKeen. All rights reserved.

itself. Since caprolactam has 6 carbon atoms, the nylon that it produces is called Nylon 6, which is nearly the same as Nylon 66 described in Section 8.1.5. The structure of Nylon 6 is shown in Figure 8.5 with the repeating unit in the brackets.

Some of the Nylon 6 characteristics:

- Outstanding balance of mechanical properties
- Outstanding toughness in equilibrium moisture content
- Outstanding chemical resistance and oil resistance.
- Outstanding wear and abrasion resistance
- Almost all grades are self-extinguishing. The flame-resistant grades are rated UL 94VO
- Outstanding long-term heat resistance (at a long-term continuous maximum temperature ranging between 80°C and 150°C)
- Grades reinforced with glass fiber and other materials offer superior elastic modulus and strength
- Offers low gasoline permeability and outstanding gas barrier properties
- Highest rate of water absorption and highest equilibrium water content (8% or more)
- Excellent surface finish even when reinforced
- Poor chemical resistance to strong acids and bases

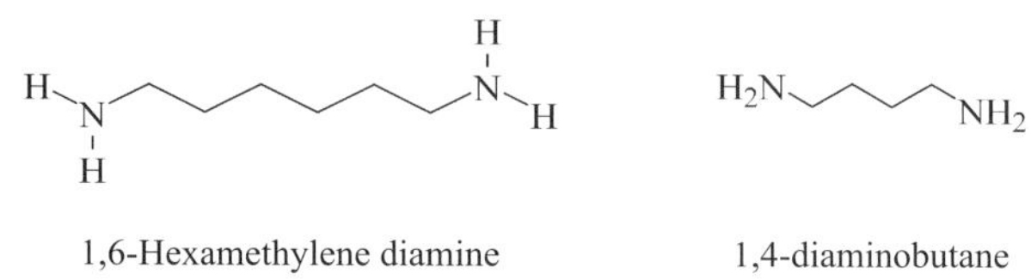

Figure 8.2 Chemical structures of diamines used to make polyamides.

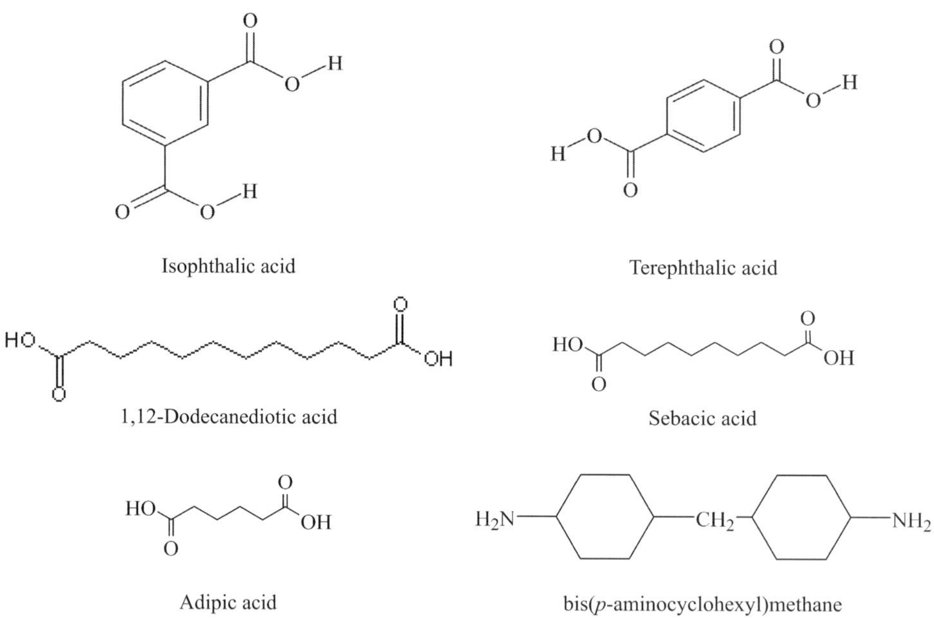

Figure 8.3 Chemical structures of diacids used to make polyamides.

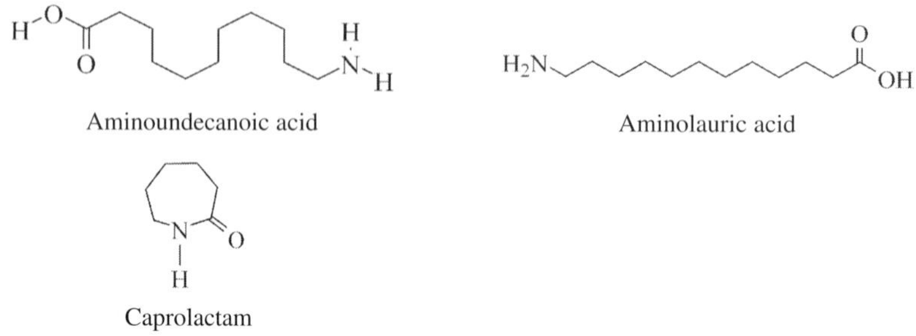

Figure 8.4 Chemical structures of amino acids used to make polyamides.

8.1.2 Nylon 11

Nylon 11 has only one monomer, aminoundecanoic acid. It has the necessary amine group on one end and the acid group on the other. It polymerizes with itself to produce the polyamide containing 11 carbons between the nitrogen of the amide groups. Its structure is shown in Figure 8.6.

Some of the Nylon 11 characteristics:

- Low water absorption for a nylon (2.5% at saturation)
- Reasonable UV resistance
- Higher strength
- Ability to accept high loading of fillers
- Better heat resistance than Nylon 12
- More expensive than Nylon 6 or Nylon 6/6
- Relatively low impact strength

8.1.3 Nylon 12

Nylon 12 has only one monomer, aminolauric acid. It has the necessary amine group on one end and the acid group on the other. It polymerizes with itself to produce the polyamide containing 12 carbons between the two nitrogen atoms of the two amide groups. Its structure is shown in Figure 8.7.

The properties of semicrystalline polyamides are determined by the concentration of amide groups in the macromolecules. Polyamide 12 has the lowest amide group concentration of all commercially available polyamides thereby substantially promoting its characteristics:

- Lowest moisture absorption (~2%): Parts show largest dimensional stability under conditions of changing humidity
- Exceptional impact and notched impact strength, even at temperatures well below the freezing point
- Good to excellent resistance against greases, oils, fuels, hydraulic fluids, various solvents, salt solutions and other chemicals
- Exceptional resistance to stress cracking, including metal parts encapsulated by injection molding or embedded
- Excellent abrasion resistance
- Low coefficient of sliding friction
- Noise and vibration damping properties
- Good fatigue resistance under high-frequency cyclical loading condition
- High processability
- Expensive
- Lowest strength and heat resistance of any polyamide unmodified generic

8.1.4 Nylon 66

The structure of Nylon 66 is shown in Figure 8.8.

Some of the Nylon 66 characteristics:

- Outstanding balance of mechanical properties
- Outstanding toughness in equilibrium moisture content

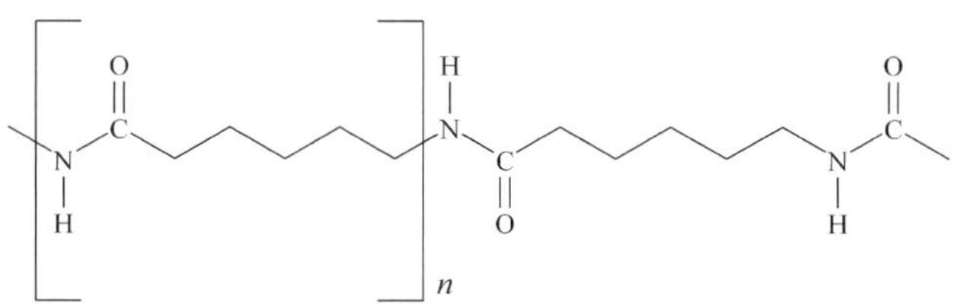

Figure 8.5 Chemical structure of Nylon 6.

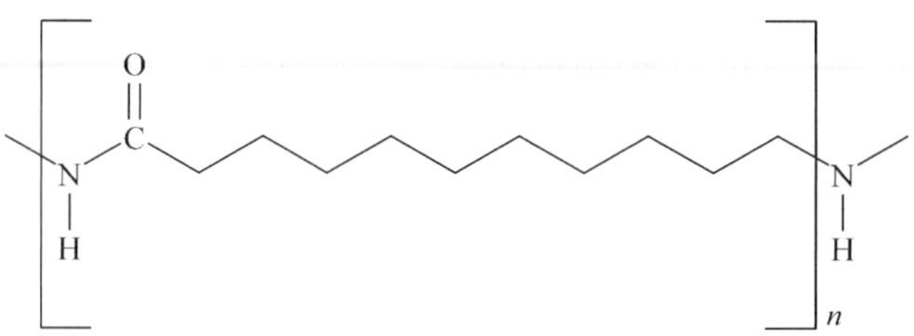

Figure 8.6 Chemical structure of Nylon 11.

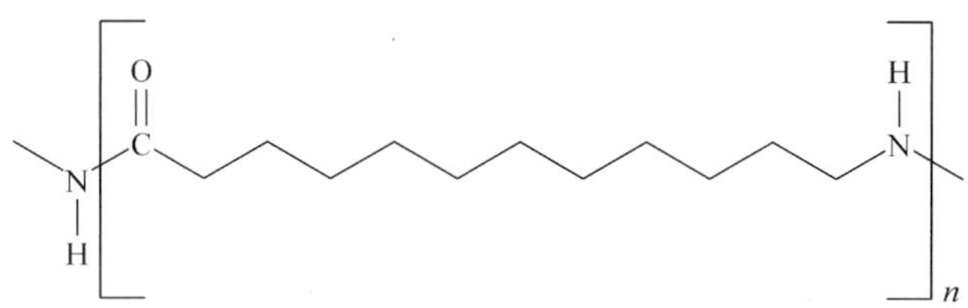

Figure 8.7 Chemical structure of Nylon 12.

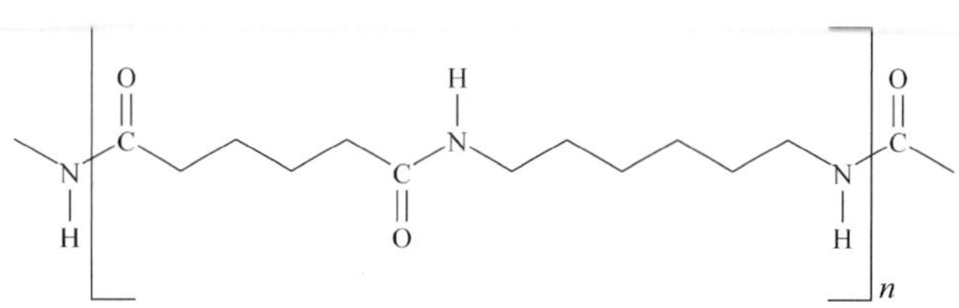

Figure 8.8 Chemical structure of Nylon 66.

- Outstanding chemical resistance and oil resistance
- Outstanding wear and abrasion resistance
- Almost all grades are self-extinguishing. The flame-resistant grades are rated UL 94VO
- Outstanding long-term heat resistance (at a long-term continuous maximum temperature ranging between 80°C and 150°C)
- Grades reinforced with glass fiber and other materials offer superior elastic modulus and strength
- Offers low gasoline permeability and outstanding gas barrier properties
- High water absorption
- Poor chemical resistance to strong acids and bases

8.1.5 Nylon 610

The structure of Nylon 610 is given in Figure 8.9. Some of the Nylon 610 characteristics:

- Outstanding suppleness and impact strength at low temperature
- Relatively low hygroscopic properties
- Outstanding flex fatigue properties

8.1.6 Nylon 612

The structure of Nylon 612 is given in Figure 8.10. Some of the Nylon 612 characteristics:

- High-impact strength
- Very good resistance to greases, oils, fuels, hydraulic fluids, water, alkalis, and saline
- Very good stress cracking resistance, even when subjected to chemical attack and when used to cover metal parts
- Low coefficients of sliding friction and high abrasion resistance, even when running dry
- Heat deflection temperature (melting point nearly 40°C higher than Nylon 12)
- Tensile and flexural strength
- Outstanding recovery at high wet strength

8.1.7 Nylon 666 or 66/6

This is the name given to copolyamides made from PA6 and PA66 building blocks. A precise structure cannot be drawn.

8.1.8 Amorphous Nylon

Amorphous nylon is designed to give no crystallinity to the polymer structure. One such amorphous nylon is shown in Figure 8.11.

The tertiary butyl group attached to the amine molecule is bulky and disrupts this molecule's ability to crystallize. This particular amorphous nylon is sometimes designated at Nylon 6-3-T. Amorphous polymers can have properties that differ significantly from crystalline types, one of which is optical transparency.

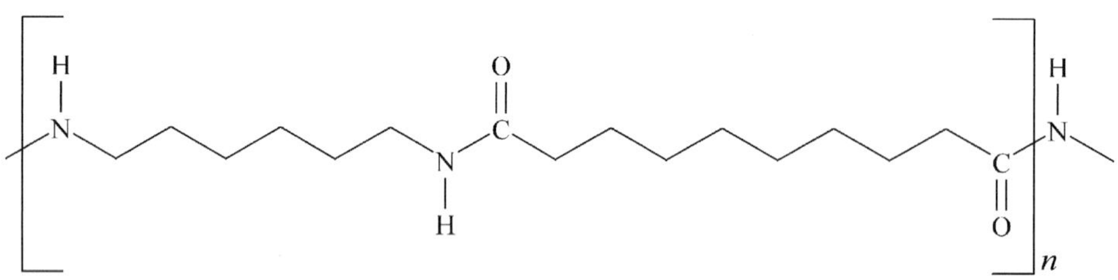

Figure 8.9 Chemical structure of Nylon 610.

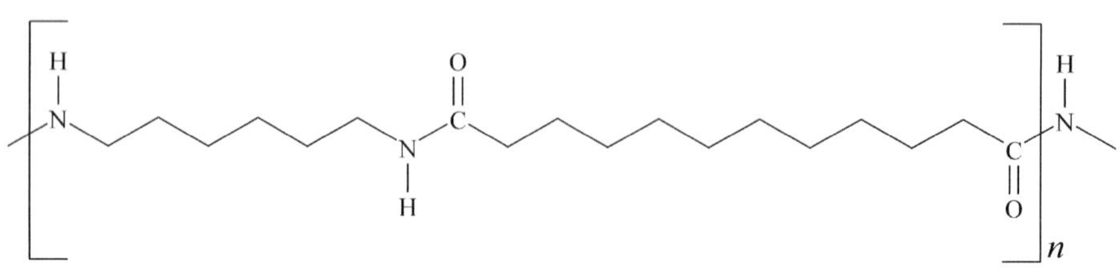

Figure 8.10 Chemical structure of Nylon 612.

Some of the amorphous nylon characteristics:

- Crystal-clear, high optical transparency
- High mechanical stability
- High heat deflection temperature
- High-impact strength
- Good chemical resistance compared to other plastics
- Good electrical properties
- Low mold shrinkage

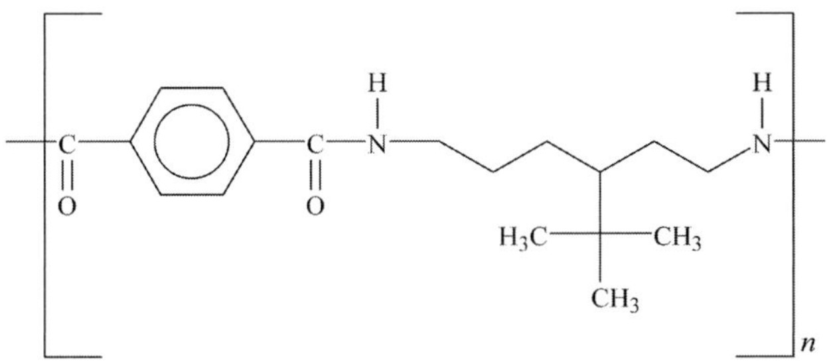

Figure 8.11 Chemical structure of amorphous nylon.

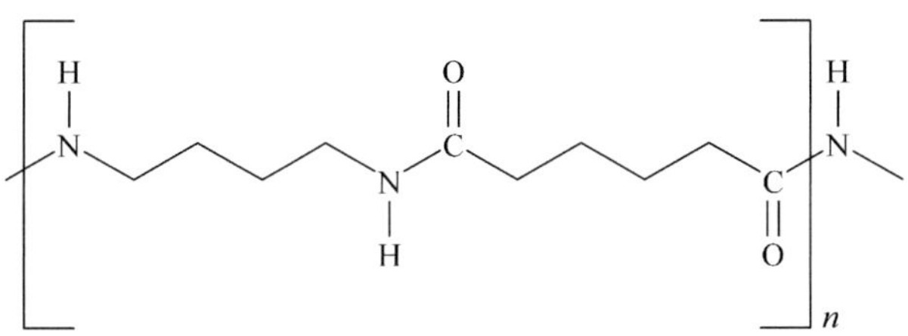

Figure 8.12 Chemical structure of Nylon 46.

8.1.9 Nylon 46

The structure of Nylon 46 is given in Figure 8.12. Some of the Nylon 46 characteristics:

- Higher heat distortion temperature than Nylon 6 or Nylon 6/6
- Higher crystallinity than Nylon 6 or Nylon 6/6
- Better chemical resistance, particularly to acidic salts
- Similar moisture absorption to Nylon 6/6, but dimensional increase is less
- High processing temperature
- Highest mechanical properties at high temperatures
- Excellent resistance to wear and low friction
- Outstanding flow for easy processing

8.1.10 Polyphthalamide (PPA)/ High-Performance Polyamide

As a member of the nylon family, it is a semi-crystalline material composed from a diacid and a diamine. However, the diacid portion contains at least 55% terephthalic acid (TPA) or isophthalic acid (IPA). TPA and IPA are aromatic components which serve to raise the melting point, glass transition temperature, and generally improve chemical resistance versus standard aliphatic nylon polymers. The structure of the polymer depends on the ratio of the diacid ingredients and the diamine used and varies from grade to grade. The polymer usually consists of mixtures of blocks of two or more different segments, four of which are shown in Figure 8.13.

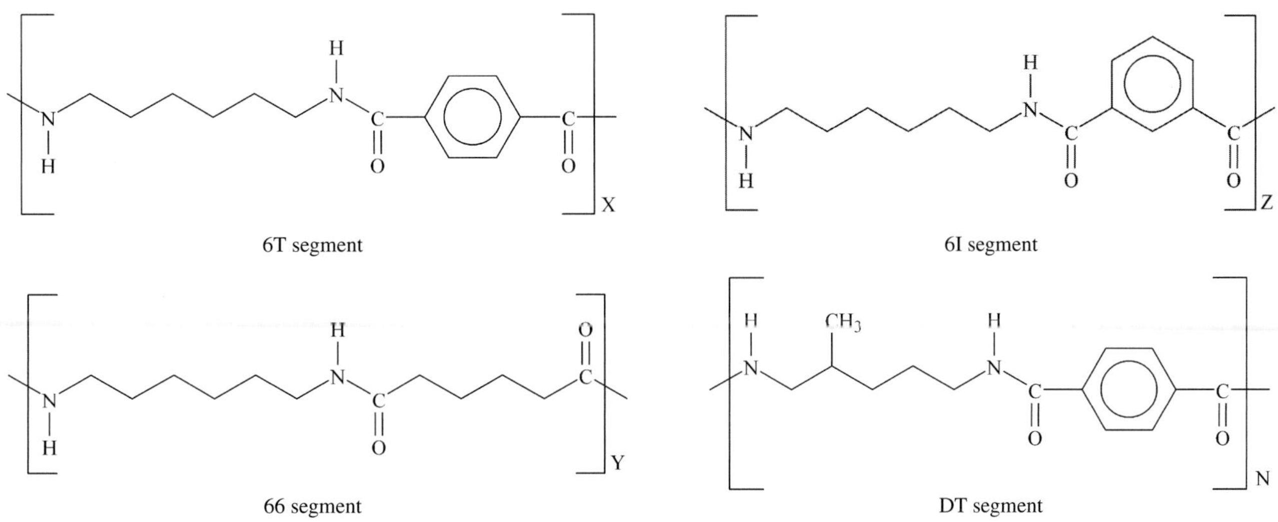

Figure 8.13 Chemical structures of block used to make PPAs.

Some of the PPA characteristics:

- Very high heat resistance
- Good chemical resistance
- Relatively low moisture absorption
- High strength or physical properties over a broad temperature range
- Not inherently flame retardant
- Requires good drying equipment
- High processing temperatures
- Good dimensional stability
- Slow rate of water absorption

Graphs of multipoint properties of polyamides as a function of temperature, moisture, and other factors are given in the following sections. Because the polyamides do absorb water, and that affects the properties, some of the data are dry, or better dry as molded. Some of the data are for conditioned specimens; they have reached equilibrium water absorption from 50% relative humidity at 23°C.

8.1.11 PAA—Polyarylamide

Another partially aromatic high-performance polyamide is polyarylamide, PAA. The primary commercial polymer, PAMXD6, is formed by the reaction of *m*-xylylenediamine and adipic acid giving the structure shown in Figure 8.14. It is a semicrystalline polymer.

- Very high rigidity
- High strength
- Very low creep
- Excellent surface finish even for a reinforced product even with a high glass fiber content
- Ease of processing

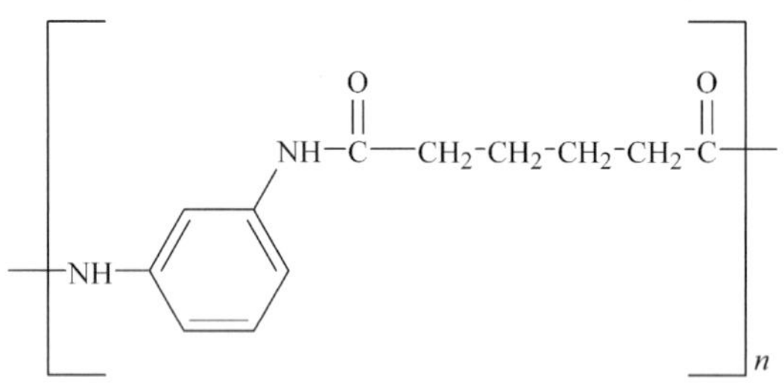

Figure 8.14 Chemical structure of PAMXD6 PAA.

8.1.12 PACM 12—Semicrystalline Polyamide

PACM 12 is a polyamide produced from bis (*p*-aminocyclohexyl)methane (54% *trans–trans*) and dodecanedioic acid. The structure is shown in Figure 8.15.

PACM 12 combines the chemical resistance of semicrystalline materials with the advantages of amorphous, UV-resistant materials. The properties of PACM 12 are:

- Crystal-clear, permanent transparency
- Superior chemical and stress cracking resistance
- High level of UV resistance
- Low water absorption compared with many other polyamides, which leaves the mechanical properties virtually unaffected
- High dimensional stability
- Balanced mechanical property profile
- High-impact resistance, even at low temperatures
- Abrasion and scratch resistance
- High glass transition temperature
- Easy processing

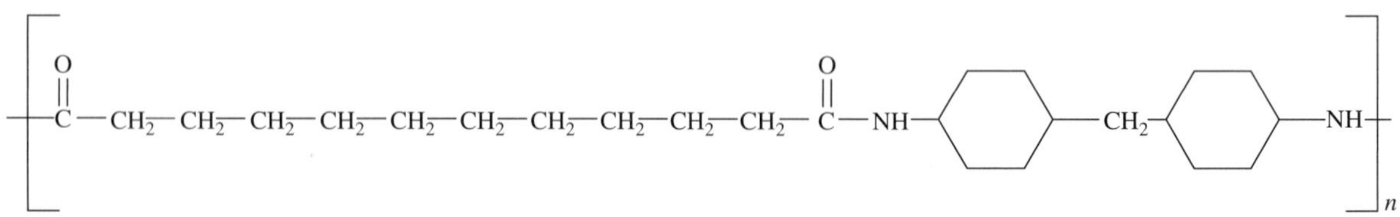

Figure 8.15 Chemical structure of PACM 12 semicrystalline polyamide.

8.2 Polyamide 6 (Nylon 6)

8.2.1 Fatigue Data

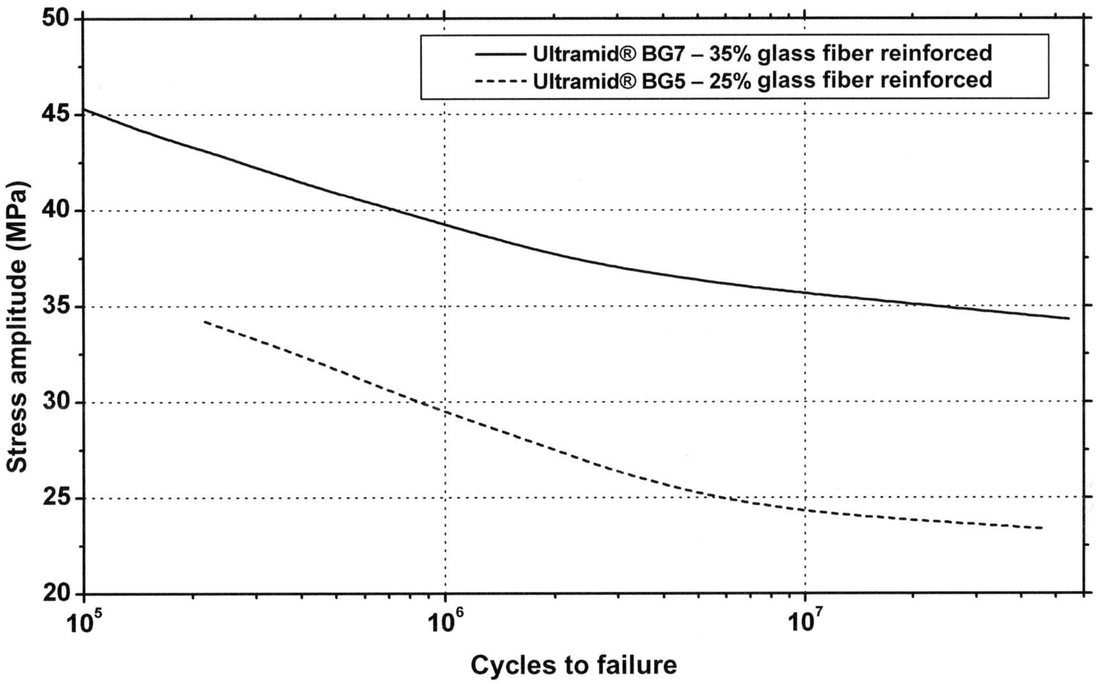

Figure 8.16 Stress amplitude vs. cycles to failure of two BASF Ultramid® glass fiber reinforced PA6 plastics.

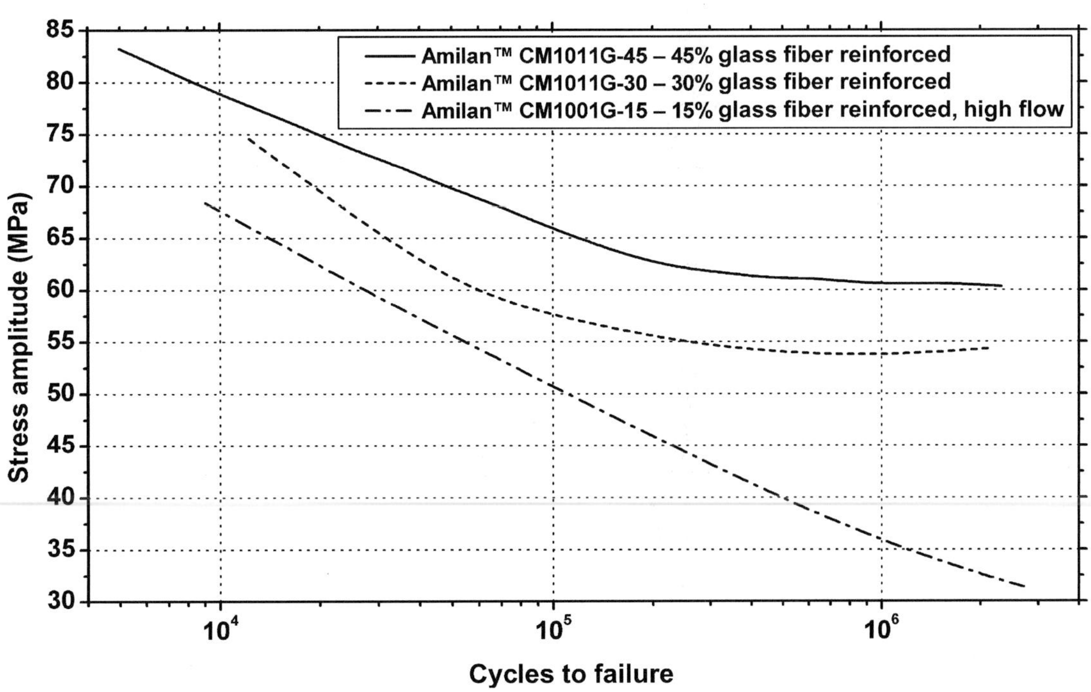

Figure 8.17 Flexural stress amplitude vs. cycles to failure of Toray Resin Company Amilan™ PA6 plastics.

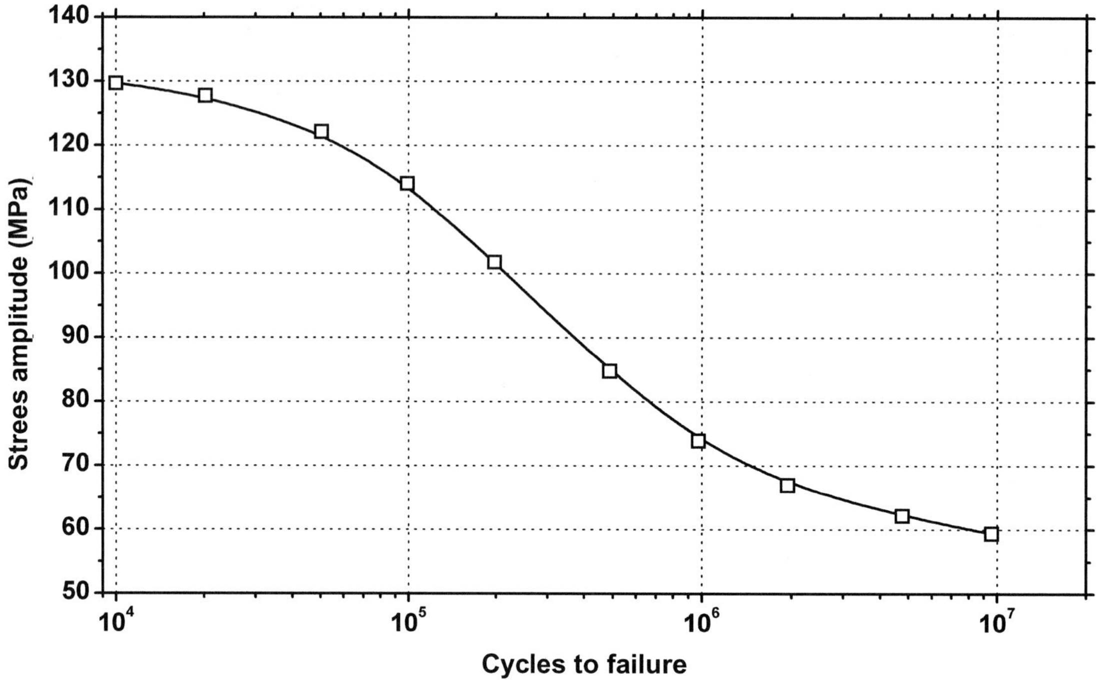

Figure 8.18 Flexural stress amplitude vs. cycles to failure of EMS-GRIVORY Grilon® PV-5 H—50% glass fiber reinforced, UV stabilized, high flow PA6.

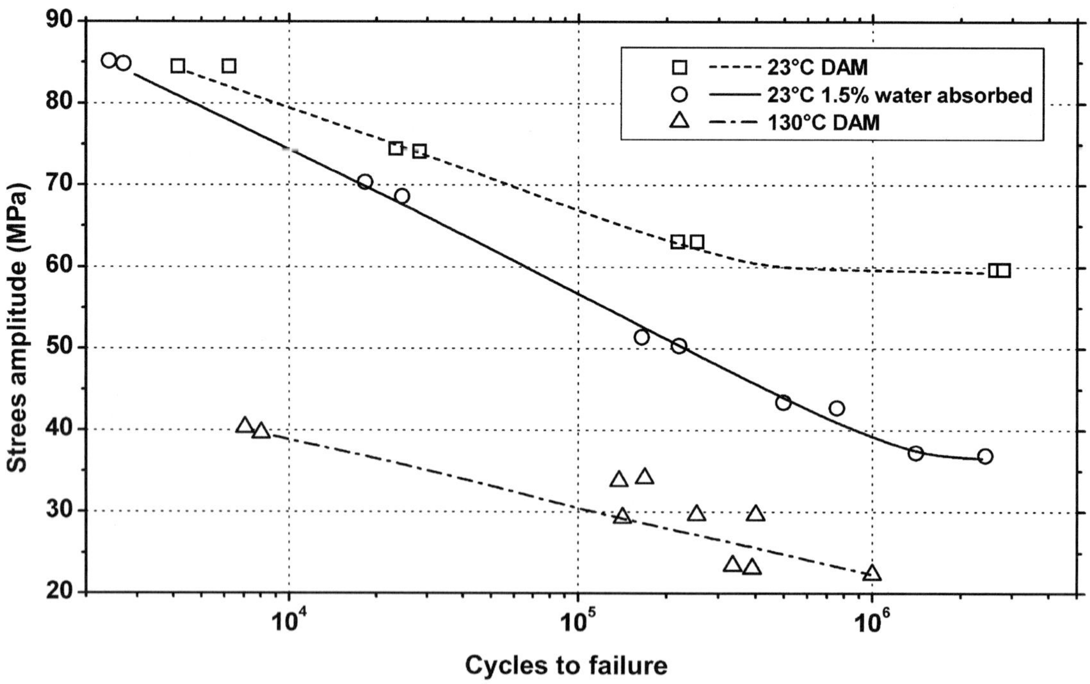

Figure 8.19 Flexural stress amplitude vs. cycles to failure under various conditions of Toray Resin Company Amilan™ CM1011G-45—45% glass fiber reinforced standard grade PA6.

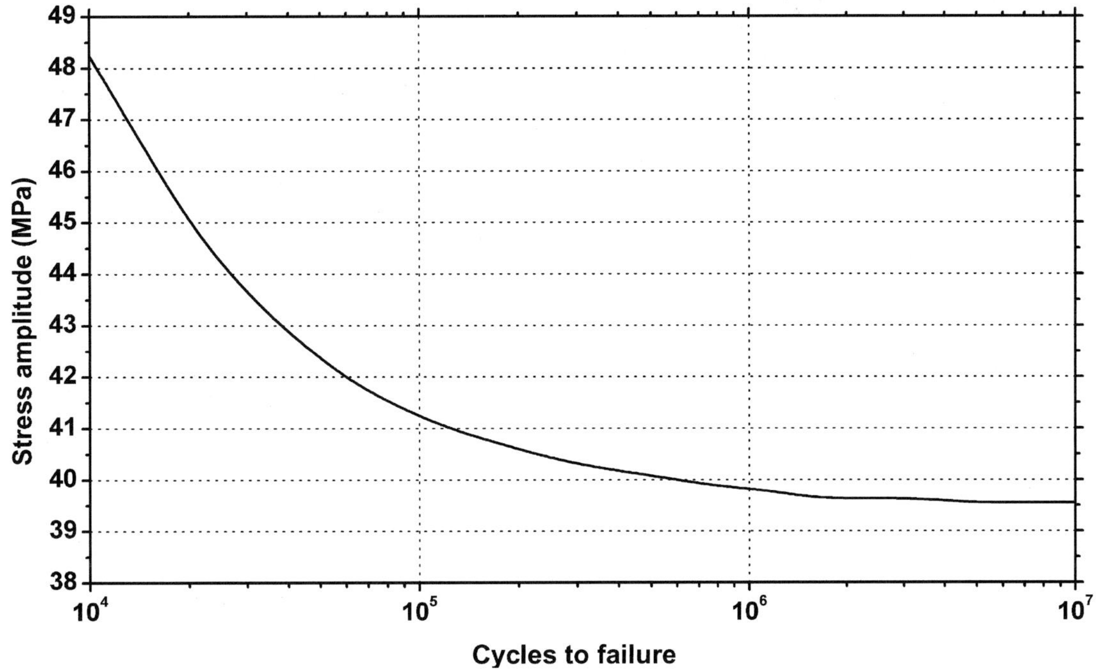

Figure 8.20 Flexural stress amplitude vs. cycles to failure under various conditions of SABIC Innovative Plastics Thermocomp® PF-1006 (PF006)—30% glass fiber reinforced PA6.

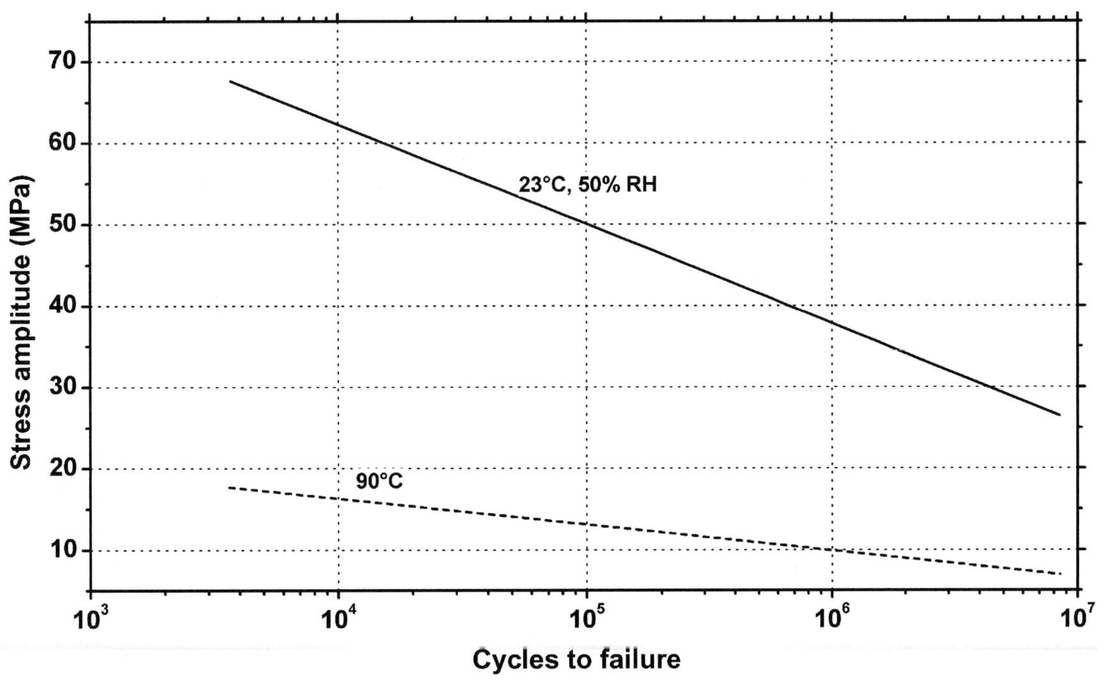

Figure 8.21 Flexural stress amplitude vs. cycles to failure and temperature of BASF Ultramid® B 3WG6—easy flow, 30% glass fiber reinforced PA6.

8.2.2 Tribology Data

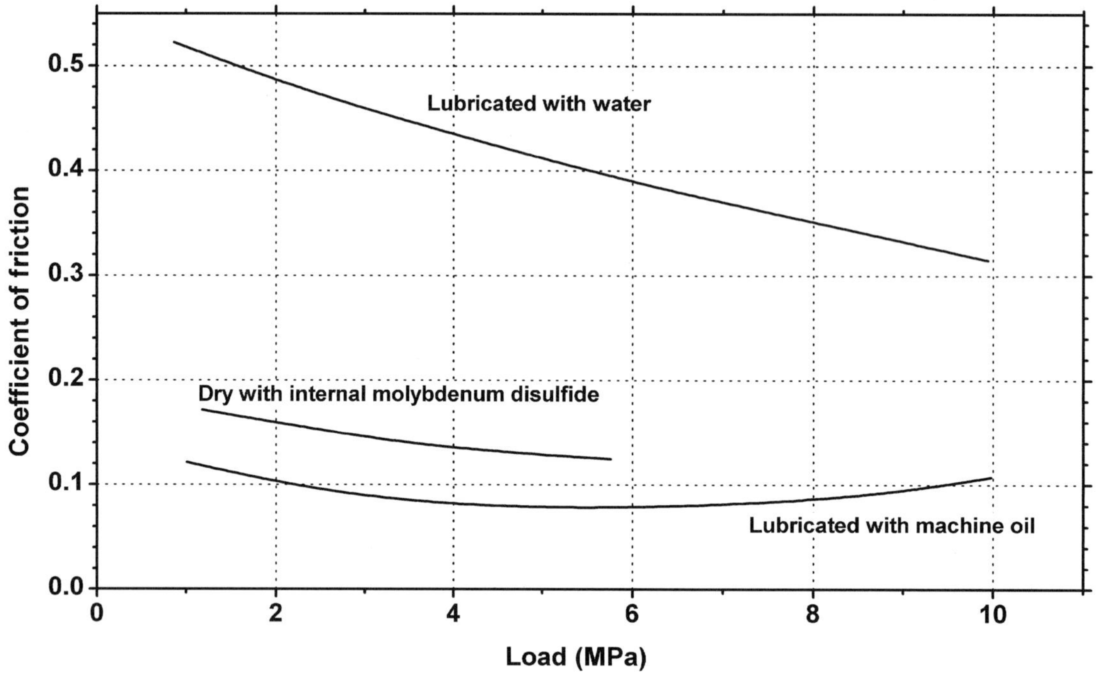

Figure 8.22 Coefficient of friction vs. load under different lubrication conditions of Toray Resin Company Amilan™ CM1021—unreinforced, medium viscosity PA6.

Table 8.2 Tribological Properties of RTP Company RTP 207A TFE 13 SI 2 HS (with Glass Fiber 40%, PTFE 13%, Silicone 2%, Heat Stabilized) vs. 1018 C Steel (Data Obtained per ASTM 3702)

PV (KPa m/s)	Load (N)	Speed (m/s)	Wear Factor × 10^{-8} (mm^3/N m)	Dynamic Coefficient of Friction
70	1.80	0.25	98	0.70
175	2.25	0.50	125	0.61
350	2.25	1.00	822	0.61

Table 8.3 Tribological Properties of RTP Company RTP 207A TFE 20 HS (with Glass Fiber 40%, PTFE 20%, Heat Stabilized) vs. 1018 C Steel (Data Obtained per ASTM 3702)

PV (KPa m/s)	Load (N)	Speed (m/s)	Wear Factor × 10^{-8} (mm^3/N m)	Dynamic Coefficient of Friction
70	1.80	0.25	92	0.10
175	2.25	0.50	77	0.17

Table 8.4 Tribological Properties of RTP Company RTP 299A × 82678 C (Proprietary Formulation) vs. 1018 C Steel (Data Obtained per ASTM 3702)

PV (KPa m/s)	Load (N)	Speed (m/s)	Wear Factor × 10^{-8} (mm^3/N m)	Dynamic Coefficient of Friction
70	1.80	0.25	663	0.60

Table 8.5 Tribological Properties of RTP Company RTP 299A × 90821 (Proprietary Formulation) vs. 1018 C Steel (Data Obtained per ASTM 3702)

PV (KPa m/s)	Load (N)	Speed (m/s)	Wear Factor × 10^{-8} (mm^3/N m)	Dynamic Coefficient of Friction
70	1.80	0.25	448	0.47

8.3 Polyamide 12 (Nylon 12)

8.3.1 Fatigue Data

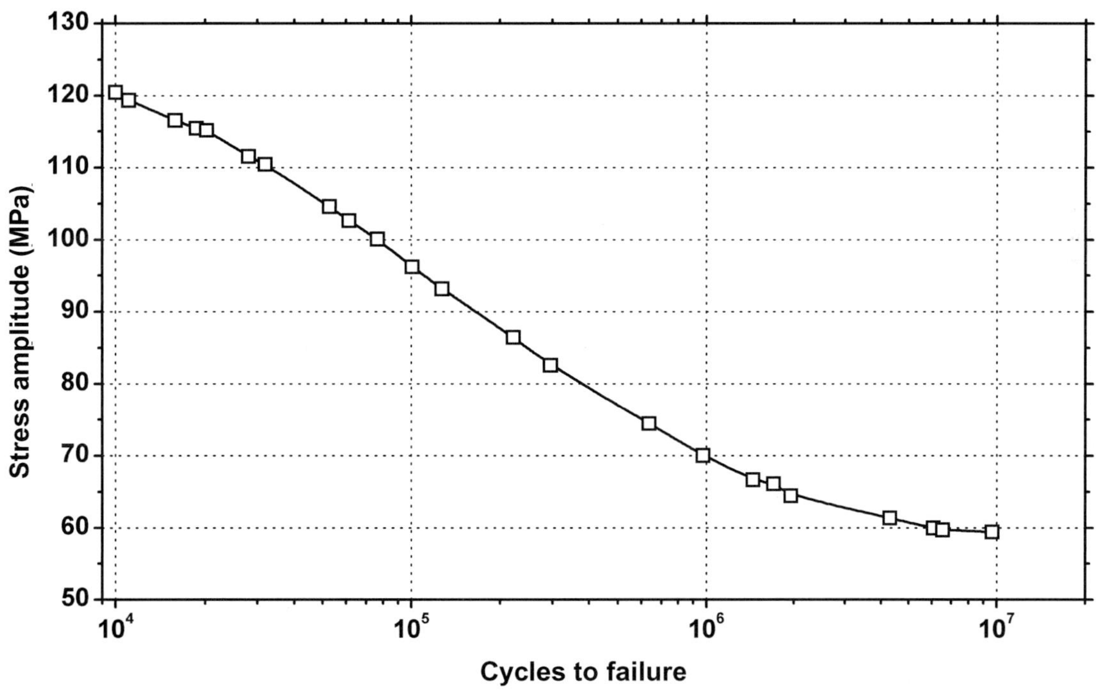

Figure 8.23 Flexural stress amplitude vs. cycles to failure of EMS-GRIVORY Grilamid® LV-5H—50% glass fiber, heat-stabilized PA12.

8.3.2 Tribology Data

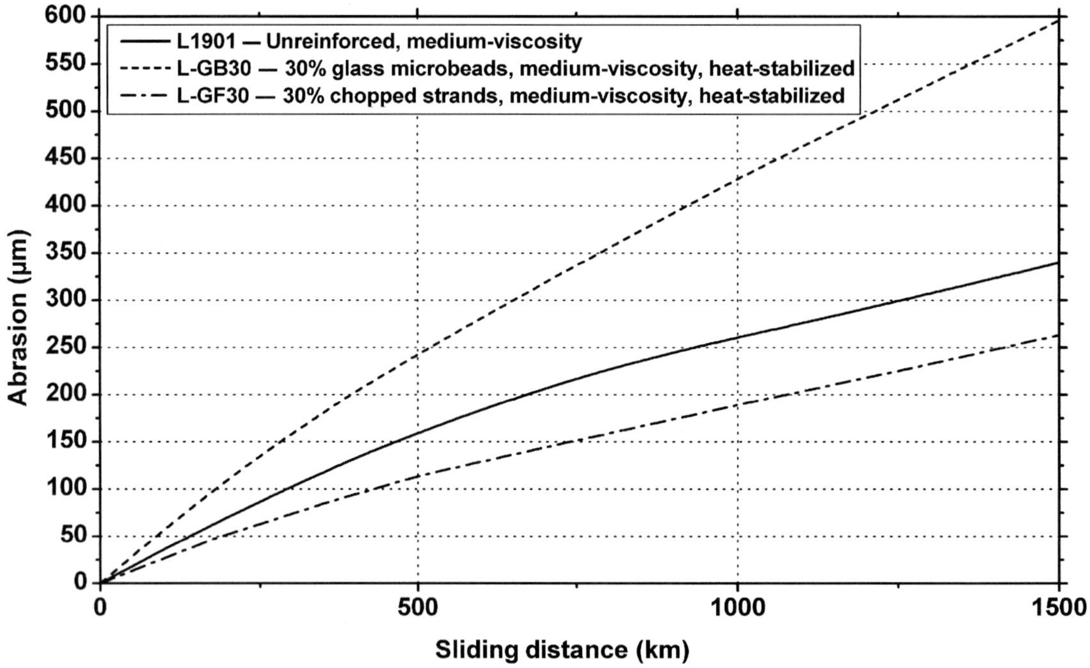

Figure 8.24 Abrasion vs. sliding distance of several Evonik Industries Vestamid® PA12 plastics.

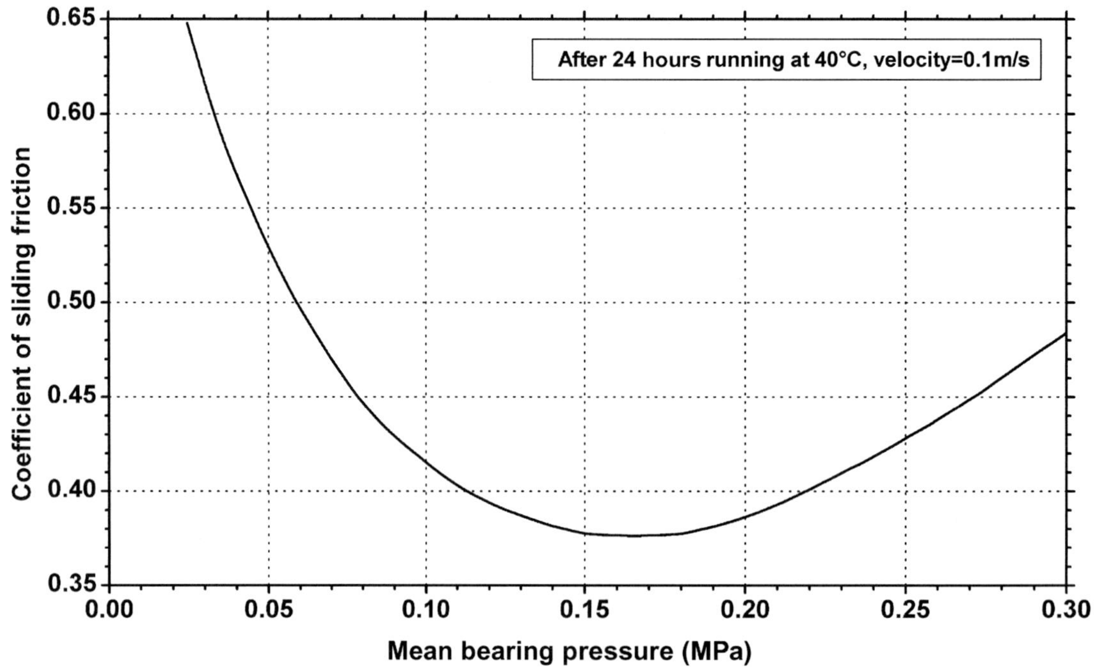

Figure 8.25 Dynamic coefficient of friction vs. bearing pressure of Evonik Industries Vestamid® L1901—unreinforced, medium viscosity PA12.

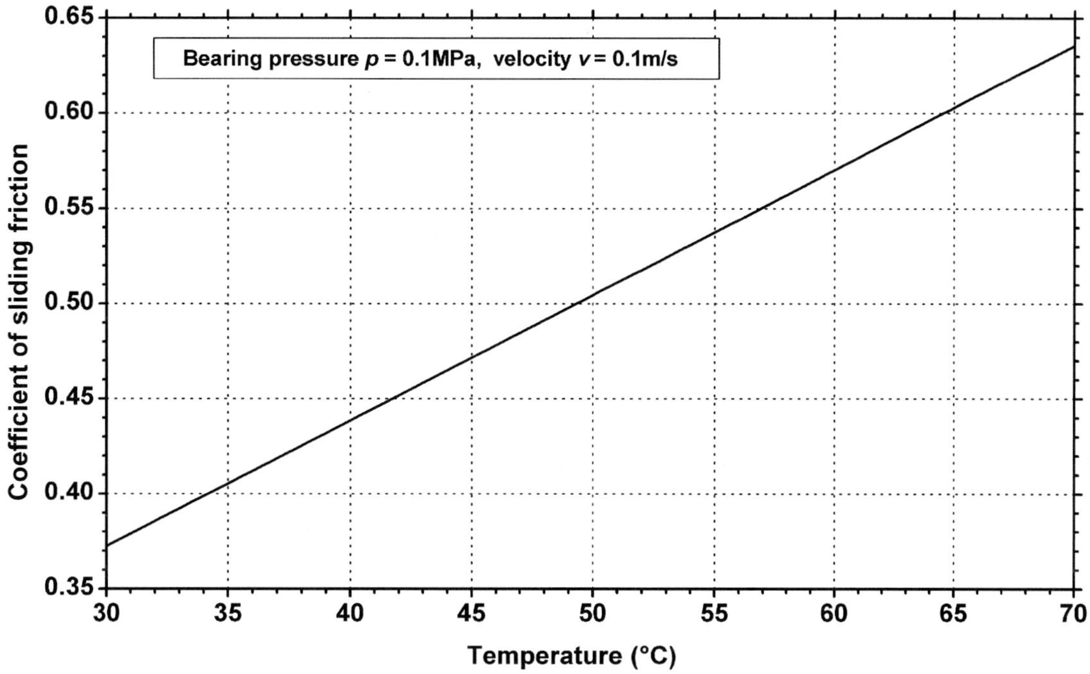

Figure 8.26 Dynamic coefficient of friction vs. bearing temperature of Evonik Industries Vestamid® L1901—Unreinforced, medium viscosity PA12.

Table 8.6 Taber Abrasion of Evonik Industries Vestamid® PA12 Plastics

Vestamid Material Code	mg/1000 cycles	mm^3/1000 cycles
L1600—Low viscosity	10–11	48
L1670—Low viscosity, heat and light stabilized with processing aid	10–11	48
L2101F—High viscosity, steam sterilizable	12–13	68
L2140—High viscosity, high heat	12–13	68
L2124—High viscosity, plasticized, heat and light stabilized, with processing aid	13–16	40
L2128—High viscosity, highly plasticized, heat and light stabilized, with processing aid	22–23	–
L1950—Medium-viscosity, heat-stabilized, molybdenum disulfide modification	12–13	39
L1930—30% Milled glass, medium viscosity, heat stabilized, with processing aid	16–19	170
L-GB30—30% Glass microbeads, medium viscosity, heat stabilized, with processing aid	14–15	120

8.4 Polyamide 66 (Nylon 66)

8.4.1 Fatigue Data

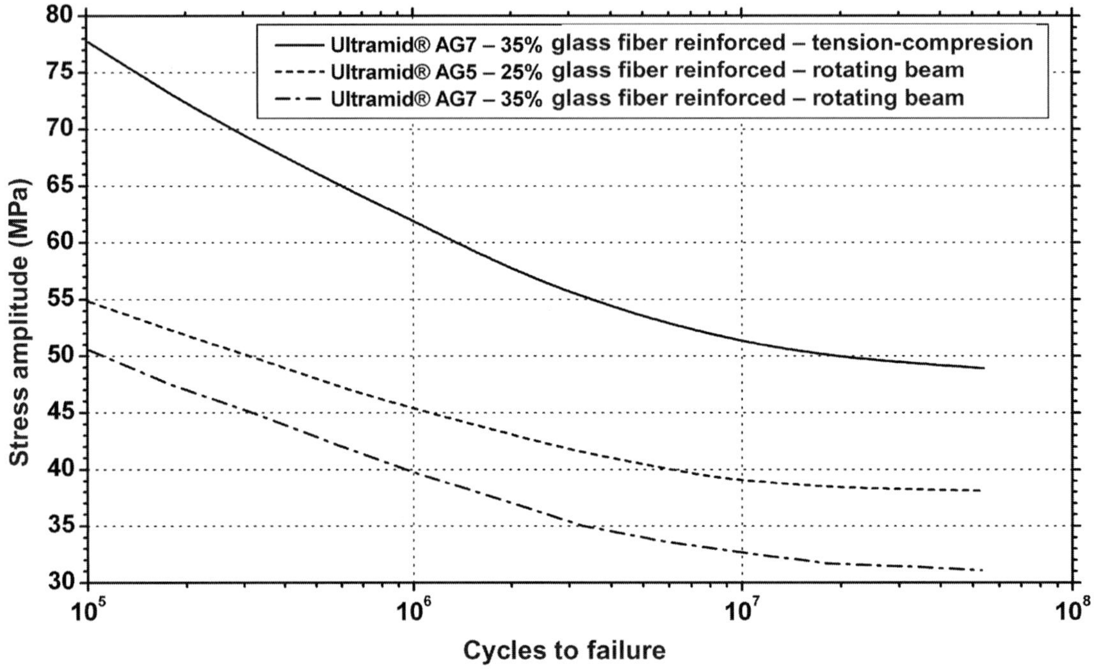

Figure 8.27 Stress amplitude vs. cycles to failure of BASF Ultramid® PA66 plastics.

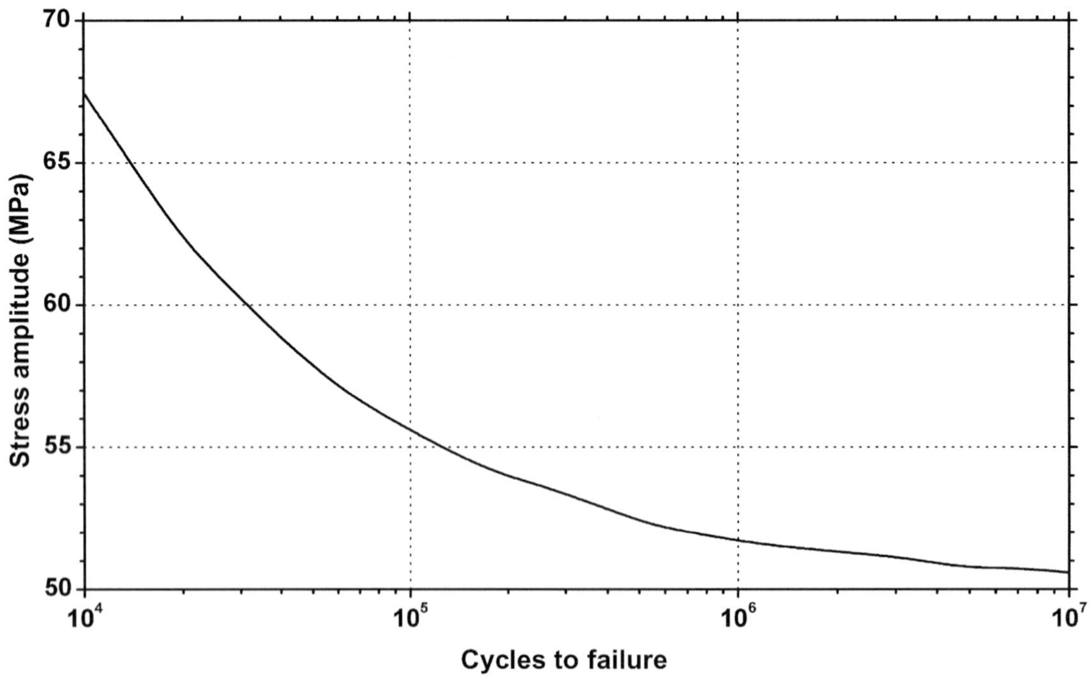

Figure 8.28 Flexural stress amplitude vs. cycles to failure of SABIC Innovative Plastics LNP Lubriloy® FR-40—40% glass fiber reinforced, lubricated PA66.

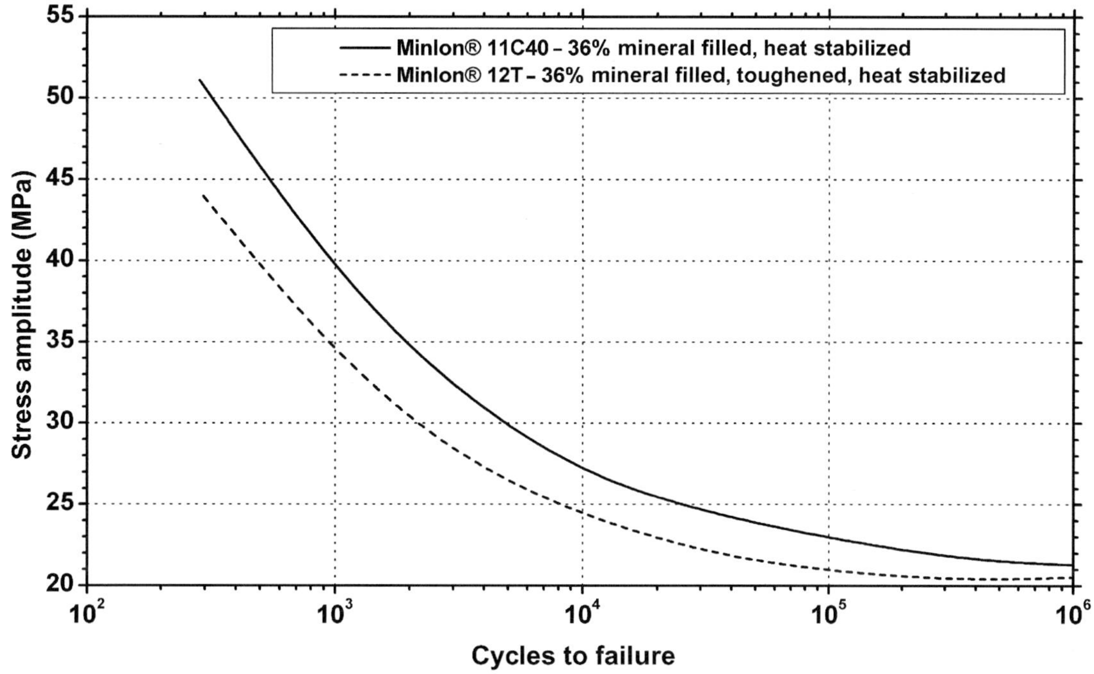

Figure 8.29 Flexural stress amplitude vs. cycles to failure at 23°C of two DuPont Engineering Plastics Minlon® mineral filled PA66 plastics.

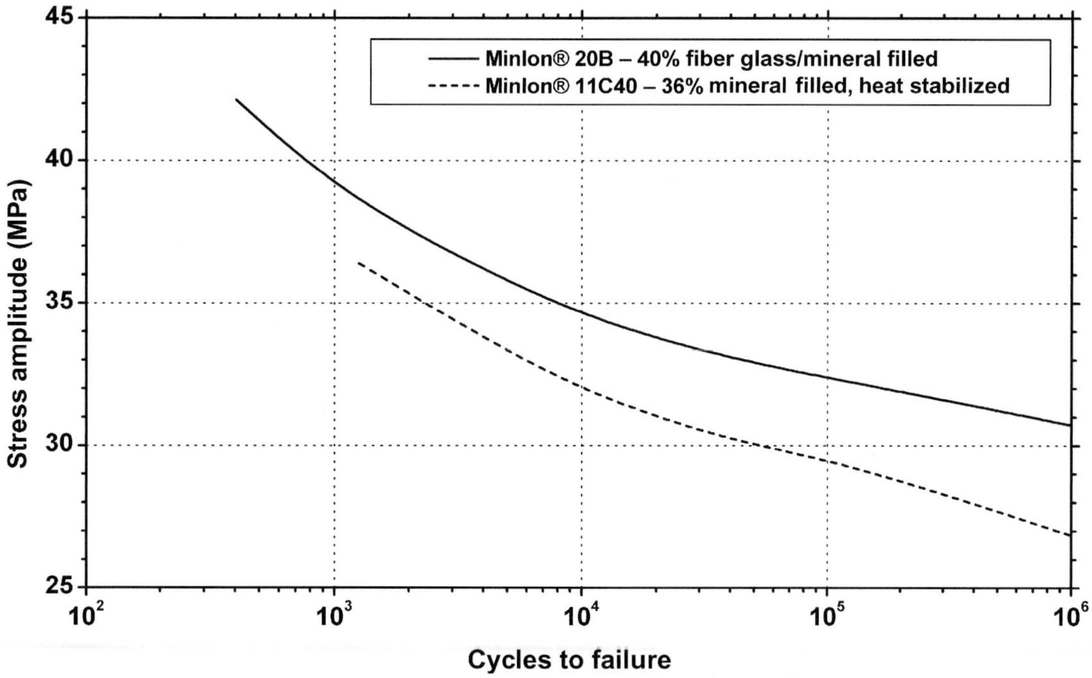

Figure 8.30 Flexural stress amplitude vs. cycles to failure at 23°C of two DuPont Engineering Plastics Minlon® filled PA66 plastics.

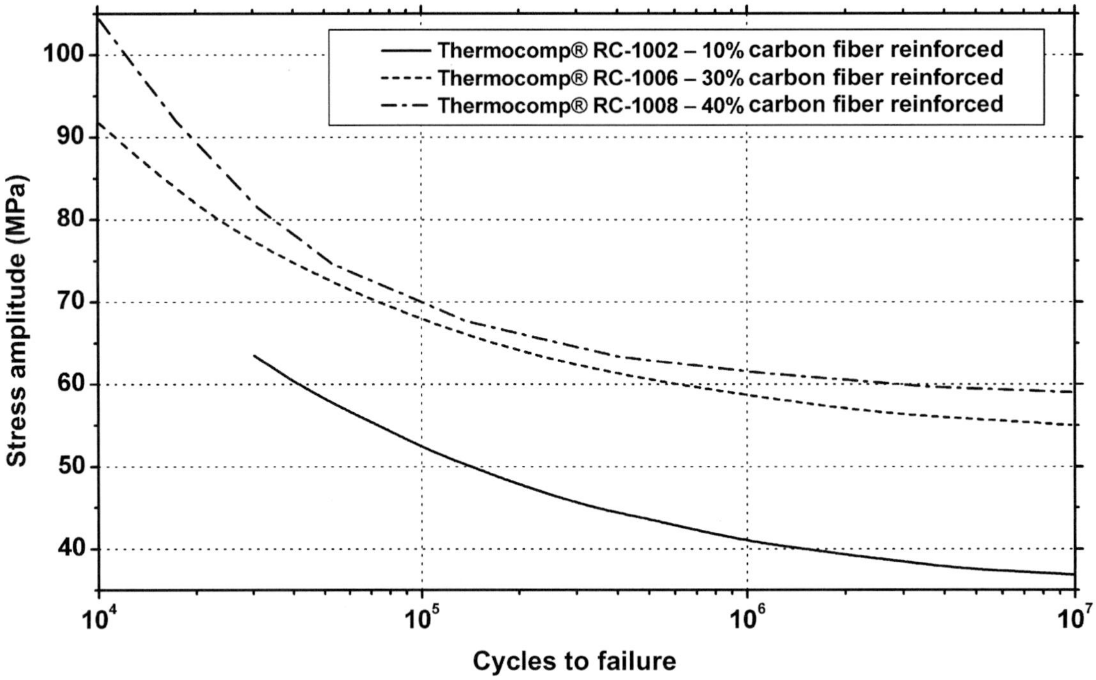

Figure 8.31 Flexural stress amplitude vs. cycles to failure at 23°C of several SABIC Innovative Plastics LNP Thermocomp® carbon fiber reinforced PA66 plastics.

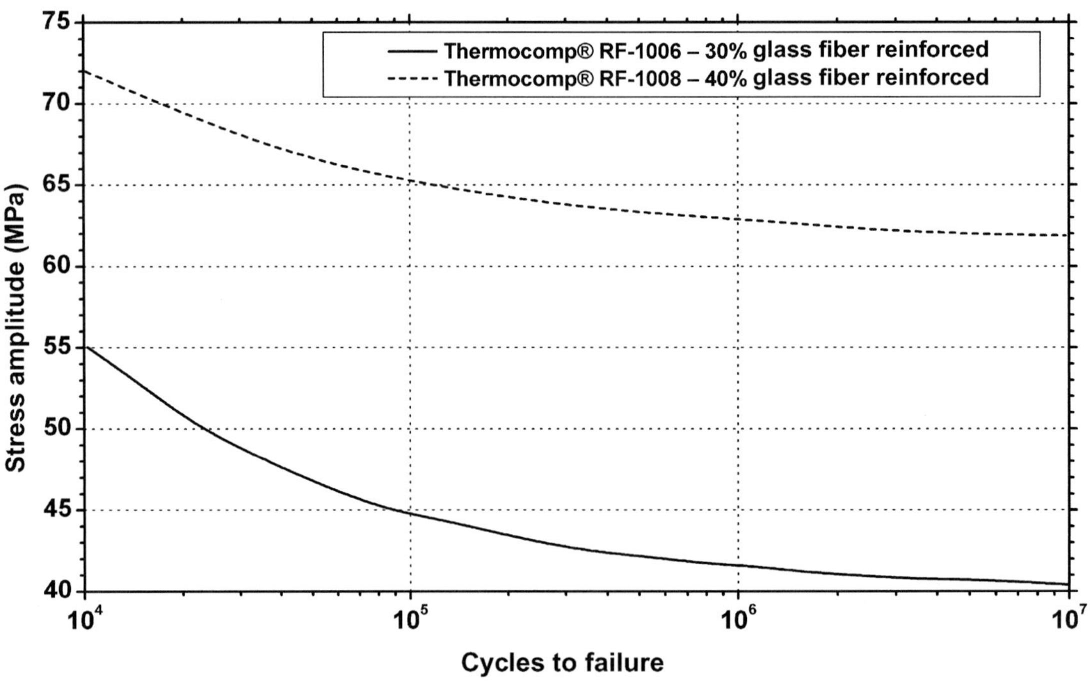

Figure 8.32 Flexural stress amplitude vs. cycles to failure at 23°C of two SABIC Innovative Plastics LNP Thermocomp® glass fiber reinforced PA66 plastics.

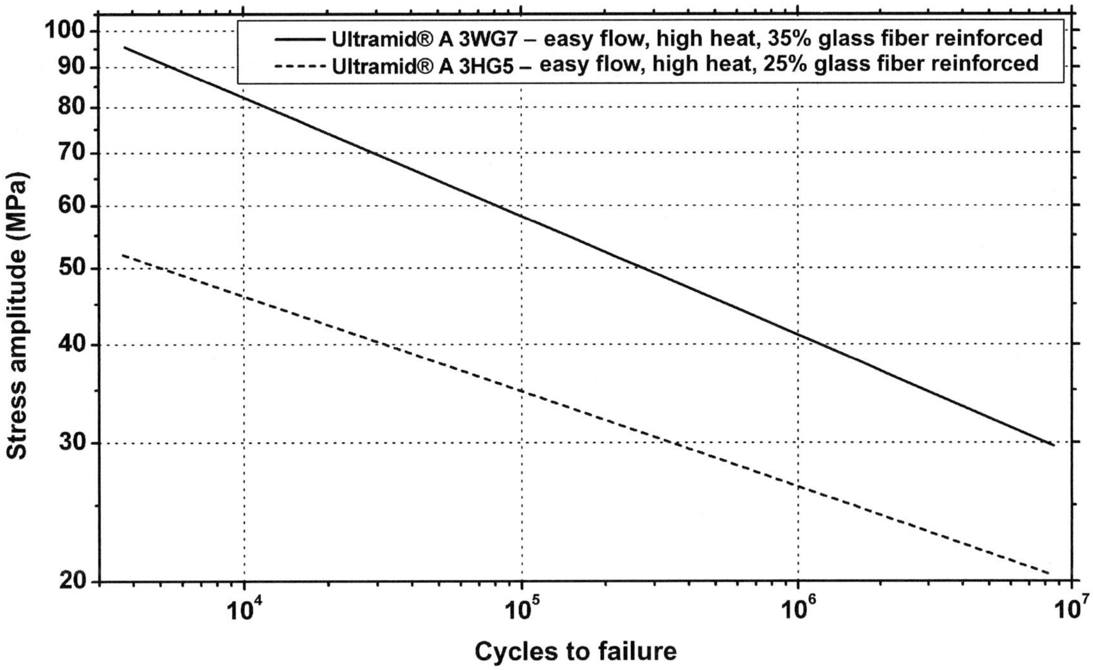

Figure 8.33 Flexural stress amplitude vs. cycles to failure at 23°C of two BASF Ultramid® glass fiber reinforced PA66 plastics.

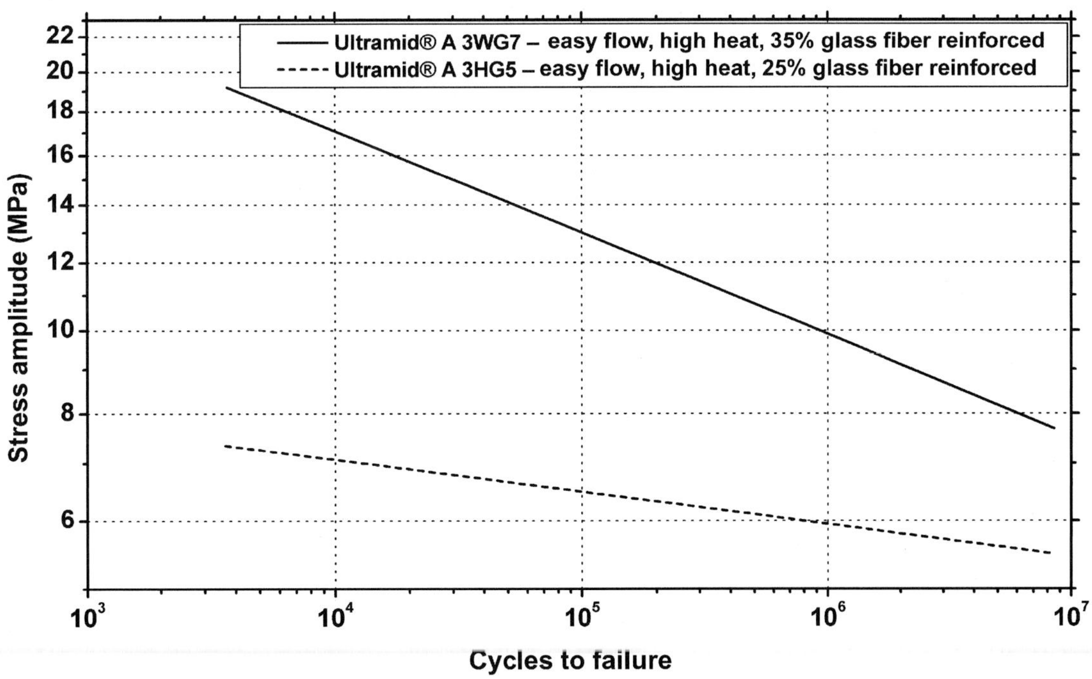

Figure 8.34 Flexural stress amplitude vs. cycles to failure at 90°C of two BASF Ultramid® glass fiber reinforced PA66 plastics.

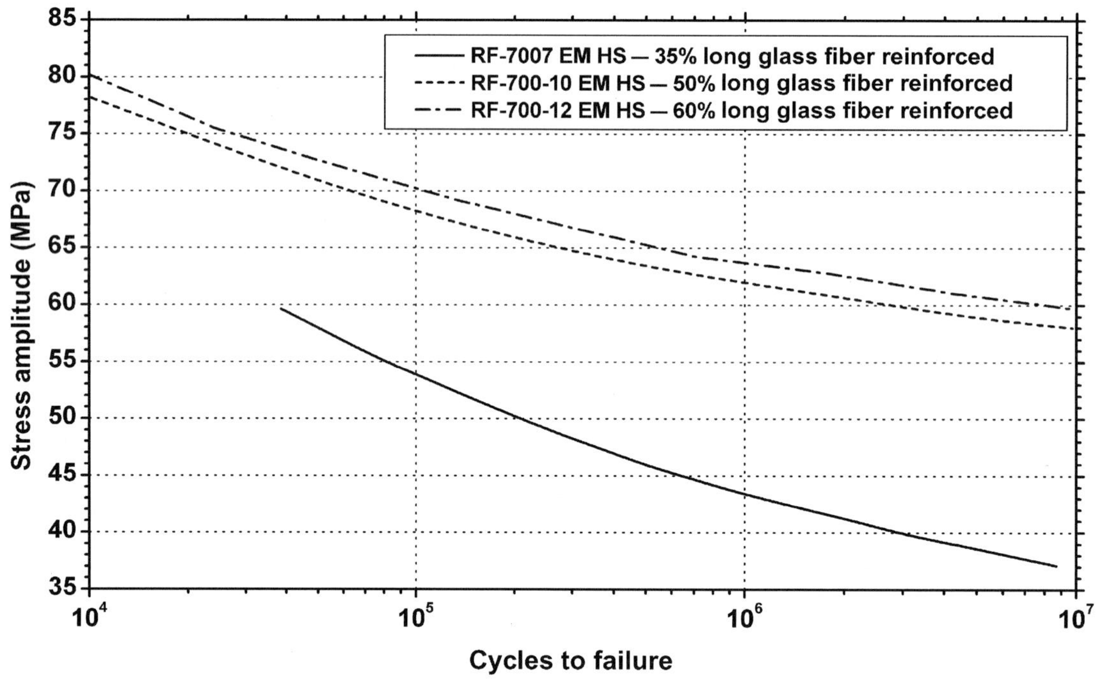

Figure 8.35 Flexural stress amplitude vs. cycles to failure at 23°C of three SABIC Innovative Plastics LNP Verton® long glass fiber reinforced PA66 plastics.

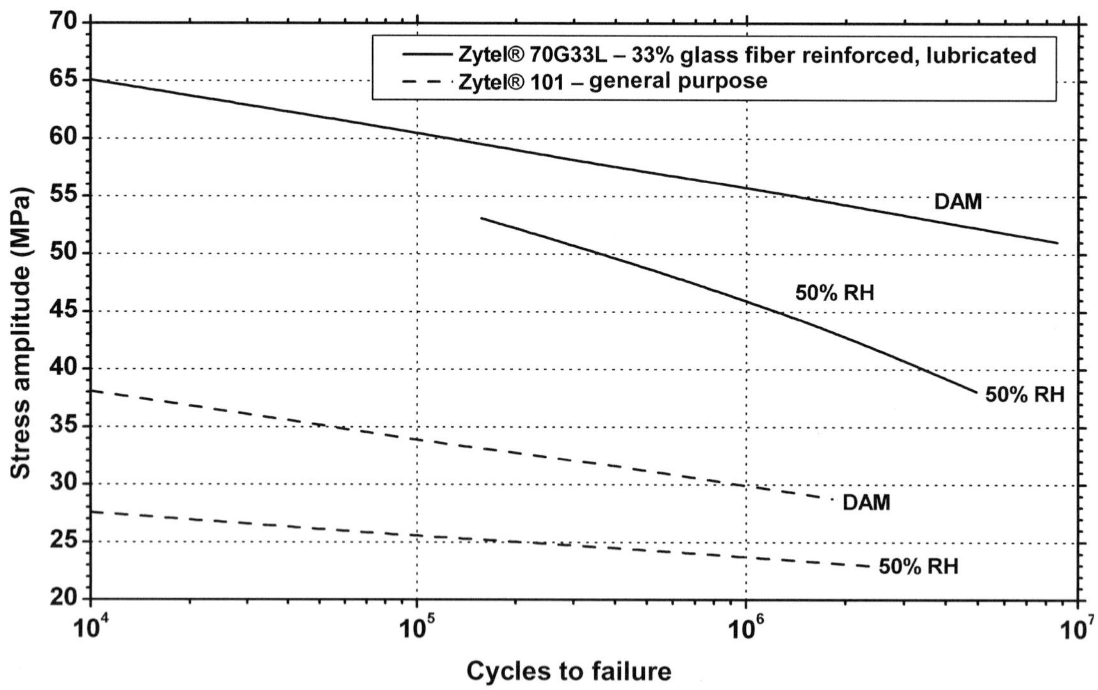

Figure 8.36 Flexural stress amplitude vs. cycles to failure at 23°C (conditioned and dry as molded) of two DuPont Engineering Plastics Zytel® PA66 plastics.

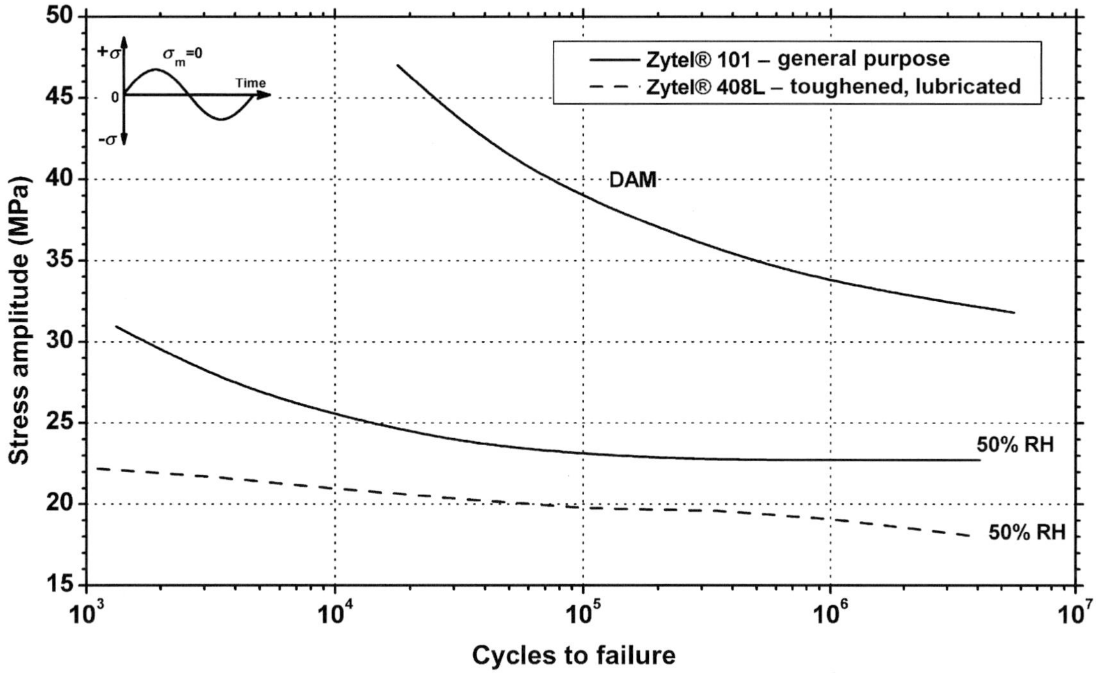

Figure 8.37 Axial tension/compression stress amplitude vs. cycles to failure at 23°C (conditioned and dry as molded) of two DuPont Engineering Plastics Zytel® PA66 plastics.

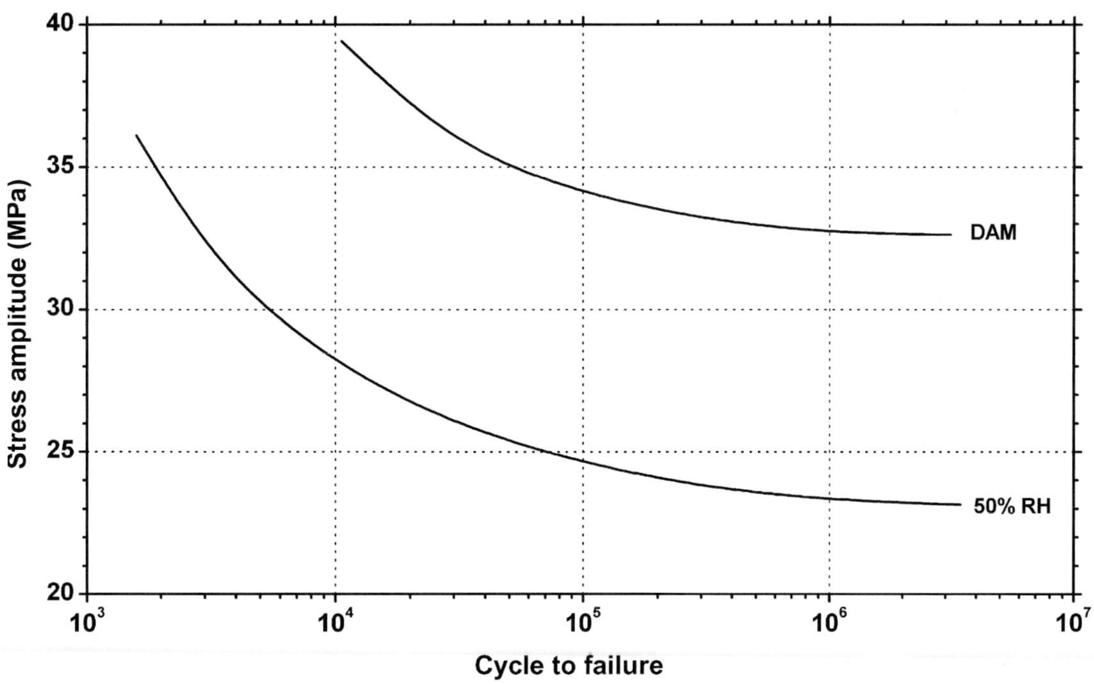

Figure 8.38 Flexural stress amplitude vs. cycles to failure at 23°C (conditioned and dry as molded) DuPont Engineering Plastics Zytel® 101—general-purpose PA66.

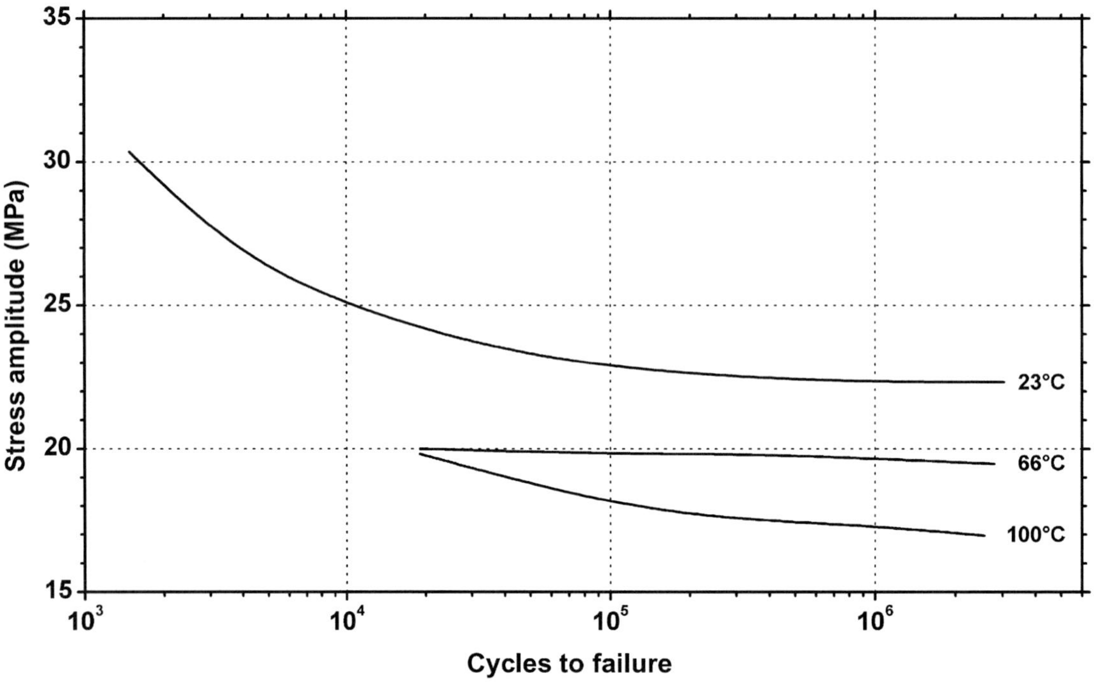

Figure 8.39 Axial stress amplitude vs. cycles to failure at different temperatures of conditioned DuPont Engineering Plastics Zytel® 101—general-purpose PA66.

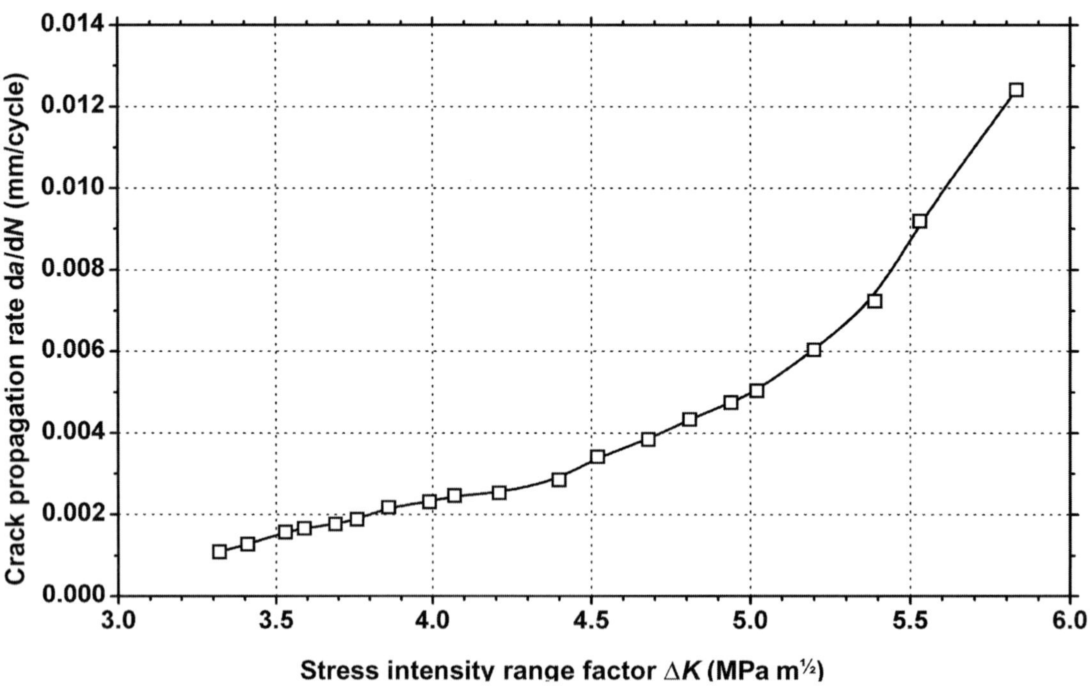

Figure 8.40 Fatigue crack propagation rate vs. stress intensity factor of DuPont Engineering Plastics Zytel® 122L PA66.

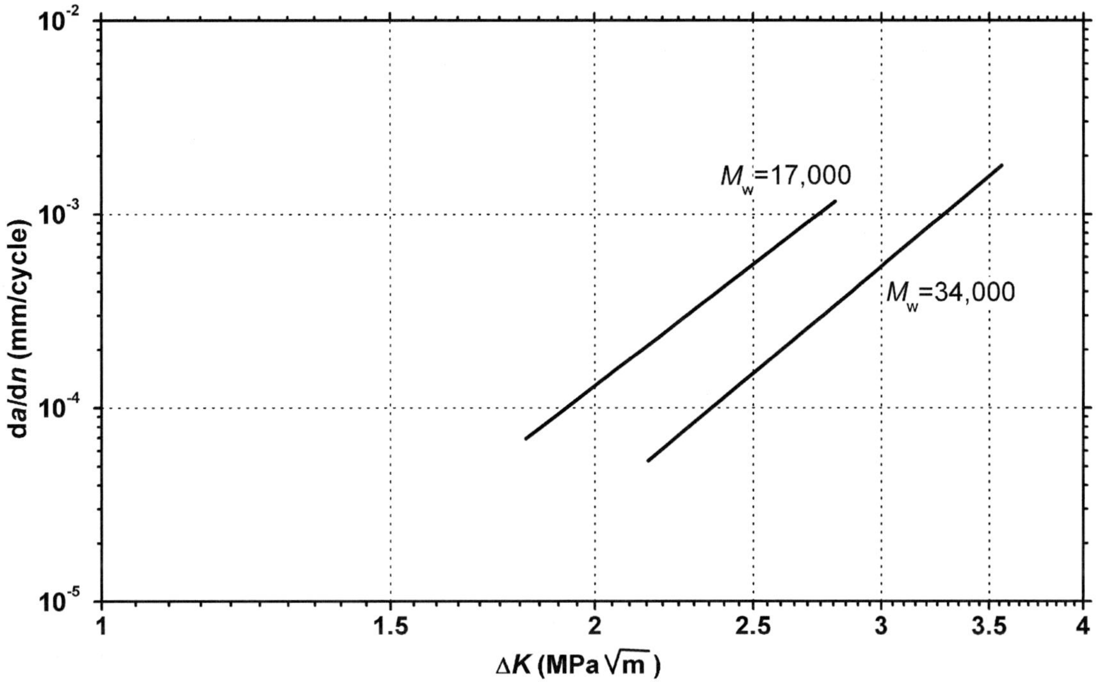

Figure 8.41 Fatigue crack propagation rate vs. stress intensity factor and molecular weight of generic PA66.

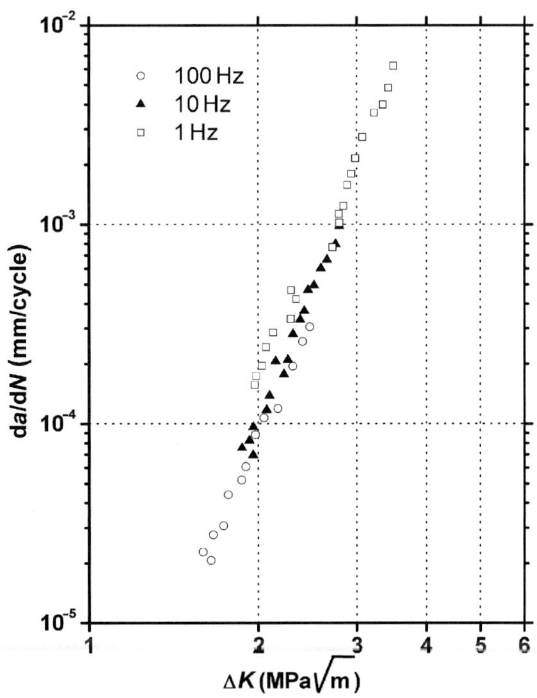

Figure 8.42 Fatigue crack propagation rate vs. stress intensity factor and cycle frequency of generic PA66.

8.4.2 Tribology Data

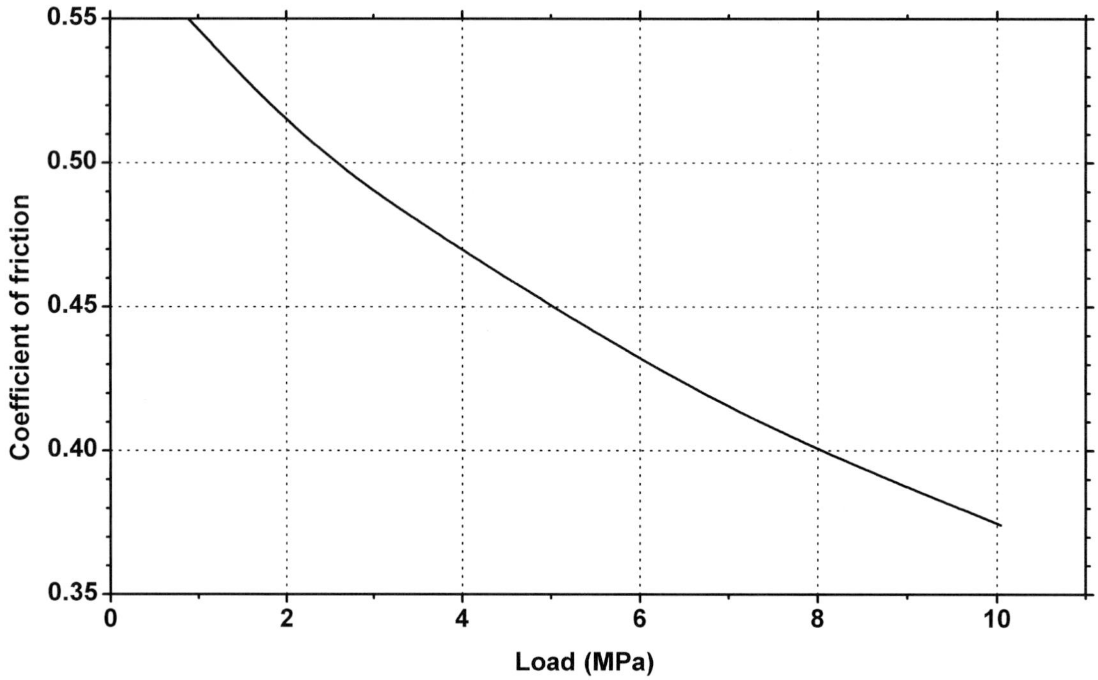

Figure 8.43 Coefficient of friction vs. load (lubricated with water) of Toray Resin Company Amilan® CM3001N—unreinforced, standard grade PA66.

Table 8.7 Tribological Properties of RTP Company RTP 200 SI 2 (with 2% Silicone) vs. 1018 C Steel (Data Obtained per ASTM 3702)

PV (KPa m/s)	Load (N)	Speed (m/s)	Wear Factor $\times 10^{-8}$ (mm^3/N m)	Dynamic Coefficient of Friction
70	1.80	0.25	1285	0.54
175	2.25	0.50	363	0.78
350	9.00	0.25	172	0.77

Table 8.8 Tribological Properties of RTP Company RTP 200 SI 2 (with 2% Silicone) vs. RTP 200 SI 2 (Data Obtained per ASTM 3702)

PV (KPa m/s)	Load (N)	Speed (m/s)	Wear Factor $\times 10^{-8}$ (mm^3/N m)	Dynamic Coefficient of Friction
17.5	0.45	0.25	820	0.41
35	0.90	0.25	609	0.15
70	1.80	0.25	7216	0.09

Table 8.9 Tribological Properties of RTP Company RTP 200 TFE 5 (with 5% PTFE) vs. 1018 C Steel (Data Obtained per ASTM 3702)

PV (KPa m/s)	Load (N)	Speed (m/s)	Wear Factor × 10^{-8} (mm^3/N m)	Dynamic Coefficient of Friction
70	1.80	0.25	1924	0.61
175	2.25	0.50	859	0.77
350	4.50	0.50	153	0.59

Table 8.10 Tribological Properties of RTP Company RTP 200 TFE 10 (with 10% PTFE) vs. 1018 C Steel (Data Obtained per ASTM 3702)

PV (KPa m/s)	Load (N)	Speed (m/s)	Wear Factor × 10^{-8} (mm^3/N m)	Dynamic Coefficient of Friction
70	1.80	0.25	878	0.42
70	1.80	0.25	1189	0.52
70	1.80	0.25	550	0.39
70	1.80	0.25	128	0.31
70	1.80	0.25	303	0.25
175	2.25	0.50	681	0.52
175	2.25	0.50	310	0.43
175	2.25	0.50	39	0.28
350	9.00	0.25	314	0.29
350	4.50	0.50	66	0.28
350	4.50	0.50	66	0.28
350	2.25	1.00	555	0.38
350	2.25	1.00	96	0.35
350	2.25	1.00	119	0.35

Table 8.11 Tribological Properties of RTP Company RTP 200 TFE 10 (with 10% PTFE) vs. RTP 200 TFE 10 (Data Obtained per ASTM 3702)

PV (KPa m/s)	Load (N)	Speed (m/s)	Wear Factor × 10^{-8} (mm^3/N m)	Dynamic Coefficient of Friction
17.5	0.45	0.25	900	0.33
70	1.80	0.25	197	0.23

Table 8.12 Tribological Properties of RTP Company RTP 200 TFE 10 SI 2 (with 10% PTFE and 2% Silicone) vs. 1018 C Steel (Data Obtained per ASTM 3702)

PV (KPa m/s)	Load (N)	Speed (m/s)	Wear Factor × 10^{-8} (mm^3/N m)	Dynamic Coefficient of Friction
17.5	0.90	0.15	144	0.14
17.5	0.45	0.25	62	0.20
17.5	0.22	0.50	237	0.24
70	1.80	0.25	446	0.25

Table 8.13 Tribological Properties of RTP Company RTP 200 TFE 20 (with 20% PTFE) vs. 1018 C Steel (Data Obtained per ASTM 3702)

PV (KPa m/s)	Load (N)	Speed (m/s)	Wear Factor × 10^{-8} (mm^3/N m)	Dynamic Coefficient of Friction
70	1.80	0.25	288	0.32
70	1.80	0.25	123	0.23
350	2.25	1.00	132	0.35
350	2.25	1.00	91	0.35
350	9.00	0.25	23	0.18

Table 8.14 Tribological Properties of RTP Company RTP 200 TFE 20 (with 20% PTFE) vs. RTP 200 TFE 20 (Data Obtained per ASTM 3702)

PV (KPa m/s)	Load (N)	Speed (m/s)	Wear Factor × 10^{-8} (mm^3/N m)	Dynamic Coefficient of Friction
17.5	0.90	0.15	156	0.60
17.5	0.22	0.50	146	0.23
35	0.90	0.25	165	0.13
70	3.60	0.15	22	0.12
70	1.80	0.25	442	0.42

Table 8.15 Tribological Properties of RTP Company RTP 200 TFE 18 SI 2 (with 18% PTFE and 2% Silicone) vs. 1018 C Steel (Data Obtained per ASTM 3702)

PV (KPa m/s)	Load (N)	Speed (m/s)	Wear Factor × 10^{-8} (mm^3/N m)	Dynamic Coefficient of Friction
70	1.80	0.25	22	0.20
70	0.90	0.50	491	0.33
70	0.90	0.50	511	0.33
175	2.25	0.50	119	0.36
350	9.00	0.25	37	0.19
350	2.25	1.00	1512	0.07

Table 8.16 Tribological Properties of RTP Company RTP 200 TFE 18 SI 2 (with 18% PTFE and 2% Silicone) vs. RTP 200 TFE 18 SI 2 (Data Obtained per ASTM 3702)

PV (KPa m/s)	Load (N)	Speed (m/s)	Wear Factor × 10^{-8} (mm^3/N m)	Dynamic Coefficient of Friction
17.5	0.45	0.25	79	0.20
17.5	0.22	0.50	79	0.08
35	0.90	0.25	25	0.16
70	1.80	0.25	128	0.02

Table 8.17 Tribological Properties of RTP Company RTP 202 TFE 15 (with 15% Glass Fiber Reinforcement and 15% PTFE) vs. 1018 C Steel (Data Obtained per ASTM 3702)

PV (KPa m/s)	Load (N)	Speed (m/s)	Wear Factor × 10^{-8} (mm^3/N m)	Dynamic Coefficient of Friction
70	1.80	0.25	429	0.44
175	2.25	0.50	75	0.50
350	9.00	0.25	460	0.27
350	2.25	1.00	263	0.34

Table 8.18 Tribological Properties of RTP Company RTP 202 TFE 15 (with 15% Glass Fiber Reinforcement and 15% PTFE) vs. RTP 202 TFE 15 (Data Obtained per ASTM 3702)

PV (KPa m/s)	Load (N)	Speed (m/s)	Wear Factor × 10^{-8} (mm^3/N m)	Dynamic Coefficient of Friction
17.5	0.45	0.25	22	0.26
70	1.80	0.25	194	0.07

Table 8.19 Tribological Properties of RTP Company RTP 202 TFE 13 SI 2 (with 15% Glass Fiber Reinforcement, 13% PTFE and 2% Silicone) vs. 1018 C Steel (Data Obtained per ASTM 3702)

PV (KPa m/s)	Load (N)	Speed (m/s)	Wear Factor × 10^{-8} (mm^3/N m)	Dynamic Coefficient of Friction
70	1.80	0.25	16	0.44
175	2.25	0.50	30	0.50
350	9.00	0.25	12	0.27
350	2.25	1.00	25	0.34

Table 8.20 Tribological Properties of RTP Company RTP 202 TFE 13 SI 2 (with 15% Glass Fiber Reinforcement, 13% PTFE, and 2% Silicone) vs. RTP 202 TFE 13 SI 2 (Data Obtained per ASTM 3702)

PV (KPa m/s)	Load (N)	Speed (m/s)	Wear Factor × 10^{-8} (mm^3/N m)	Dynamic Coefficient of Friction
17.5	0.45	0.25	18	0.12
70	1.80	0.25	62	0.14

Table 8.21 Tribological Properties of RTP Company RTP 205 TFE 15 (Glass Fiber 30%, PTFE 15%) vs. 1018 C Steel (Data Obtained per ASTM 3702)

PV (KPa m/s)	Load (N)	Speed (m/s)	Wear Factor × 10^{-8} (mm^3/N m)	Dynamic Coefficient of Friction
70	1.80	0.25	84	0.50
70	0.90	0.50	103	0.42
70	0.45	1.00	134	0.50
175	4.50	0.25	95	0.53
175	2.25	0.50	199	0.77
175	1.12	1.00	307	0.42
350	8.99	0.25	262	0.42
350	4.50	0.50	351	0.46
350	2.25	1.00	546	0.52

Table 8.22 Tribological Properties of RTP Company RTP 282 TFE 15 (Carbon Fiber 15%, PTFE 15%) vs. 1018 C Steel (Data Obtained per ASTM 3702)

PV (KPa m/s)	Load (N)	Speed (m/s)	Wear Factor × 10^{-8} (mm^3/N m)	Dynamic Coefficient of Friction
70	1.80	0.25	89	0.23
175	2.25	0.50	52	0.18
350	9.00	0.25	167	0.27
350	2.25	1.02	43	—

Table 8.23 Tribological Properties of RTP Company RTP 282 TFE 15 (Carbon Fiber 15%, PTFE 15%) vs. RTP 282 TFE 15 (Data Obtained per ASTM 3702)

PV (KPa m/s)	Load (N)	Speed (m/s)	Wear Factor × 10^{-8} (mm^3/N m)	Dynamic Coefficient of Friction
17.5	0.45	0.25	667	0.52
70	1.80	0.25	7	0.40

Table 8.24 Tribological Properties of RTP Company RTP 282 TFE 13 SI 2 (Carbon Fiber 15%, PTFE 13%, and Silicone 2%) vs. 1018 C Steel (Data Obtained per ASTM 3702)

PV (KPa m/s)	Load (N)	Speed (m/s)	Wear Factor × 10^{-8} (mm^3/N m)	Dynamic Coefficient of Friction
70	1.80	0.25	354	0.25
175	2.25	0.50	140	0.31
350	9.00	0.25	226	0.55
350	2.25	1.00	386	0.70

Table 8.25 Tribological Properties of RTP Company RTP 282 TFE 13 SI 2 (Carbon Fiber 15%, PTFE 13%, and Silicone 2%) vs. RTP 282 TFE 13 SI 2 (Data Obtained per ASTM 3702)

PV (KPa m/s)	Load (N)	Speed (m/s)	Wear Factor × 10^{-8} (mm^3/N m)	Dynamic Coefficient of Friction
17.5	0.45	0.25	124	0.08
70	1.80	0.25	227	0.22
175	2.25	0.50	2412	0.25

Table 8.26 Tribological Properties of RTP Company RTP 285 TFE 13 SI 2 (Carbon Fiber 30%, PTFE 13%) vs. 1018 C Steel (Data Obtained per ASTM 3702)

PV (KPa m/s)	Load (N)	Speed (m/s)	Wear Factor × 10^{-8} (mm^3/N m)	Dynamic Coefficient of Friction
70	1.80	0.25	72	0.35
70	0.90	0.50	76	0.35
70	0.45	1.00	121	0.31
175	4.50	0.25	151	0.34
175	2.25	0.50	106	0.31
175	1.15	1.00	134	0.28
350	9.00	0.25	169	0.59
350	4.50	0.50	220	0.74
350	2.25	1.00	189	0.64

Table 8.27 Tribological Properties of RTP Company RTP 200 AR 15 TFE 15 (Aramid Fiber 15%, PTFE 13%) vs. 1018 C Steel (Data Obtained per ASTM 3702)

PV (KPa m/s)	Load (N)	Speed (m/s)	Wear Factor × 10^{-8} (mm^3/N m)	Dynamic Coefficient of Friction
70	1.80	0.25	224	0.39
70	0.90	0.50	246	0.44
70	0.45	1.00	55	0.57
175	4.50	0.25	108	0.38
175	2.25	0.50	64	0.44
175	1.15	1.00	89	0.60
350	9.00	0.25	298	0.39
350	4.50	0.50	48	0.37
35	2.25	1.00	69	0.37

Table 8.28 Tribological Properties of Polyamide 66 Resins

Trade or Common Name	Supplier	Material Note	Test Temperature (°C)	Mating Surface	Pressure (MPa)	Sliding Velocity (m/min)	PV (MPa m/min)	Test Method	Coefficient of Friction Static	Coefficient of Friction, Kinetic	Wear Factor *K* (10^{-8} mm^3/Nm)	Wear Factor K Mating Surface (10^{-8} mm^3/Nm)
R1000	SABIC	Unmodified	23	Carbon steel; surface finish: 0.3–0.4 μm; 18–20 Rockwell C	0.28	15.2	4.3	Thrust washer	0.2	0.28	403	
R1000	SABIC	Unmodified	23	PC; unmodified	0.28	15.2	4.3	Thrust washer	0.06	0.05	108778	1964053
R1000	SABIC	Unmodified	23	PC (30% glass fiber, 15% PTFE)	0.28	15.2	4.3	Thrust washer	0.15	0.18	2820	20
R1000	SABIC	Unmodified	23	POM (unmodified)	0.28	15.2	4.3	Thrust washer	0.06	0.07	151	111
R1000	SABIC	Unmodified	23	PA66 (unmodified)	0.28	15.2	4.3	Thrust washer	0.06	0.07	5036	2216
R1000	SABIC	Unmodified	23	PA66 (30% glass fiber)	0.28	15.2	4.3	Thrust washer	0.07	0.08	44317	705
R1006	SABIC	30% glass fiber	23	AISI 440 stainless steel; surface finish: 1.3–1.8 μm	0.28	15.2	4.3	Thrust washer	0.15	0.18	99	0
R1006	SABIC	30% glass fiber	23	POM (20% PTFE)	0.28	15.2	4.3	Thrust washer	0.05	0.06	60	77
R1006	SABIC	30% glass fiber	23	PA66 (20% PTFE)	0.28	15.2	4.3	Thrust washer	0.05	0.07	50	81
R1006	SABIC	30% glass fiber	23	PA66 (30% glass fiber)	0.28	15.2	4.3	Thrust washer	0.12	0.12	1209	1209
R1006	SABIC	30% glass fiber	23	PA66 (30% glass fiber, 15% PTFE)	0.28	15.2	4.3	Thrust washer	0.07	0.09	705	806
R1006	SABIC	30% glass fiber	23	Steel AISI 1141; surface finish: 0.2–0.3 μm	0.28	15.2	4.3	Thrust washer	0.16	0.21	286	4

(Continued)

Table 8.28 (Continued)

Trade or Common Name	Supplier	Material Note	Test Temperature (°C)	Mating Surface	Pressure (MPa)	Sliding Velocity (m/min)	PV (MPa m/min)	Test Method	Coefficient of Friction Static	Coefficient of Friction, Kinetic	Wear Factor *K* (10^{-8} mm^3/Nm)	Wear Factor K Mating Surface (10^{-8} mm^3/Nm)
R1006	SABIC	30% glass fiber	23	Steel AISI 1141; surface finish: 0.3–0.4 μm	0.28	15.2	4.3	Thrust washer	0.25	0.31	151	2
R1006	SABIC	30% glass fiber	23	Steel AISI 1141; surface finish: 1.3–1.8 μm	0.28	15.2	4.3	Thrust washer	0.22	0.28	201	2
R1006	SABIC	30% glass fiber	23	AISI 304 stainless steel; surface finish: 0.2–0.4 μm	0.28	15.2	4.3	Thrust washer	0.12	0.22	66	1
R1006	SABIC	30% glass fiber	23	AISI 304 stainless steel; surface finish: 1.3–1.8 μm	0.28	15.2	4.3	Thrust washer	0.1	0.18	91	0
R1006	SABIC	30% glass fiber	23	AISI 440 stainless steel; surface finish: 0.2–0.4 μm	0.28	15.2	4.3	Thrust washer	0.2	0.22	97	1
R1006	SABIC	30% glass fiber	23	70/30 brass; surface finish: 0.2–0.4 μm	0.28	15.2	4.3	Thrust washer	0.17	0.22	107	85
R1006	SABIC	30% glass fiber	23	70/30 brass; surface finish: 1.3–1.8 μm	0.28	15.2	4.3	Thrust washer	0.16	0.19	105	34
R1006	SABIC	30% glass fiber	23	2024 aluminum; surface finish: 0.2–0.3 μm	0.28	15.2	4.3	Thrust washer	0.18	0.2		
R1006	SABIC	30% glass fiber	23	2024 aluminum; surface finish: 0.3–0.4 μm	0.28	15.2	4.3	Thrust washer	0.15	0.2	806	534
R1006	SABIC	30% glass fiber	23	2024 aluminum; surface finish: 1.3–1.8 μm	0.28	15.2	4.3	Thrust washer	0.16	0.21	4029	201

R1006	SABIC	30% glass fiber	23	PC 30% glass fiber	0.28	15.2	4.3	Thrust washer	0.16	0.27	26187	37267
R1006HS	SABIC	30% glass fiber	23	Carbon steel; surface finish: 0.3–0.4 μm; 18–20 Rockwell C	0.28	15.2	4.3	Thrust washer	0.25	0.31	151	
RAL4022	SABIC	10% aramid fiber, 10% PTFE	23	Carbon steel; surface finish: 0.3–0.4 μm; 18–20 Rockwell C	0.28	15.2	4.3	Thrust washer	0.12	0.13	26	
RAL4022	SABIC	10% aramid fiber, 10% PTFE	23	Steel AISI 1141; surface finish: 0.2–0.3 μm	0.28	15.2	4.3	Thrust washer	0.18	0.21	36	1
RAL4022	SABIC	10% aramid fiber, 10% PTFE	23	Steel AISI 1141; surface finish: 0.3–0.4 μm	0.28	15.2	4.3	Thrust washer	0.12	0.13	26	0
RAL4022	SABIC	10% aramid fiber, 10% PTFE	23	Steel AISI 1141; surface finish: 1.3–1.8 μm	0.28	15.2	4.3	Thrust washer	0.11	0.12	28	0
RAL4022	SABIC	10% aramid fiber, 10% PTFE	23	AISI 304 stainless steel; surface finish: 0.2–0.4 μm	0.28	15.2	4.3	Thrust washer	0.08	0.11	36	0
RAL4022	SABIC	10% aramid fiber, 10% PTFE	23	AISI 304 stainless steel; surface finish: 1.3–1.8 μm	0.28	15.2	4.3	Thrust washer	0.07	0.11	79	1
RAL4022	SABIC	10% aramid fiber, 10% PTFE	23	AISI 440 stainless steel; surface finish: 0.2–0.4 μm	0.28	15.2	4.3	Thrust washer	0.1	0.13	36	1
RAL4022	SABIC	10% aramid fiber, 10% PTFE	23	AISI 440 stainless steel; surface finish: 1.3–1.8 μm	0.28	15.2	4.3	Thrust washer	0.08	0.12	46	1
RAL4022	SABIC	10% aramid fiber, 10% PTFE	23	70/30 brass; surface finish: 0.2–0.4 μm	0.28	15.2	4.3	Thrust washer	0.12	0.15	32	0
RAL4022	SABIC	10% aramid fiber, 10% PTFE	23	70/30 brass; surface finish: 1.3–1.8 μm	0.28	15.2	4.3	Thrust washer	0.1	0.14	38	1

(*Continued*)

Table 8.28 (Continued)

Trade or Common Name	Supplier	Material Note	Test Temperature (°C)	Mating Surface	Pressure (MPa)	Sliding Velocity (m/min)	PV (MPa m/min)	Test Method	Coefficient of Friction Static	Coefficient of Friction, Kinetic	Wear Factor *K* (10^{-8} mm^3/Nm)	Wear Factor K Mating Surface (10^{-8} mm^3/Nm)
RAL4022	SABIC	10% aramid fiber, 10% PTFE	23	2024 aluminum; surface finish: 0.2–0.3 μm	0.28	15.2	4.3	Thrust washer	0.1	0.17	1007	16
RAL4022	SABIC	10% aramid fiber, 10% PTFE	23	2024 aluminum; surface finish: 0.3–0.4 μm	0.28	15.2	4.3	Thrust washer	0.11	0.16	97	8
RAL4022	SABIC	10% aramid fiber, 10% PTFE	23	2024 aluminum; surface finish: 1.3–1.8 μm	0.28	15.2	4.3	Thrust washer	0.09	0.17	258	10
RAL4022	SABIC	10% aramid fiber, 10% PTFE	23	PA66 (30% carbon fiber, 15% PTFE)	0.28	15.2	4.3	Thrust washer	0.1	0.13	604	81
RAL4022	SABIC	10% aramid fiber, 10% PTFE	23	PA66 (30% carbon fiber, 15% PTFE)	0.28	15.2	4.3	Thrust washer	0.1	0.13	604	81
RC1004	SABIC	20% carbon fiber	23	Carbon steel; surface finish: 0.3–0.4 μm; 18–20 Rockwell C	0.28	15.2	4.3	Thrust washer	0.16	0.2	81	
RC1006	SABIC	30% carbon fiber	23	Carbon steel; surface finish: 0.3–0.4 μm; 18–20 Rockwell C	0.28	15.2	4.3	Thrust washer	0.16	0.2	40	
RC1006	SABIC	30% carbon fiber	23	Steel AISI 1141; surface finish: 0.2–0.3 μm	0.28	15.2	4.3	Thrust washer	0.13	0.14	73	3
RC1006	SABIC	30% carbon fiber	23	Steel AISI 1141; surface finish: 0.3–0.4 μm	0.28	15.2	4.3	Thrust washer	0.16	0.2	40	1
RC1006	SABIC	30% carbon fiber	23	Steel AISI 1141; surface finish: 1.3–1.8 μm	0.28	15.2	4.3	Thrust washer	0.17	0.21	60	1

RC1006	SABIC	30% carbon fiber	23	AISI 304 stainless steel; surface finish: 0.2–0.4 μm	0.28	15.2	4.3	Thrust washer	0.11	0.21	48	0
RC1006	SABIC	30% carbon fiber	23	AISI 304 stainless steel; surface finish: 1.3–1.8 μm	0.28	15.2	4.3	Thrust washer	0.11	0.17	68	0
RC1006	SABIC	30% carbon fiber	23	AISI 440 stainless steel; surface finish: 0.2–0.4 μm	0.28	15.2	4.3	Thrust washer	0.08	0.28	83	1
RC1006	SABIC	30% carbon fiber	23	AISI 440 stainless steel; surface finish: 1.3–1.8 μm	0.28	15.2	4.3	Thrust washer	0.13	0.16	101	0
RC1006	SABIC	30% carbon fiber	23	70/30 brass; surface finish: 0.2–0.4 μm	0.28	15.2	4.3	Thrust washer	0.21	0.21	81	68
RC1006	SABIC	30% carbon fiber	23	70/30 brass; surface finish: 1.3–1.8 μm	0.28	15.2	4.3	Thrust washer	0.18	0.18	68	38
RC1006	SABIC	30% carbon fiber	23	PC (30% glass fiber, 15% PTFE)	0.28	15.2	4.3	Thrust washer	0.09	0.1	262	222
RC1006	SABIC	30% carbon fiber	23	PA66 (unmodified)	0.28	15.2	4.3	Thrust washer	0.26	0.16	2820	3425
RC1006	SABIC	30% carbon fiber	23	PA66 (30% glass fiber)	0.28	15.2	4.3	Thrust washer	0.09	0.18	604	91
RC1006	SABIC	30% carbon fiber	23	PA66 (30% glass fiber, 15% PTFE)	0.28	15.2	4.3	Thrust washer	0.11	0.12	262	60
RC1006	SABIC	30% carbon fiber	23	PA66 (30% carbon fiber, 15% PTFE)	0.28	15.2	4.3	Thrust washer	0.06	0.11	181	201
RC1008	SABIC	40% carbon fiber	23	RCL-4536 (30% carbon fiber, 13% PTFE, 2% silicone)	0.28	15.2	4.3	Thrust washer	0.12	0.14	91	101
RC-1008	SABIC	40% carbon fiber	23	Carbon steel; surface finish: 0.3–0.4 μm; 18–20 Rockwell C	0.28	15.2	4.3	Thrust washer	0.13	0.18	28	
RCL4036	SABIC	30% carbon fiber, 15% PTFE	23	AISI 304 stainless steel; surface finish: 1.3–1.8 μm	0.28	15.2	4.3	Thrust washer	0.16	0.2	34	1

(*Continued*)

Table 8.28 (Continued)

Trade or Common Name	Supplier	Material Note	Test Temperature (°C)	Mating Surface	Pressure (MPa)	Sliding Velocity (m/min)	PV (MPa m/min)	Test Method	Coefficient of Friction Static	Coefficient of Friction, Kinetic	Wear Factor *K* (10^{-8} mm^3/Nm)	Wear Factor K Mating Surface (10^{-8} mm^3/Nm)
RCL4036	SABIC	30% carbon fiber, 15% PTFE	23	AISI 440 stainless steel; surface finish: 0.2–0.4 μm	0.28	15.2	4.3	Thrust washer	0.1	0.23	32	0
RCL4036	SABIC	30% carbon fiber, 15% PTFE	23	PA66 (30% glass fiber)	0.28	15.2	4.3	Thrust washer	0.08	0.11	1612	161
RCL4036	SABIC	30% carbon fiber, 15% PTFE	23	PA66 (30% carbon fiber, 15% PTFE)	0.28	15.2	4.3	Thrust washer	0.1	0.11	161	181
RCL4036	SABIC	30% carbon fiber, 15% PTFE	23	2024 aluminum; surface finish: 1.3–1.8 μm	0.28	15.2	4.3	Thrust washer	0.12	0.13	498	304
RCL4036	SABIC	30% carbon fiber, 15% PTFE	23	PA66 (30% glass fiber, 15% PTFE)	0.28	15.2	4.3	Thrust washer	0.08	0.08	151	40
RCL4036	SABIC	30% carbon fiber, 15% PTFE	23	Carbon steel; surface finish: 0.3–0.4 μm; 18–20 Rockwell C	0.28	15.2	4.3	Thrust washer	0.11	0.15	20	
RCL4036	SABIC	30% carbon fiber, 15% PTFE	23	Steel AISI 1141; surface finish: 0.2–0.3 μm	0.28	15.2	4.3	Thrust washer	0.12	0.15	54	2
RCL4036	SABIC	30% carbon fiber, 15% PTFE	23	Steel AISI 1141; surface finish: 0.3–0.4 μm	0.28	15.2	4.3	Thrust washer	0.11	0.15	26	1
RCL4036	SABIC	30% carbon fiber, 15% PTFE	23	Steel AISI 1141; surface finish: 1.3–1.8 μm	0.28	15.2	4.3	Thrust washer	0.13	0.16	30	1
RCL4036	SABIC	30% carbon fiber, 15% PTFE	23	AISI 304 stainless steel; surface finish: 0.2–0.4 μm	0.28	15.2	4.3	Thrust washer	0.17	0.22	30	0

RCL4036	SABIC	30% carbon fiber, 15% PTFE	23	AISI 440 stainless steel; surface finish: 1.3–1.8 μm	0.28	15.2	4.3	Thrust washer	0.15	0.21	91	1
RCL4036	SABIC	30% carbon fiber, 15% PTFE	23	70/30 brass; surface finish: 0.2–0.4 μm	0.28	15.2	4.3	Thrust washer	0.15	0.15	36	10
RCL4036	SABIC	30% carbon fiber, 15% PTFE	23	70/30 brass; surface finish: 1.3–1.8 μm	0.28	15.2	4.3	Thrust washer	0.13	0.14	26	12
RCL4036	SABIC	30% carbon fiber, 15% PTFE	23	2024 aluminum; surface finish: 0.2–0.3 μm	0.28	15.2	4.3	Thrust washer	0.13	0.14	5036	1209
RCL4036	SABIC	30% carbon fiber, 15% PTFE	23	2024 aluminum; surface finish: 0.3–0.4 μm	0.28	15.2	4.3	Thrust washer	0.12	0.12	353	212
RCL4536	SABIC	30% carbon fiber, 13% PTFE, 2% silicone	23	Carbon steel; surface finish: 0.3–0.4 μm; 18–20 Rockwell C	0.28	15.2	4.3	Thrust washer	0.1	0.11	12	
RCL4536	SABIC	30% carbon fiber, 13% PTFE, 2% silicone	23	RCL-4536 (30% carbon fiber, 13% PTFE, 2% silicone)	0.28	15.2	4.3	Thrust washer	0.11	0.15	50	60
RF100-10	SABIC	50% glass fiber	23	Carbon steel; surface finish: 0.3–0.4 μm; 18–20 Rockwell C	0.28	15.2	4.3	Thrust washer	0.28	0.35	121	
RF1002	SABIC	10% glass fiber	23	Carbon steel; surface finish: 0.3–0.4 μm; 18–20 Rockwell C	0.28	15.2	4.3	Thrust washer	0.21	0.28	161	
RF1004	SABIC	20% glass fiber	23	Carbon steel; surface finish: 0.3–0.4 μm; 18–20 Rockwell C	0.28	15.2	4.3	Thrust washer	0.23	0.3	161	
RF1008	SABIC	40% glass fiber	23	Carbon steel; surface finish: 0.3–0.4 μm; 18–20 Rockwell C	0.28	15.2	4.3	Thrust washer	0.26	0.33	141	

(*Continued*)

Table 8.28 (Continued)

Trade or Common Name	Supplier	Material Note	Test Temperature (°C)	Mating Surface	Pressure (MPa)	Sliding Velocity (m/min)	PV (MPa m/min)	Test Method	Coefficient of Friction Static	Coefficient of Friction, Kinetic	Wear Factor *K* (10^{-8} mm^3/Nm)	Wear Factor K Mating Surface (10^{-8} mm^3/Nm)
RFL4036	SABIC	30% glass fiber, 15% PTFE	23	Carbon steel; surface finish: 0.3–0.4 μm; 18–20 Rockwell C	0.28	15.2	4.3	Thrust washer	0.19	0.26	32	
RFL4036	SABIC	30% glass fiber, 15% PTFE	23	Steel AISI 1141; surface finish: 0.2–0.3 μm	0.28	15.2	4.3	Thrust washer	0.2	0.26	60	2
RFL4036	SABIC	30% glass fiber, 15% PTFE	23	Steel AISI 1141; surface finish: 0.3–0.4 μm	0.28	15.2	4.3	Thrust washer	0.19	0.26	32	1
RFL4036	SABIC	30% glass fiber, 15% PTFE	23	Steel AISI 1141; surface finish: 1.3–1.8 μm	0.28	15.2	4.3	Thrust washer	0.17	0.2	32	1
RFL4036	SABIC	30% glass fiber, 15% PTFE	23	AISI 304 stainless steel; surface finish: 0.2–0.4 μm	0.28	15.2	4.3	Thrust washer	0.17	0.18	24	0
RFL4036	SABIC	30% glass fiber, 15% PTFE	23	AISI 304 stainless steel; surface finish: 1.3–1.8 μm	0.28	15.2	4.3	Thrust washer	0.13	0.15	26	0
RFL4036	SABIC	30% glass fiber, 15% PTFE	23	AISI 440 stainless steel; surface finish: 0.2–0.4 μm	0.28	15.2	4.3	Thrust washer	0.14	0.21	44	0
RFL4036	SABIC	30% glass fiber, 15% PTFE	23	AISI 440 stainless steel; surface finish: 1.3–1.8 μm	0.28	15.2	4.3	Thrust washer	0.13	0.15	32	0
RFL4036	SABIC	30% glass fiber, 15% PTFE	23	70/30 brass; surface finish: 0.2–0.4 μm	0.28	15.2	4.3	Thrust washer	0.18	0.15	42	24
RFL4036	SABIC	30% glass fiber, 15% PTFE	23	70/30 brass; surface finish: 1.3–1.8 μm	0.28	15.2	4.3	Thrust washer	0.16	0.15	36	24

RFL4036	SABIC	30% glass fiber, 15% PTFE	23	2024 aluminum; surface finish: 0.2–0.3 μm	0.28	15.2	4.3	Thrust washer	0.15	0.18	4532	3022
RFL4036	SABIC	30% glass fiber, 15% PTFE	23	2024 aluminum; surface finish: 0.3–0.4 μm	0.28	15.2	4.3	Thrust washer	0.15	0.18	645	353
RFL4036	SABIC	30% glass fiber, 15% PTFE	23	2024 aluminum; surface finish: 1.3–1.8 μm	0.28	15.2	4.3	Thrust washer	0.14	0.19	3626	151
RFL4036	SABIC	30% glass fiber, 15% PTFE	23	POM (20% PTFE)	0.28	15.2	4.3	Thrust washer	0.05	0.06	50	64
RFL4036	SABIC	30% glass fiber, 15% PTFE	23	PA66 (20% PTFE)	0.28	15.2	4.3	Thrust washer	0.05	0.06	30	50
RFL4036	SABIC	30% glass fiber, 15% PTFE	23	PA66 (30% glass fiber)	0.28	15.2	4.3	Thrust washer	0.07	0.1	705	1007
RFL4036	SABIC	30% glass fiber, 15% PTFE	23	PA66 (30% glass fiber, 15% PTFE)	0.28	15.2	4.3	Thrust washer	0.11	0.12	201	232
RFL4036	SABIC	30% glass fiber, 15% PTFE	23	PA66 (30% carbon fiber, 15% PTFE)	0.28	15.2	4.3	Thrust washer	0.1	0.15	60	282
RFL4036	SABIC	30% glass fiber, 15% PTFE	93	Cold rolled steel; surface finish: 0.3–0.4 μm; 22 Rockwell C	0.28	15.2	4.3	Thrust washer	0.29	0.24	121	
RFL4036	SABIC	30% glass fiber, 15% PTFE	149	Cold rolled steel; surface finish: 0.3–0.4 μm; 22 Rockwell C	0.28	15.2	4.3	Thrust washer	0.36	0.32	604	
RFL4036	SABIC	30% glass fiber, 15% PTFE	204	Cold rolled steel; surface finish: 0.3–0.4 μm; 22 Rockwell C	0.28	15.2	4.3	Thrust washer	0.37	0.4	1410	
RFL-4036	SABIC	30% carbon fiber, 15% PTFE	23	POM (30% glass fiber, 15% PTFE)	0.28	15.2	4.3	Thrust washer	0.05	0.07	121	56

(Continued)

Table 8.28 (Continued)

Trade or Common Name	Supplier	Material Note	Test Temperature (°C)	Mating Surface	Pressure (MPa)	Sliding Velocity (m/min)	PV (MPa m/min)	Test Method	Coefficient of Friction Static	Coefficient of Friction, Kinetic	Wear Factor K (10^{-8} mm^3/Nm)	Wear Factor K Mating Surface (10^{-8} mm^3/Nm)
RFL4216	SABIC	30% glass fiber, <5% MoS2	23	Carbon steel; surface finish: 0.3–0.4 μm; 18–20 Rockwell C	0.28	15.2	4.3	Thrust washer	0.24	0.31	151	
RFL4218	SABIC	40% glass fiber, <5% MoS2	23	Carbon steel; surface finish: 0.3–0.4 μm; 18–20 Rockwell C	0.28	15.2	4.3	Thrust washer	0.26	0.33	141	
RFL4416	SABIC	30% glass fiber, 2% silicone	23	Carbon steel; surface finish: 0.3–0.4 μm; 18–20 Rockwell C	0.28	15.2	4.3	Thrust washer	0.19	0.26	131	
RFL4536	SABIC	30% glass fiber, 13% PTFE, 2% silicone	23	Carbon steel; surface finish: 0.3–0.4 μm; 18–20 Rockwell C	0.28	15.2	4.3	Thrust washer	0.12	0.14	18	
RFL4536	SABIC	30% glass fiber, 13% PTFE, 2% silicone	23	Steel AISI 1141; surface finish: 0.2–0.3 μm	0.28	15.2	4.3	Thrust washer	0.2	0.26	40	2
RFL4536	SABIC	30% glass fiber, 13% PTFE, 2% silicone	23	Steel AISI 1141; surface finish: 0.3–0.4 μm	0.28	15.2	4.3	Thrust washer	0.18	0.2	18	1
RFL4536	SABIC	30% glass fiber, 13% PTFE, 2% silicone	23	Steel AISI 1141; surface finish: 1.3–1.8 μm	0.28	15.2	4.3	Thrust washer	0.16	0.19	40	2
RFL4536	SABIC	30% glass fiber, 13% PTFE, 2% silicone	23	AISI 304 stainless steel; surface finish: 0.2–0.4 μm	0.28	15.2	4.3	Thrust washer	0.17	0.2	30	0

RFL4536	SABIC	30% glass fiber, 13% PTFE, 2% silicone	23	AISI 304 stainless steel; surface finish: 1.3–1.8 μm	0.28	15.2	4.3	Thrust washer	0.15	0.18	52	1
RFL4536	SABIC		23	AISI 440 stainless steel; surface finish: 0.2–0.4 μm	0.28	15.2	4.3		0.4	59	4	203
RFL4536	SABIC	30% glass fiber, 13% PTFE, 2% silicone	23	AISI 440 stainless steel; surface finish: 1.3–1.8 μm	0.28	15.2	4.3	Thrust washer	0.1	0.16	36	1
RFL4536	SABIC	30% glass fiber, 13% PTFE, 2% silicone	23	70/30 brass; surface finish: 0.2–0.4 μm	0.28	15.2	4.3	Thrust washer	0.18	0.17	40	24
RFL4536	SABIC	30% glass fiber, 13% PTFE, 2% silicone	23	70/30 brass; surface finish: 1.3–1.8 μm	0.28	15.2	4.3	Thrust washer	0.19	0.18	36	24
RFL4616	SABIC	30% glass fiber, 2% silicone	23	Carbon steel; surface finish: 0.3–0.4 μm; 18–20 Rockwell C	0.28	15.2	4.3	Thrust washer	0.14	0.15	201	
RL4010	SABIC	5% PTFE	23	Carbon steel; surface finish: 0.3–0.4 μm; 18–20 Rockwell C	0.28	15.2	4.3	Thrust washer	0.13	0.2	161	
RL4040	SABIC	20% PTFE	23	Carbon steel; surface finish: 0.3–0.4 μm; 18–20 Rockwell C	0.28	15.2	4.3	Thrust washer	0.1	0.18	24	
RL4040	SABIC	20% PTFE	23	Steel AISI 1141; surface finish: 0.2–0.3 μm	0.28	15.2	4.3	Thrust washer	0.05	0.1	32	0
RL4040	SABIC	20% PTFE	23	Steel AISI 1141; surface finish: 0.3–0.4 μm	0.28	15.2	4.3	Thrust washer	0.1	0.14	24	0
RL4040	SABIC	20% PTFE	23	Steel AISI 1141; surface finish: 1.3–1.8 μm	0.28	15.2	4.3	Thrust washer	0.11	0.13	48	0

(*Continued*)

Table 8.28 (Continued)

Trade or Common Name	Supplier	Material Note	Test Temperature (°C)	Mating Surface	Pressure (MPa)	Sliding Velocity (m/min)	PV (MPa m/min)	Test Method	Coefficient of Friction Static	Coefficient of Friction, Kinetic	Wear Factor *K* (10^{-8} mm^3/Nm)	Wear Factor K Mating Surface (10^{-8} mm^3/Nm)
RL4040	SABIC	20% PTFE	23	AISI 304 stainless steel; surface finish: 0.2–0.4 μm	0.28	15.2	4.3	Thrust washer	0.05	0.09	14	0
RL4040	SABIC	20% PTFE	23	AISI 304 stainless steel; surface finish: 1.3–1.8 μm	0.28	15.2	4.3	Thrust washer	0.04	0.09	26	1
RL4040	SABIC	20% PTFE	23	AISI 440 stainless steel; surface finish: 0.2–0.4 μm	0.28	15.2	4.3	Thrust washer	0.1	0.12	14	0
RL4040	SABIC	20% PTFE	23	AISI 440 stainless steel; surface finish: 1.3–1.8 μm	0.28	15.2	4.3	Thrust washer	0.08	0.11	24	0
RL4040	SABIC	20% PTFE	23	70/30 brass; surface finish: 0.2–0.4 μm	0.28	15.2	4.3	Thrust washer	0.06	0.09	16	0
RL4040	SABIC	20% PTFE	23	70/30 brass; surface finish: 1.3–1.8 μm	0.28	15.2	4.3	Thrust washer	0.05	0.09	42	1
RL4040	SABIC	20% PTFE	23	2024 aluminum; surface finish: 0.2–0.3 μm	0.28	15.2	4.3	Thrust washer	0.07	0.09	504	20
RL4040	SABIC	20% PTFE	23	2024 aluminum; surface finish: 0.3–0.4 μm	0.28	15.2	4.3	Thrust washer	0.06	0.09	52	12
RL4040	SABIC	20% PTFE	23	2024 aluminum; surface finish: 1.3–1.8 μm	0.28	15.2	4.3	Thrust washer	0.08	0.1	212	12
RL4040	SABIC	20% PTFE	23	PC 30% glass fiber	0.28	15.2	4.3	Thrust washer	0.08	0.12	302	302
RL4040	SABIC	20% PTFE	23	PC (30% glass fiber, 15% PTFE)	0.28	15.2	4.3	Thrust washer	0.06	0.07	46	26

RL4040	SABIC	20% PTFE	23	POM (20% PTFE)	0.28	15.2	4.3	Thrust washer	0.03	0.04	91	60
RL4040	SABIC	20% PTFE	23	PA66 (20% PTFE)	0.28	15.2	4.3	Thrust washer	0.05	0.08	71	60
RL4040	SABIC	20% PTFE	23	PA66 (30% glass fiber)	0.28	15.2	4.3	Thrust washer	0.09	0.09	30	10
RL4040	SABIC	20% PTFE	23	PA66 (30% glass fiber, 15% PTFE)	0.28	15.2	4.3	Thrust washer	0.06	0.06	60	30
RL4040	SABIC	20% PTFE	23	PA66 (30% carbon fiber, 15% PTFE)	0.28	15.2	4.3	Thrust washer	0.05	0.06	40	30
RL4040FR (94VO)	SABIC	Flame retardant, 20% PTFE	23	Carbon steel; surface finish: 0.3–0.4 μm; 18–20 Rockwell C	0.28	15.2	4.3	Thrust washer	0.12	0.19	50	
RL4310	SABIC	5% graphite	23	Carbon steel; surface finish: 0.3–0.4 μm; 18–20 Rockwell C	0.28	15.2	4.3	Thrust washer	0.15	0.2	111	
RL4410	SABIC	2% silicone	23	Carbon steel; surface finish: 0.3–0.4 μm; 18–20 Rockwell C	0.28	15.2	4.3	Thrust washer	0.09	0.09	81	
RL4410	SABIC	2% silicone	23	PC; unmodified	0.28	15.2	4.3	Thrust washer	0.08	0.08	822	60
RL4410	SABIC	2% silicone	23	PC (30% glass fiber, 15% PTFE)	0.28	15.2	4.3	Thrust washer	0.1	0.14	68490	40
RL4410	SABIC	2% silicone	23	POM (30% glass fiber, 15% PTFE)	0.28	15.2	4.3	Thrust washer	0.1	0.11	101	1894
RL4410	SABIC	2% silicone	23	PA66 (30% glass fiber)	0.28	15.2	4.3	Thrust washer	0.09	0.14	705	71
RL4410	SABIC	2% silicone	23	PA66 (30% carbon fiber, 15% PTFE)	0.28	15.2	4.3	Thrust washer	0.1	0.13	403	60
RL4530	SABIC	13% PTFE, 2% silicone	23	PC; unmodified	0.28	15.2	4.3	Thrust washer	0.06	0.06	62	64
RL4530	SABIC	13% PTFE, 2% silicone	23	PC (30% glass fiber, 15% PTFE)	0.28	15.2	4.3	Thrust washer	0.06	0.06	282	40
RL4530	SABIC	13% PTFE, 2% silicone	23	PA66 (30% glass fiber)	0.28	15.2	4.3	Thrust washer	0.06	0.07	856	60

(Continued)

Table 8.28 (Continued)

Trade or Common Name	Supplier	Material Note	Test Temperature (°C)	Mating Surface	Pressure (MPa)	Sliding Velocity (m/min)	PV (MPa m/min)	Test Method	Coefficient of Friction Static	Coefficient of Friction, Kinetic	Wear Factor *K* (10^{-8} mm^3/Nm)	Wear Factor K Mating Surface (10^{-8} mm^3/Nm)
RL4530	SABIC	13% PTFE, 2% silicone	23	PA66 (30% glass fiber, 15% PTFE)	0.28	15.2	4.3	Thrust washer	0.06	0.06	121	40
RL4530	SABIC	13% PTFE, 2% silicone	23	PA66 (30% carbon fiber, 15% PTFE)	0.28	15.2	4.3	Thrust washer	0.1	0.1	81	40
RL4540	SABIC	18% PTFE, 2% silicone	23	Carbon steel; surface finish: 0.3–0.4 μm; 18–20 Rockwell C	0.28	15.2	4.3	Thrust washer	0.06	0.08	12	
RL4610	SABIC	2% silicone	23	Carbon steel; surface finish: 0.3–0.4 μm; 18–20 Rockwell C	0.28	15.2	4.3	Thrust washer	0.19	0.19	312	
RL4730	SABIC	13% PTFE, 2% silicone	23	Carbon steel; surface finish: 0.3–0.4 μm; 18–20 Rockwell C	0.28	15.2	4.3	Thrust washer	0.11	0.18	50	
Ultramid A3K	BASF	High flow, heat stabilized		Steel, Cr 6/800/ HV; surface finish: 0.15–0.2 μm	1	30	29.9	pin on disk		0.45–0.6		
Ultramid A3K	BASF	High flow, heat stabilized		Steel, Cr 6/800/ HV; surface finish: 2.0–2.6 μm	1	30	29.9	pin on disk		0.4–0.53		
Ultramid A3R	BASF	Noiseless bearings; high flow; PE modified; stabilized		Steel, Cr 6/800/ HV; surface finish: 0.15–0.2 μm	1	30	29.9	pin on disk		0.32–0.42		
Ultramid A3R	BASF	Noiseless bearings; high flow; PE modified; stabilized		Steel, Cr 6/800/ HV; surface finish: 2.0–2.6 μm	1	30	29.9	pin on disk		0.4–0.5		

Ultramid A3WC6	BASF	High flow, heat stabilized; 30% carbon fiber		Steel, Cr 6/800/ HV; surface finish: 0.15–0.2 μm	1	30	29.9	pin on disk		0.4–0.5		
Ultramid A3WC6	BASF	High flow, heat stabilized; 30% carbon fiber		Steel, Cr 6/800/ HV; surface finish: 2.0–2.6 μm	1	30	29.9	pin on disk		0.4–0.5		
Ultramid A3WG6	BASF	High flow, heat stabilized; 30% glass fiber		Steel, Cr 6/800/ HV; surface finish: 0.15–0.2 μm	1	30	29.9	pin on disk		0.6–0.7		
Ultramid A3WG6	BASF	High flow, heat stabilized; 30% glass fiber		Steel, Cr 6/800/ HV; surface finish: 2.0–2.6 μm	1	30	29.9	pin on disk		0.55–0.65		
Ultramid A4	BASF	Moderate flow		Steel, Cr 6/800/ HV; surface finish: 0.15–0.2 μm	1	30	29.9	pin on disk		0.45–0.6		
Ultramid A4	BASF	Moderate flow		Steel, Cr 6/800/ HV; surface finish: 2.0–2.6 μm	1	30	29.9	pin on disk		0.4–0.53		
Verton RF-700-1OHS	SABIC	50% glass fiber	23	Carbon steel; surface finish: 0.3–0.4 μm; 18–20 Rockwell C	0.28	15.2	4.3	Thrust washer	0.26	0.32	60	
Verton RF-7007HS	SABIC	35% glass fiber	23	Carbon steel; surface finish: 0.3–0.4 μm; 18–20 Rockwell C	0.28	15.2	4.3	Thrust washer	0.24	0.3	81	
	DuPont		23		1.72	3	5.3	Thrust washer		0.435	1847	
	DuPont	33% glass fiber	23		1.72	3	5.3	Thrust washer		0.42	854	
	DuPont	20% aramid fiber	23		1.72	3	5.3	Thrust washer		0.39	481	
	DuPont		23		0.28	15.2	4.3	Thrust washer		0.574	1464	
	DuPont	33% glass fiber	23		0.28	15.2	4.3	Thrust washer		0.476	276	

8.5 Polyamide 610 (Nylon 610)

8.5.1 Fatigue Data

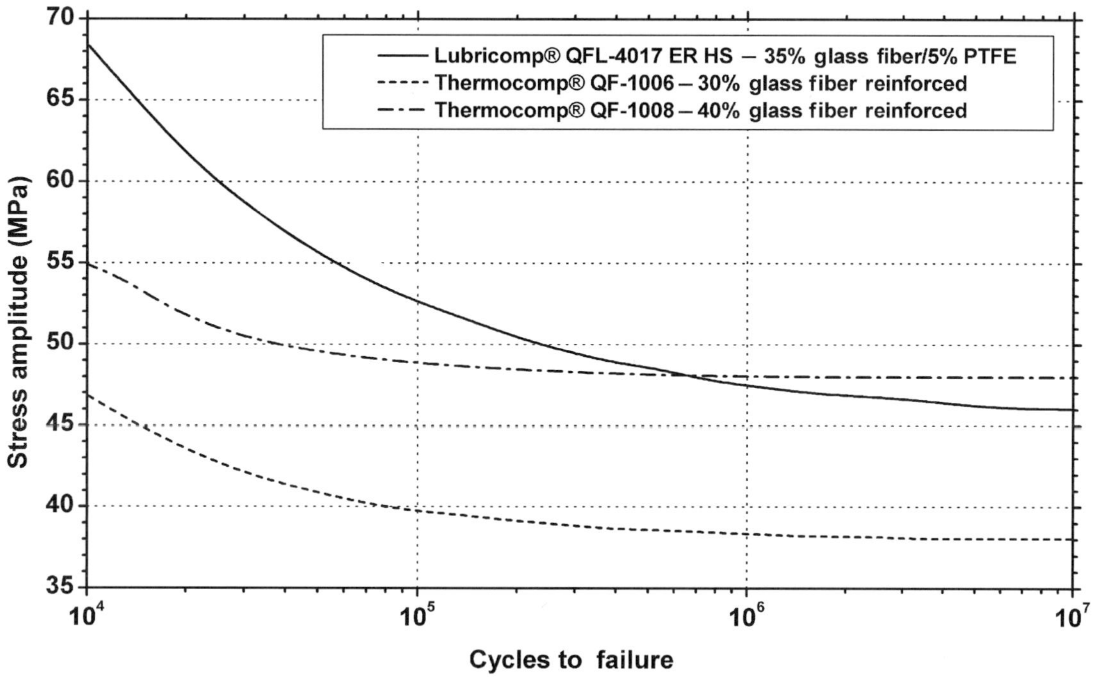

Figure 8.44 Flexural stress amplitude vs. cycles to failure of several SABIC Innovative Plastics PA610 plastics.

8.5.2 Tribology Data

Table 8.29 Tribological Properties of RTP Company RTP 299B × 89491 A (Proprietary Formulation) vs. 1018 C Steel (Data Obtained per ASTM 3702)

PV (KPa m/s)	Load (N)	Speed (m/s)	Wear Factor × 10^{-8} (mm^3/N m)	Dynamic Coefficient of Friction
70	1.80	0.25	740	0.52
175	2.25	0.50	531	1.13

8.6 Polyamide 612 (Nylon 612)

8.6.1 Fatigue Data

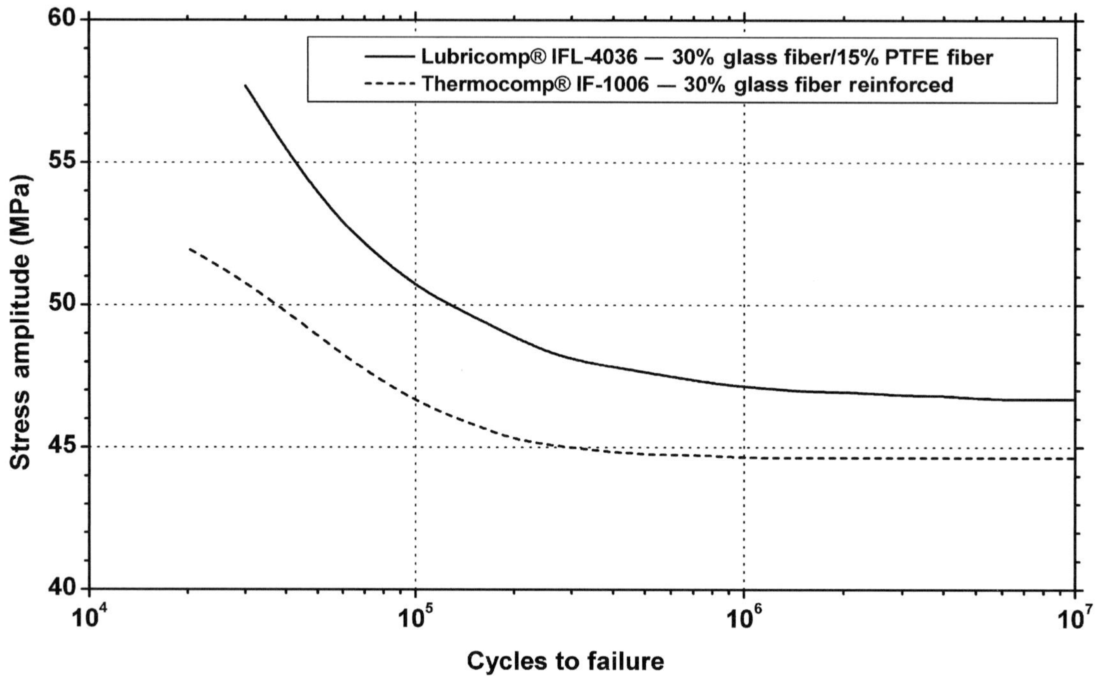

Figure 8.45 Flexural stress amplitude vs. cycles to failure of two SABIC Innovative Plastics PA612 plastics.

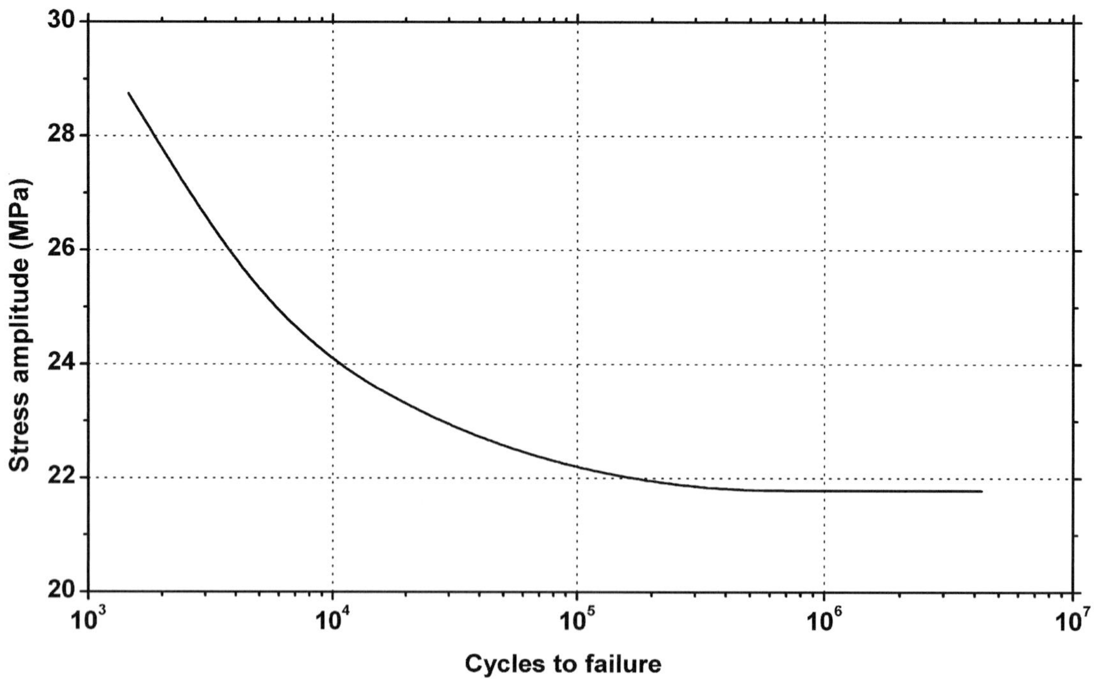

Figure 8.46 Axial stress amplitude vs. cycles to failure at 23°C of conditioned DuPont Engineering Polymers Zytel® 158L NC010—general-purpose, lubricated, higher melt viscosity PA612.

8.6.2 Tribology Data

Table 8.30 Tribological Properties of RTP Company RTP 200D TFE 10 (10% PTFE) vs. 1018 C Steel (Data Obtained per ASTM 3702)

PV (KPa m/s)	Load (N)	Speed (m/s)	Wear Factor × 10^{-8} (mm^3/N m)	Dynamic Coefficient of Friction
70	1.80	0.25	686	0.42
175	2.25	0.50	571	0.89
350	9.00	0.25	520	0.77

Table 8.31 Tribological Properties of RTP Company RTP 200D TFE 10 (10% PTFE) vs. RTP 200D TFE 10 (Data Obtained per ASTM 3702)

PV (KPa m/s)	Load (N)	Speed (m/s)	Wear Factor × 10^{-8} (mm^3/N m)	Dynamic Coefficient of Friction
17.5	0.45	0.25	1097	0.38
35	0.90	0.25	533	0.63

Table 8.32 Tribological Properties of RTP Company RTP 200D TFE 20 (20% PTFE) vs. 1018 C Steel (Data Obtained per ASTM 3702)

PV (KPa m/s)	Load (N)	Speed (m/s)	Wear Factor × 10^{-8} (mm^3/N m)	Dynamic Coefficient of Friction
70	1.80	0.25	311	0.27
175	2.25	0.50	218	0.30
350	9.00	0.25	68	0.23

Table 8.33 Tribological Properties of RTP Company RTP 200D TFE 20 (20% PTFE) vs. RTP 200D TFE 10 (Data Obtained per ASTM 3702)

PV (KPa m/s)	Load (N)	Speed (m/s)	Wear Factor × 10^{-8} (mm^3/N m)	Dynamic Coefficient of Friction
17.5	0.45	0.25	154	0.30
70	1.80	0.25	251	0.22

Table 8.34 Tribological Properties of RTP Company RTP 200D TFE 18 SI 2 (18% PTFE, 2% Silicone) vs. 1018 C Steel (Data Obtained per ASTM 3702)

PV (KPa m/s)	Load (N)	Speed (m/s)	Wear Factor × 10^{-8} (mm^3/N m)	Dynamic Coefficient of Friction
70	1.80	0.25	187	0.18
175	2.25	0.50	79	0.20
350	9.00	0.25	13	0.08

Table 8.35 Tribological Properties of RTP Company RTP 200D TFE 18 SI 2 (18% PTFE, 2% Silicone) vs. RTP 200D TFE 18 SI 2 (Data Obtained per ASTM 3702)

PV (KPa m/s)	Load (N)	Speed (m/s)	Wear Factor × 10^{-8} (mm^3/N m)	Dynamic Coefficient of Friction
17.5	0.45	0.25	86	0.14
70	1.80	0.25	105	0.11

Table 8.36 Tribological Properties of RTP Company RTP 202D TFE 15 (15% Glass Fiber, 15% PTFE) vs. 1018 C Steel (Data Obtained per ASTM 3702)

PV (KPa m/s)	Load (N)	Speed (m/s)	Wear Factor × 10^{-8} (mm^3/N m)	Dynamic Coefficient of Friction
70	1.80	0.25	244	0.28
175	2.25	0.50	141	0.37
350	9.00	0.25	219	0.24
350	2.25	1.00	173	0.25

Table 8.37 Tribological Properties of RTP Company RTP 202D TFE 15 (15% Glass Fiber, 15% PTFE) vs. RTP 202D TFE 15 (Data Obtained per ASTM 3702)

PV (KPa m/s)	Load (N)	Speed (m/s)	Wear Factor × 10^{-8} (mm^3/N m)	Dynamic Coefficient of Friction
17.5	0.45	0.25	86	0.14
70	1.80	0.25	105	0.11

Table 8.38 Tribological Properties of RTP Company RTP 282D TFE 15 (15% Carbon Fiber, 15% PTFE) vs. 1018 C Steel (Data Obtained per ASTM 3702)

PV (KPa m/s)	Load (N)	Speed (m/s)	Wear Factor × 10^{-8} (mm^3/N m)	Dynamic Coefficient of Friction
70	1.80	0.25	143	0.41
175	2.25	0.50	5	0.35
350	9.00	0.25	109	0.56
350	2.25	1.00	270	0.45

Table 8.39 Tribological Properties of RTP Company RTP 282D TFE 15 (15% Carbon Fiber, 15% PTFE) vs. RTP 282D TFE 15 (Data Obtained per ASTM 3702)

PV (KPa m/s)	Load (N)	Speed (m/s)	Wear Factor × 10^{-8} (mm^3/N m)	Dynamic Coefficient of Friction
17.5	0.45	0.25	434	0.13
70	1.80	0.25	10	0.12
175	2.25	0.50	3638	0.12

Table 8.40 Tribological Properties of RTP Company RTP 285D TFE 15 (30% Carbon Fiber, 15% PTFE) vs. 1018 C Steel (Data Obtained per ASTM 3702)

PV (KPa m/s)	Load (N)	Speed (m/s)	Wear Factor × 10^{-8} (mm^3/N m)	Dynamic Coefficient of Friction
70	1.80	0.25	46	0.30
175	2.25	0.50	90	0.31
350	9.00	0.25	156	0.54
350	2.25	1.00	7	0.42

Table 8.41 Tribological Properties of RTP Company RTP 285D TFE 15 (30% Carbon Fiber, 15% PTFE) vs. RTP 285D TFE 15 (Data Obtained per ASTM 3702)

PV (KPa m/s)	Load (N)	Speed (m/s)	Wear Factor × 10^{-8} (mm^3/N m)	Dynamic Coefficient of Friction
17.5	0.45	0.25	64	0.26
70	1.80	0.25	94	0.19
175	2.25	0.50	174	0.28

8.7 Polyamide 666 (Nylon 666)

8.7.1 Fatigue Data

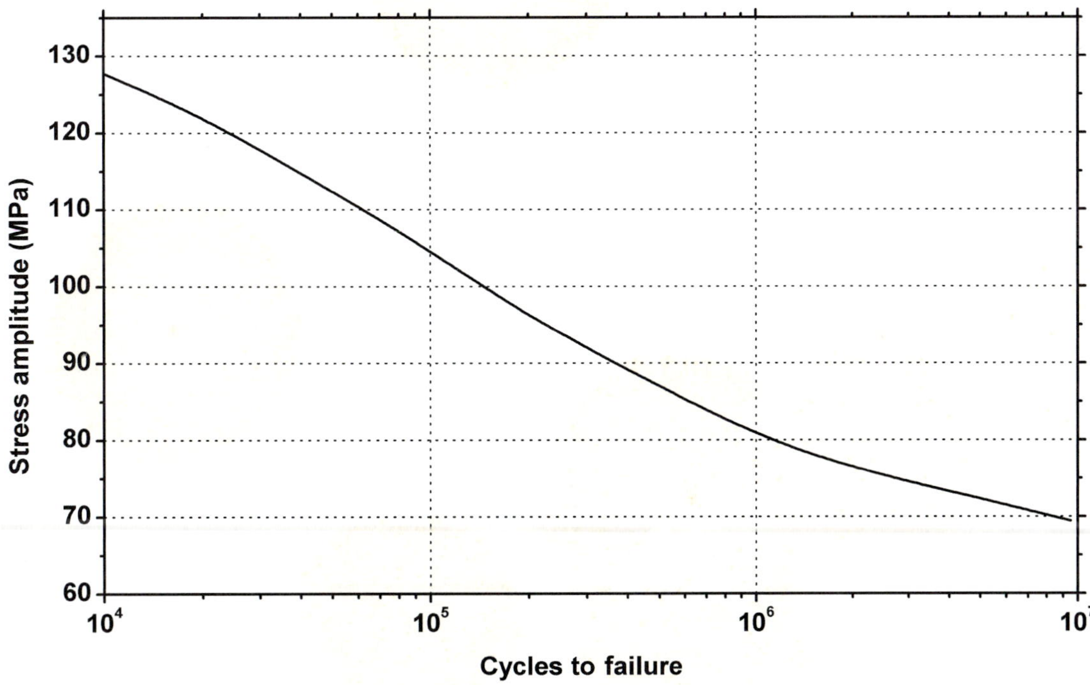

Figure 8.47 Flexural stress amplitude vs. cycles to failure of conditioned EMS-GRIVORY Grivory® GV-5H—50% glass fiber reinforced, normal viscosity, heat-stabilized PA666.

8.8 Amorphous Polyamide

8.8.1 Fatigue Data

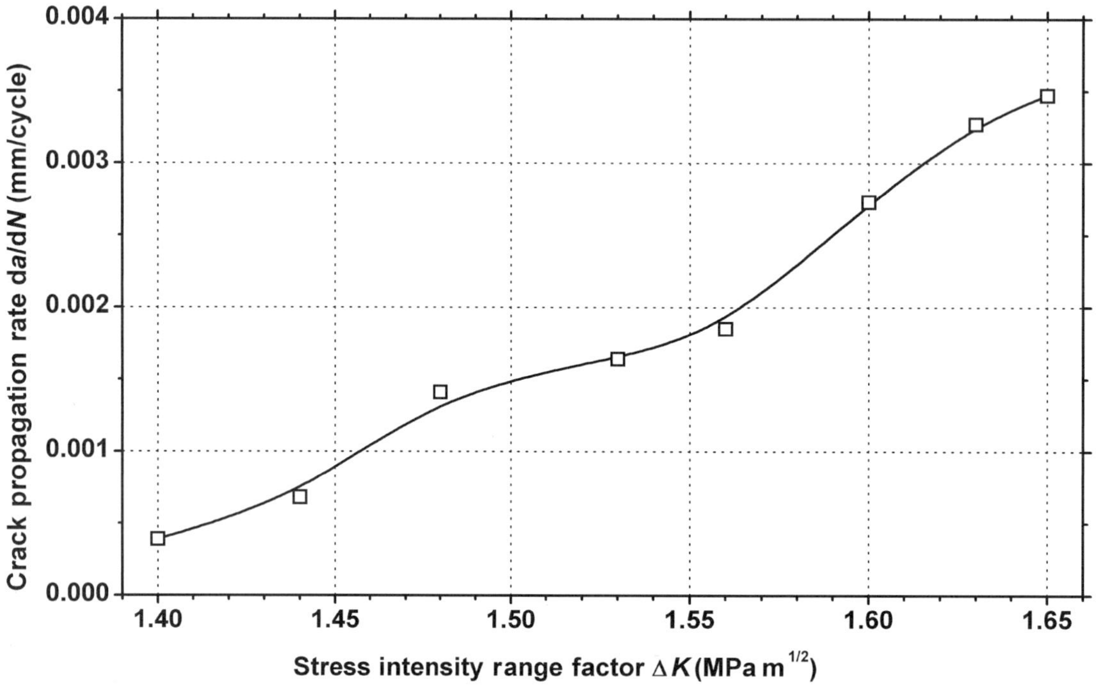

Figure 8.48 Fatigue crack propagation rate vs. stress intensity factor of Evonik Industries Trogamid® T5000—standard grade amorphous polyamide.

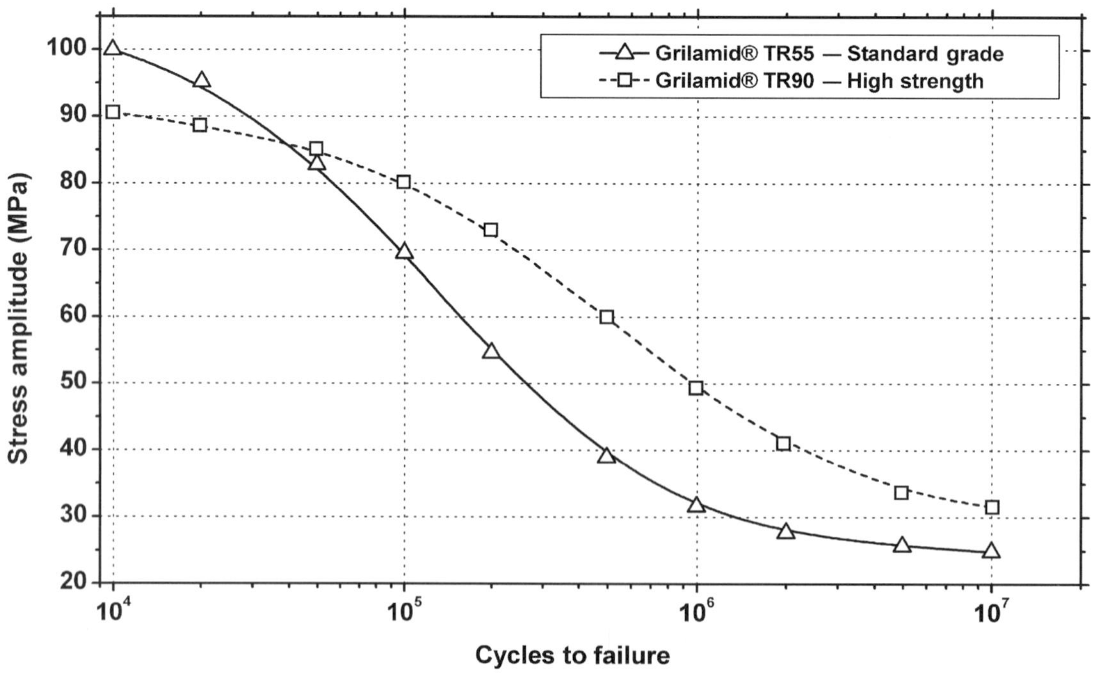

Figure 8.49 Flexural stress amplitude vs. cycles to failure of two EMS-GRIVORY Grilamid® amorphous polyamide plastics.

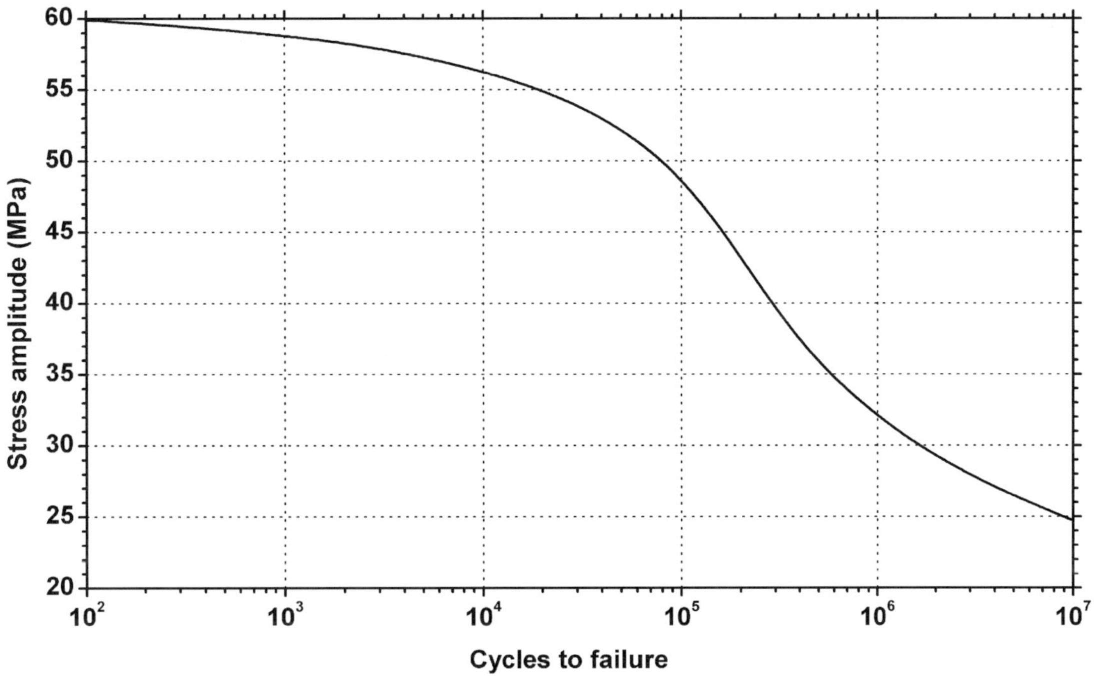

Figure 8.50 Flexural stress amplitude vs. cycles to failure of Evonik Industries Trogamid® T5000—standard grade amorphous polyamide.

8.9 Polyamide 46 (Nylon 46)

8.9.1 Fatigue Data

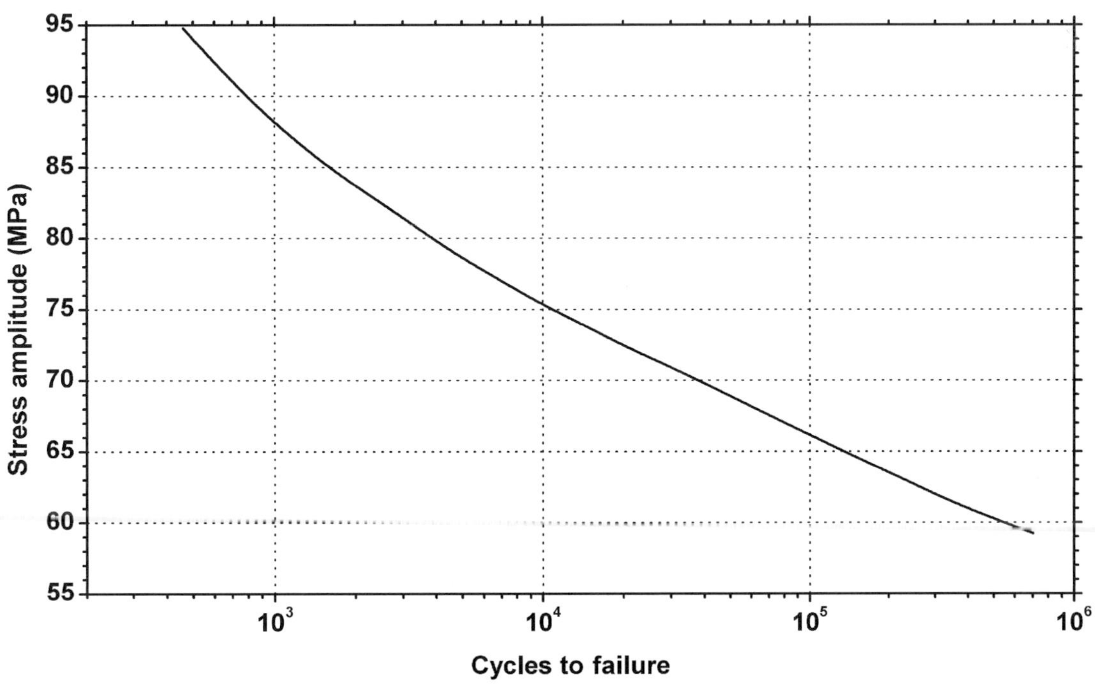

Figure 8.51 Stress amplitude vs. cycles to failure at 140°C and 8 Hz of DSM Engineering Plastics Stanyl® TE200F6—30% glass fiber reinforced, heat-stabilized PA46.

8.10 PPA/High-Performance Polyamide

8.10.1 Fatigue Data

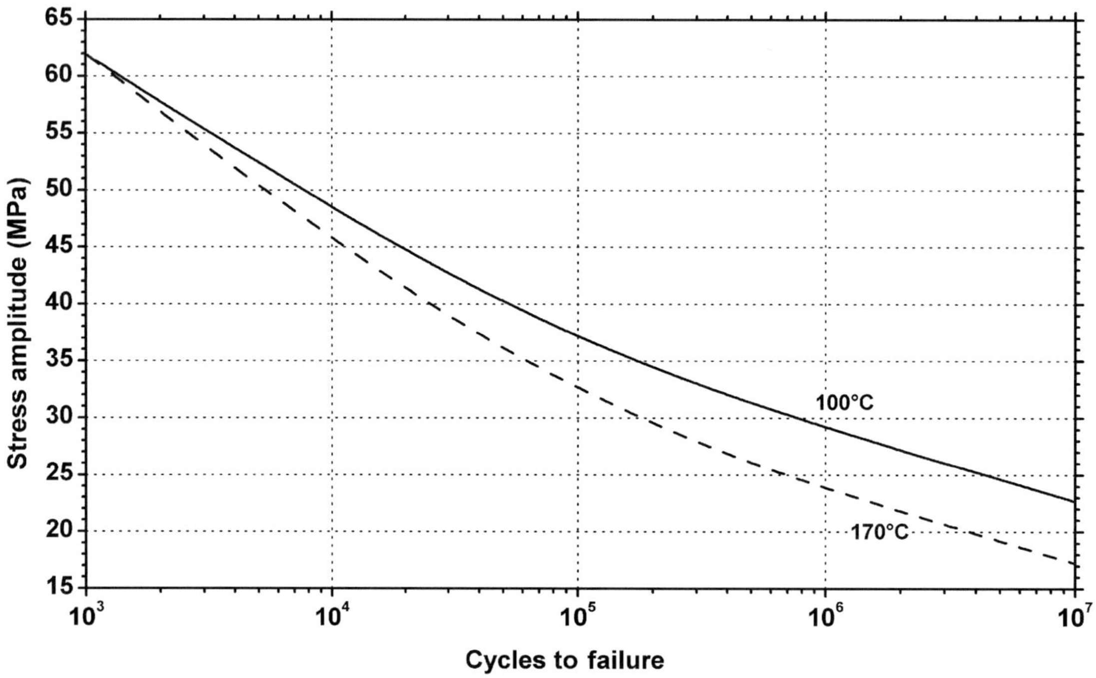

Figure 8.52 Flexural stress amplitude vs. cycles to failure and temperature of Solvay Amodel® A-1145 HS—45% glass fiber reinforced, heat-stabilized PAA.

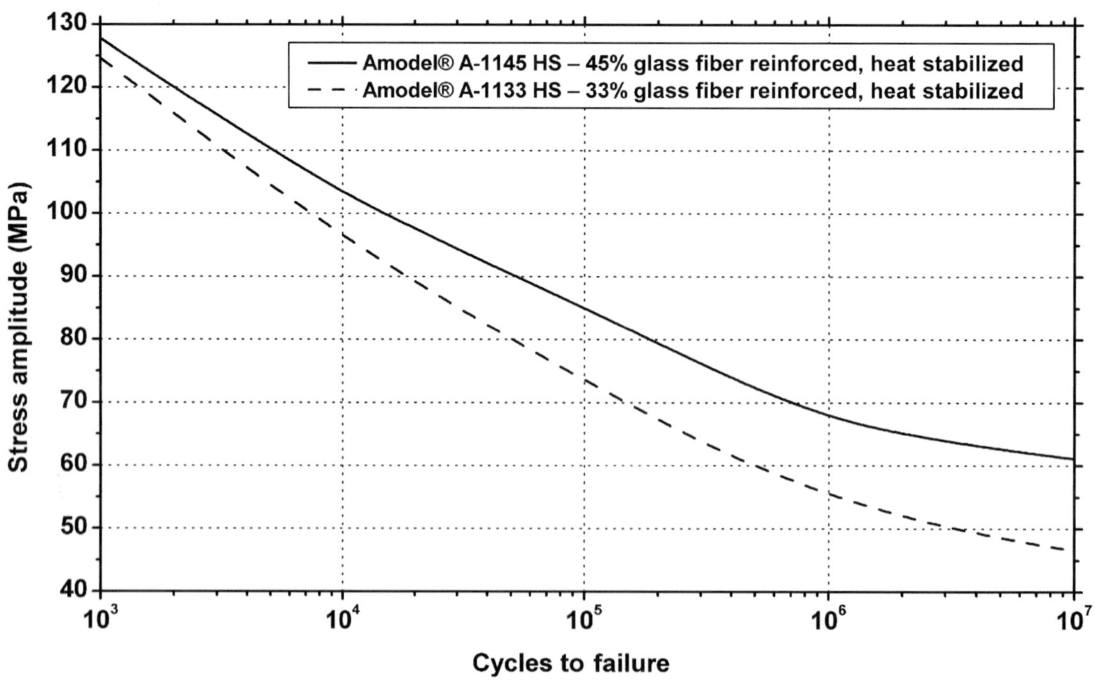

Figure 8.53 Flexural stress amplitude vs. cycles to failure at 23°C and 32 Hz of Solvay Amodel® glass fiber reinforced, heat-stabilized PAA plastics.

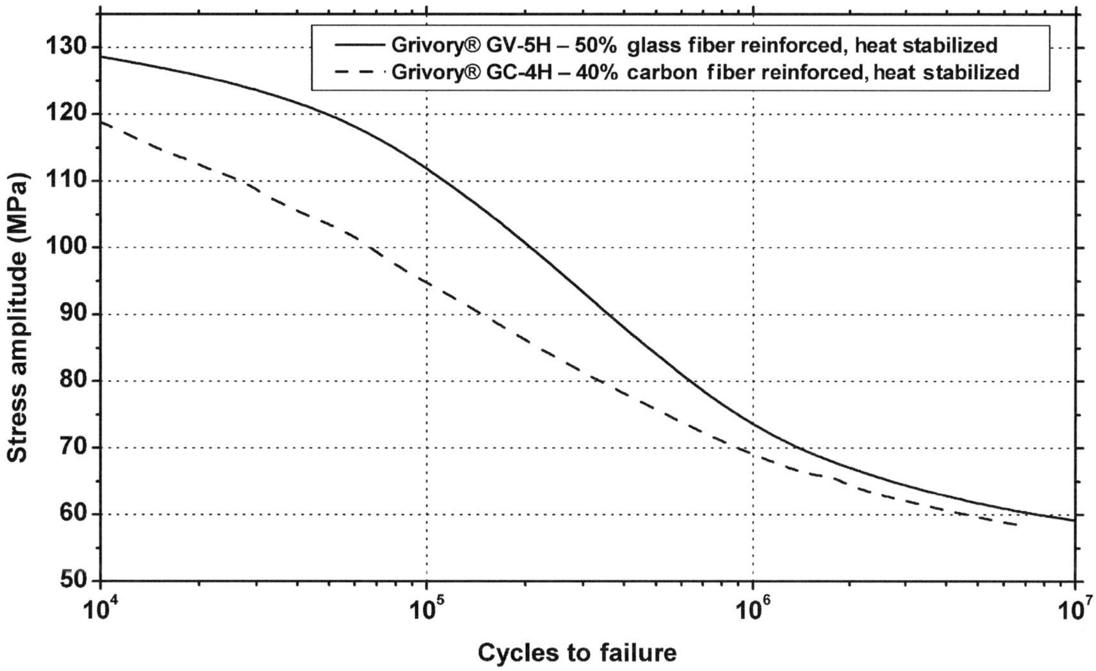

Figure 8.54 Flexural stress amplitude vs. cycles to failure at 23°C and 8 Hz of conditioned EMS-GRIVORY Grivory® fiber reinforced PAA plastics.

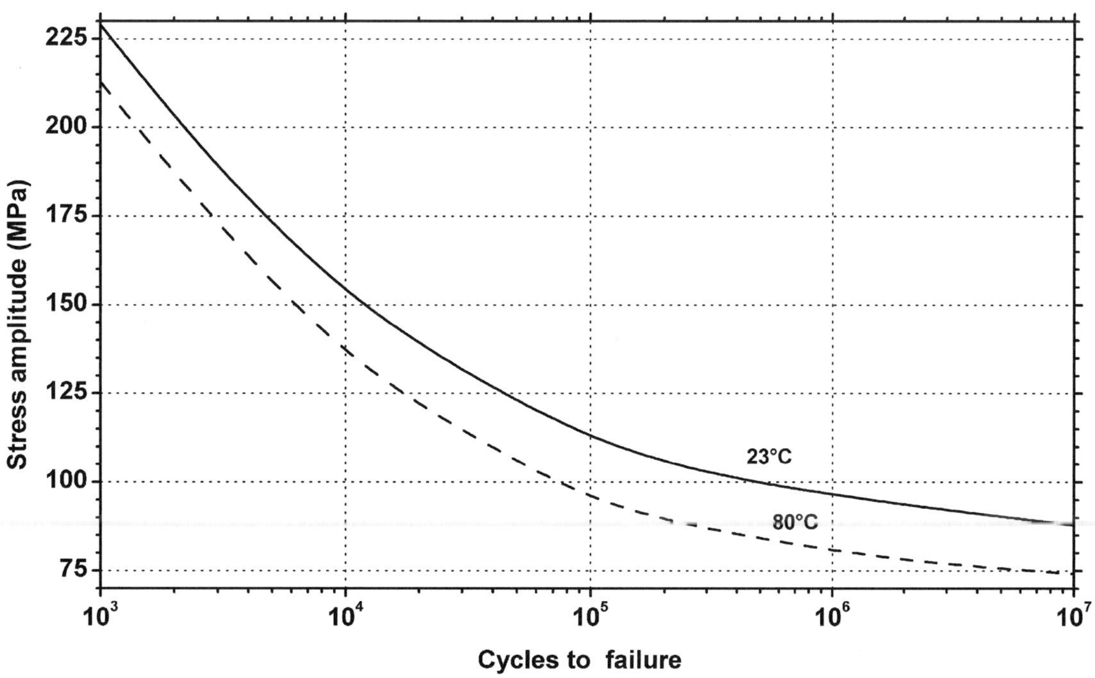

Figure 8.55 Flexural stress amplitude vs. cycles to failure at 23°C and 80°C of EMS-GRIVORY Grivory® HTV-5H1—50% glass fiber reinforced, heat-stabilized (PA6T/6I) PAA.

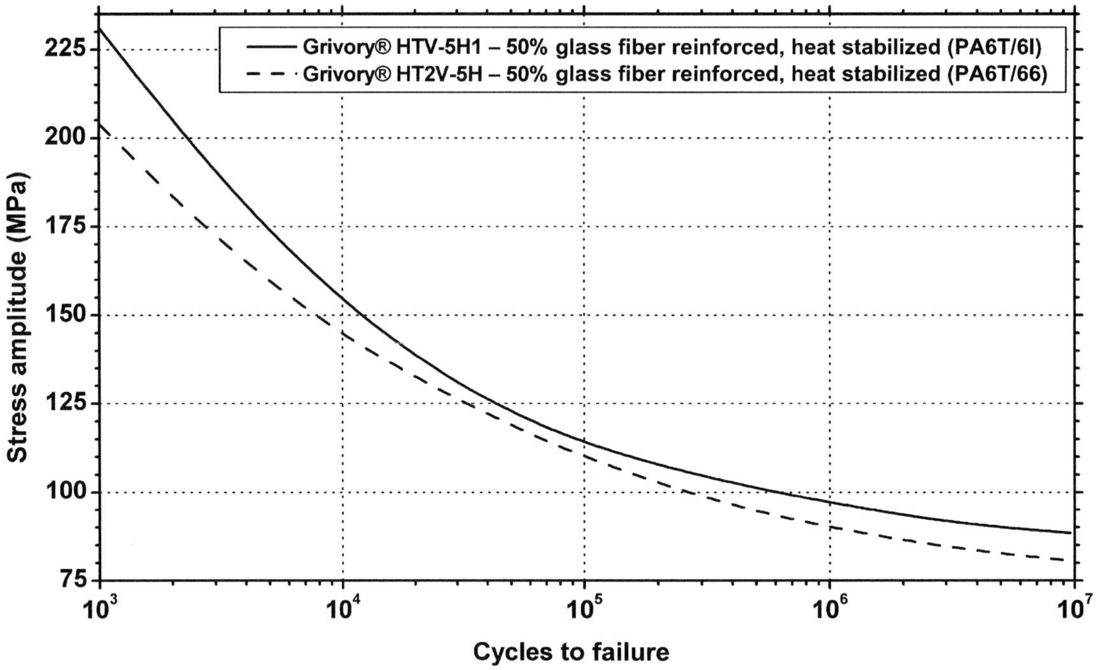

Figure 8.56 Flexural stress amplitude vs. cycles to failure at 23°C of EMS-GRIVORY Grivory® 50% glass fiber reinforced, heat-stabilized PAA.

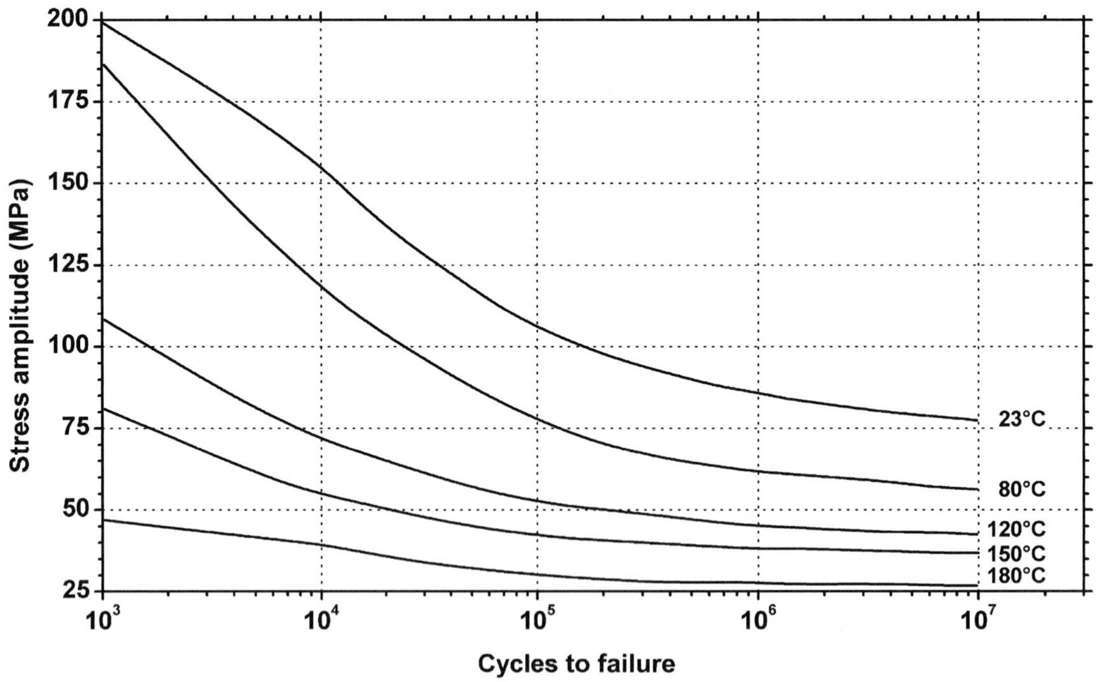

Figure 8.57 Flexural stress amplitude vs. cycles to failure at various temperatures of EMS-GRIVORY Grivory® HTV-6H1—60% glass fiber reinforced, heat-stabilized (PA6T/6I) PAA.

8.11 Polyarylamide

8.11.1 Fatigue Data

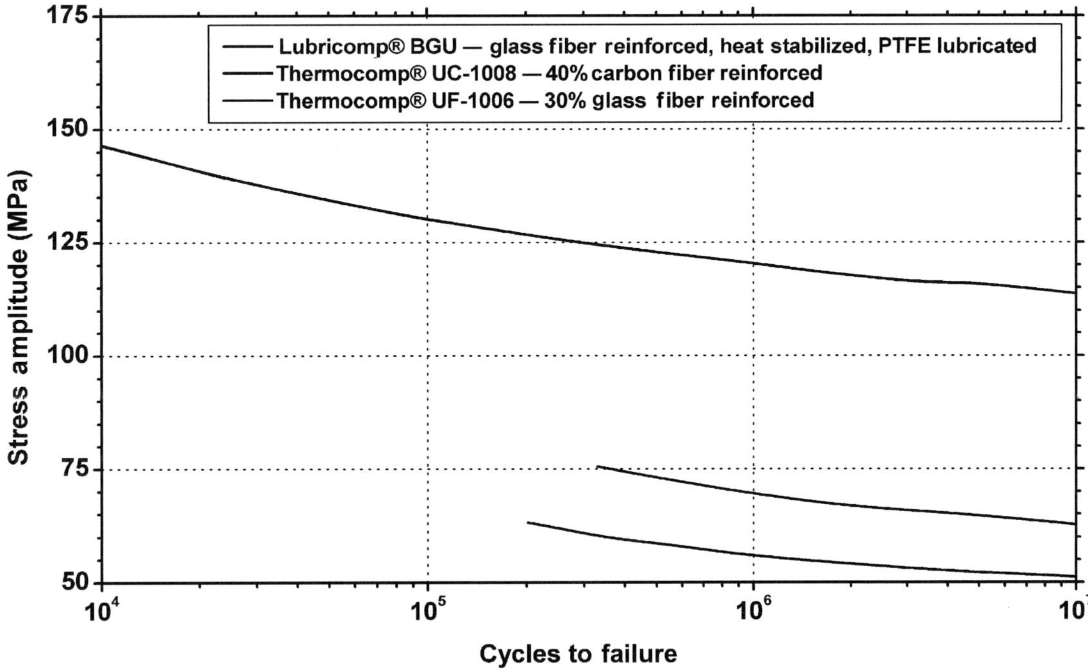

Figure 8.58 Flexural stress amplitude vs. cycles to failure at 23°C of several SABIC Innovative Plastics PAA plastics.

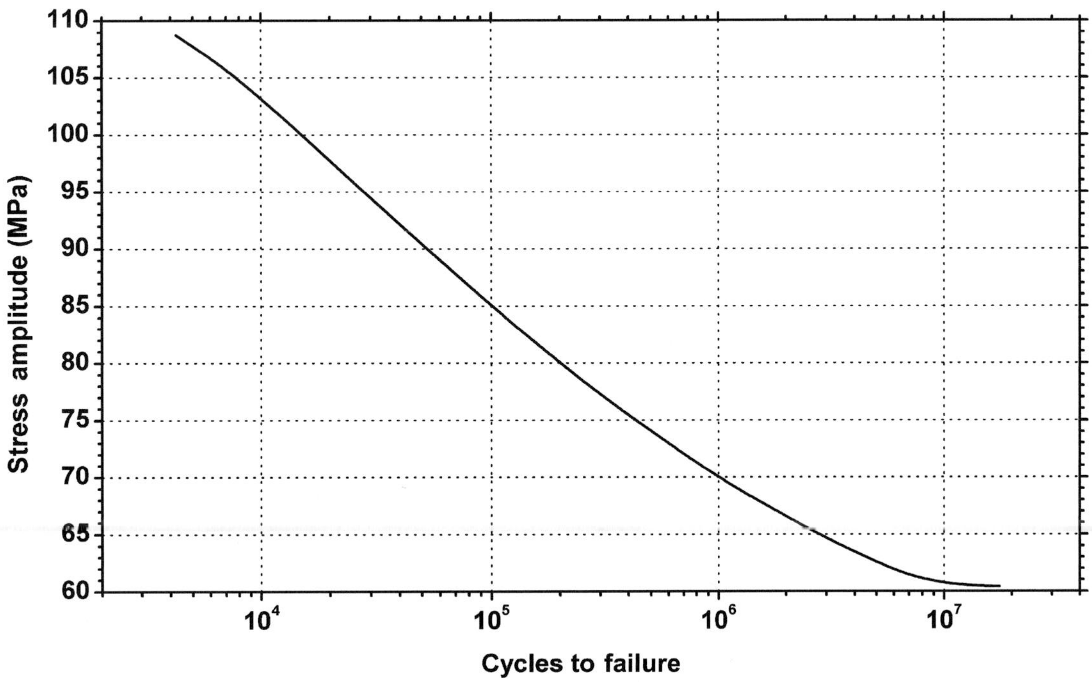

Figure 8.59 Flexural stress amplitude vs. cycles to failure at 23°C of Solvay IXEF® 1022—50% glass fiber reinforced PAA.

Table 8.42 Tribological Properties of Solvay IXEF® PAA Plastics

	Dynamic Coefficient of Friction	Taber Abrasion CS17 Wheel, 1 kg mg/1000 revolutions	Taber Abrasion H22 Wheel, 1 kg mg/1000 revolutions
IXEF® 1002 (30% glass fiber)	0.36–0.45		
IXEF® 1022 (50% glass fiber)	0.40–0.53	16	53

8.12 Semicrystalline Polyamide (PACM 12)

8.12.1 Tribology Data

Table 8.43 Abrasion Resistance of Degussa Trogamid Transparent Polyamides

Property	Test Method	Unit	TROGAMID CX7323 (Medium Viscosity)	TROGAMID T5000
Abrasion resistance	DIN 53754	mg/100 revolutions	18	23

9 Polyolefins and Acrylics

9.1 Background

In organic chemistry, an alkene, also called an olefin, is a chemical compound containing at least one carbon-to-carbon double bond. The simplest alkenes, with only one double bond and no other functional groups, form a homologous series of hydrocarbons with the general formula C_nH_{2n}. The two simplest alkenes of this series are ethylene and propylene. When these are polymerized, they form polyethylene (PE) and polypropylene (PP), which are two of the plastics discussed in this chapter. A slightly more complex alkene is 4-methylpentene-1, the basis of poly(methyl pentene), known under the trade name of TPX™. If one of the hydrogens on the ethylene molecule is changed to chlorine, the molecule is called vinyl chloride, the basis of polyvinyl chloride, commonly called PVC. Acrylic polymers are also polymerized through the carbon–carbon double bond. Methyl methacrylate is the monomer used to make poly(methyl methacrylate).

The structures of these monomers are shown in Figure 9.1, with polymer structures shown in Figure 9.2. The copolymer structure using the norbornene monomer is shown later in Figure 9.5.

9.1.1 Polyethylene

PE can be made in a number of ways. The way it is produced can affect its physical properties. It can also have very small amounts of comonomers, which will alter its structure and properties.

The basic types or classifications of PE, according to the ASTM 1248, are:

- Ultralow-density PE (ULDPE), polymers with densities ranging from 0.890 to 0.905 g/cm^3, contains comonomer
- Very low-density PE (VLDPE), polymers with densities ranging from 0.905 to 0.915 g/cm^3, contains comonomer

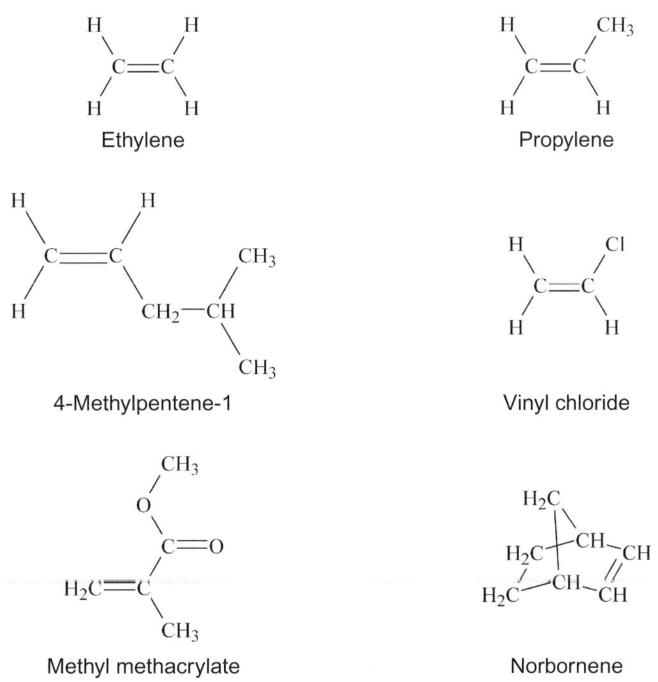

Figure 9.1 Chemical structures of monomers used to make polyolefins.

Copyright © 2010 Laurence W. McKeen. All rights reserved.

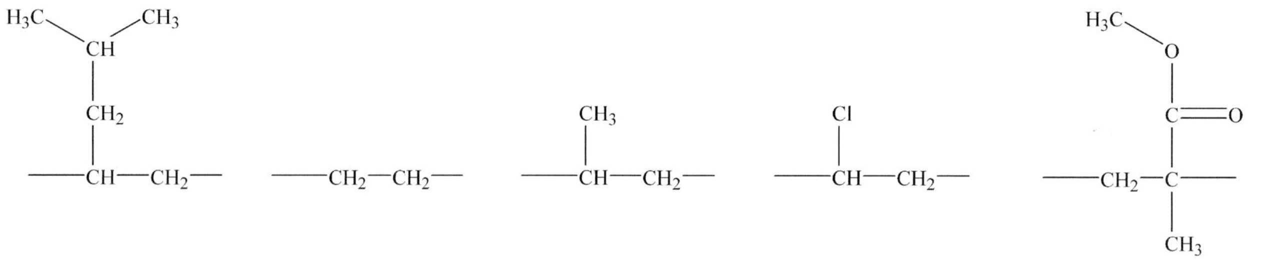

Figure 9.2 Structures of polyolefin polymers.

- Linear low-density PE (LLDPE), polymers with densities ranging from 0.915 to 0.935 g/cm^3, contains comonomer
- Low-density PE (LDPE), polymers with densities ranging from about 0.915 to 0.935 g/cm^3
- Medium-density PE (MDPE), polymers with densities ranging from 0.926 to 0.940 g/cm^3, may or may not contain comonomer
- High-density PE (HDPE), polymers with densities ranging from 0.940 to 0.970 g/cm^3, may or may not contain comonomer

Figure 9.3 shows the differences graphically. The differences in the branches in terms of number and length affect the density and melting points of some of the types.

Branching affects the crystallinity. A diagram of a representation of the crystal structure of PE is shown in Figure 9.4. One can imagine how branching in the polymer chain can disrupt the crystalline regions. The crystalline regions are the highly ordered areas in the shaded rectangles of Figure 9.4. A high degree of branching would reduce the size of the crystalline regions, which leads to lower crystallinity.

9.1.2 Cross-linked PE

A modification of HDPE is called cross-linked PE (PEX). It is a form of PE with cross-links and it is commonly abbreviated as PEX or XLPE. The HDPE has undergone a chemical or physical reaction that causes the molecular structure of the PE chains to link together as described in Section 3.3 and Figure 3.11. This reaction creates a three-dimensional structure which has superior resistance to high temperature and pressure. PEX is primarily used in tubing. There are three primary commercial methods for producing PEX tubing:

- Peroxide (or Engel) method (PEX-a)—Cross-links during extrusion while the polymer is molten
- Silane method (PEX-b)—A chemical cross-linking method in which reactive silane groups are grafted onto the polymer backbone
- Radiation or electron-beam method (PEX-c)—The formed tubing passes through a radiation chamber that generates the cross-links

The details are beyond the scope of this book.

9.1.3 Polypropylene

The three main types of PP generally available:

1. Homopolymers are made in a single reactor with propylene and catalyst. It is the stiffest of the three propylene types and has the highest tensile strength at yield. In the natural state (no colorant added), it is translucent and has excellent see-through or contact clarity with liquids. In comparison to the other two types, it has less impact resistance, especially below 0°C.
2. Random copolymers (homophasic copolymers) are made in a single reactor with a small amount of ethylene (<5%) added which disrupts the crystallinity of the polymer allowing this type to be the clearest. It is also the most flexible with the lowest tensile strength of the three. It has better room temperature impact than homopolymer but shares the same relatively poor impact resistance at low temperatures.
3. Impact copolymers (heterophasic copolymers), also known as block copolymers, are made in a two reactor system where the homopolymer matrix is made in the first reactor and then transferred to the second reactor where ethylene and

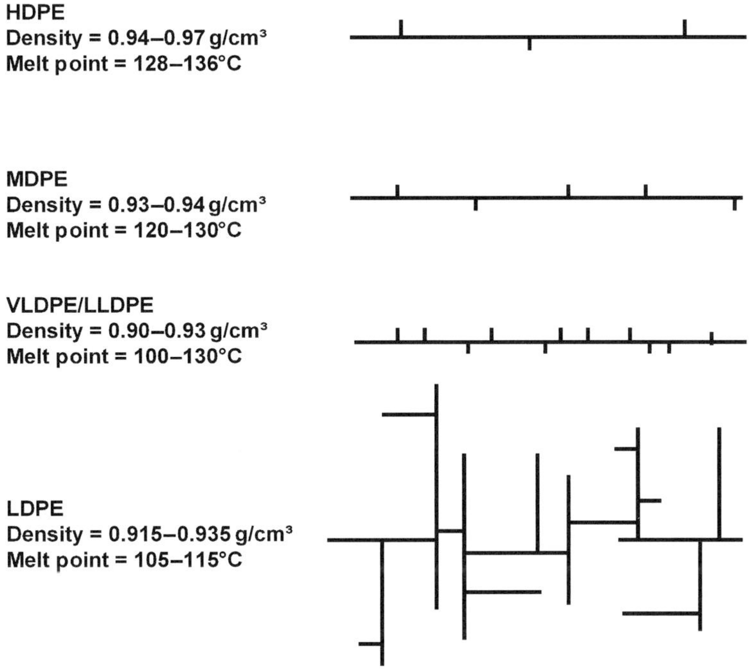

Figure 9.3 Graphical depictions of PE types.

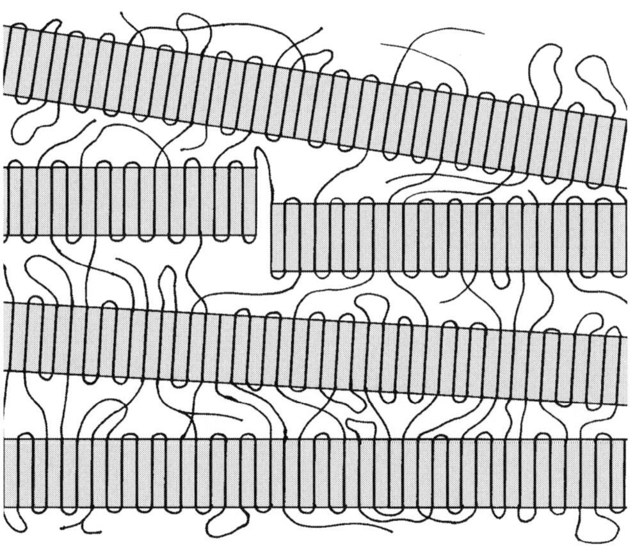

Figure 9.4 Graphical diagram of PE crystal structure.

propylene are polymerized to create ethylene propylene rubber (EPR) in the form of microscopic nodules dispersed in the homopolymer matrix phase. These nodules impart impact resistance both at ambient and cold temperatures to the compound. This type has intermediate stiffness and tensile strength and is quite cloudy. In general, the more ethylene monomer added, the greater the impact resistance with correspondingly lower stiffness and tensile strength.

9.1.4 Polymethyl Pentene

4-Methylpentene-1-based polyolefin is manufactured and marketed solely by Mitsui Chemicals, Inc under the trade name TPX™. This lightweight, functional polymer displays a unique combination of physical properties and characteristics due to its distinctive molecular structure, which includes a bulky side chain as shown in Figure 9.2. Polymethyl pentene (PMP) possesses many characteristics inherent in traditional polyolefins such as excellent electrical insulating properties and strong hydrolysis resistance. Moreover, it features low dielectric, superb clarity, transparency, gas permeability, heat and chemical resistance and release qualities.

It can be used for extruded and film products, injection molded, and blow molded application items, including:

- paper coatings and baking cartons
- release film and release paper
- high-frequency films
- microwave cookware
- food packaging such as gas permeable packages for fruit and vegetables
- LED molds

9.1.5 Ultrahigh Molecular Weight PE

Thermoplastic ultrahigh molecular weight PE (UHMWPE) is also known as high-modulus PE (HMPE) or high-performance PE (HPPE). It has extremely long chains, with molecular weight numbering in the millions (usually between 3.1 and 5.67 million). The high molecular weight leads to very good packing of the chains into the crystal structure. This makes UHMWPE a very tough material, with the highest impact strength of any thermoplastic presently made. It is highly resistant to corrosive chemicals, with the exception of oxidizing acids. It has extremely low moisture absorption and is highly resistant to abrasion. Its coefficient of friction is significantly lower than that of nylon and acetal.

9.1.6 Rigid Polyvinyl Chloride

PVC is a flexible or rigid material that is chemically nonreactive. Rigid PVC is easily machined, heat formed, welded, and even solvent cemented. PVC can also be machined using standard metal working tools and finished to close tolerances and finishes without great difficulty. PVC resins are normally mixed with other additives such as impact modifiers and stabilizers, providing hundreds of PVC-based materials with a variety of engineering properties.

There are three broad classifications for rigid PVC compounds: Type II, CPVC, and Type I. Type II differs from Type I due to greater impact values, but lower chemical resistance. CPVC has greater high-temperature resistance. These materials are considered "unplasticized," because they are less flexible than the plasticized formulations. PVC has a broad range of applications, from high-volume construction related products to simple electric wire insulation and coatings.

9.1.7 Cyclic Olefin Copolymer

Cyclic olefin copolymer (COC) is an amorphous polyolefin made by reaction of ethylene and norbornene in varying ratios. Its structure is given in Figure 9.5. The properties can be customized by changing the ratio of the monomers found in the polymer. Being amorphous it is transparent. Other performance benefits include:

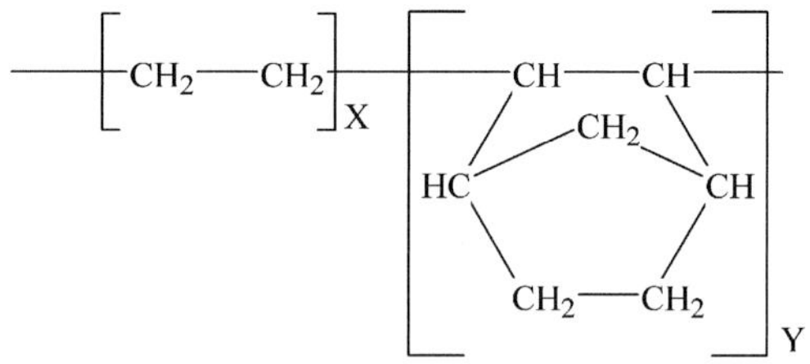

Figure 9.5 Chemical structure of COCs.

- low density
- extremely low water absorption
- excellent water vapor barrier properties
- high rigidity, strength, and hardness
- variable heat deflection temperature up to 170°C
- very good resistance to acids and alkalis

9.1.8 Polyacrylics

While a large number of acrylic polymers are manufactured, polymethyl methacrylate PMMA is by far the most common. Nearly everyone has heard of Plexiglas®. PMMA has two very distinct properties that set the products apart from others. First it is optically clear and colorless. It has a light transmission of 92%. The 4% reflection loss at each surface is unavoidable. Second its surface is extremely hard. They are also highly weather resistant.

9.1.9 Other Olefin Acrylic Polymers

There are a large number of acrylic/olefin copolymers manufactured. One of the best known is the copolymer of ethylene and methacrylic acid forming a polymer known as EMAA. This is more commonly known by its trade name Surlyn® made by DuPont. Generally, there is little multipoint data publicly available for these polymers – so they are not included in this book.

9.2 Polyethylene

9.2.1 Fatigue Data

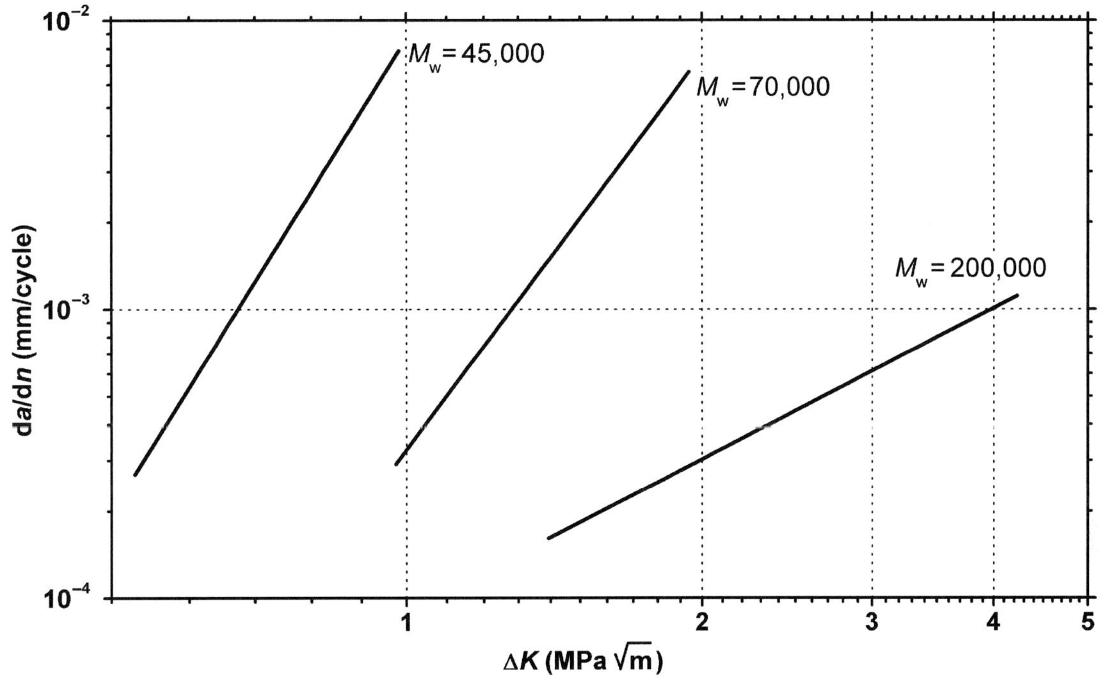

Figure 9.6 Fatigue crack propagation vs. stress intensity factor and molecular weight of generic high-density PE.

9.2.2 Tribology Data

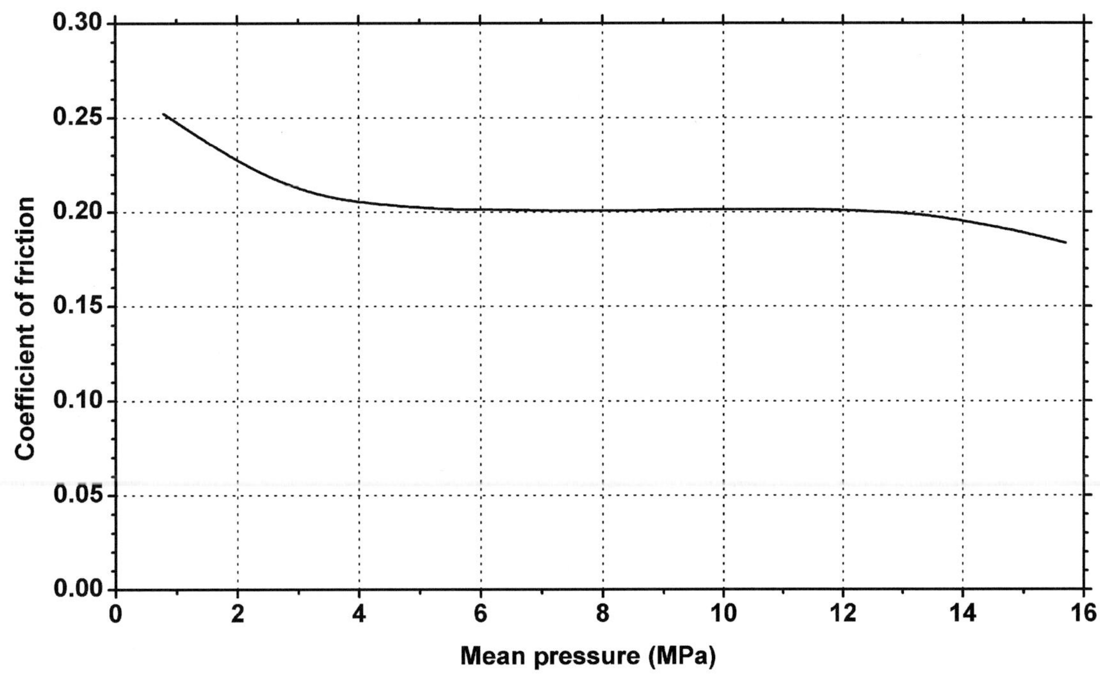

Figure 9.7 Dynamic coefficient of friction vs. pressure of LyondellBasell Industries Polyolefins Lupolen® PE.

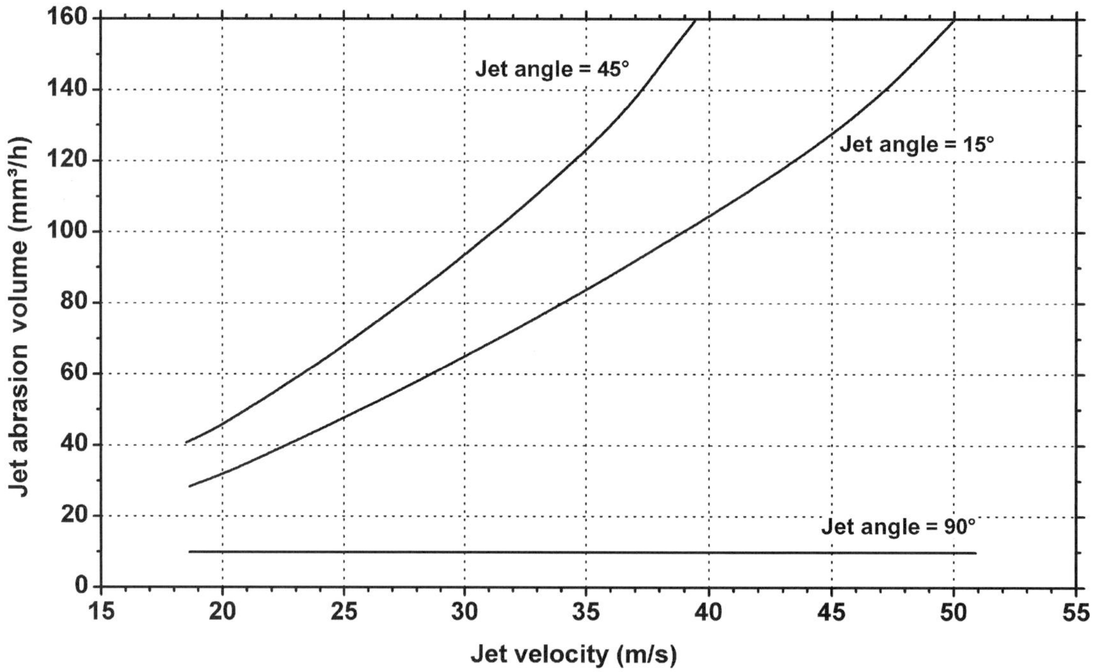

Figure 9.8 Jet abrasion volume vs. jet velocity of LyondellBasell Industries Polyolefins Lupolen® PE.

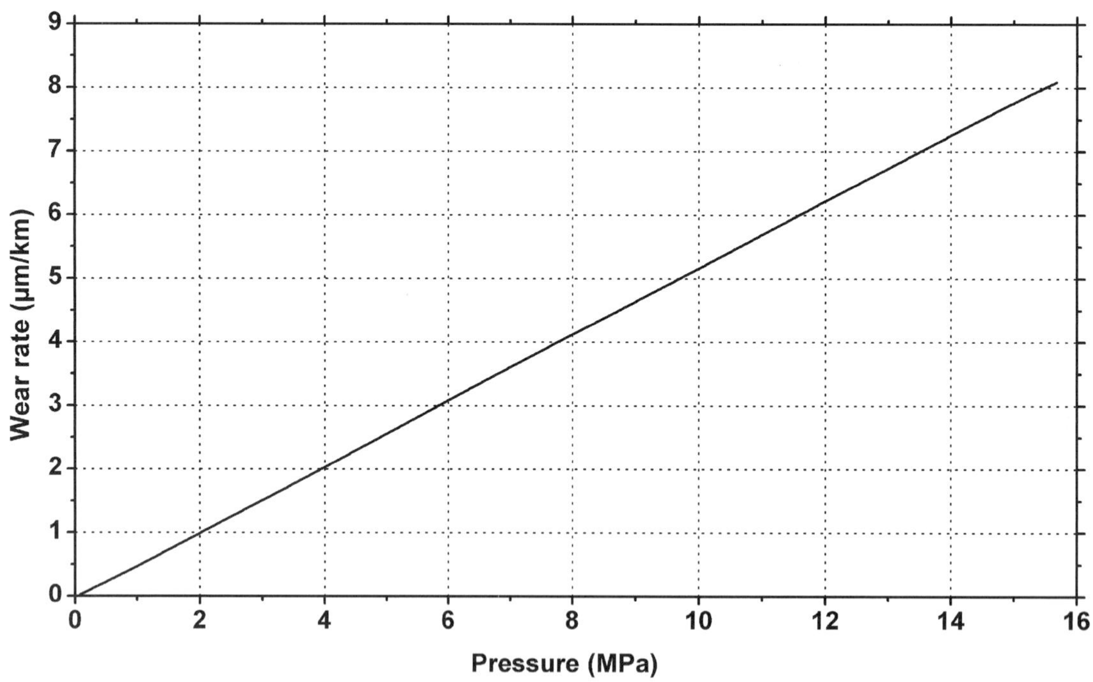

Figure 9.9 Wear rate vs. mean pressure by pin-on-disk of LyondellBasell Industries Polyolefins Lupolen® PE.

9.3 Polypropylene

9.3.1 Fatigue Data

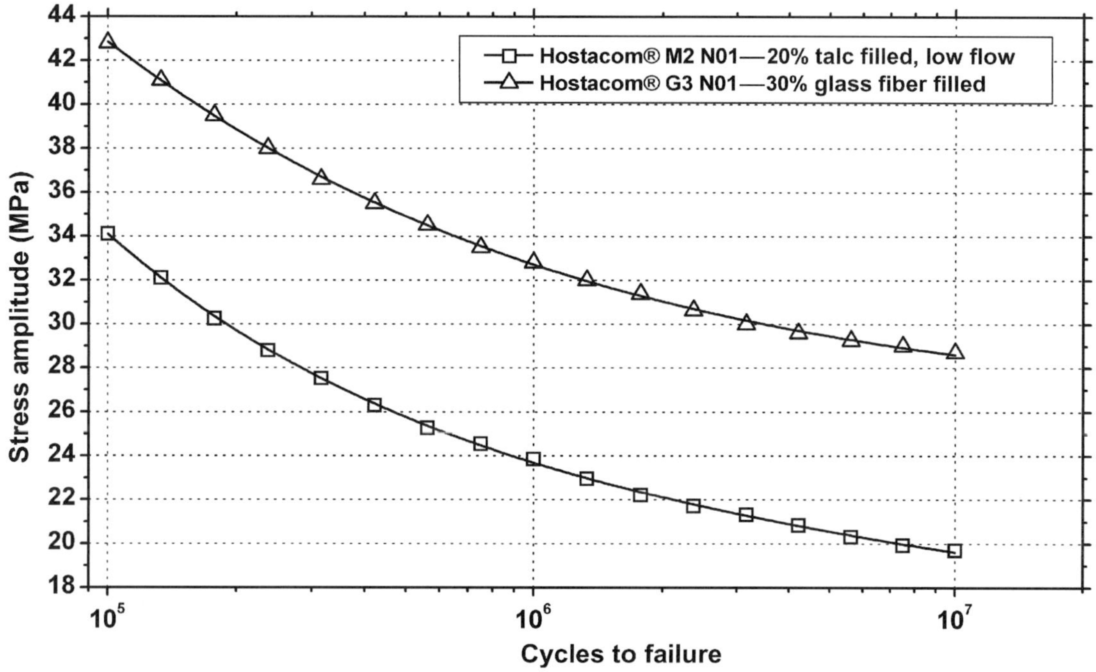

Figure 9.10 Flexural stress amplitude vs. cycles to failure at 23°C and 10 Hz of LyondellBasell Industries Polyolefins Hostacom® PP plastics.

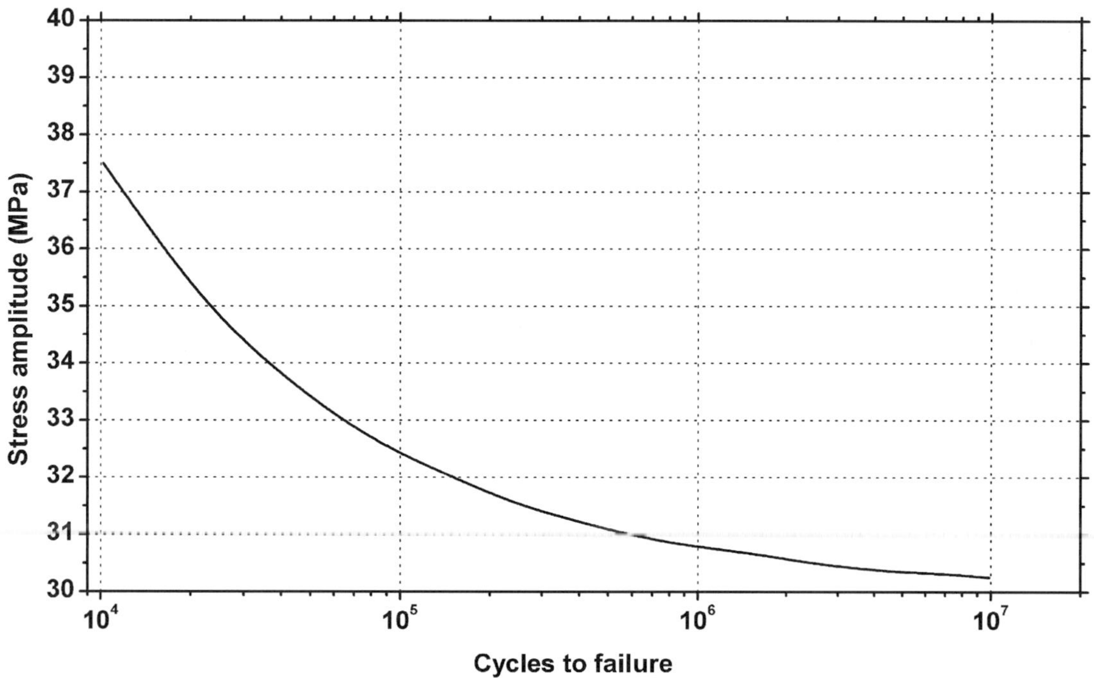

Figure 9.11 Flexural stress amplitude vs. cycles to failure at 23°C of SABIC Innovative Plastics Thermocomp® MF-1006—30% glass fiber reinforced PP.

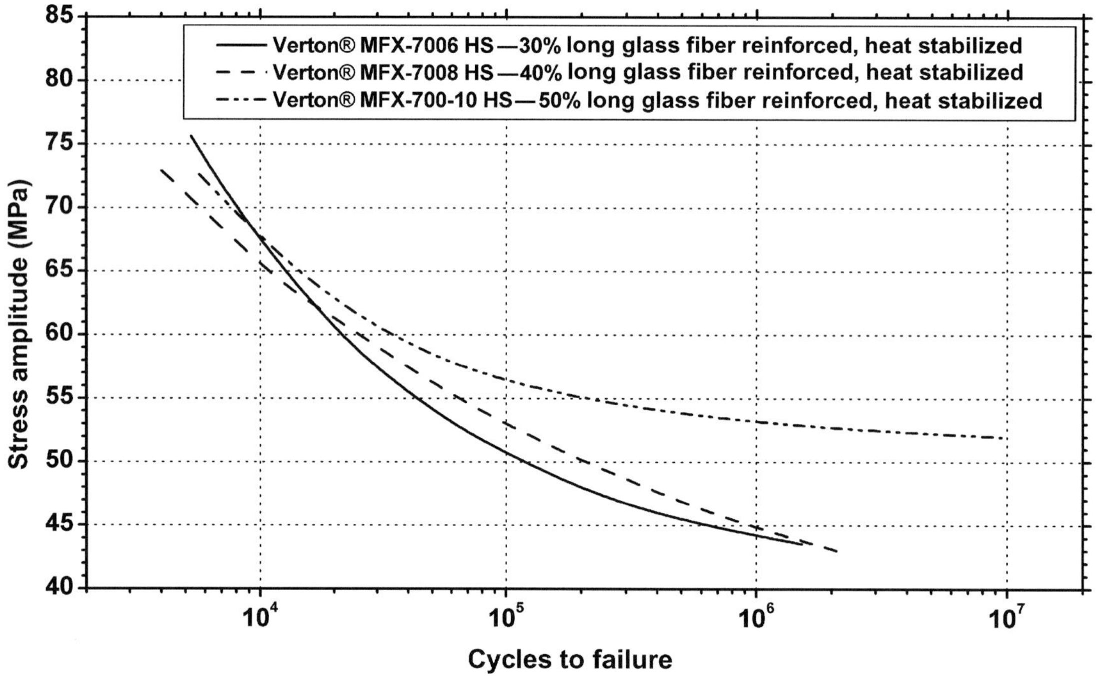

Figure 9.12 Flexural stress amplitude vs. cycles to failure at 23°C of several SABIC Innovative Plastics Verton®—long glass fiber reinforced PP plastics.

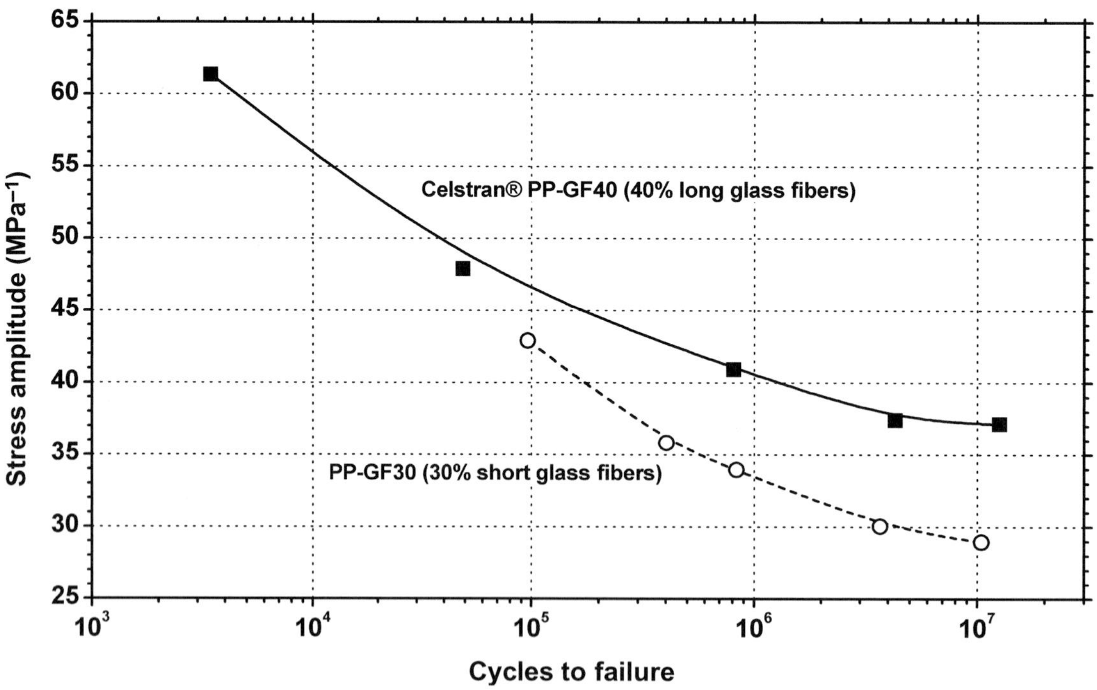

Figure 9.13 Flexural stress amplitude vs. cycles to failure of two Ticona glass fiber reinforced PP plastics.

9.4 Ultrahigh-Molecular-Weight PE

9.4.1 Fatigue Data

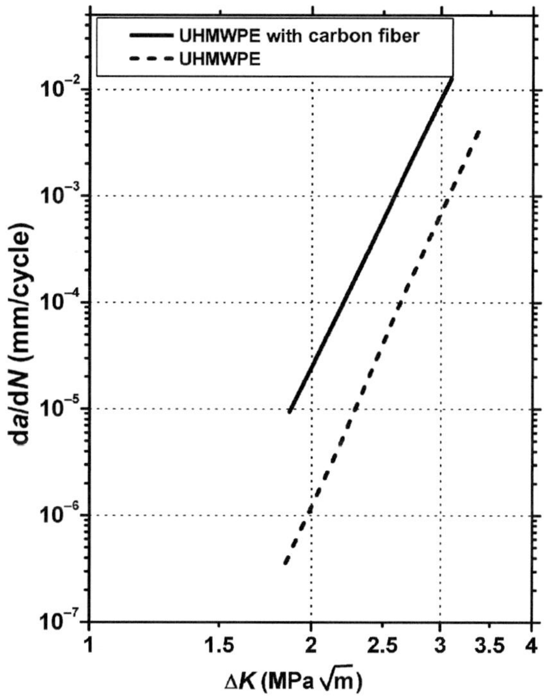

Figure 9.14 Fatigue crack propagation vs. stress intensity factor of two UHMWPE plastics.

9.4.2 Tribology Data

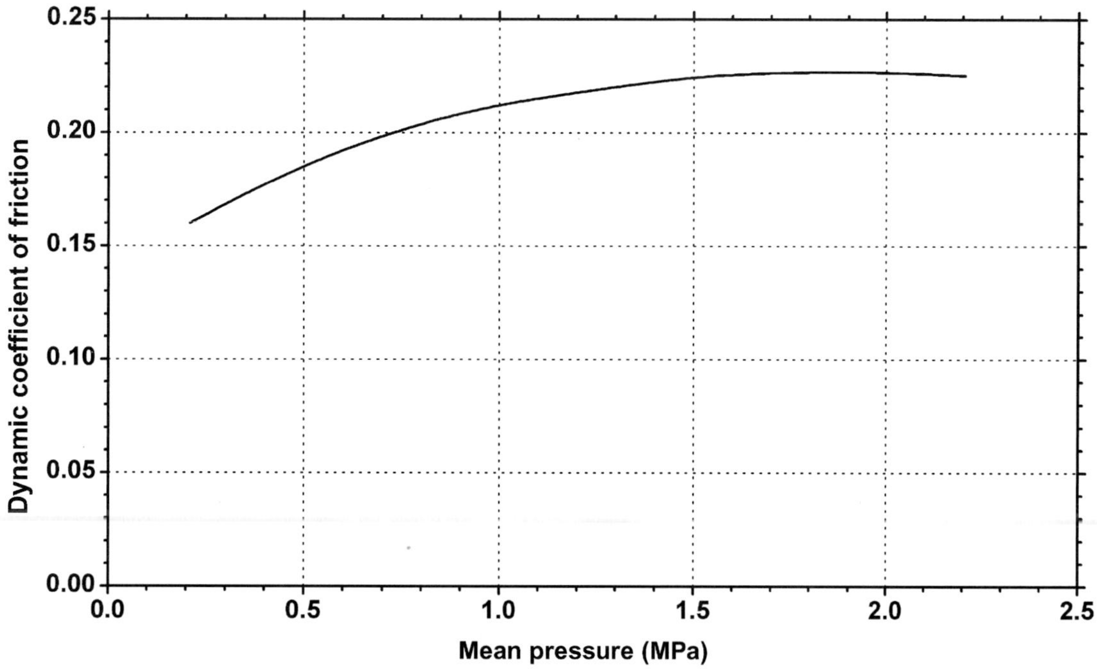

Figure 9.15 Dynamic coefficient of friction vs. pressure at a velocity of 10 m/min of Ticona Engineering Polymers GUR® UHMWPE.

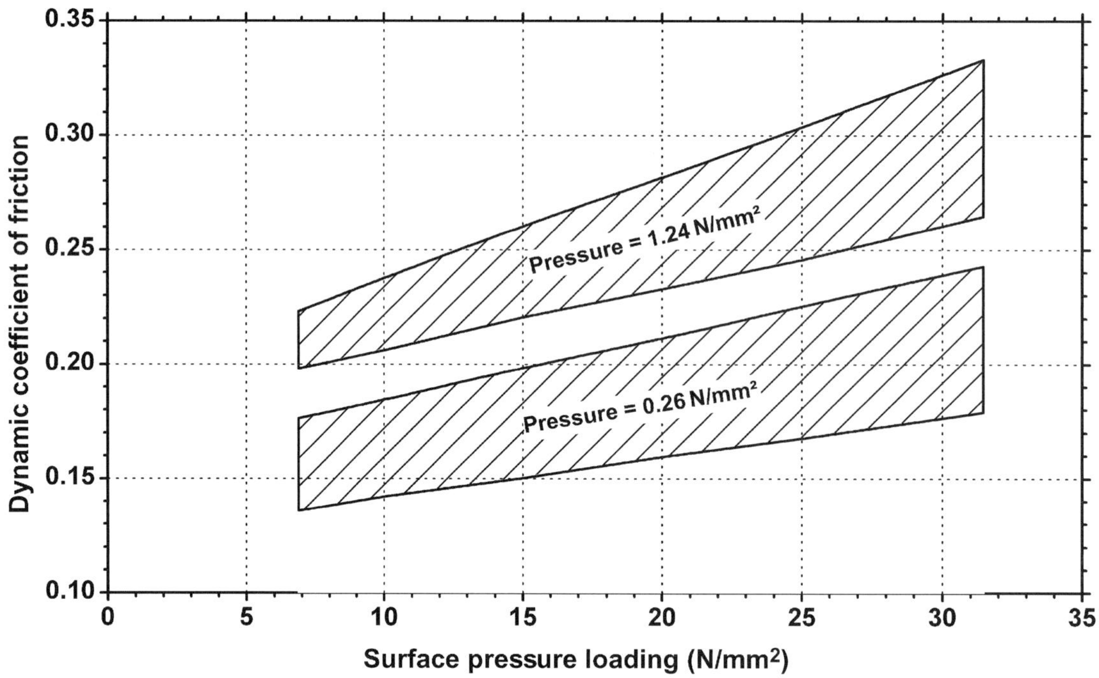

Figure 9.16 Dynamic coefficient of friction vs. sliding speed and pressure of Ticona Engineering Polymers GUR® UHMWPE.

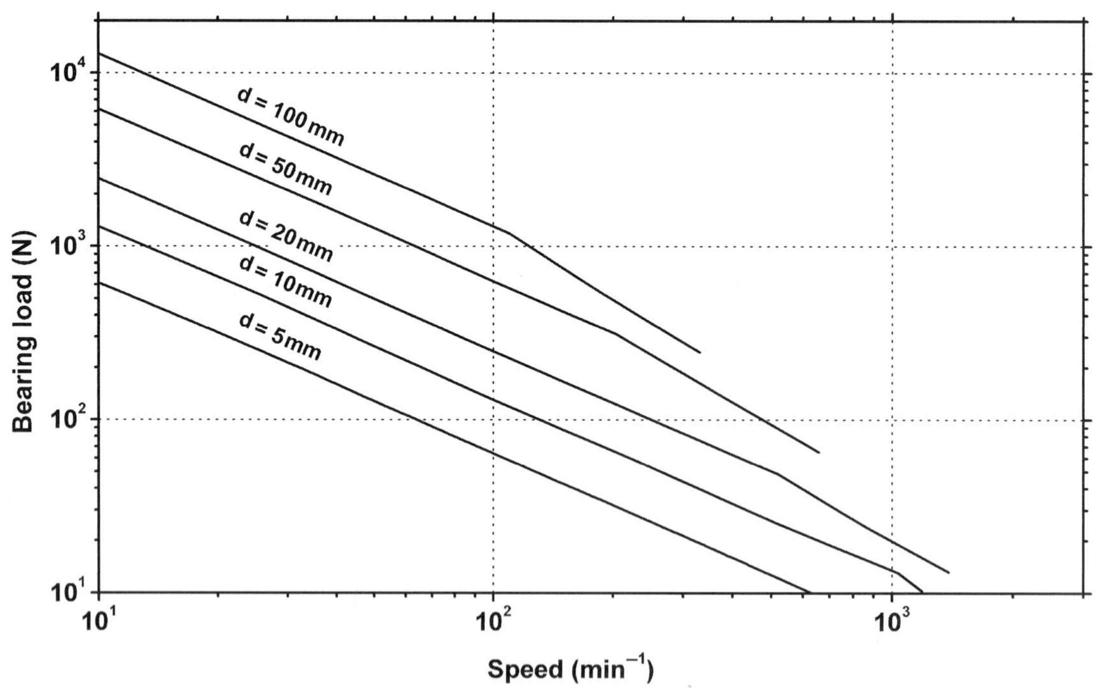

Figure 9.17 Permissible unlubricated bearing load vs. sliding speed for bearings made of Ticona Engineering Polymers GUR® UHMWPE.

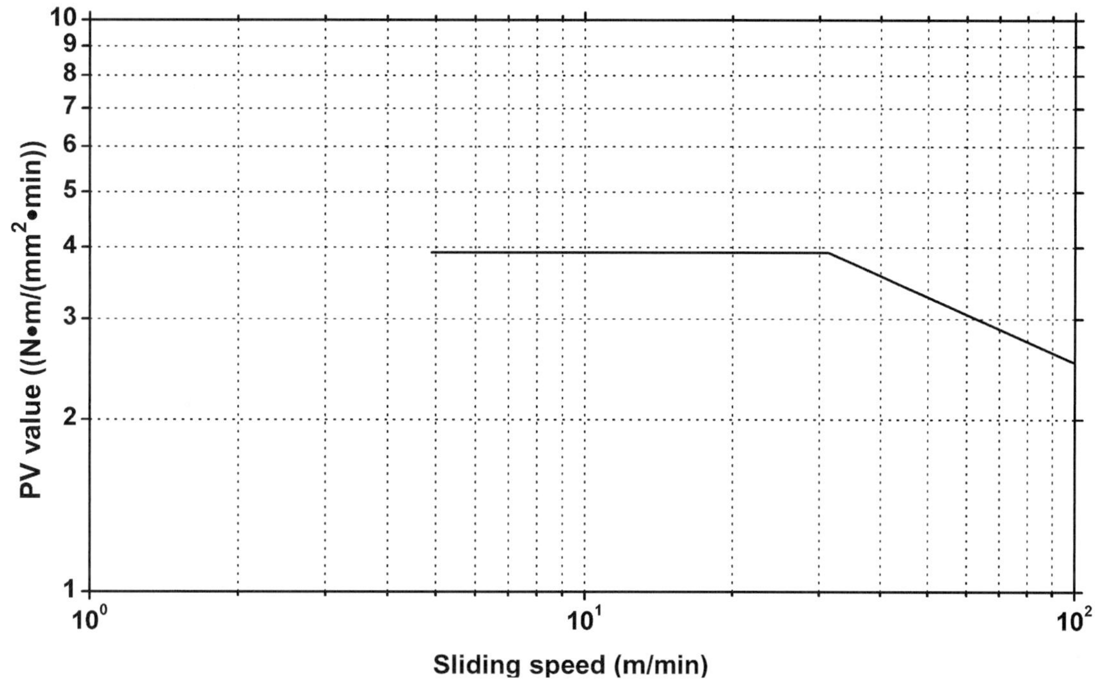

Figure 9.18 PV load limit vs. sliding speed for bearings made of Ticona Engineering Polymers GUR® UHMWPE.

9.5 Polyvinyl Chloride

9.5.1 Fatigue Data

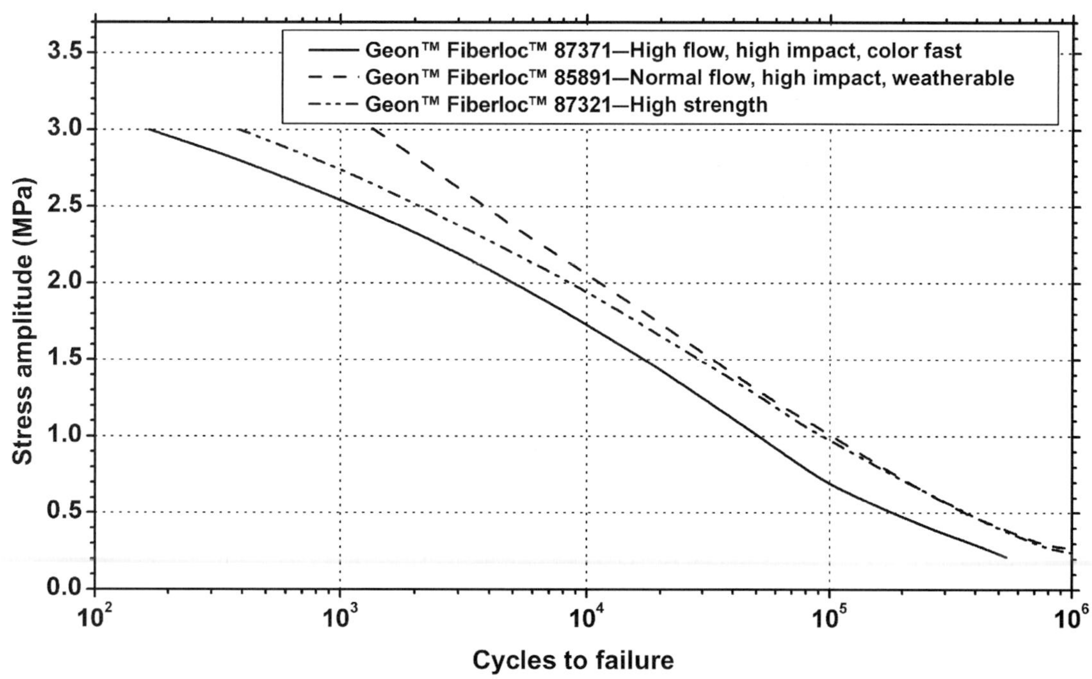

Figure 9.19 Flexural stress amplitude vs. cycles to failure at 23°C of several PolyOne Corporation Geon™ Fiberloc™ PVC plastics.

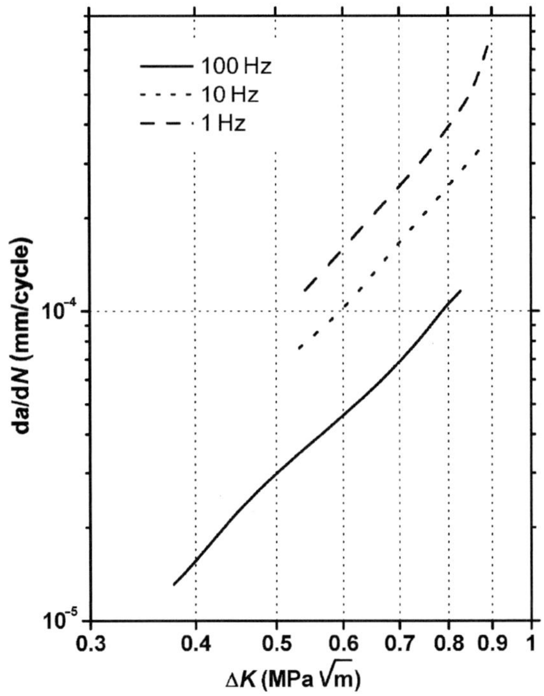

Figure 9.20 Fatigue crack propagation rate vs. stress intensity factor and cycle frequency of generic PVC.

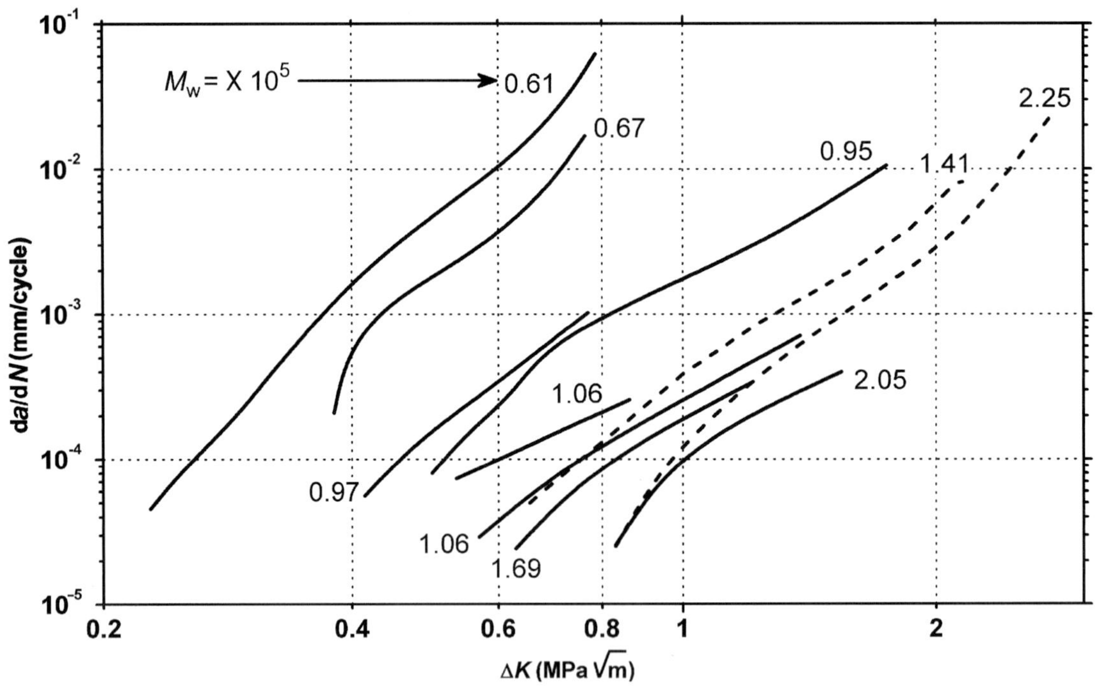

Figure 9.21 Fatigue crack propagation rate vs. stress intensity factor and molecular weight of generic PVC.

9.6 Acrylics

9.6.1 Fatigue Data

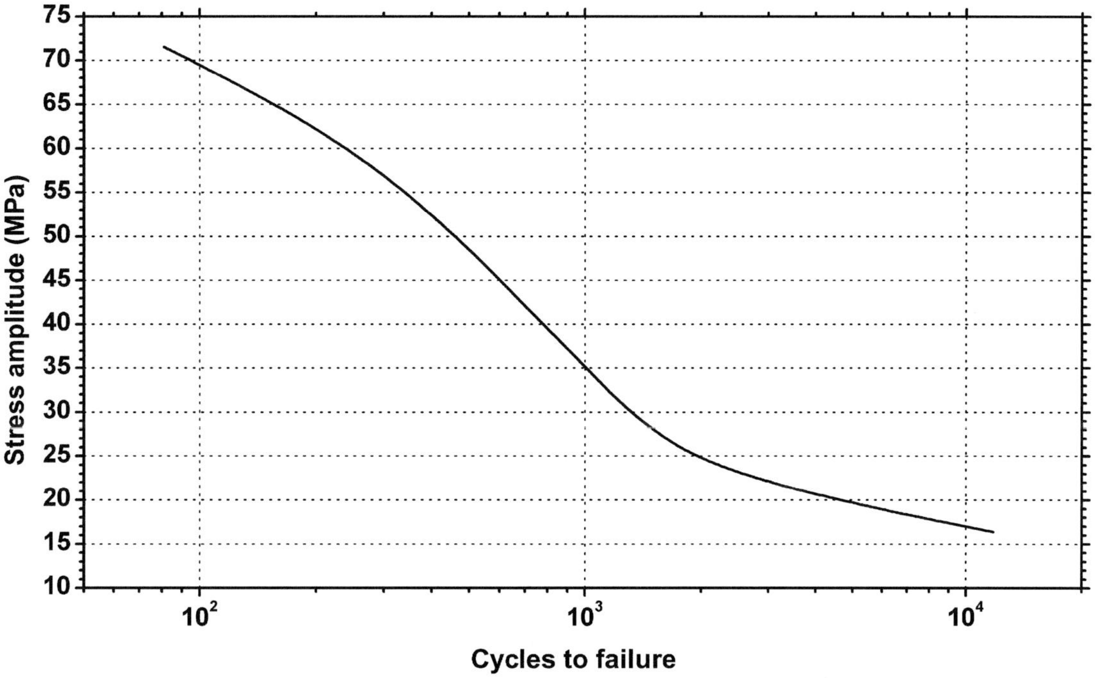

Figure 9.22 Flexural stress amplitude vs. cycles to failure at 23°C and 65% relative humidity of Lucite International Inc Diakon™ CMG302—general-purpose, high-heat-resistant acrylic.

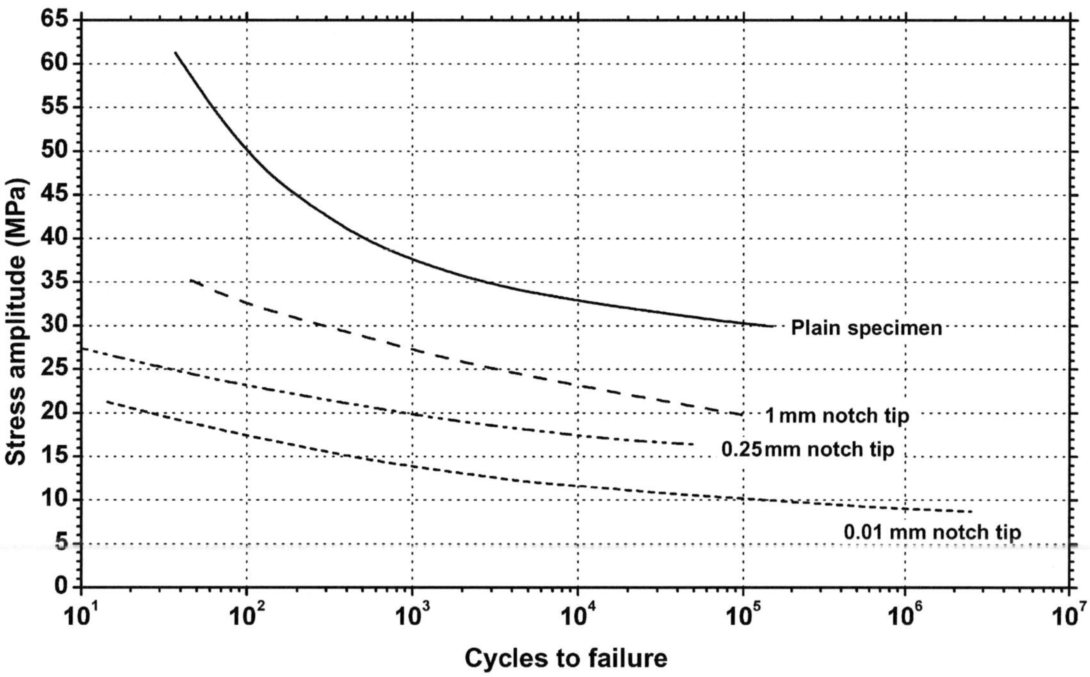

Figure 9.23 Tension/compression stress amplitude vs. cycles to failure at 20°C and 5 Hz at different notch sizes of generic acrylic.

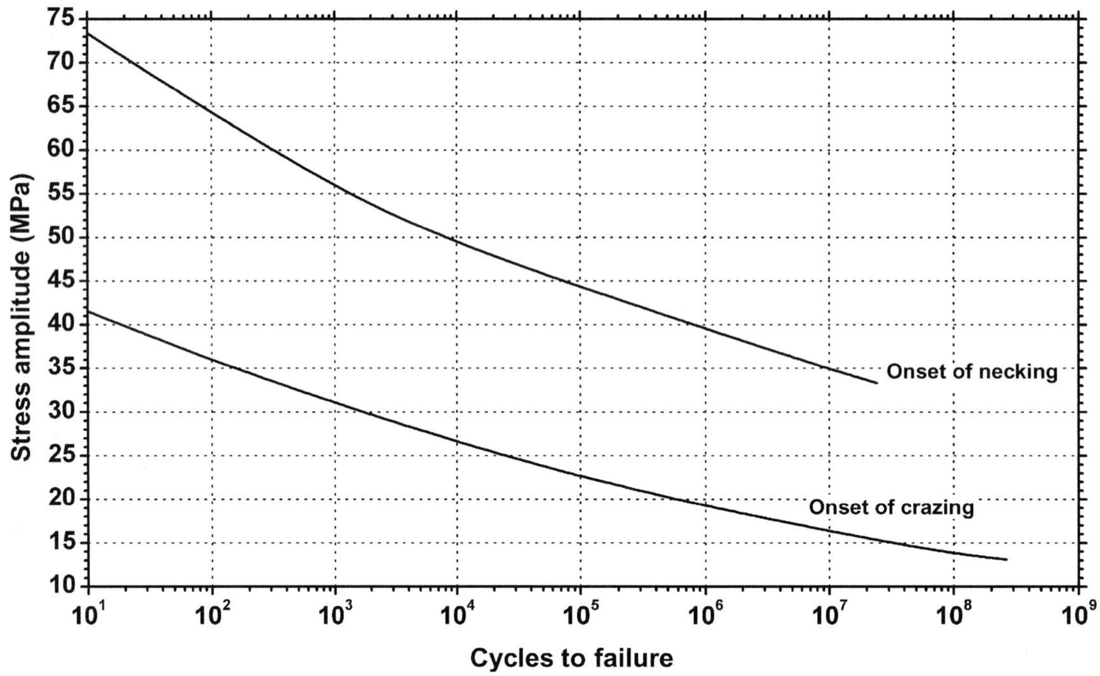

Figure 9.24 Stress amplitude vs. cycles to failure at 23°C and 65% RH by type of failure of Lucite International Inc Diakon™ CMG302—general-purpose, high-heat-resistant acrylic.

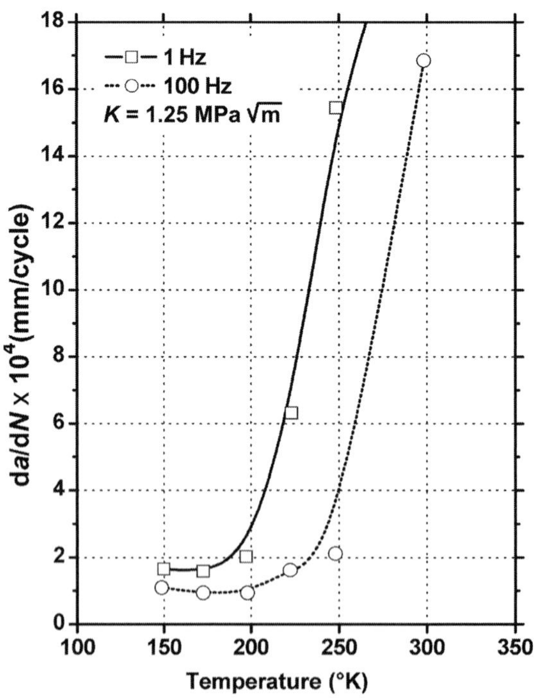

Figure 9.25 Fatigue crack propagation rate vs. temperature and cycle frequency of generic acrylic.

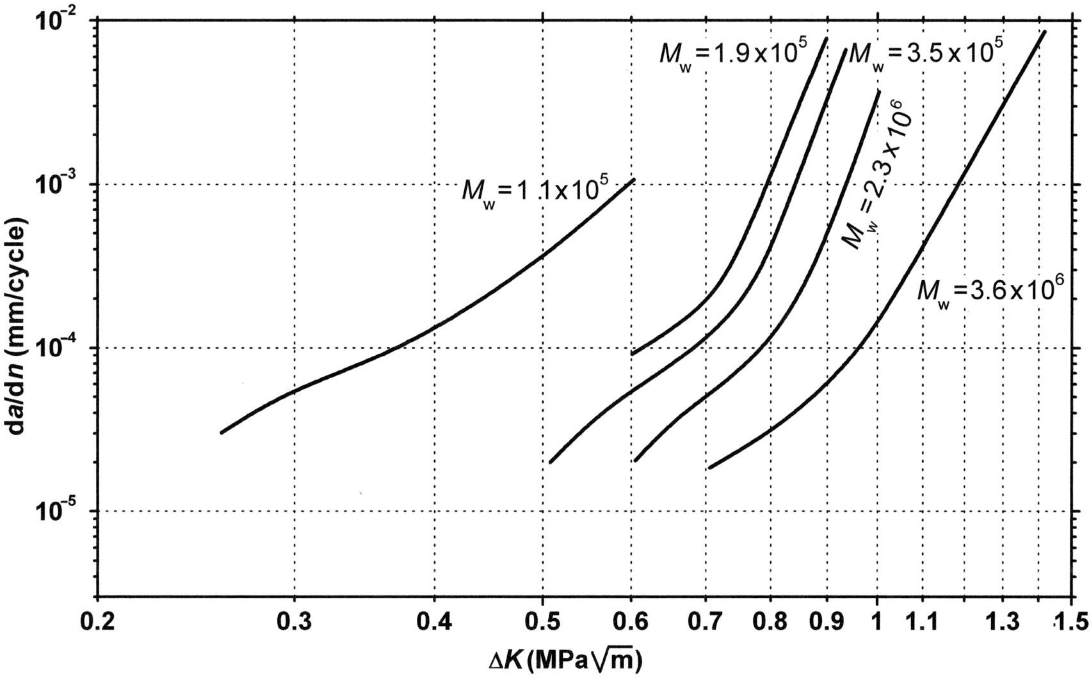

Figure 9.26 Fatigue crack propagation rate vs. stress intensity factor and molecular weight of generic acrylic.

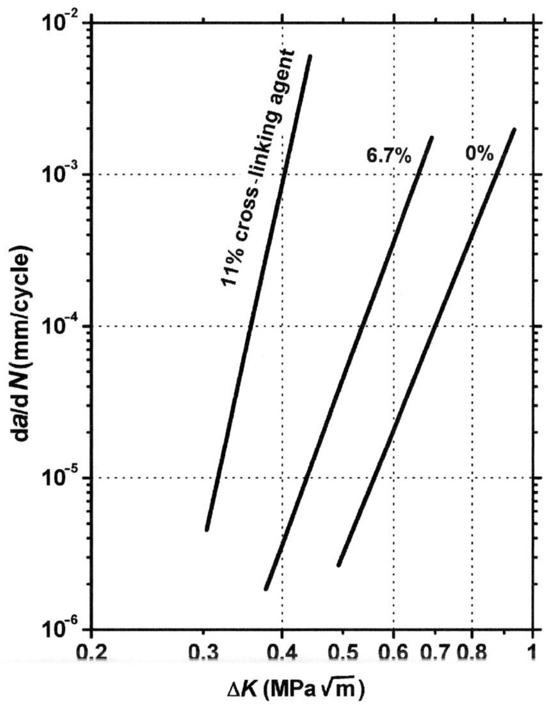

Figure 9.27 Fatigue crack propagation rate vs. stress intensity factor and the amount of cross-linking agent of generic acrylic with a molecular weight of 3.7×10^5.

10 Thermoplastic Elastomers

10.1 Background

Thermoplastic elastomers (TPEs) have two big advantages over the conventional thermoset (vulcanized) elastomers. Those are ease and speed of processing. Other advantages of TPEs are recyclability of scrap, lower energy costs for processing, and the availability of standard, uniform grades (not generally available in thermosets).

TPEs are molded or extruded on standard plastics-processing equipment in considerably shorter cycle times than those required for compression or transfer molding of conventional rubbers. They are made by copolymerizing two or more monomers, using either block or graft polymerization techniques. One of the monomers provides the hard, or crystalline, polymer segment that functions as a thermally stable component; the other monomer develops the soft or amorphous segment, which contributes the elastomeric or rubbery characteristic.

Physical and chemical properties can be controlled by varying the ratio of the monomers and the length of the hard and soft segments. Block techniques create long-chain molecules that have various or alternating hard and soft segments. Graft polymerization methods involve attaching one polymer chain to another as a branch. The properties that are affected by each phase can be generalized:

"Hard phase"—plastic properties:

1. Processing temperatures
2. Continuous use temperature
3. Tensile strength
4. Tear strength
5. Chemical and fluid resistance
6. Adhesion to inks, adhesives, and over-molding substrates

"Soft phase"—elastomeric properties:

1. Lower service temperature limits
2. Hardness
3. Flexibility
4. Compression set and tensile set

Three high-performance types of TPEs make up this chapter.

10.1.1 Thermoplastic Polyurethane Elastomers

Urethanes are a reaction product of a diisocyanate and long- and short-chain polyether, polyester, or caprolactone glycols. The polyols and the short-chain diols react with the diisocyanates to form linear polyurethane molecules. This combination of diisocyanate and short-chain diol produces the rigid or hard segment. The polyols form the flexible or soft segment of the final molecule. Figure 10.1 shows the molecular structure in schematic form.

The properties of the resin depend on the nature of the raw materials, the reaction conditions, and the ratio of the starting raw materials. The polyols used have a significant influence on certain properties of the thermoplastic polyurethane. Polyether and polyester polyols are both used to produce many products.

The polyester-based thermoplastic polyurethane elastomers (TPUs) have the following characteristic features:

- Good oil/solvent resistance
- Good UV resistance
- Abrasion resistance
- Good heat resistance
- Mechanical properties

The polyether-based TPUs have the following characteristic features:

- Fungus resistance
- Low-temperature flexibility
- Excellent hydrolytic stability
- Acid/base resistance

Copyright © 2010 Laurence W. McKeen. All rights reserved.

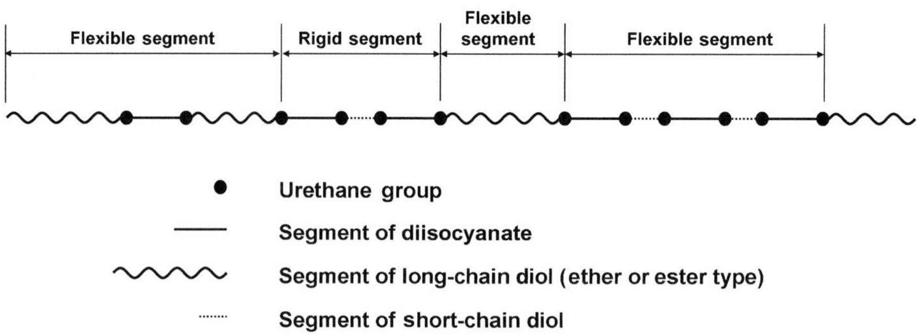

Figure 10.1 Molecular structure of a TPU.

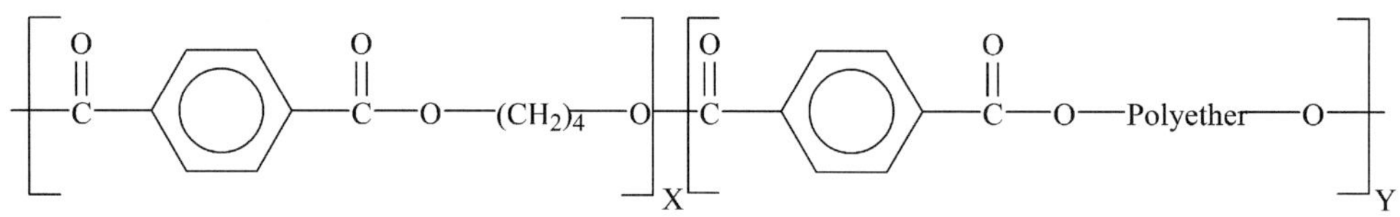

Figure 10.2 Molecular structure of Ticona Riteflex® TPE or COPE.

In addition to the basic components described above, most resin formulations contain additives to facilitate production and processability. Other additives can also be included such as:

- Demolding agents
- Flame retardants
- Heat/UV stabilizers
- Plasticizers

The polyether types are slightly more expensive and have better hydrolytic stability and low-temperature flexibility than the polyester types.

10.1.2 Thermoplastic Copolyester Elastomers

Thermoplastic copolyester elastomers (TPE-E or COPE) are block copolymers. The chemical structure of one such elastomer is shown in Figure 10.2.

These TPEs are generally tougher over a broader temperature range than the urethanes described in Section 10.1.1. Also, they are easier and more forgiving in processing.

- Excellent abrasion resistance
- High tensile, compressive, and tear strength
- Good flexibility over a wide range of temperatures
- Good hydrolytic stability
- Resistance to solvents and fungus attack
- Selection of a wide range of hardness

In these polyester TPEs, the hard polyester segments can crystallize, giving the polymer some of the attributes of semicrystalline thermoplastics, most particularly better solvent resistance than ordinary rubbers, but also better heat resistance. Above the melting temperature of the crystalline regions, these TPEs can have low viscosity and can be molded easily in thin sections and complex structures. Properties of thermoplastic polyester elastomers can be fine-tuned over a range by altering the ratio of hard to soft segments.

In DuPont Hytrel® polyester TPEs, the resin is a block copolymer. The hard phase is polybutylene terephthalate (PBT). The soft segments are long-chain polyether glycols.

10.1.3 Thermoplastic Polyether Block Amide Elastomers

Polyether block amides are plasticizer-free TPEs. The soft segment is the polyether and the hard segment is the polyamide (nylon). For example, Arkema PEBAX® 33 series products are based on Nylon 12 (see Section 8.1.3) and polytetramethylene glycol segments (PTMG). They are easy to process by injection molding and profile or film extrusion. Often they can be easily melt-blended with other polymers, and many compounders will provide custom products by doing this. Their chemistry allows them to achieve a wide range of physical and mechanical properties by varying the monomeric block types and ratios.

- Light weight
- Great flexibility (extensive range)
- Resiliency
- Very good dynamic properties
- High strength
- Outstanding impact resistance properties at low temperature
- Easy processing
- Good resistance to most chemicals

10.1.4 Styrenic Block Copolymer TPEs

Styrenic block copolymer (SBS) TPEs are multiphase compositions in which the phases are chemically bonded by block copolymerization (see Section 3.2). At least one of the phases is a hard styrenic polymer. This styrenic phase may become fluid when the TPE composition is heated. Another phase is a softer elastomeric material that is rubber like at room temperature. The polystyrene blocks act as cross-links, tying the elastomeric chains together in a three-dimensional network. SBS TPEs have no commercial applications when the product is just a pure polymer. They must be compounded with other polymers, oils, fillers, and additives to have any commercial value.

10.1.5 Polyolefin TPE

Polyolefin TPE (TPO) materials are defined as compounds of various polyolefin polymers, semicrystalline thermoplastics, and amorphous elastomers. Most TPOs are composed of polypropylene and a copolymer of ethylene and propylene called ethylene–propylene rubber (EPR). A common rubber of this type is called EPDM rubber, which has a small amount of a third monomer, a diene (two carbon–carbon double bonds in it). The diene monomer leaves a small amount of unsaturation in the polymer chain that can be used for sulfur cross-linking. Like most TPEs, TPO products are composed of hard and soft segments. TPO compounds include fillers, reinforcements, lubricants, heat stabilizers, antioxidants, UV stabilizers, colorants, and processing aids.

10.1.6 Elastomeric Alloy-Thermoplastic Vulcanizate

Vulcanized elastomeric alloys are TPEs composed of mixtures of two or more polymers that have received a proprietary treatment. Elastomeric alloy-thermoplastic vulcanizates (EA-TPVs) are a category of TPEs made of a rubber and plastic polymer mixture in which the rubber phase is highly vulcanized. The plastic phase of an EA-TPV is a polypropylene, and the rubber phase is an ethylene–propylene elastomer.

The vulcanization of the rubber phase of an EA-TPV results in various property improvements such as insoluble in rubber solvents and reduced swelling in some solvents. The vulcanization offers other property improvements such as:

- increase in tensile strength and modulus
- decrease in compression set
- decrease in swelling caused by oils
- the retention of properties at temperatures below 200°F (93°C)
- fatigue resistance

11 Fluoropolymers

11.1 Background

The following sections will briefly explain the structures and properties between the various fluoropolymers. It is important to keep in mind there are variations of most of these polymers. The most common variation is the molecular weight, which will affect the melting point somewhat, and the viscosity of the polymer above its melt point, properties that are important in determining processing conditions and use.

Traditionally, a fluoropolymer or fluoroplastic is defined as a polymer consisting of carbon (C) and fluorine (F). Sometimes these are referred to as perfluoropolymers to distinguish them from partially fluorinated polymers, fluoroelastomers, and other polymers that contain fluorine in their chemical structure. For example, fluorosilicone and fluoroacrylate polymers are not referred to as fluoropolymers.

11.1.1 Polytetrafluoroethylene

Polytetrafluoroethylene polymer (PTFE) is an example of a linear fluoropolymer. Its structure in simplistic form is shown in Figure 11.1.

Formed by the polymerization of tetrafluoroethylene (TFE), the ($-CF_2-CF_2-$) groups repeat many thousands of times. The fundamental properties of fluoropolymers evolve from the atomic structure of fluorine and carbon and their covalent bonding in specific chemical structures. The backbone is formed of carbon–carbon bonds and the pendant groups are carbon–fluorine bonds. Both are extremely strong bonds. The basic properties of PTFE stem from these two very strong chemical bonds. The size of the fluorine atom allows the formation of a uniform and continuous covering around the carbon–carbon bonds and protects them from chemical attack, thus imparting chemical resistance and stability to the molecule. PTFE is rated for use up to (260°C). PTFE does not dissolve in any known solvent. The fluorine sheath is also responsible for the low surface energy (18 dynes/cm) and low coefficient of friction (0.05–0.8, static) of PTFE. Another attribute of the uniform fluorine sheath is the electrical inertness (or nonpolarity) of the PTFE molecule. Electrical fields impart only slight polarization in this molecule, so volume and surface resistivity are high.

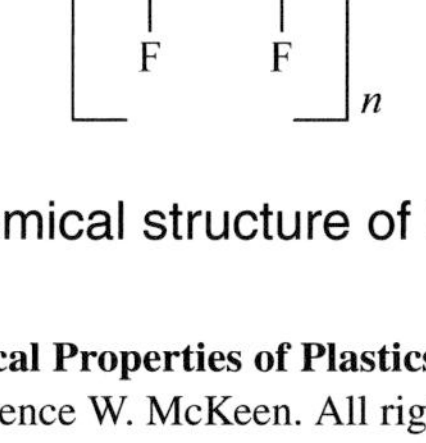

Figure 11.1 Chemical structure of PTFE.

The PTFE molecule is simple and is quite ordered and so it can align itself with other molecules or other portions of the same molecule. Disordered regions are called *amorphous* regions. This is important because polymers with high crystallinity require more energy to melt. In other words they have higher melting points. When this happens it forms what is called a crystalline region. Crystalline polymers have a substantial fraction of their mass in the form of parallel, closely packed molecules. High-molecular-weight PTFE resins have high crystallinity and therefore high-melting points, typically as high as 320–342°C (608–648°F). The crystallinity of as-polymerized PTFE is typically 92–98%. Further, the viscosity in the molten state (called melt creep viscosity) is so high that high-molecular-weight PTFE particles do not flow even at temperatures above its melting point. They sinter much like powdered metals; they stick to each other at the contact points and combine into larger particles.

PTFE is called a *homopolymer*, a polymer made from a single monomer. Recently many PTFE manufacturers have added minute amounts of other monomers to their PTFE polymerizations to produce alternate grades of PTFE designed for specific applications. Generally, polymers made from two monomers are called copolymers, but fluoropolymer manufacturers call these grades modified homopolymer when the copolymer is used at less than 1% by weight. DuPont grades of this type are called *Teflon®*

Copyright © 2010 Laurence W. McKeen. All rights reserved.

NXT Resins. These modified granular PTFE materials retain the exceptional chemical, thermal, anti-stick, and low-friction properties of conventional PTFE resin, but offer some improvements:

- Weldability
- Improved permeation resistance
- Less creep
- Smoother, less porous surfaces
- Better high-voltage insulation

The copolymers described in the next sections contain significantly more of the non-TFE monomers.

11.1.2 Polyethylene Chlorotrifluoroethylene

Polyethylene chlorotrifluoroethylene (E-CTFE) is a copolymer of ethylene and chlorotrifluoroethylene. Figure 11.2 shows the molecular structure of E-CTFE.

This simplified structure shows the ratio of the monomers being 1–1 and strictly alternating, which is the desirable proportion. Commonly known by the trade name, Halar®, E-CTFE is an expensive, melt processable, semi-crystalline, whitish semi-opaque thermoplastic with good chemical resistance, and barrier properties. It also has good tensile and creep properties and good high frequency electrical characteristics. Applications include chemically resistant linings, valve and pump components, barrier films, and release/vacuum bagging films.

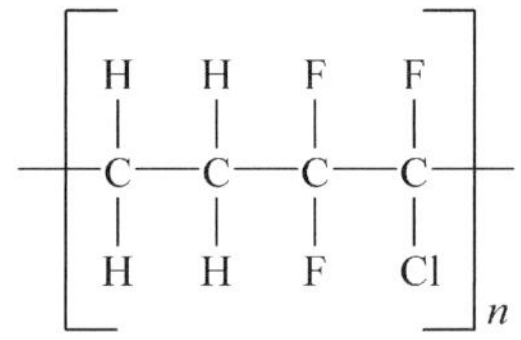

Figure 11.2 Chemical structure of E-CTFE.

11.1.3 Polyethylene Tetrafluoroethylene

Polyethylene tetrafluoroethylene (ETFE) is a copolymer of ethylene and TFE. The basic molecular structure of ETFE is shown in Figure 11.3.

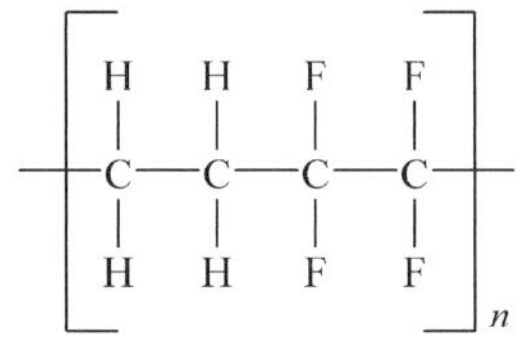

Figure 11.3 Chemical structure of ETFE.

This depicted structure shows alternating units of TFE and ethylene. While this can be readily made, many grades of ETFE vary the ratio of the two monomers slightly to optimize properties for specific end uses.

ETFE is a fluoroplastic with excellent electrical and chemical properties. It also has excellent mechanical properties. ETFE is especially suited for uses requiring high mechanical strength, chemical, thermal, and/or electrical properties. The mechanical properties of ETFE are superior to those of PTFE and FEP. ETFE has:

- Excellent resistance to extremes of temperature, ETFE has a working temperature range of −200–150°C.
- Excellent chemical resistance.
- Mechanical strength ETFE is good with excellent tensile strength and elongation and has superior physical properties compared to most fluoropolymers.
- With low smoke and flame characteristics, ETFE is rated 94V-0 by the Underwriters Laboratories Inc. It is odorless and nontoxic.
- Outstanding resistance to weather and aging.
- Excellent dielectric properties.
- Nonstick characteristics.

11.1.4 Fluorinated Ethylene Propylene

If one of the fluorine atoms on TFE is replaced with a trifluoromethyl group ($-CF_3$) then the new monomer is called hexafluoropropylene (HFP). Polymerization of monomers (HFP) and TFE yield a different fluoropolymer, fluorinated ethylene propylene, called FEP. The number of HFP groups is typically 13% by weight or less and its structure is shown in Figure 11.4.

The effect of using HFP is to put a "bump" along the polymer chain. This bump disrupts the crystallization of the FEP, which has atypical as-polymerized crystallinity of 70% versus 92–98% for PTFE. It also lowers its melting point. The reduction of the

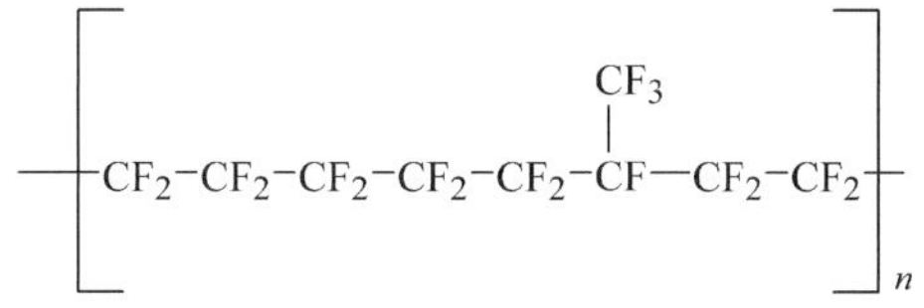

Figure 11.4 Chemical structure of FEP.

Table 11.1 PFA Comonomers

Comonomer	Structure
Perfluoromethyl vinyl ether (MVE)	$CF_2CF–O–CF_3$
Perfluoroethyl vinyl ether (EVE)	$CF_2CF–O–CF_2–CF_3$
Perfluoropropyl vinyl ether (PVE)	$CF_2CF–O–CF_2–CF_2–CF_3$

melting point depends on the amount of trifluoromethyl groups added and secondarily on the molecular weight. Most FEP resins melt around 274°C (525°F), although lower melting points are possible. Even high-molecular-weight FEP will melt and flow. The high chemical resistance, low surface energy, and good electrical insulation properties of PTFE are retained.

11.1.5 Perfluoro Alkoxy

Making a more dramatic change in the side-group than that done in making FEP, chemists put a perfluoro alkoxy (PFA) group on the polymer chain. This group is signified as $–O–R_f$, where R_f can be any number of totally fluorinated carbons. The most common comonomer is perfluoropropyl ($–O–CF_2–CF_2–CF_3$). However, other common comonomers are shown in Table 11.1.

The polymers based on propyl vinyl ether (PVE) are called PFA and the perfluoroalkylvinylether group is typically added at 3.5% or less. When the comonomer is methyl vinyl ether (MVE) the polymer is called MFA. A structure of PFA is shown in Figure 11.5.

The large side group reduces the crystallinity drastically. The melting point is generally between 305°C and 310°C (581–590°F) depending on the molecular weight. The melt viscosity is also dramatically dependent on the molecular weight. Since PFA is still perfluorinated as with FEP the high chemical resistance, low surface energy, and good electrical insulation properties are retained.

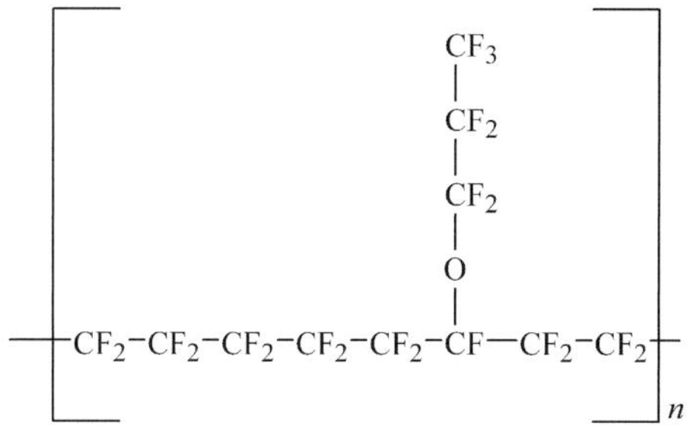

Figure 11.5 Chemical structure of PFA.

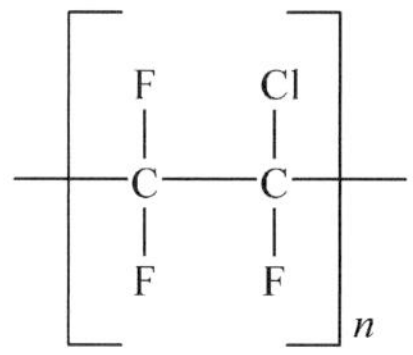

Figure 11.6 Chemical structure of PCTFE.

11.1.6 Polychlorotrifluoroethylene

CTFE is a homopolymer of chlorotrifluoroethylene, characterized by the following structure shown in Figure 11.6.

The addition of the one chlorine atom contributes to lowering the melt viscosity to permit extrusion and injection molding. It also contributes to the transparency, the exceptional flow, and the rigidity characteristics of the polymer. Fluorine is responsible for its chemical inertness and zero moisture absorption. Therefore, PCTFE has unique properties. Its resistance to cold flow, dimensional stability, rigidity, low gas permeability, and low moisture absorption is superior to any other fluoropolymer. It can be used at low temperatures.

11.1.7 Polyvinylidene Fluoride

The polymers made from 1,1-di-fluoro-ethene (or vinylidene fluoride) are known as PVDF—polyvinylidene fluoride. They are resistant to oils and fats, water and steam, and gas and odors, making them of particular value for the food industry. PVDF is known for its exceptional chemical stability and excellent resistance to UV radiation. It is used chiefly in the production and coating of equipment used in aggressive environments, and where high levels of mechanical and thermal resistance are required. It has also been used in architectural applications as a coating on metal siding where it provides exceptional resistance to environmental exposure. The chemical structure of PVDF is shown in Figure 11.7.

One of the trade names of PVDF is KYNAR®. The alternating CH_2 and CF_2 groups along the polymer chain provide a unique polarity that influences its solubility and electric properties. At elevated temperatures PVDF can be dissolved in polar solvents such as organic esters and amines. This selective solubility offers a way to prepare corrosion resistant coatings for chemical process equipment and long-life architectural finishes on building panels.

Key attributes of PVDF include:

- Mechanical strength and toughness
- High abrasion resistance
- High thermal stability
- High dielectric strength
- High purity
- Readily melt processable
- Resistant to most chemicals and solvents
- Resistant to UV and nuclear radiation
- Resistant to weathering
- Resistant to fungi

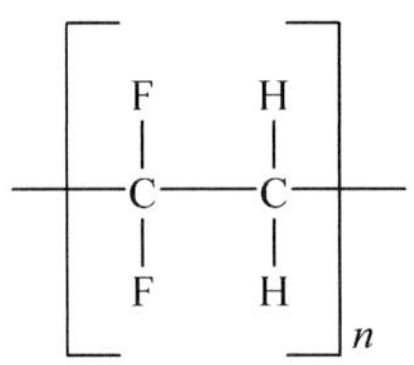

Figure 11.7 Chemical structure of PVDF.

- Low permeability to most gases and liquids
- Low flame and smoke characteristics

11.1.8 THV™

THV™ is a polymer of TFE, HFP, and vinylidene fluoride. It is made by 3M Dyneon. It has the following properties:

- Low processing temperature
- Bonds to elastomers and hydrocarbon plastics
- Good flexibility
- Permeation resistance
- Excellent clarity and light transmission

11.1.9 HTE (Hexafluoropropylene–Tetrafluoroethylene–Ethylene copolymer)

HTE is a polymer of HFP, TFE, and ethylene. It is made by 3M Dyneon. It has the following properties:

- Broad processing range
- Very good chemical resistance
- Permeation resistance
- High light transmission in visible and UV
- Excellent dimensional stability and toughness
- Good electrical properties

11.1.10 Fluoroplastic Melting Points

Table 11.2 Melting Point Ranges of Various Fluoroplastics

Fluoroplastic	Melting Point (°C)
Polytetrafluoroethylene (PTFE)	320–340
Polyethylene chlorotrifluoroethylene (ECTFE)	240
Polyethylene tetrafluoroethylene (ETFE)	255–280
Fluorinated ethylene propylene (FEP)	260–270
Propyl Perfluoro alkoxy (PFA)	302–310
Methyl Perfluoro alkoxy (MFA)	280–290
Polychlorotrifluoroethylene (PCTFE)	210–212
Polyvinylidene fluoride (PVDF)	155–170
THV™	115–235
HTE	155–215

11.2 Polytetrafluoroethylene

11.2.1 Fatigue Data

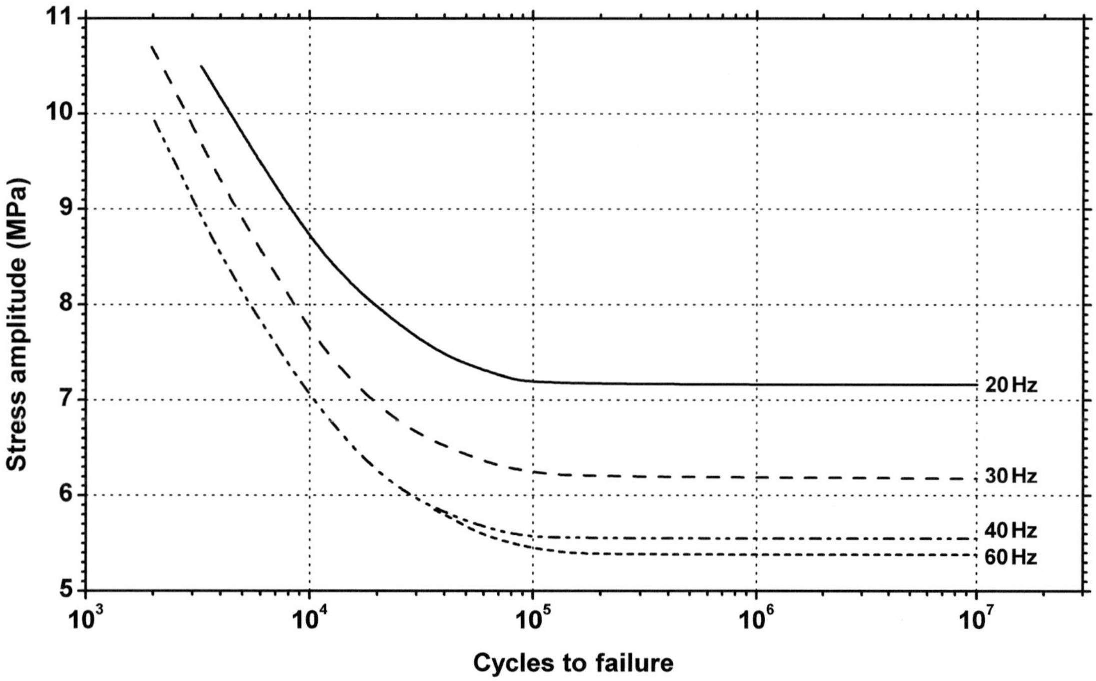

Figure 11.8 Flexural stress amplitude vs. cycles to failure at 23°C and different cycle frequencies of generic PTFE.

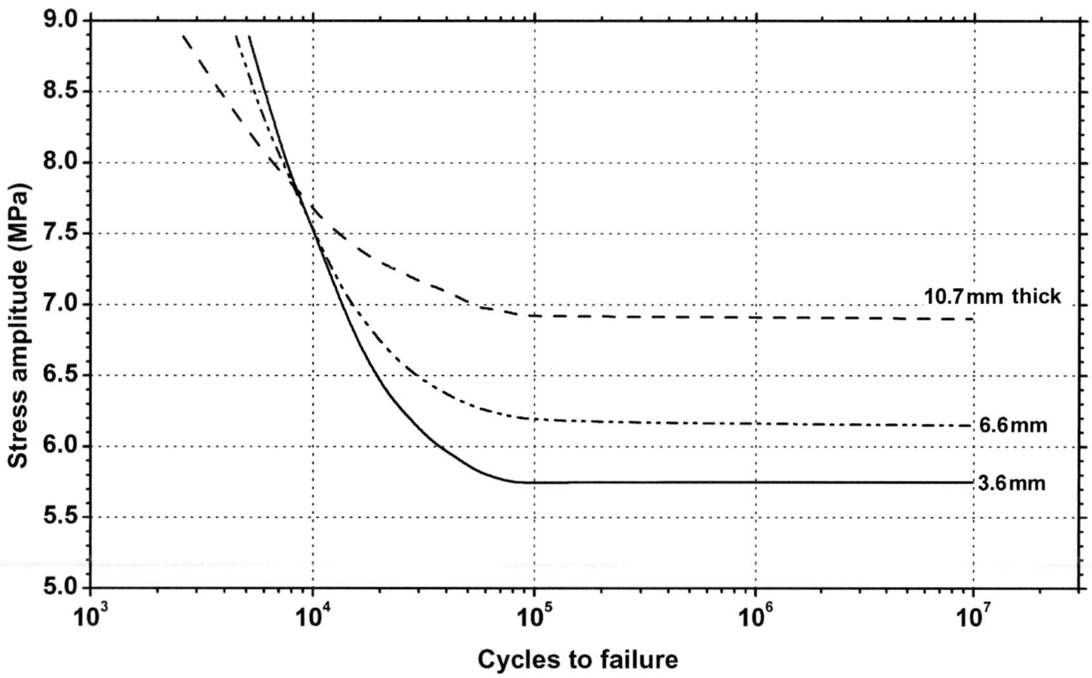

Figure 11.9 Flexural stress amplitude vs. cycles to failure at 23°C and 30 Hz at different thicknesses of generic PTFE.

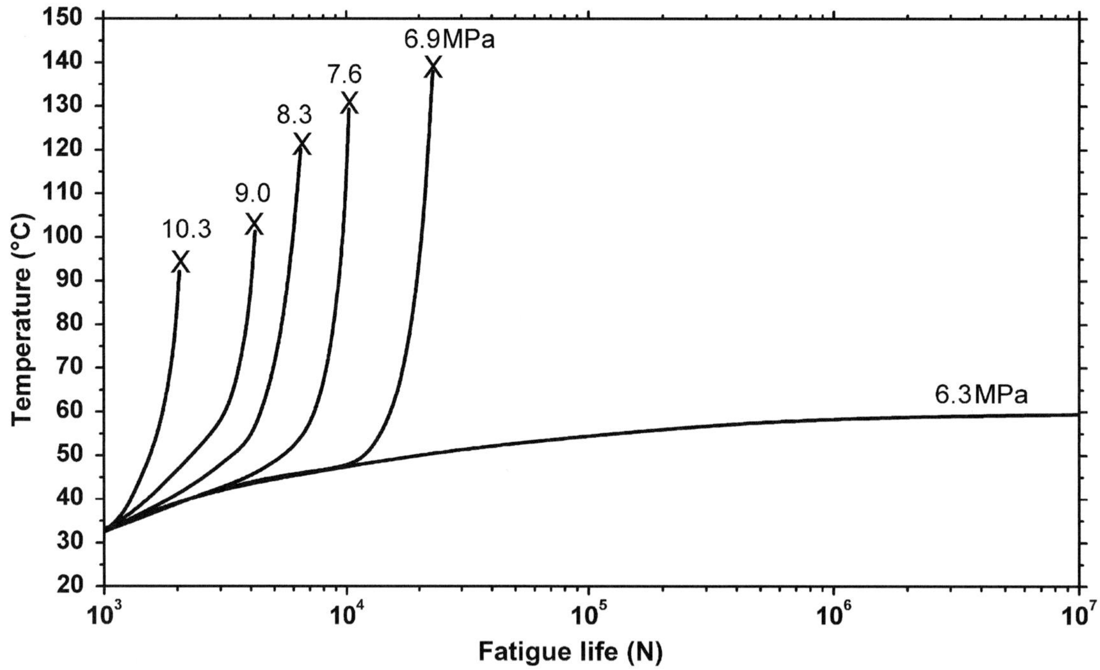

Figure 11.10 Temperature rise vs. fatigue cycles at 30 Hz at different stress levels of generic PTFE (X denoted fatigue failure).

11.2.2 Tribology Data

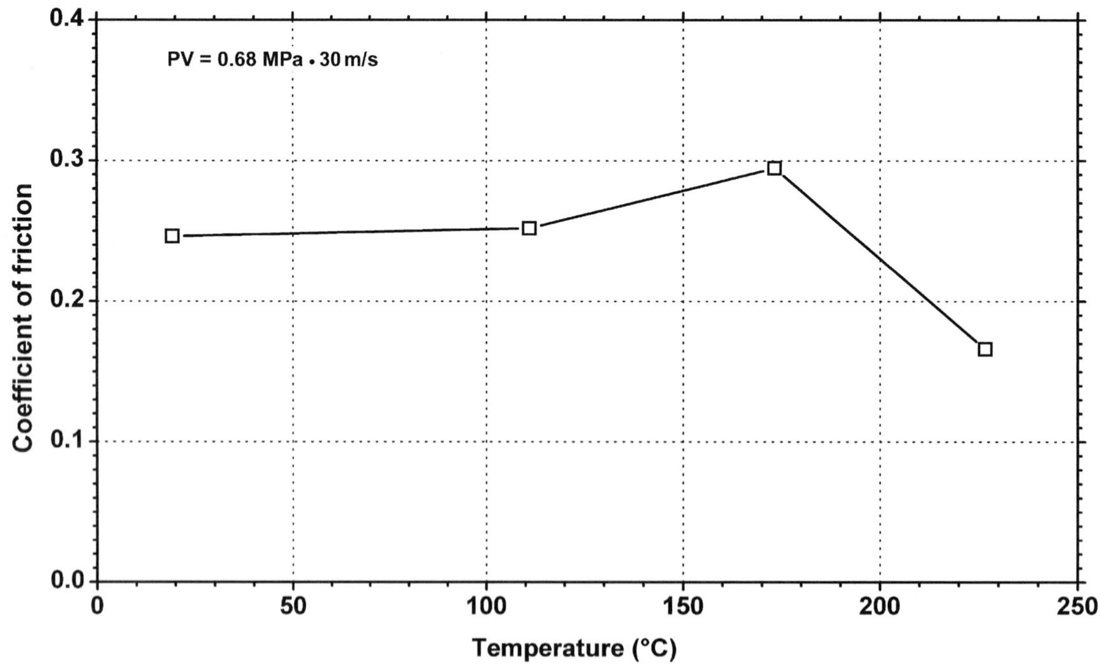

Figure 11.11 Dynamic coefficient of friction vs. temperature of generic PTFE filled with 25% carbon.

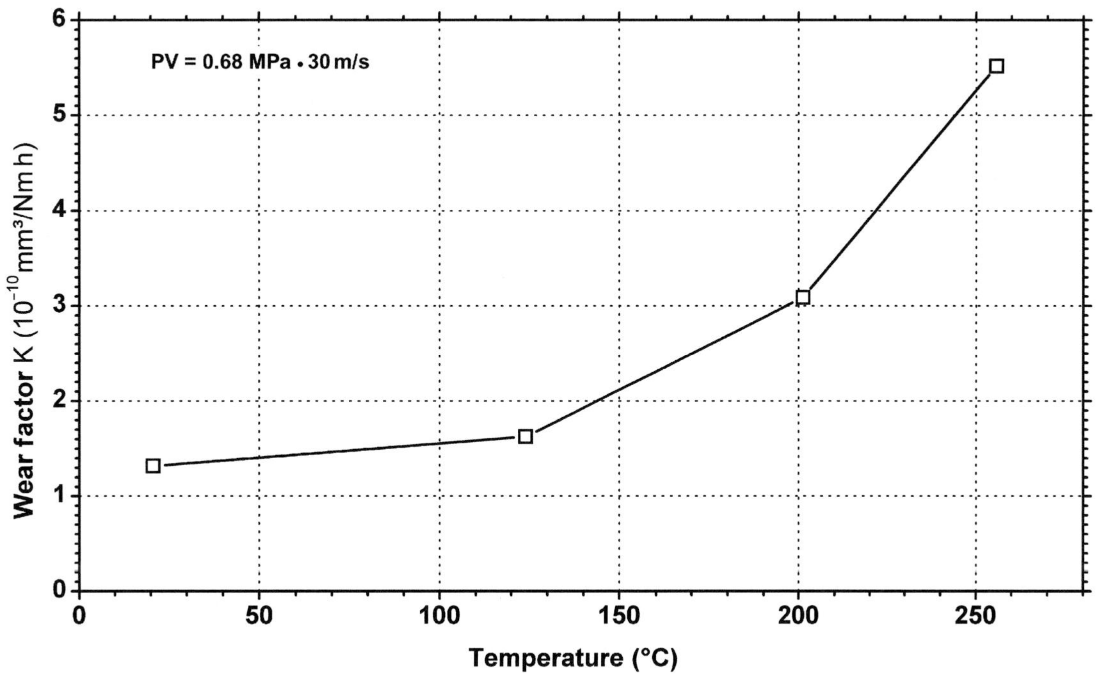

Figure 11.12 Wear factor vs. temperature of generic PTFE filled with 25% carbon.

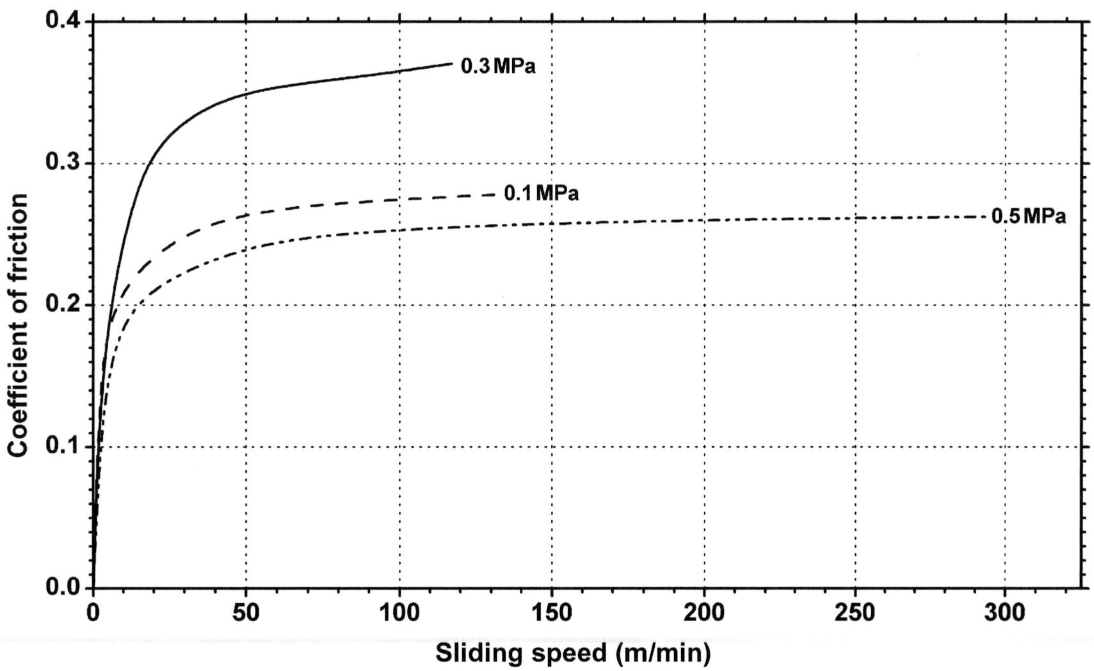

Figure 11.13 Dynamic coefficient of friction vs. sliding speed and different pressures of DuPont Teflon® PTFE.

Table 11.3 PV and Wear Performance of DuPont Teflon® PTFE Filled Compounds by Weight Percent

Property	PTFE	PTFE (15% glass fiber)	PTFE (25% glass fiber)	PTFE (15% graphite)	PTFE (60% bronze)	PTFE (20% glass, 5% graphite)	PTFE (15% glass, 5% MoS2)	PTFE (25% carbon)
PV Limit at 10 ft/min	42	350	350	350	525	385	385	490
PV Limit at 100 ft/min	63	438	455	595	648	525	490	701
PV Limit at 1,000 ft/min	88	525	560	981	771	771	613	1051
PV for 0.005 in. radial wear in 1000 hours (nonlubricated)	1	109	175	53	291	116	193	151
Wear factor × 10^{-8} (mm^3/N m)	5036	32	20	68	12	30	18	23

11.3 Polyethylene Chlorotrifluoroethylene

11.3.1 Tribology Data

Table 11.4 Solvay Solexis Halar® Tribology Properties

Property	Test Method	Unit	Standard Copolymers	Terpolymer (Halar® 600)	Halar® 902
Abrasion resistance	TABER	mg/1000 rev	5	5	5
Friction coefficient: static	ASTM D1894	–	0.1–0.2	0.2	0.1–0.2
Dynamic	ASTM D1894	–	0.1–0.2	0.2	0.1–0.2

11.4 Polyethylene Tetrafluoroethylene

11.4.1 Fatigue Data

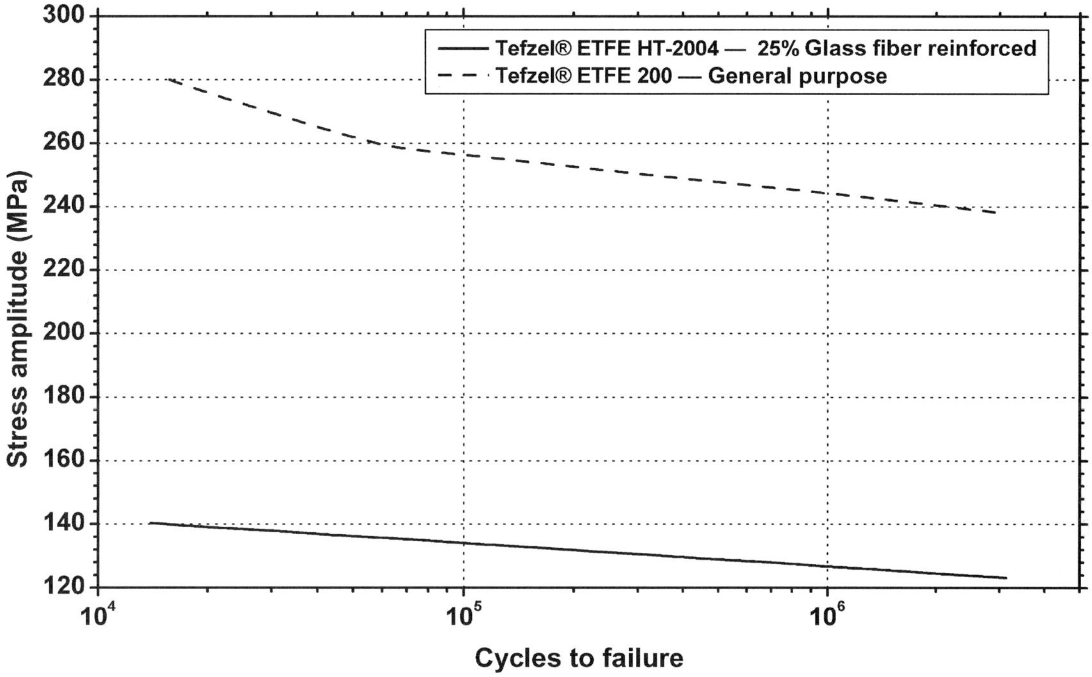

Figure 11.14 Flexural stress amplitude vs. cycles to failure at 23°C, 50% relative humidity and 1800 Hz of two DuPont Tefzel® ETFE plastics.

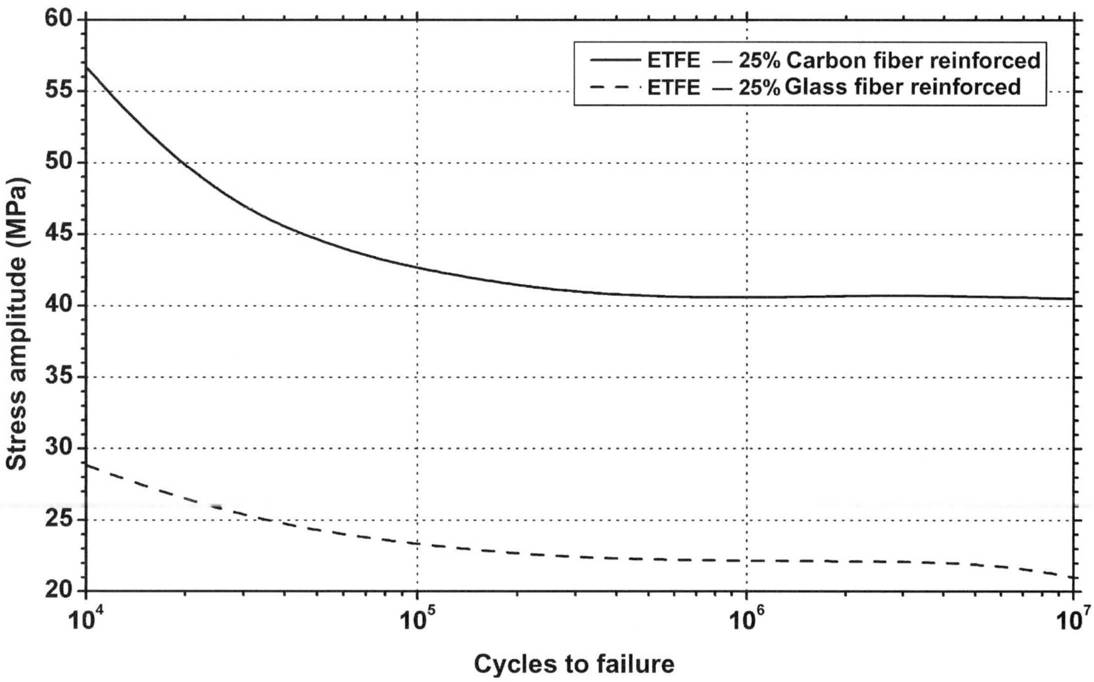

Figure 11.15 Flexural stress amplitude vs. cycles to failure at 23°C and 30 Hz of two fiber reinforced ETFE plastics.

11.4.2 Tribology Data

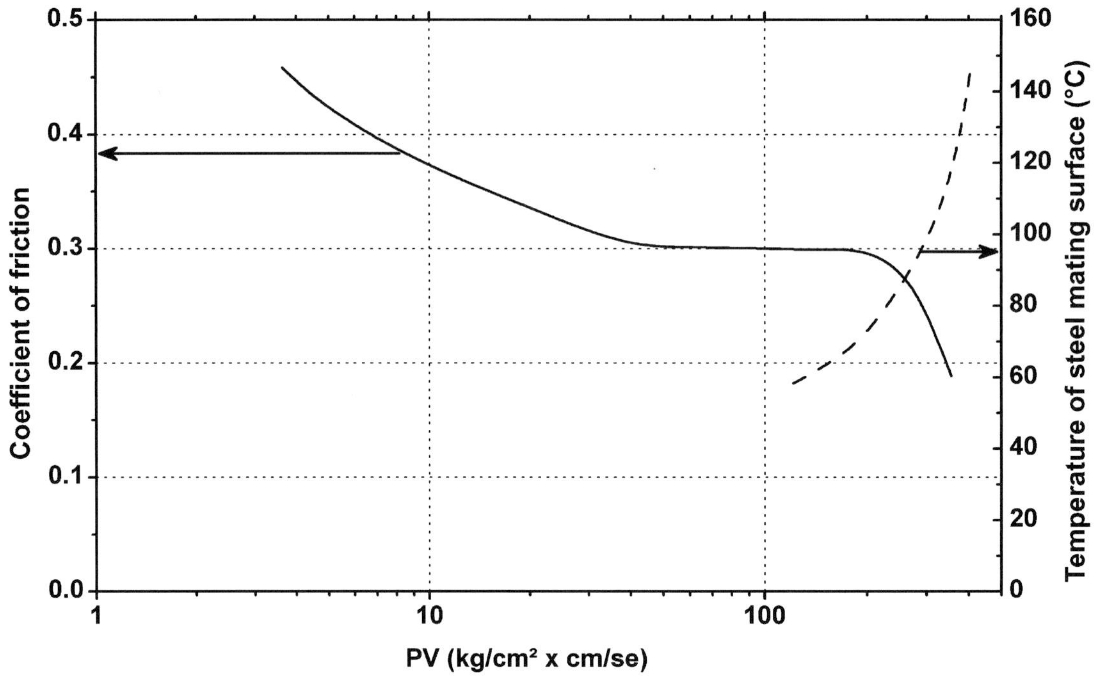

Figure 11.16 The frictional behavior of DuPont Tefzel® HT-2004—25% glass fiber reinforced ETFE (thrust bearing tester, unlubricated).

Table 11.5 Static Coefficient of Friction of DuPont™ Tefzel® HT-2004—25% Glass Fiber Reinforced ETFE

Pressure (MPa)	Static Coefficient of Friction
0.069	0.51
0.345	0.38
0.69	0.31
3.45	0.34

Table 11.6 DuPont™ Tefzel® HT-2004—25% Glass Fiber Reinforced ETFE Bearing Wear Rate

Mating Surface	Pressure (MPa)	Velocity (cm/s)	Tefzel®	Metal
Steel[a]	6.9	2.5	16	4
Steel[a]	6.9	2.5	16	4
Steel[a]	6.9	2.5	16	4
Steel[a]	6.9	2.5	16	4
Steel[a]	6.9	10.2	FAIL	–
Aluminum[b]	2.07	5.1	1,220	1,220
Aluminum[b]	0.69	25.4	480	390

Thrust bearing tester, no lubricant, ambient air temperature, metal finish 16 microinches (406 nm).
[a]Steel mating surface AISI 1018.
[b]Aluminum mating surface LM24M (English).

11.5 Fluorinated Ethylene Propylene

11.5.1 Tribology Data

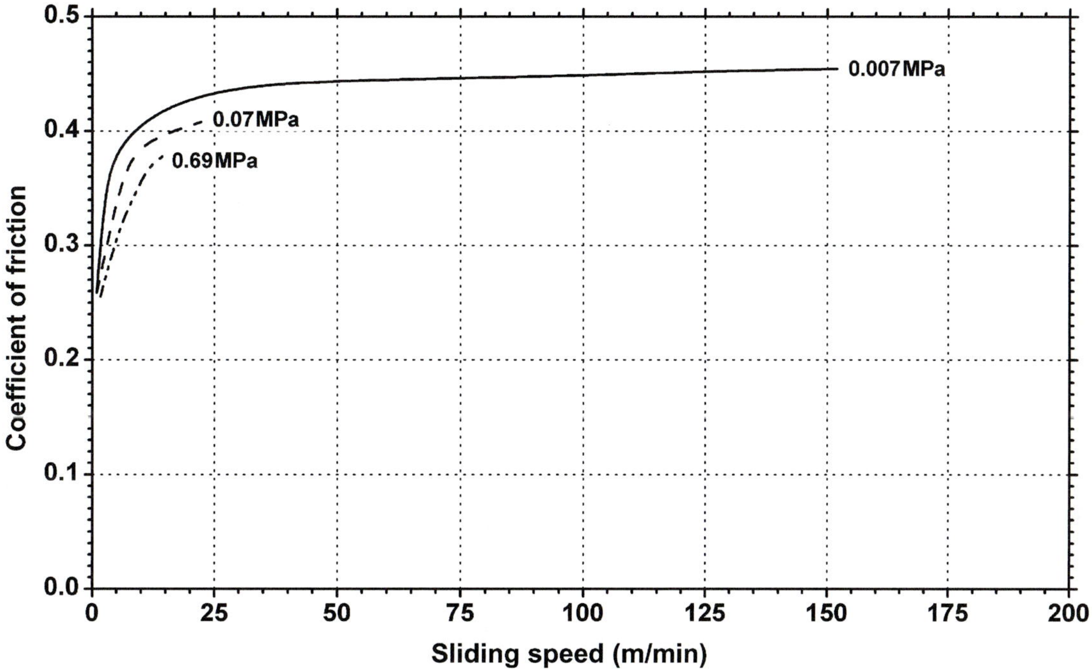

Figure 11.17 Dynamic coefficient of friction vs. sliding speed and different pressures of DuPont Teflon® FEP.

Table 11.7 PV and Wear Factors of DuPont Teflon® FEP

Property	FEP	FEP (15% glass fiber)	FEP (10% bronze by volume)
PV Limit at 10 ft/min	21	158	315
PV Limit at 100 ft/min	28	350	420
PV Limit at 1000 ft/min	35	280	350
PV for 0.005 in. radial wear in 1000 hours (nonlubricated)	<0.35	58	175
Wear Factor $\times 10^{-8}$ (mm³/N m)	>10000	60	20

11.6 Perfluoro Alkoxy

11.6.1 Fatigue Data

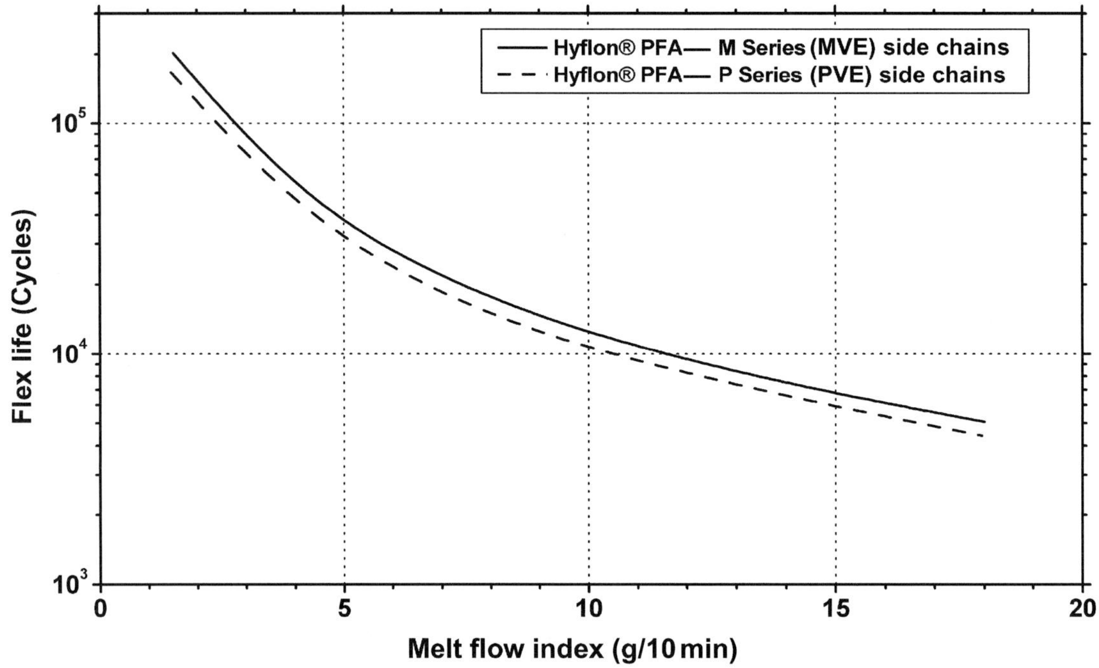

Figure 11.18 MIT flex life vs. melt flow index of two Solvay Solexis Hyflon® PFA type plastics.

11.6.2 Tribology Data

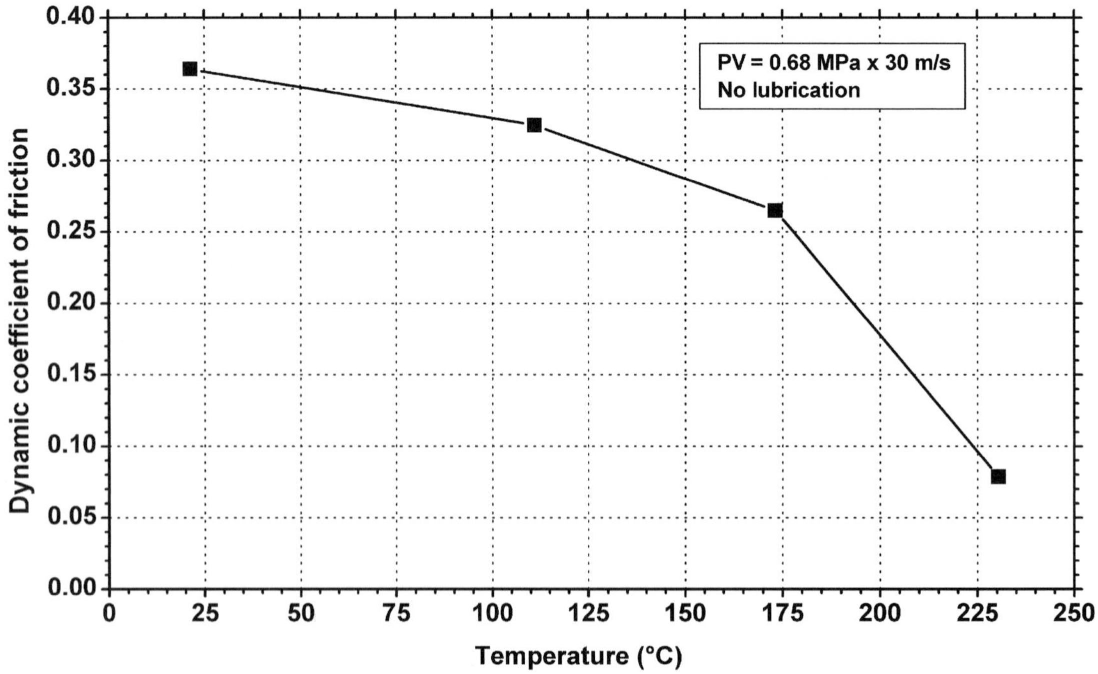

Figure 11.19 Dynamic coefficient of friction vs. temperature of DuPont Engineering Polymers Vespel® CR-6100—30% carbon fiber reinforced PFA.

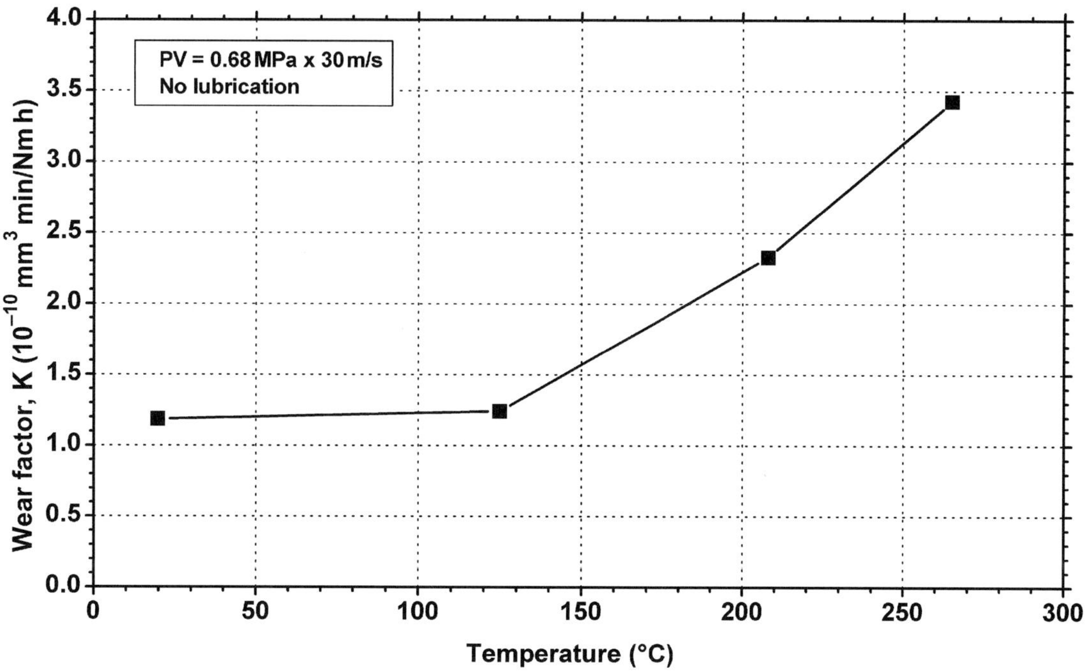

Figure 11.20 Wear factor vs. temperature of DuPont Engineering Polymers Vespel® CR-6100—30% carbon fiber reinforced PFA.

Table 11.8 Flex Life Properties of Solvay Solexis PFA products

	Test Method	SI Units	M620	M640	M720	P220	P420	P450
Flex Life (0.3 mm film)	ASTM D2176	$\times 10^3$ cycles	70–100	4–6	–	–	90–120	4–6

Table 11.9 Tribology Properties of DuPont Engineering Polymers Vespel® CR-6100 (Unlubricated Tri-pin-on AISI Carbon Steel Disc Finished to 16 microinches (0.4 micrometers): 400 psi (8.9 MPa))

Wear rate at 25 ft/min (cm/h)	68.8
Wear rate at 50 ft/min (cm/h)	189.0
Dynamic COF at 25 ft/min	0.20
Dynamic COF at 50 ft/min	0.29
Limiting PV (MPa-m/s)	>5.4

11.7 Polyvinylidene Fluoride

11.7.1 Fatigue Data

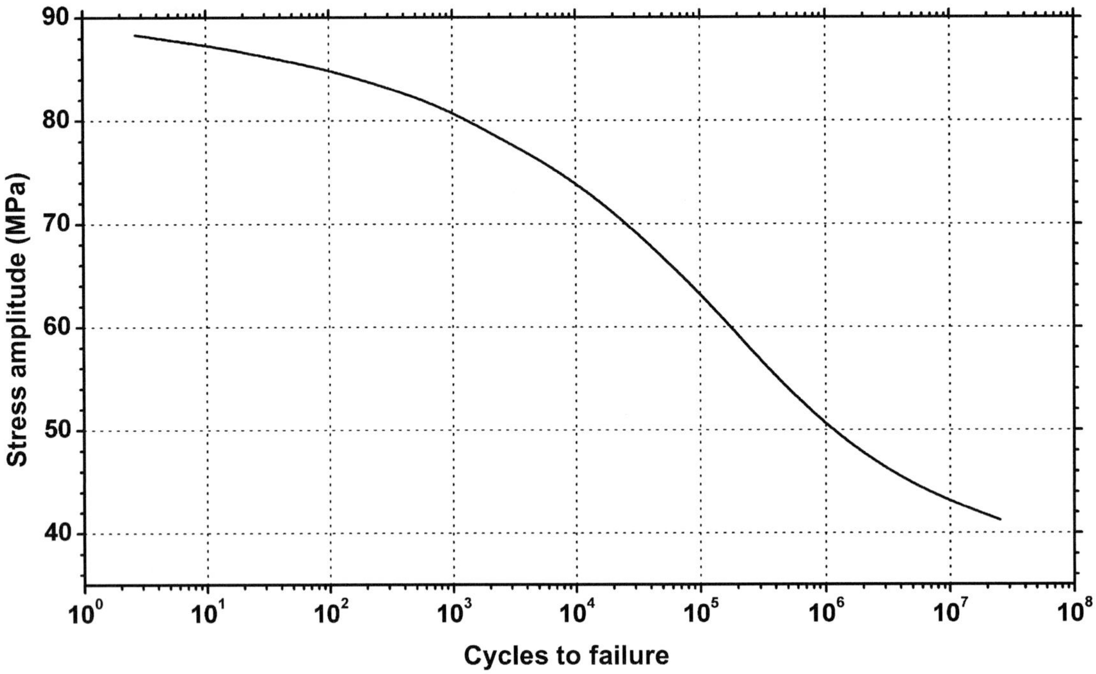

Figure 11.21 Tensile stress amplitude vs. cycles to failure at 25°C of Solvay Solexis Solef® PVDF (notched specimens, 0.4 mm on each side).

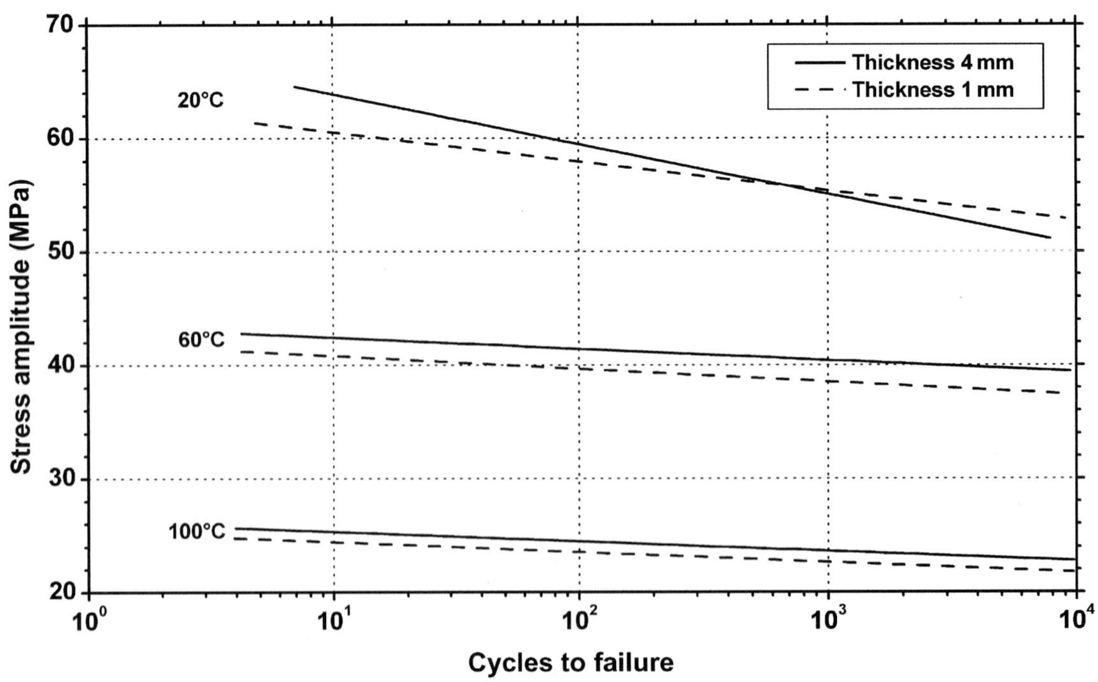

Figure 11.22 Tensile stress amplitude vs. cycles to failure at various temperatures and thicknesses of Solvay Solexis Solef® 1010—general purpose homopolymer molding and extrusion (unnotched specimens).

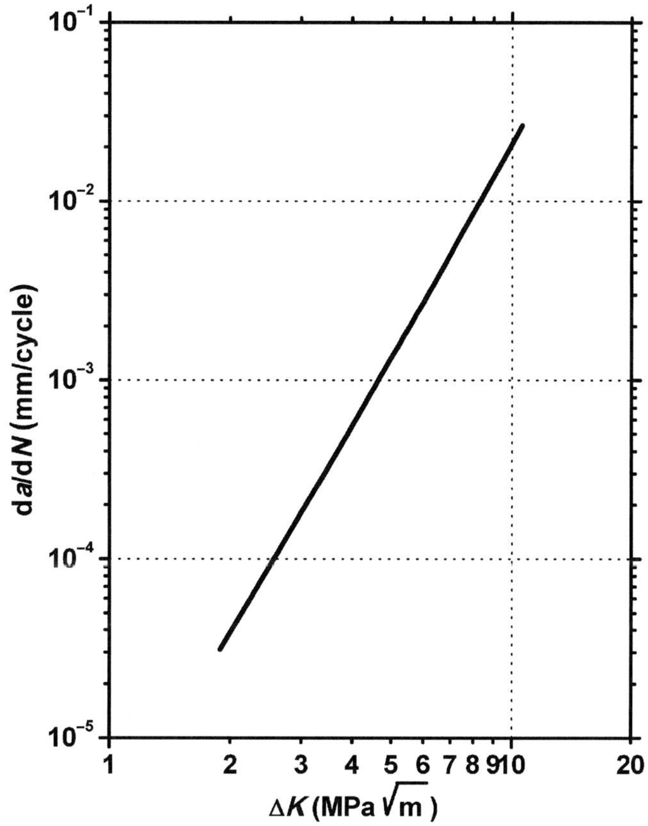

Figure 11.23 Fatigue crack propagation vs. stress intensity factor of generic PVDF.

11.7.2 Tribology Data

Table 11.10 Taber Abrasion Properties of Arkema Kynar® and Kynar Flex® Fluoropolymers

Test Method	Units	Kynar® 710	Kynar® 460	Kynar Flex® 2500	Kynar Flex® 2750-01	Kynar Flex® 2800-00	Kynar Flex® 2850-00
Taber Abrasion (CS 17 1000 g)	mg/1000 cycles	5–9	7–9	28–33	21–25	16–19	6–9
		Kynar Flex® 3120-10	**Kynar Flex® 3120-50**	**Flame Retardant Kynar Flex® 2850-02**	**Flame Retardant Kynar Flex® 2900-04**	**Flame Retardant Kynar Flex® 2950-05**	**Flame Retardant Kynar Flex® 3120-15**
CS 17 1000 g	mg/1000 cycles	16–19	16–19	6–9	16–19	21–25	16–19

Table 11.11 Tribology Properties of Arkema Kynar® and Kynar Flex® Fluoropolymers

Mechanical Properties	Standard/ Conditions	Units	460	1000 Series	700 Series	370	2500
Taber abrasion	CS-17 1000 g	Mg per 1000 cycles	7–9	5–9	5–9	–	28–33
Coefficient of friction-static vs. steel	ASTM D1894 (23°C)		0.23	0.22	0.20	0.18	0.49
Coefficient of friction—dynamic vs. steel	ASTM D1894 (23°C)		0.17	0.15	0.14	0.12	0.54
Mechanical Properties	Standard/ Conditions	Units	2750/2950	2800/2900	2850	3120	
Taber abrasion	CS-17 1000 g	Mg per 1000 cycles	21–25	16–19	6–9	16–19	
Coefficient of friction-static vs. steel	ASTM D1894 (23°C)		0.55	0.33	0.26	0.31	
Coefficient of friction-dynamic vs. steel	ASTM D1894 (23°C)		0.54	0.33	0.19	0.30	

12 High-Temperature Polymers

12.1 Background

This section contains information and multipoint properties for several high-temperature, high-performance plastics. They might be classified or been appropriate to include in another chapter, but they are grouped in this chapter because of their performance levels.

12.1.1 Polyetheretherketone

Polyetheretherketones (PEEK) are also referred to as polyketones. The most common structure is given in Figure 12.1.

PEEK is a thermoplastic with extraordinary mechanical properties. The Young's modulus of elasticity is 3.6 GPa and its tensile strength is 170 MPa. PEEK is partially crystalline, melts at around 350°C, and is highly resistant to thermal degradation. The material is also resistant to both organic and aqueous environments, and is used in bearings, piston parts, pumps, compressor plate valves, and cable insulation applications. It is one of the few plastics compatible with ultrahigh vacuum applications. In summary, the properties of PEEK include:

- outstanding chemical resistance
- outstanding wear resistance
- outstanding resistance to hydrolysis
- excellent mechanical properties
- outstanding thermal properties
- very good dielectric strength, volume resistivity, tracking resistance
- excellent radiation resistance

12.1.2 Polyethersulfone

Polyethersulfone (PES) is an amorphous polymer and a high-temperature engineering thermoplastic. Even though PES has high-temperature performance, it can be processed on conventional plastics processing equipment. Its chemical structure is shown in Figure 12.2. PES has an outstanding ability to withstand exposure to elevated temperatures in air and water for prolonged periods.

Because PES is amorphous, mold shrinkage is low and is suitable for applications requiring close tolerances and little dimensional change over a wide temperature range. Its properties include:

- excellent thermal resistance—T_g 224°C
- outstanding mechanical, electrical, flame and chemical resistance
- very good hydrolytic and sterilization resistance

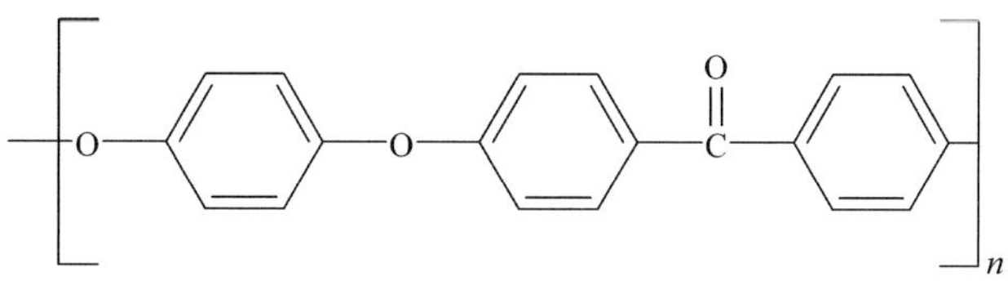

Figure 12.1 The structure of PEEK.

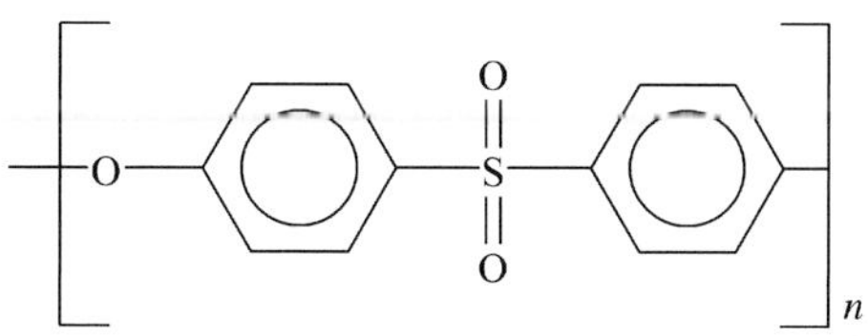

Figure 12.2 The structure of PES.

Copyright © 2010 Laurence W. McKeen. All rights reserved.

- good optical clarity
- processed by all conventional techniques

12.1.3 Polyphenylene Sulfide

Polyphenylene sulfide (PPS) is a semicrystalline material. It offers an excellent balance of properties, including high-temperature resistance, chemical resistance, flowability, dimensional stability, and electrical characteristics. PPS must be filled with fibers and fillers to overcome its inherent brittleness. Because of its low viscosity, PPS can be molded with high loadings of fillers and reinforcements. Because of its outstanding flame resistance, PPS is ideal for high-temperature electrical applications. It is unaffected by all industrial solvents. The structure of PPS is shown in Figure 12.3.

There are several variants to regular PPS that may be talked about by suppliers or may be seen in the literature. These are:

- Regular PPS is of "modest" molecular weight. Materials of this type are often used in coating products.
- Cured PPS is PPS that has been heated to high temperature, above 300°C, in the presence of air or oxygen. The oxygen causes some cross-linking and chain extension called oxidative cross-linking. This results in some thermoset-like properties such as improved thermal stability, dimensional stability, and improved chemical resistance.

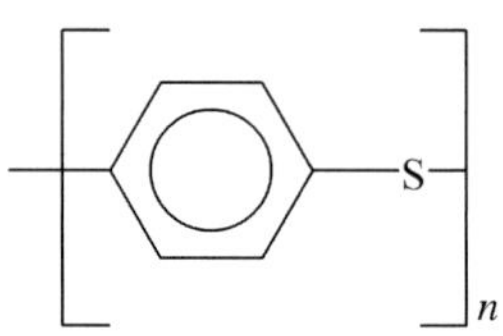

Figure 12.3 The structure of polyphenylene sulfide (PPS).

- High-molecular-weight (HMW) linear PPS has a molecular weight about double of that of regular PPS. The higher molecular weight improves elongation and impact strength.
- High-molecular weight (HMW) branched PPS has higher molecular weight than regular PPS, but it also has polymer chain branches along the main molecule backbone. This provides improved mechanical properties.

PPS properties are summarized:

- Continuous use temperature of 220°C
- Excellent dimensional properties
- Transparent
- Improved impact strength and toughness as compared to PES
- Excellent hydrolytic stability
- High stress cracking resistance
- Good chemical resistance
- Good surface release properties
- Expected continuous temperature of 180°C

12.1.4 Polysulfone

Polysulfone (PSU) is a rigid, strong, tough, high-temperature amorphous thermoplastic. The structure of PSU is shown in Figure 12.4.

Its properties summarized:

- High thermal stability
- High toughness and strength
- Good environmental stress crack resistance
- Inherent fire resistance
- Transparence

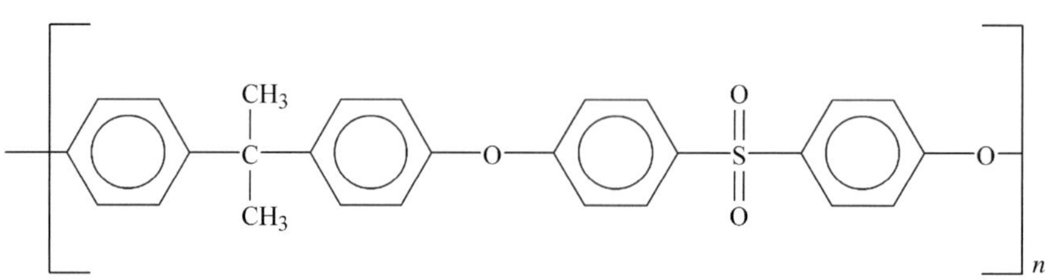

Figure 12.4 The structure of PSU.

12.1.5 Polyphenylsulfone

Polyphenylsulfone (PPSU) is a rigid, strong, tough, high-temperature amorphous thermoplastic. It has a high heat deflection temperature of 405°F (207°C); it can withstand continuous exposure to heat and still absorb tremendous impact without cracking or breaking. It is inherently flame retardant and offers exceptional resistance to bases and other chemicals. The structure of PPSU is shown in Figure 12.5.

Its properties summarized:

- 207°C HDT
- Superior toughness
- Exceptional hydrolytic stability
- Good chemical resistance
- Transparent

12.1.6 Polybenzimidazole

Polybenzimidazole (PBI) is a unique and highly stable linear heterocyclic polymer. The chemical structure is shown in Figure 12.6. PBI exhibits excellent thermal stability, resistance to chemicals, acid and base hydrolysis, and temperature resistance. PBI can withstand temperatures as high as 430°C, and in short bursts, to 760°C. PBI does not burn and maintains its properties as low as −196°C.

Ideally suited for its application in extreme environments, PBI can be formed into stock shapes and subsequently machined into high precision finished parts. Since PBI does not have a melt point, moldings from virgin PBI polymer can only be formed in a high-temperature, high-pressure compression molding process.

PBI is highly resistant to deformation, and has low hysteresis loss and high elastic recovery. PBI exhibits ductile failure, and may be compressed to over 50% strain without fracture. Celazole® PBI has the highest compressive strength of any thermoplastic or thermosetting resin at 400 MPa. There is no weight loss or change in compressive strength of Celazole® PBI exposed to 260°C in air for 500 hours. At 371°C, no weight or strength change takes place for 100 hours. In spite of these unusual properties, PBI is usually blended with other plastics, particularly polyesters and PEEK.

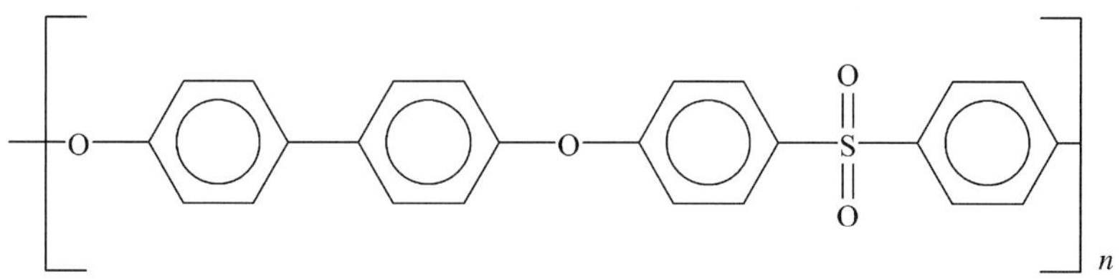

Figure 12.5 The structure of PPSU.

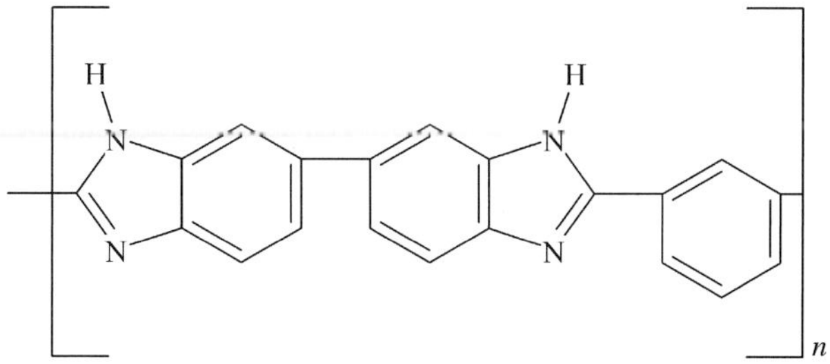

Figure 12.6 The structure of PBI.

12.2 Polyetheretherketone

12.2.1 Fatigue Data

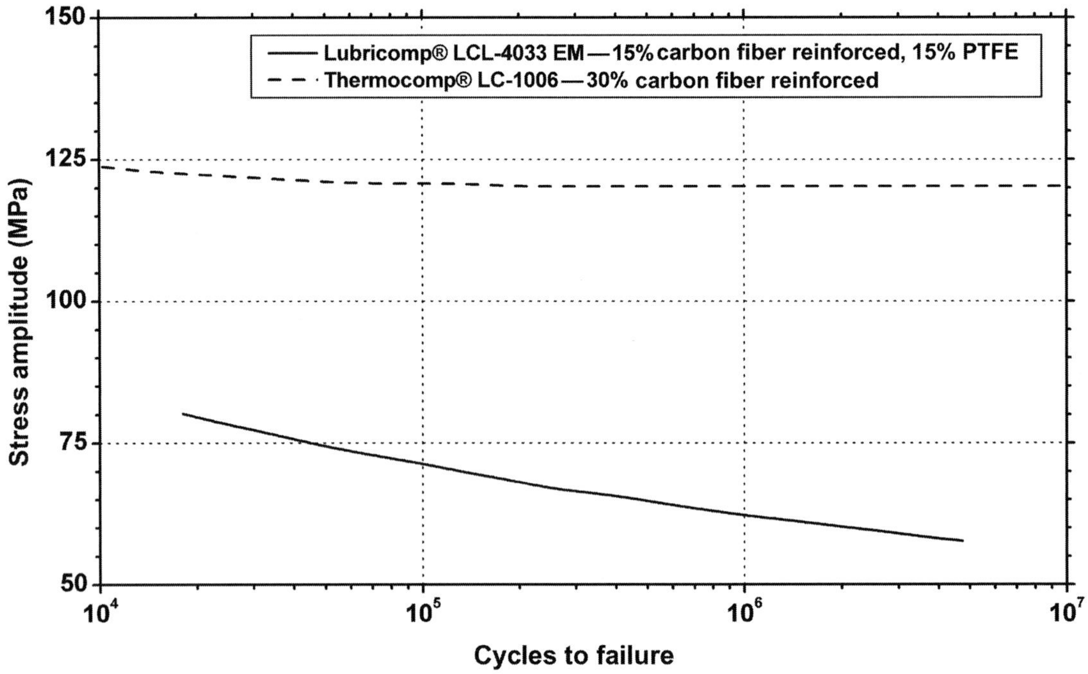

Figure 12.7 Flexural stress amplitude vs. cycles to failure at 23°C of two carbon fiber reinforced SABIC Innovative Plastics PEEK plastics.

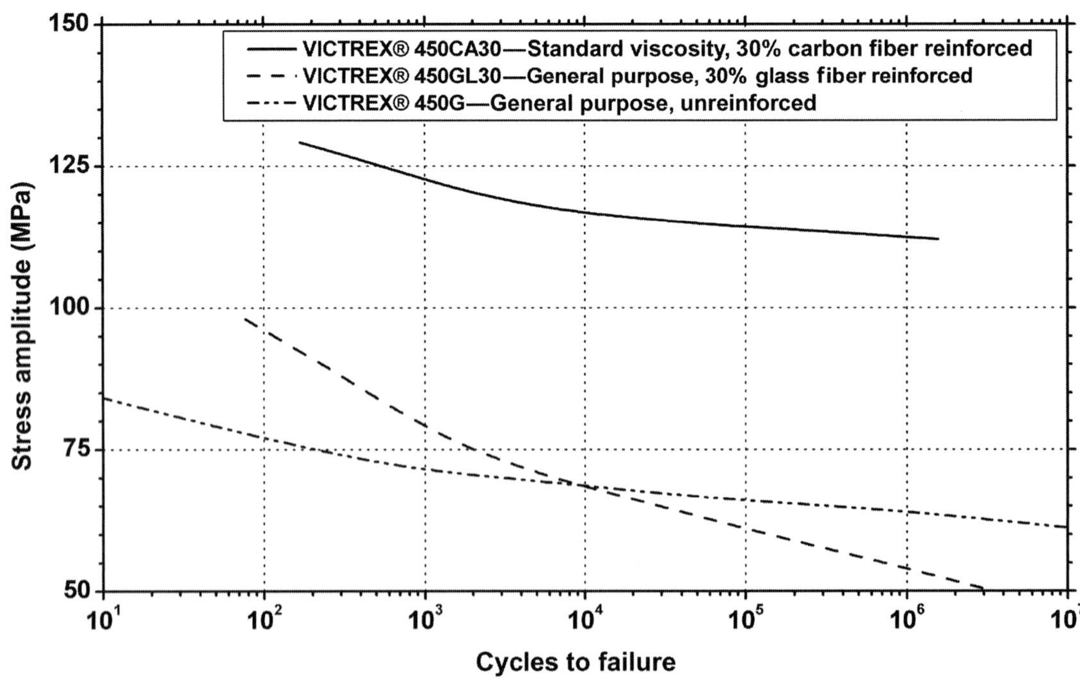

Figure 12.8 Stress amplitude vs. cycles to failure at 23°C and 0.5 Hz of several Victrex® PEEK plastics.

12.2.2 Tribology Data

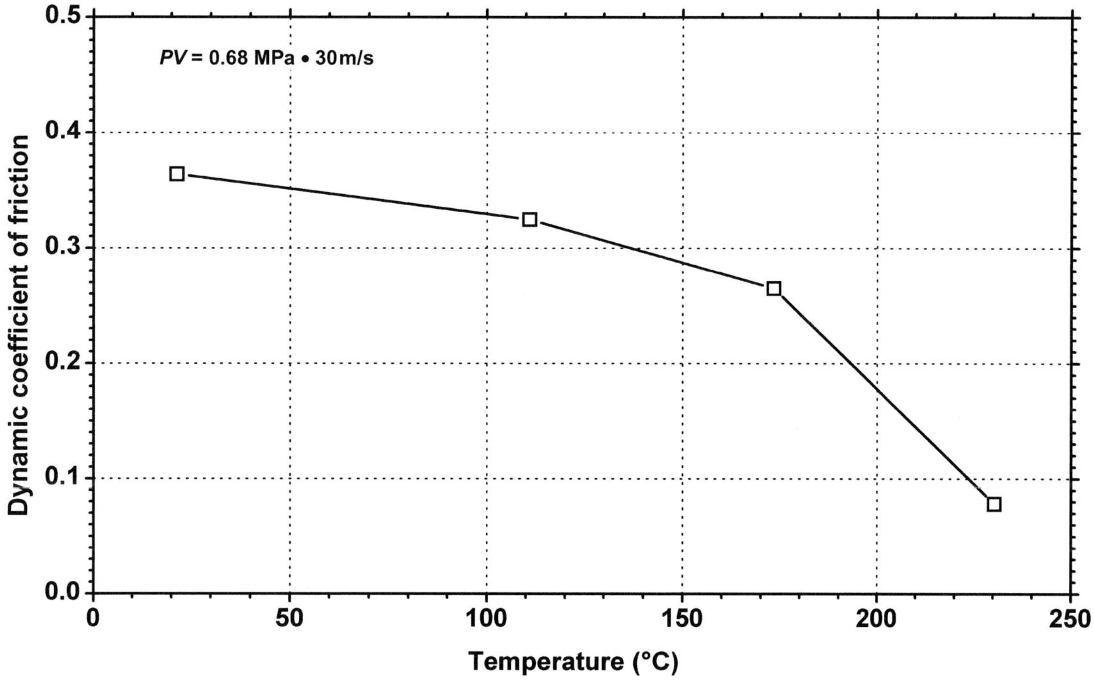

Figure 12.9 Dynamic coefficient of friction vs. temperature of Greene, Tweed & Co. Arlon® 1260—carbon fiber reinforced PEEK.

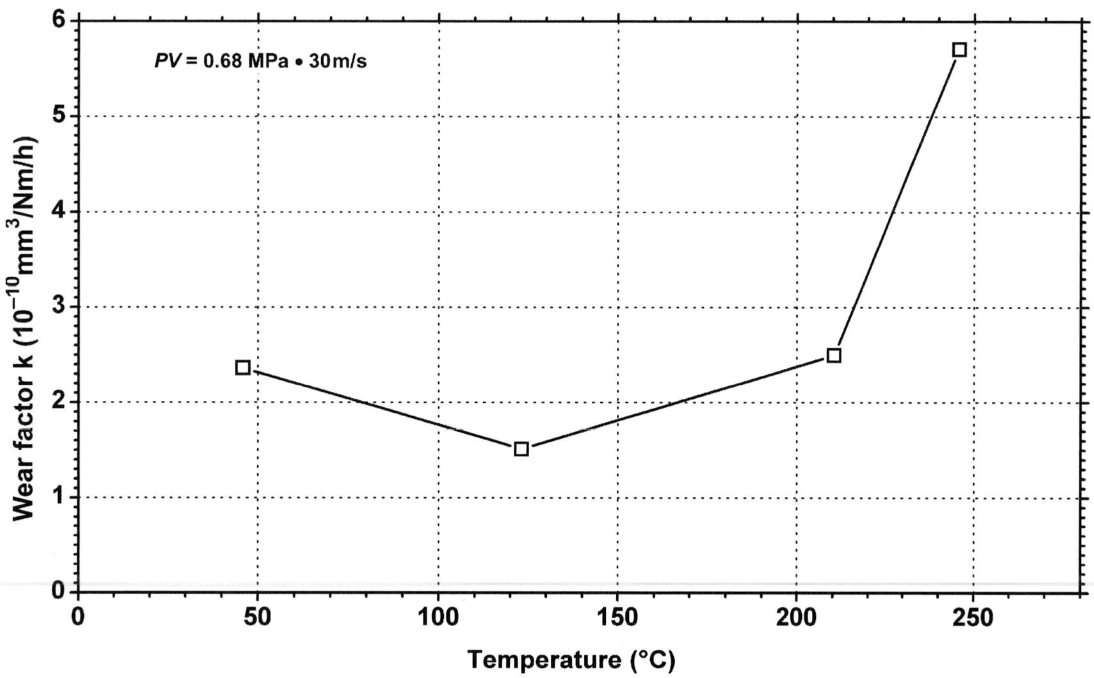

Figure 12.10 Wear factor vs. temperature of Greene, Tweed & Co. Arlon® 1260—carbon fiber reinforced PEEK.

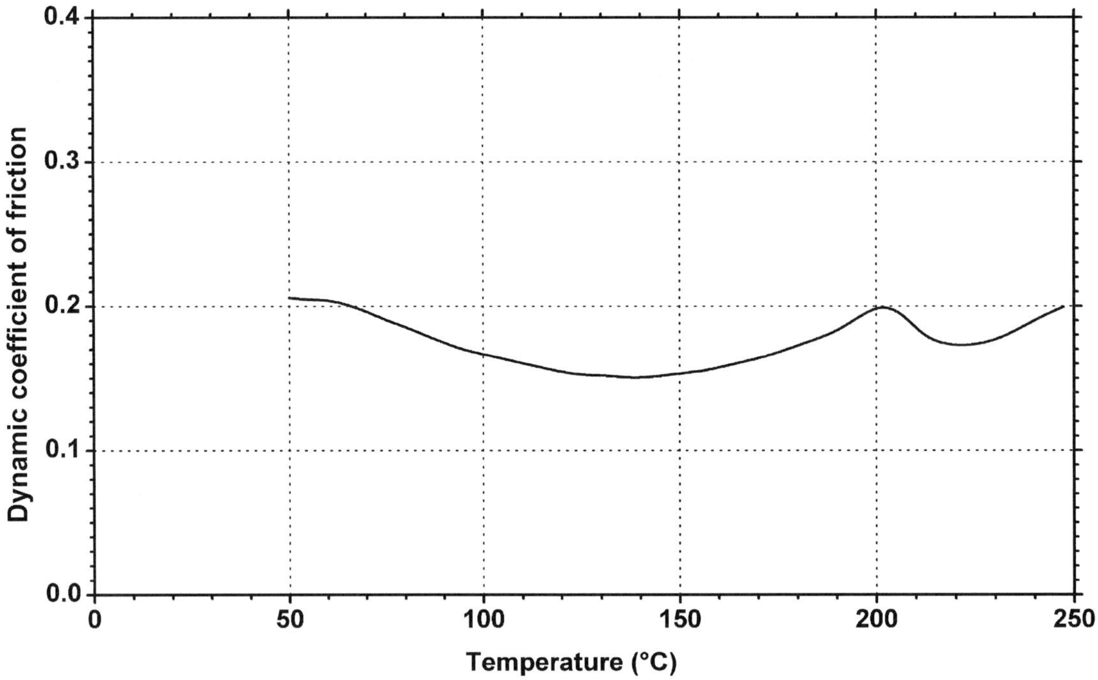

Figure 12.11 Dynamic coefficient of friction vs. temperature of Victrex plc Victrex® 450FC30—lubricated, 30% carbon fiber/PTFE PEEK.

Table 12.1 Comparative Tribological Data of Victrex plc Victrex® PEEK Plastics (with $v = 183$ m/min)

Material	20°C				200°C			
	Load (kg)	*Lpv* (MPa)	μ^a	Wear Rate[b] (μm min^{-1})	Load (kg)	*Lpv* (MPa)	μ^a	Wear Rate[b] (μm min^{-1})
Victrex® 450FC30 450FC30—Lubricated, 30% carbon fiber/PTFE	40	794	0.17	3.2	40	622	0.14	132
Victrex® 450G—general-purpose grade	8	145	0.58	7.5	8	147	0.51	150
Victrex® 450CA30—30% carbon fiber	22	376	0.28	3.8	13	445	0.25	—

[a]Average of the coefficient of friction at *Lpv* and 50% *Lpv*.
[b]Wear rate at 50% *Lpv*.

Table 12.2 Friction Coefficients and Wear Rates of PEEKs and Their Composites[a]

Material Type	Sliding Condition	Friction Coefficient	Relative Error (%)	Wear Rate (10^{-6} mm³/Nm)	Relative Error (%)
MA, low-molecular-weight PEEK					
	1 MPa, 1 m/s	0.34	3.28	19.69	13.87
	2 MPa, 1 m/s	0.38	1.36	25.85	21.57
	4 MPa, 1 m/s	0.41	5.63	18.10	23.18
MB, medium-molecular weight PEEK					
	1 MPa, 1 m/s	0.37	16.10	12.41	33.93
	2 MPa, 1 m/s	0.39	1.18	16.37	21.60
	4 MPa, 1 m/s	0.39	9.49	14.14	1.70
MC, high-molecular weight PEEK					
	1 MPa, 1 m/s	0.37	13.14	11.59	21.63
	2 MPa, 1 m/s	0.41	3.44	13.00	35.62
	4 MPa, 1 m/s	0.42	5.55	22.30	15.62
MB FC30, medium-molecular weight PEEK, with 10 wt% silicone carbide fiber, 9.1 vol% graphite and PTFE					
	1 MPa, 1 m/s	0.35	11.93	0.51	7.64
	2 MPa, 1 m/s	0.41	3.30	0.75	6.53
	4 MPa, 1 m/s	0.37	18.87	0.99	13.36
MC FC30, high-molecular weight PEEK, with 10 wt% silicone carbide fiber, 9.1 vol% graphite and PTFE					
	1 MPa, 1 m/s	0.36	12.05	0.60	13.16
	2 MPa, 1 m/s	0.41	10.14	0.75	5.24
	4 MPa, 1 m/s	0.27	1.80	0.83	3.36

[a]Zhang G, Schlarb AK, Correlation of the tribological behaviors with the mechanical properties of poly-etherether-ketones (PEEKs) with different molecular weights and their fiber filled composites. *Wear* 2008, doi: 10.1016/j.wear.2008.07.004.

Table 12.3 Tribological Properties of RTP Company RTP 2200 LF TFE 15 (PTFE 15%, Low Flow) vs. 1018 C Steel (Data Obtained per ASTM 3702)

PV (KPa m/s)	Load (N)	Speed (m/s)	Wear Factor × 10^{-8} (mm³/N m)	Dynamic Coefficient of Friction
70	1.80	0.25	246	0.26
175	2.25	0.50	212	0.50
350	2.25	1.00	483	0.36

Table 12.4 Tribological Properties of RTP Company RTP 2200 LF TFE 20 (PTFE 20%, Low Flow) vs. 1018 C Steel (Data Obtained per ASTM 3702)

PV (KPa m/s)	Load (N)	Speed (m/s)	Wear Factor × 10^{-8} (mm³/N m)	Dynamic Coefficient of Friction
70	1.80	0.25	166	0.22
175	2.25	0.50	119	0.39
175	2.25	0.50	166	0.34
350	2.25	1.00	399	0.37
350	2.25	1.00	396	0.36

Table 12.5 Tribological Properties of RTP Company RTP 2205 TFE 15 (Glass Fiber 30%, PTFE 15%) vs. 1018 C Steel (Data Obtained per ASTM 3702)

PV (KPa m/s)	Load (N)	Speed (m/s)	Wear Factor × 10^{-8} (mm^3/N m)	Dynamic Coefficient of Friction
70	1.80	0.25	149	0.48
70	0.90	0.50	462	0.44
70	0.45	1.00	241	0.38
175	4.50	0.25	143	0.47
175	2.25	0.50	123	0.46
175	1.15	1.00	386	0.44
350	9.00	0.25	147	0.32
350	4.50	0.50	249	0.40
350	2.25	1.00	251	0.44

Table 12.6 Tribological Properties of RTP Company RTP 2200 AR 15 TFE 15 (Aramid Fiber 30%, PTFE 15%) vs. 1018 C Steel (Data Obtained per ASTM 3702)

PV (KPa m/s)	Load (N)	Speed (m/s)	Wear Factor × 10^{-8} (mm^3/N m)	Dynamic Coefficient of Friction
70	1.80	0.25	12	0.28
70	0.90	0.50	30	0.24
70	0.45	1.00	48	0.27
175	4.50	0.25	26	0.28
175	2.25	0.50	30	0.24
175	1.15	1.00	44	0.20
350	9.00	0.25	84	0.30
350	4.50	0.50	72	0.27
350	2.25	1.00	40	0.31

Table 12.7 Tribological Properties of RTP Company RTP 2285 TFE 15 (Carbon Fiber 30%, PTFE 15%) vs. 1018 C Steel (Data Obtained per ASTM 3702)

PV (KPa m/s)	Load (N)	Speed (m/s)	Wear Factor × 10^{-8} (mm^3/N m)	Dynamic Coefficient of Friction
70	1.80	0.25	94	0.37
70	0.90	0.50	92	0.36
70	0.45	1.00	127	0.42
175	4.50	0.25	129	0.33
175	2.25	0.50	94	0.37
175	1.15	1.00	84	—
350	9.00	0.25	90	0.65
350	4.50	0.50	123	0.73
350	2.25	1.00	88	0.61

Table 12.8 Tribological Properties of RTP Company RTP 2299 × 57352 A (Proprietary Formula) vs. 1018 C Steel (Data Obtained per ASTM 3702)

PV (KPa m/s)	Load (N)	Speed (m/s)	Wear Factor × 10^{-8} (mm^3/N m)	Dynamic Coefficient of Friction
70	1.80	0.25	12	0.33
70	0.45	1.00	32	0.27
175	2.25	0.50	48	0.57
350	9.00	0.25	48	0.44
350	2.25	1.00	58	0.46

12.3 Polyethersulfone

12.3.1 Fatigue Data

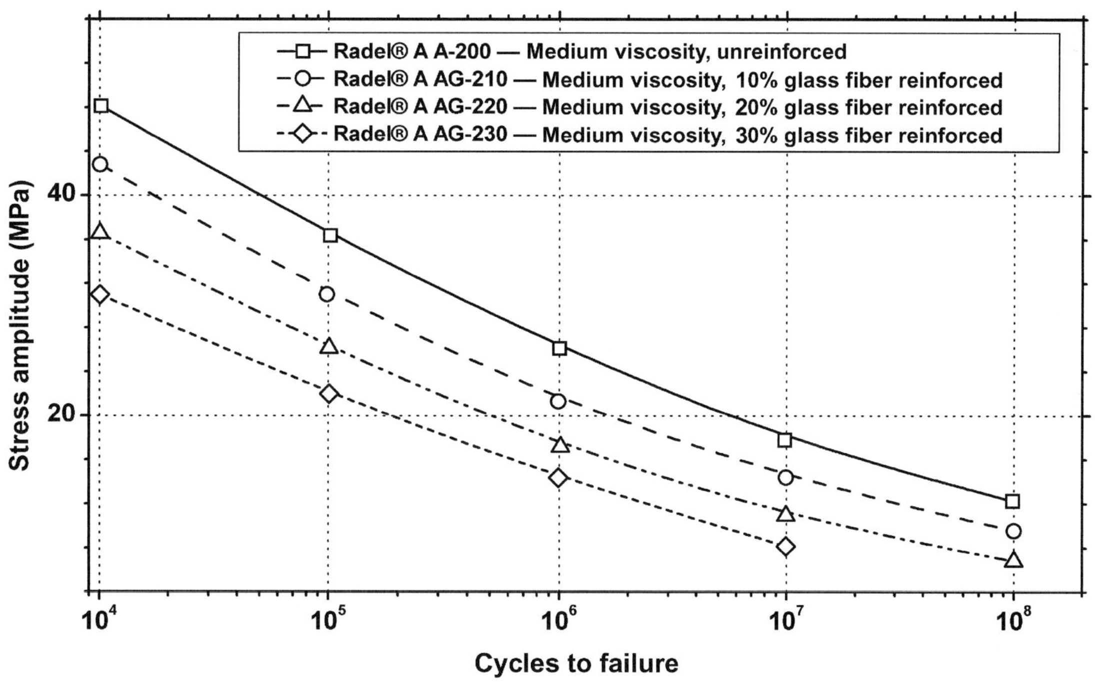

Figure 12.12 Flexural stress amplitude vs. cycles to failure at 23°C and 30 Hz of several Solvay Advanced Polymers, L.L.C. Radel® PES plastics.

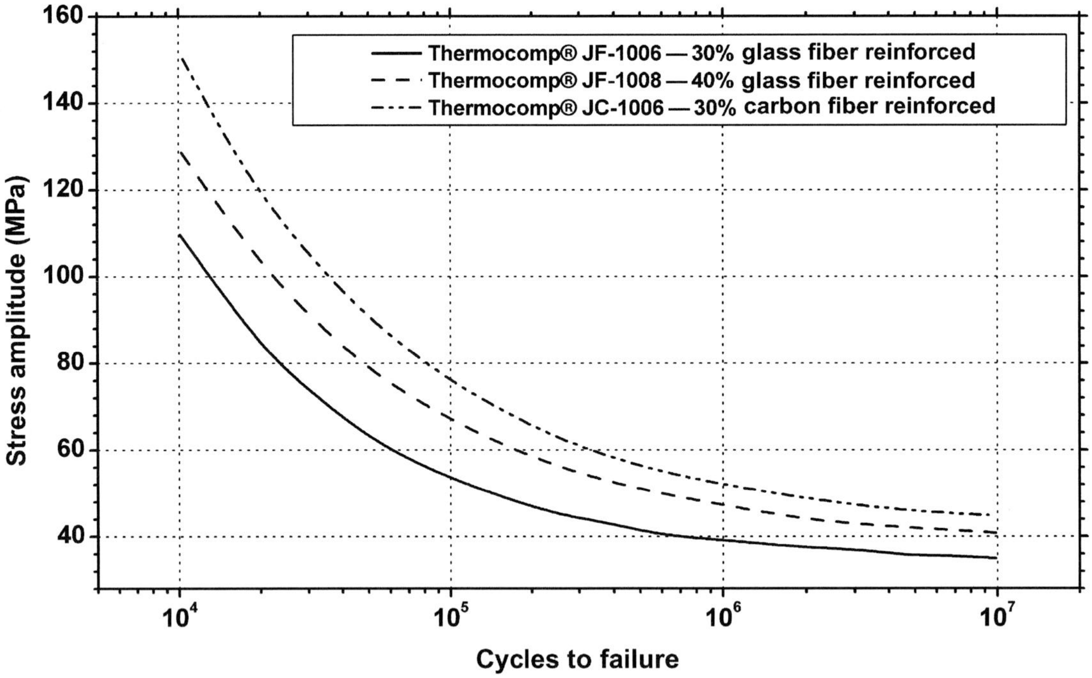

Figure 12.13 Flexural stress amplitude vs. cycles to failure at 23°C and 30 Hz of several SABIC Innovative Plastics Thermocomp® PES plastics.

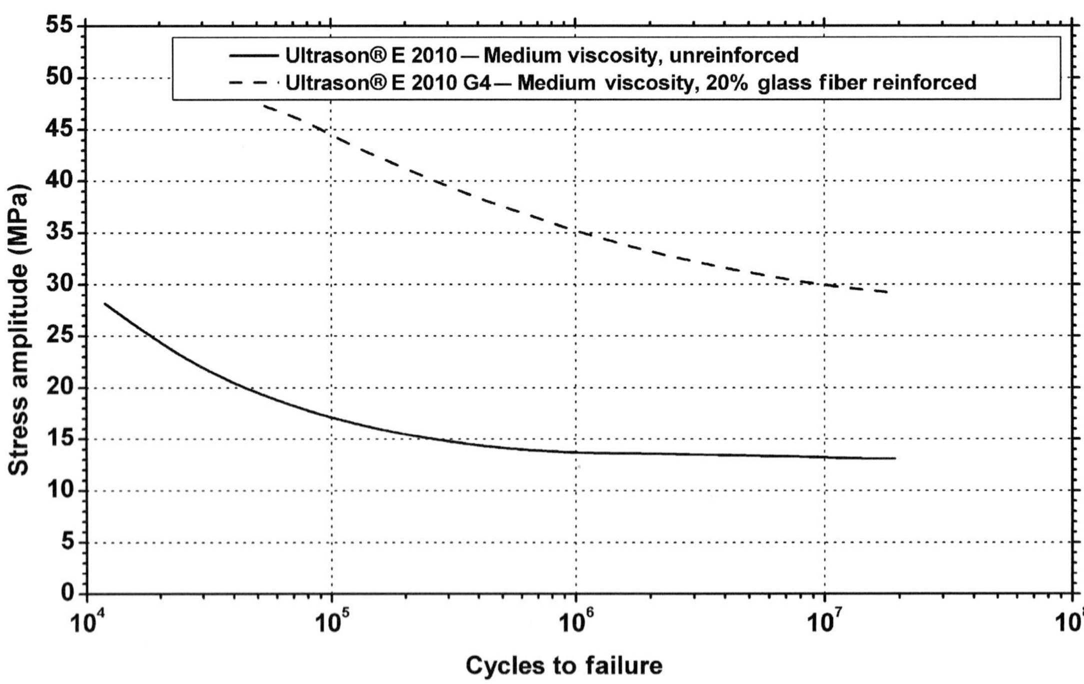

Figure 12.14 Flexural stress amplitude vs. cycles to failure at 23°C and 15 Hz of two BASF Ultrason® PES plastics.

12.3.2 Tribology Data

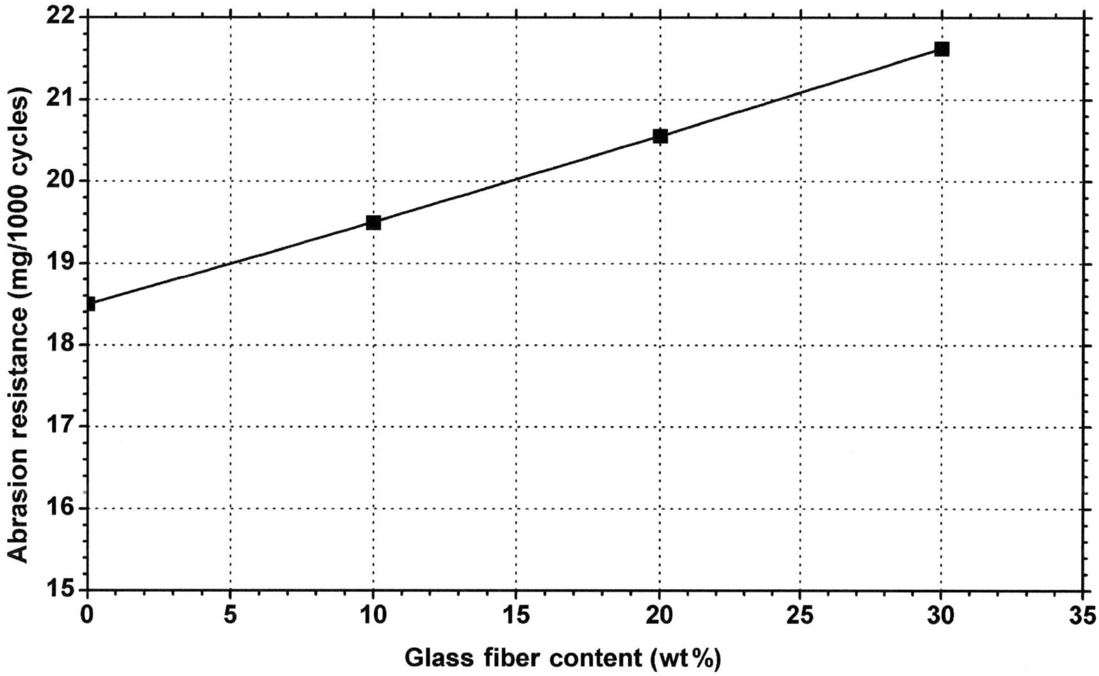

Figure 12.15 Taber abrasion loss vs. glass fiber content of Solvay Radel® A PES plastics.

Table 12.9 Tribology Data for Mitsui Chemicals, Inc. PES Plastics by Suzuki Friction/Wear Test (Ring on Disk Configuration) (Testing Conditions: P = 1 MPa (P = 0.5 MPa Only for SNG2020R), V = 10 m/min, T = 30 min)

Mating Surface	Test	Units	FO-10D	SGF2030	SGF2040	SGN2020R
Stainless steel #304	Kinetic coefficient of friction	—	0.19	0.25–0.40	0.30–0.40	0.30–0.40
Stainless steel #304	Wear	mg	9	3	3	38
Aluminum	Kinetic coefficient of friction	—	0.17	0.15–0.25	0.15–0.35	0.30–0.50
Aluminum	Wear	mg	7	7	7	117

Mitsui Chemicals, Inc. PES FO-10D—Low-friction/low-wear grade with fluorocarbon resin added. Mitsui Chemicals, Inc. PES SGF2030—Low-friction/low-wear grade + 20% glass fiber with fluorocarbon resin added. Mitsui Chemicals, Inc. PES SGF2040—Low-friction/low-wear grade + 30% glass fiber with fluorocarbon resin added. Mitsui Chemicals, Inc. PES SGN2020R—High-flowability grade + 20% glass fiber, for injection molding and improved mold release.

Table 12.10 Friction Coefficient and Wear Intensity for BASF Ultrason® E PES Plastics (Tribological System: Peg-and-Disk Apparatus, Pressure (P) = 1.0 MPa, Rubbing Velocity (V) = 0.5 m/s, Temperature of Rubbing Surfaces 40°C, Mating Material: Steel 100 Cr 6 700 Hv 10, No Lubricant)

Ultrason® Grade	Dynamic Friction Coefficient μ	Wear Intensity S (μm/km)	Mean Roughness RZ (μm)
E 2010 (unreinforced)	0.62	<1000	2.5
E 2010 G4 (20% glass fiber)	0.54	5	2.5
E 2010 G6 (30% glass fiber)	0.54	5.4	2.5
KR 4113	0.27	0.26	2.5

12.4 Polyphenylene Sulfide

12.4.1 Fatigue Data

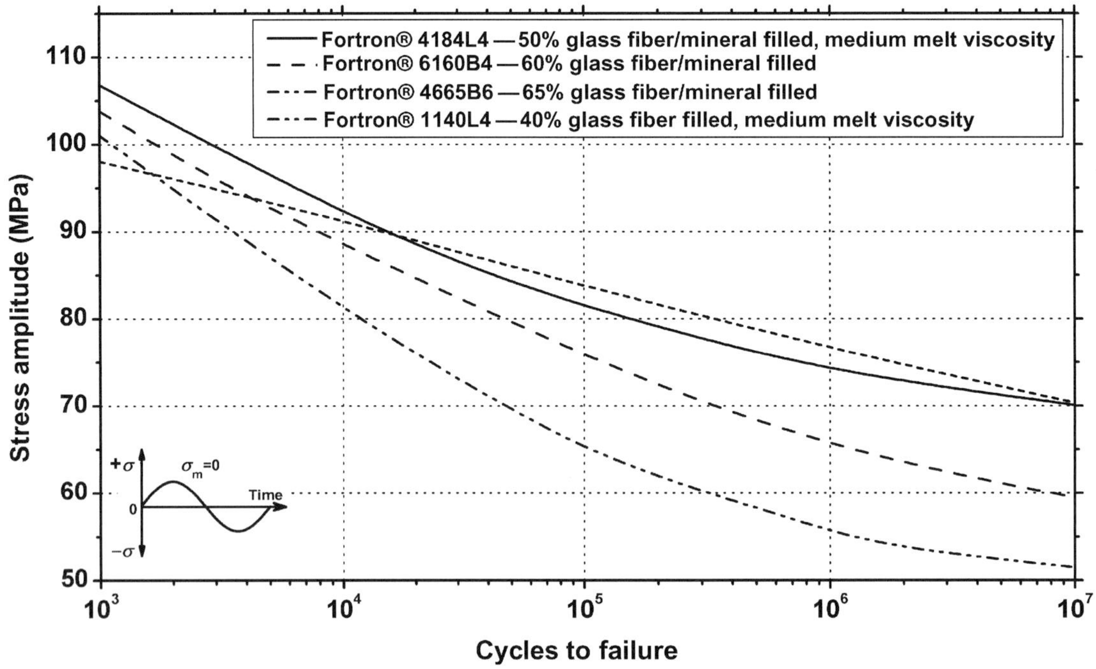

Figure 12.16 Flexural stress amplitude vs. cycles to failure at 23°C and 10 Hz of several fiber reinforced Ticona Fortron® PPS plastics.

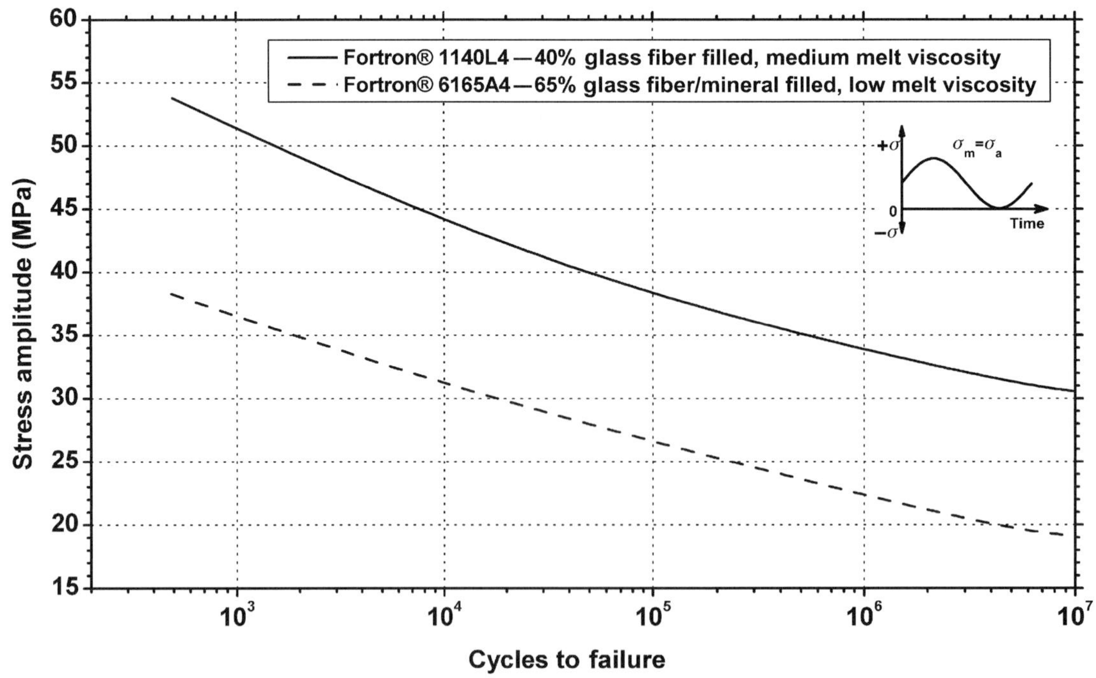

Figure 12.17 Tensile stress amplitude vs. cycles to failure at 23°C and 5 Hz of two fiber reinforced Ticona Fortron® PPS plastics.

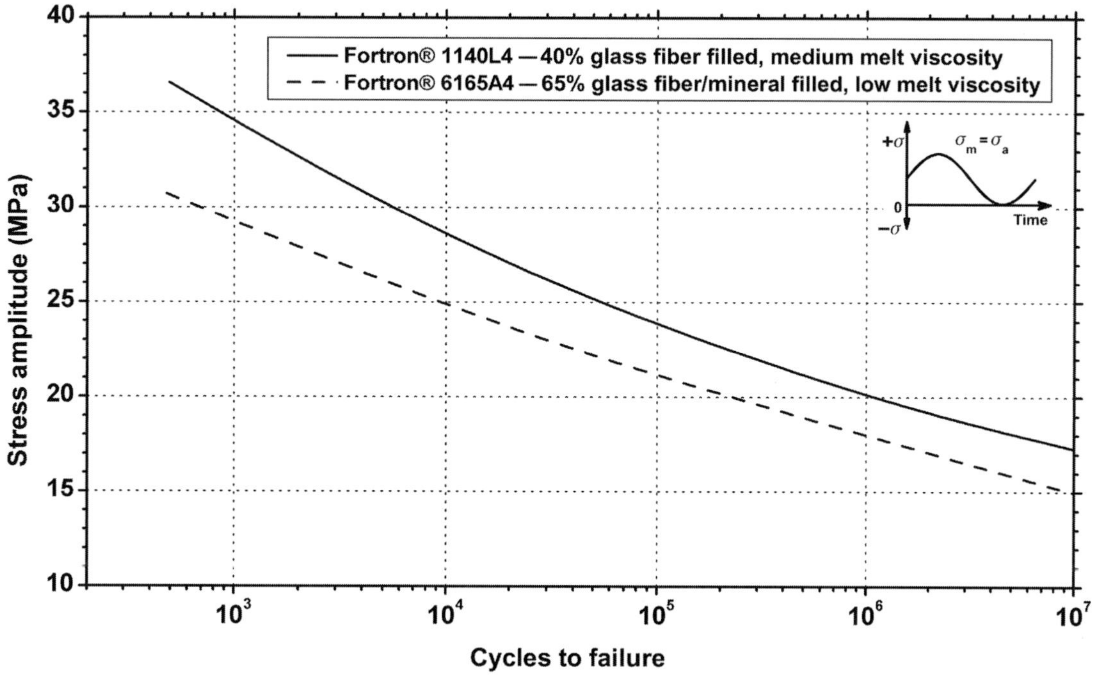

Figure 12.18 Tensile stress amplitude vs. cycles to failure at 90°C and 5 Hz of two fiber reinforced Ticona Fortron® PPS plastics.

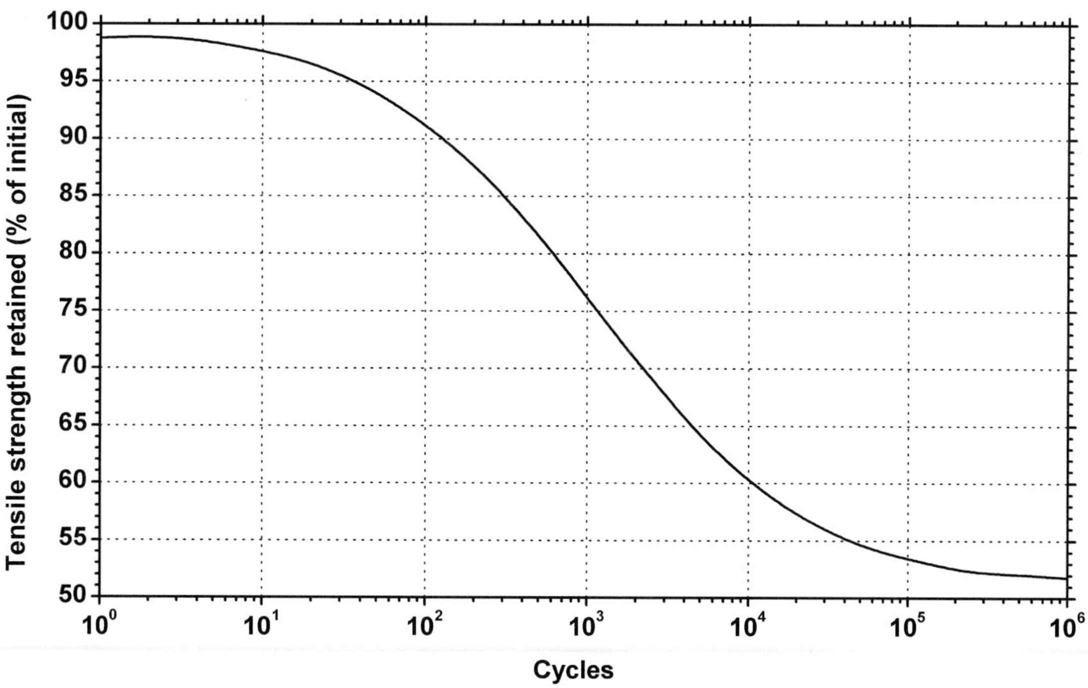

Figure 12.19 Tensile strength retained vs. cycles to failure at 23°C and 10 Hz of Chevron Phillips Chemical Ryton® A-200 PPS.

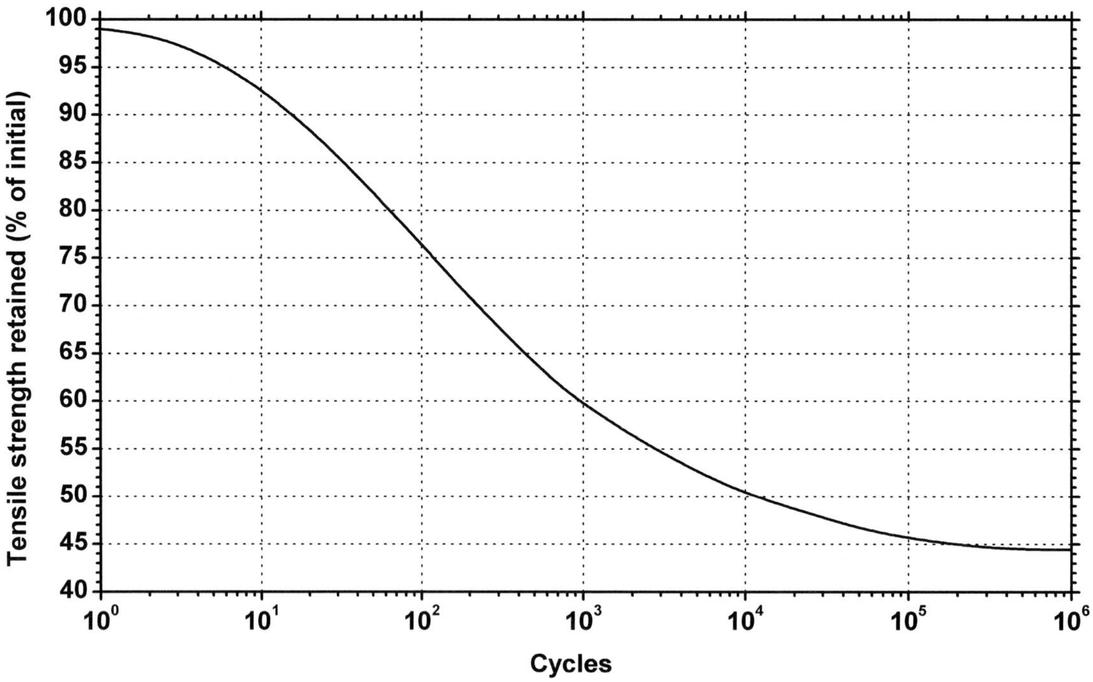

Figure 12.20 Tensile strength retained vs. cycles to failure at 23°C and 10 Hz of Chevron Phillips Chemical Ryton® R-4 02XT—40% glass fiber filled PPS.

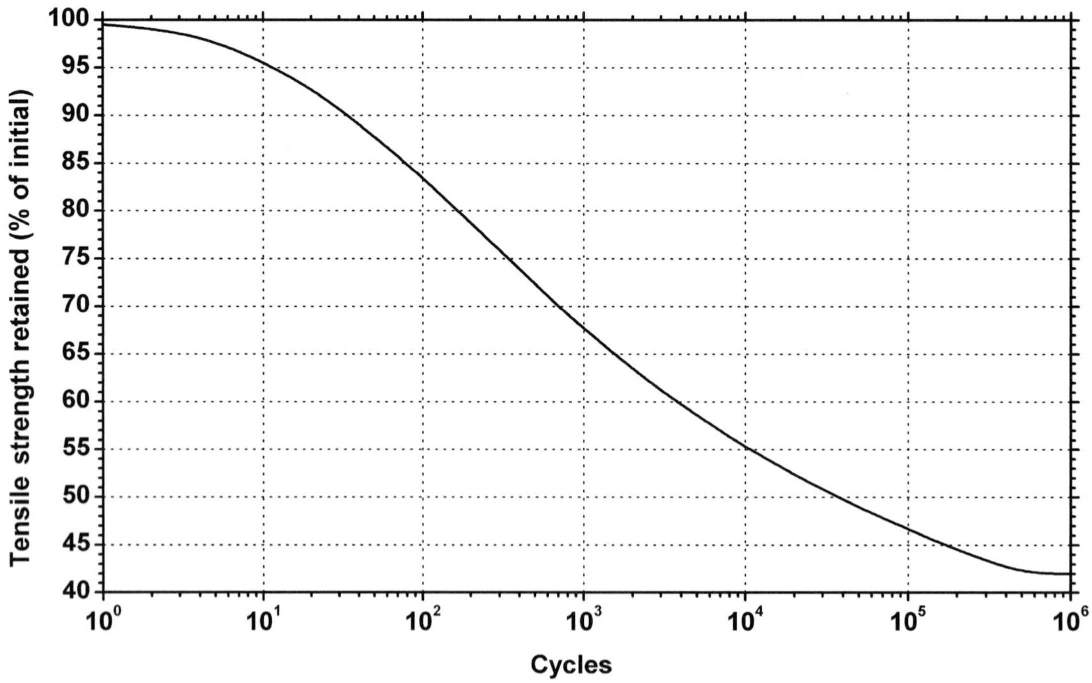

Figure 12.21 Tensile strength retained vs. cycles to failure at 23°C and 10 Hz of Chevron Phillips Chemical Ryton® R-4 04—40% glass fiber filled PPS.

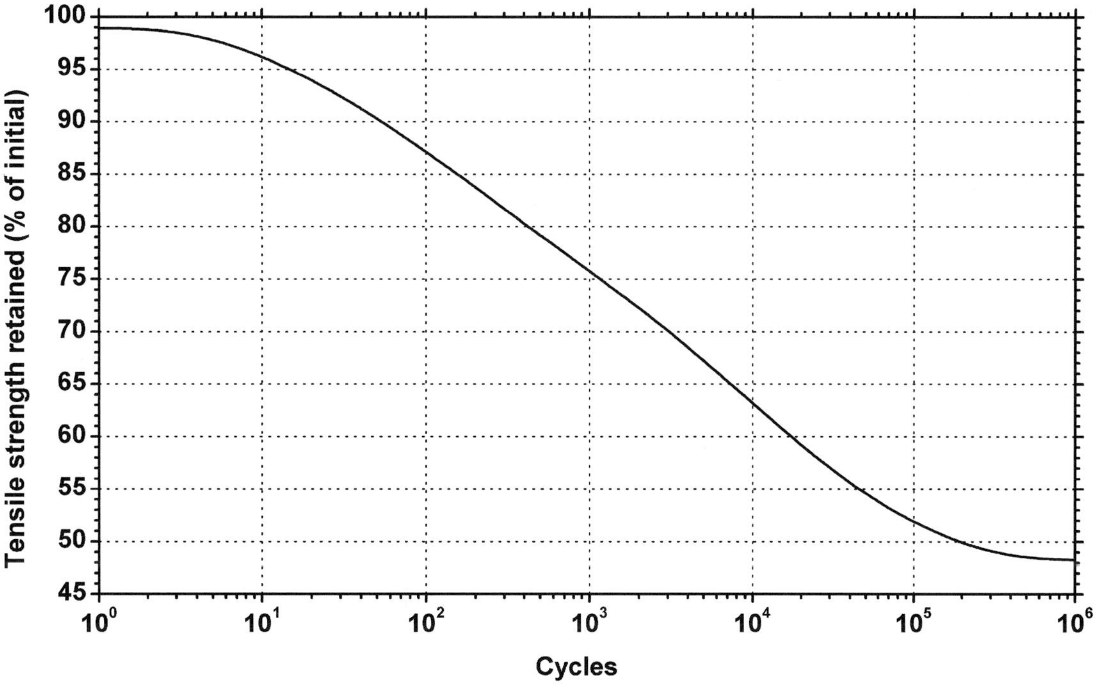

Figure 12.22 Tensile strength retained vs. cycles to failure at 23°C and 10 Hz of Chevron Phillips Chemical Ryton® R-7—65% glass fiber/mineral filled PPS.

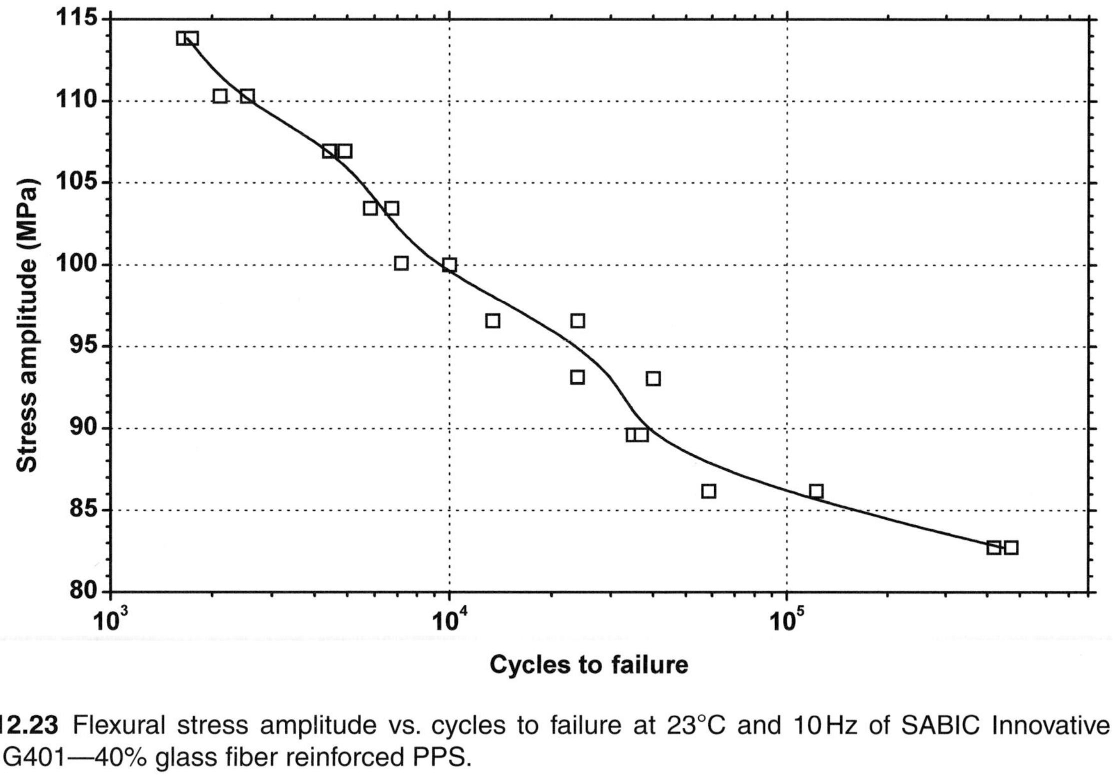

Figure 12.23 Flexural stress amplitude vs. cycles to failure at 23°C and 10 Hz of SABIC Innovative Plastics Supec® G401—40% glass fiber reinforced PPS.

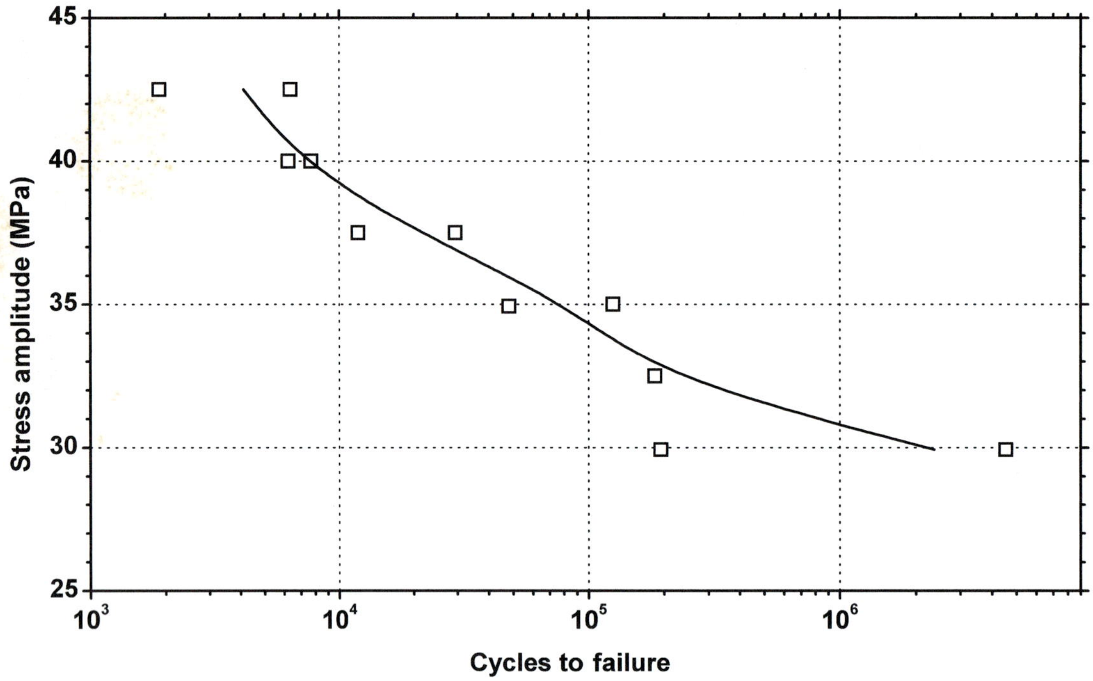

Figure 12.24 Flexural stress amplitude vs. cycles to failure at 23°C and 10 Hz of SABIC Innovative Plastics Supec® G620—40% glass fiber reinforced PPS.

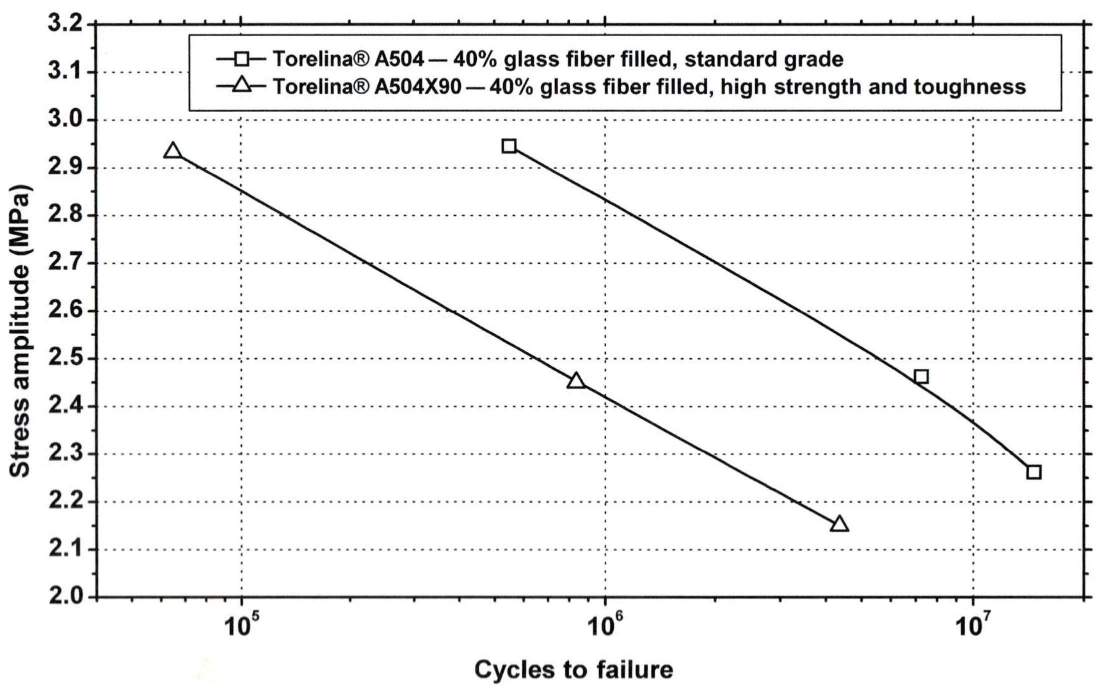

Figure 12.25 Stress amplitude vs. cycles to failure at 110°C and 2000 Hz of two Toray Resin Company Torelina® PPS plastics.

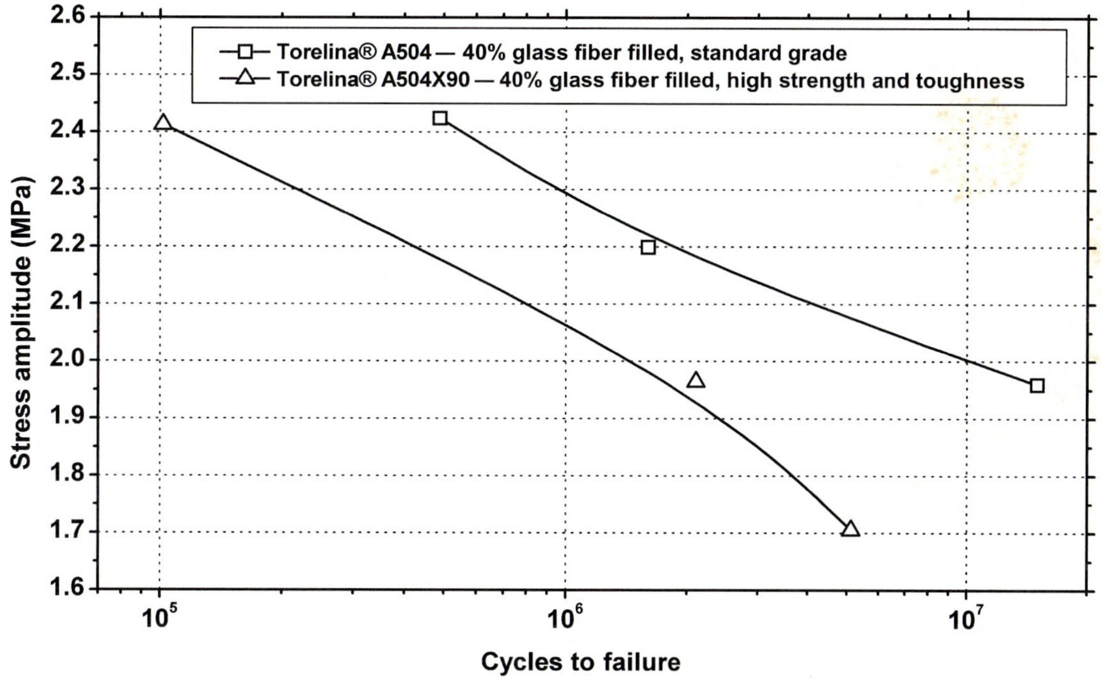

Figure 12.26 Stress amplitude vs. cycles to failure at 160°C and 2000 Hz of two Toray Resin Company Torelina® PPS plastics.

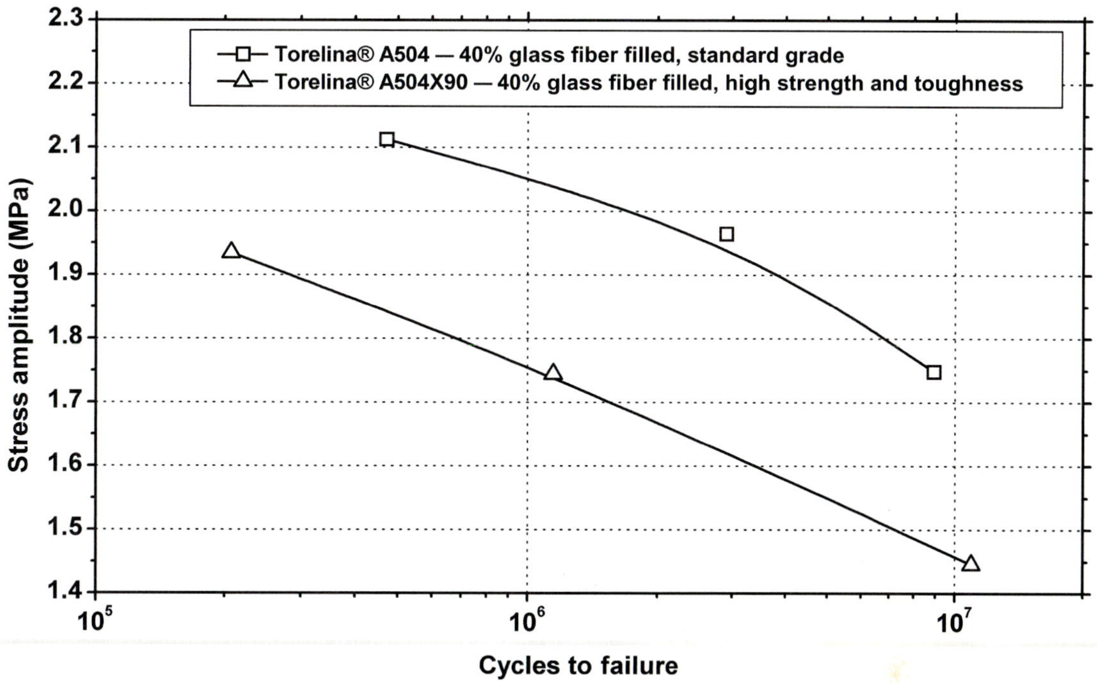

Figure 12.27 Stress amplitude vs. cycles to failure at 180°C and 2000 Hz of two Toray Resin Company Torelina® PPS plastics.

12.4.2 Tribology Data

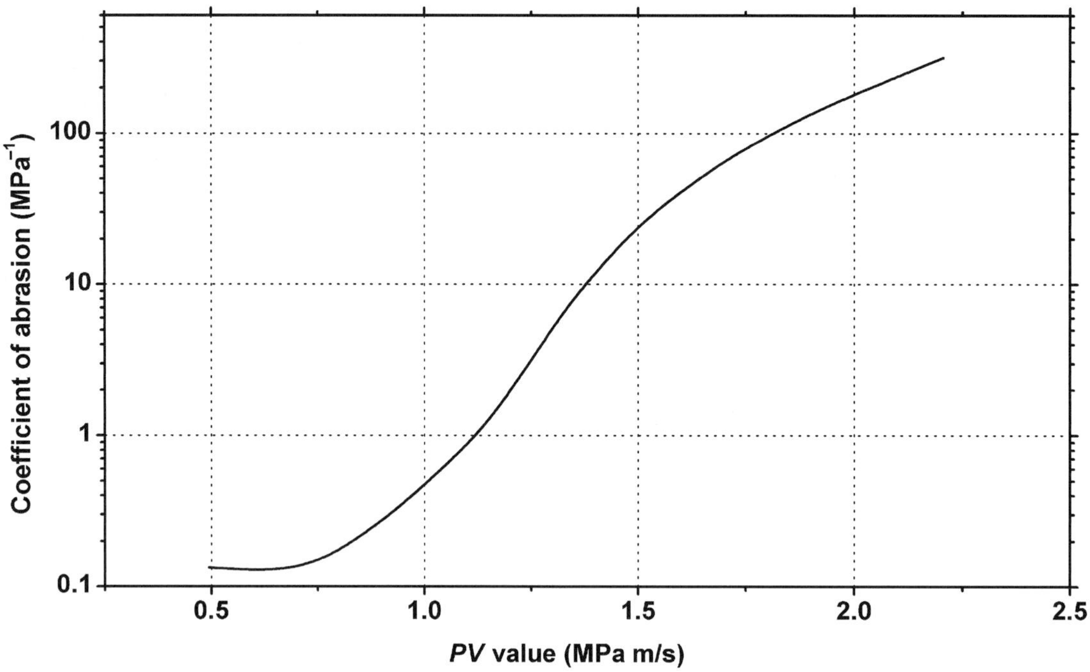

Figure 12.28 Coefficient of abrasion (against SCM21 steel) vs. PV value of Toray Resin Company Torelina® A504—40% glass fiber filled, standard grade PPS.

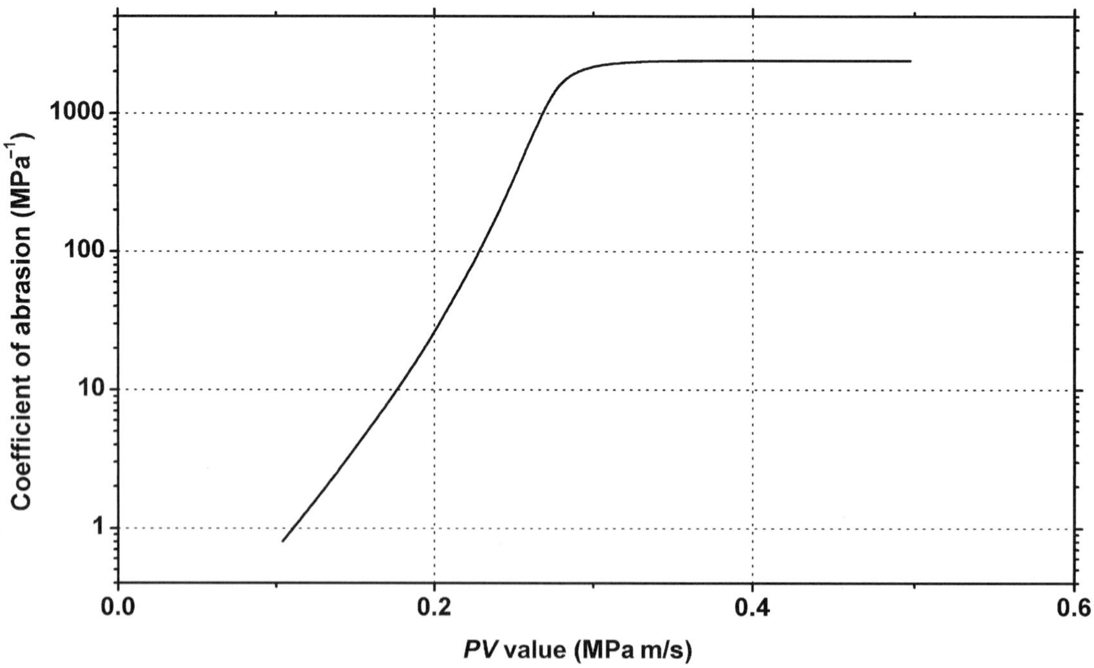

Figure 12.29 Coefficient of abrasion (against itself) vs. PV value of two Toray Resin Company Torelina® A504—40% glass fiber filled, standard grade PPS.

Table 12.11 Taber Abrasion of Chevron Phillips Chemical Ryton® Plastics

Ryton PPS Grade	Abrasion Wheel	Shore D Hardness	Weight Loss (g/1000 revolutions)
R-4 (40% glass fiber filled)	CS-10	89	0.070
R-7 (65% glass fiber/mineral filled)	CS-17	89	0.034
R-7	CS-17	—	0.068
A-200	CS-17	—	0.023

Table 12.12 Coefficient of Friction (Against Steel) of Chevron Phillips Chemical Ryton R-4 40% Glass Fiber Filled PPS

Ryton PPS Grade	Static	100 rpm, 29 ft/min	190 rpm, 55 ft/min
R-4	0.50	0.55	0.53

12.5 Polysulfone

12.5.1 Fatigue Data

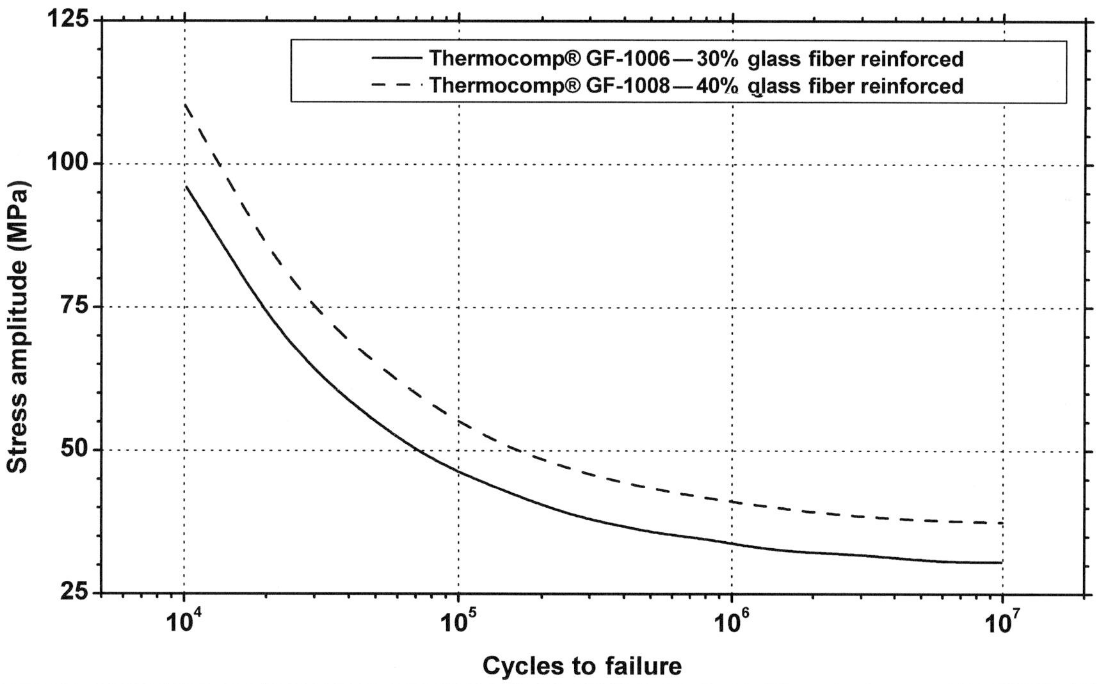

Figure 12.30 Flexural stress amplitude vs. cycles to failure at 23°C of two glass fiber reinforced SABIC Innovative Plastics Thermocomp® PSU plastics.

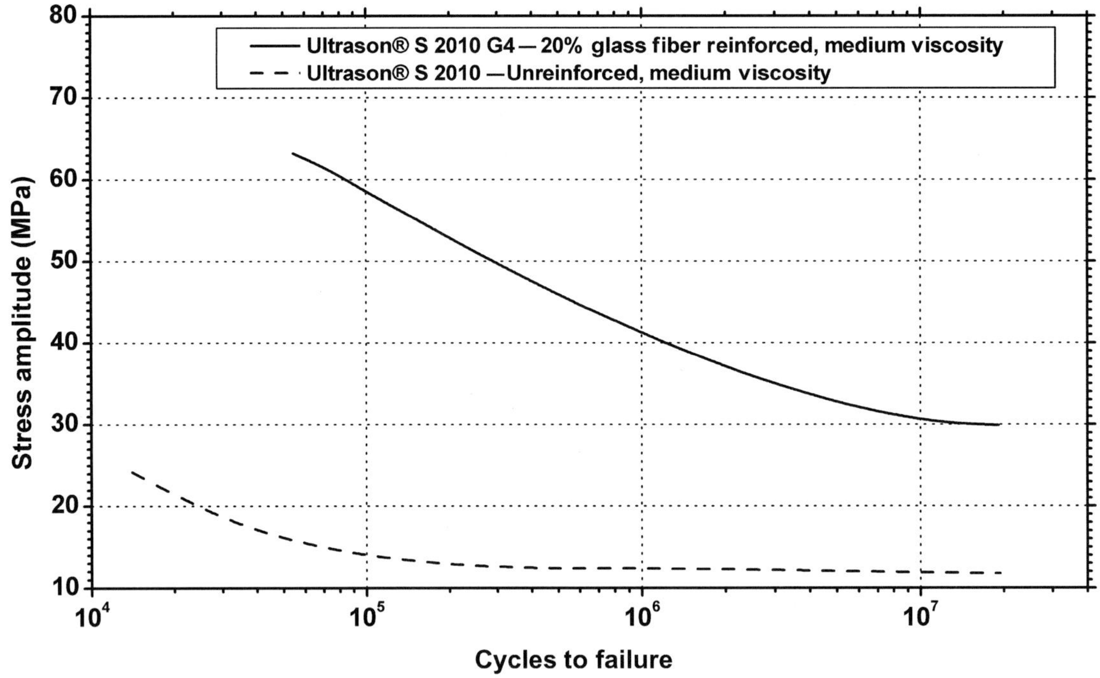

Figure 12.31 Flexural stress amplitude vs. cycles to failure at 23°C and 15 Hz of two BASF Ultrason® PSU plastics.

12.5.2 Tribology Data

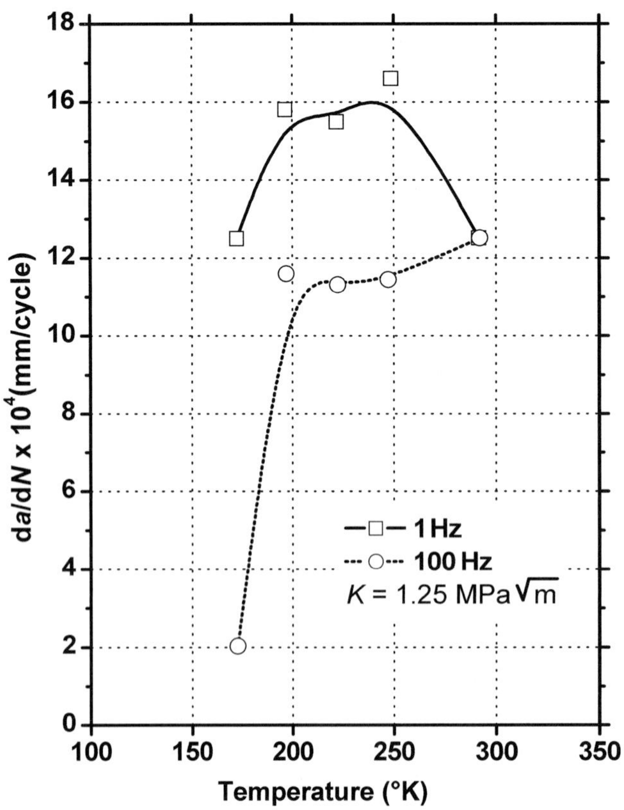

Figure 12.32 Fatigue crack propagation rate vs. temperature and cycle frequency of generic PSU.

Table 12.13 Friction Coefficient and Wear Intensity for BASF Ultrason® S PSU Plastics (Tribological System: Peg-and-Disk Apparatus, Pressure (P) = 1.0 MPa, Rubbing Velocity (V) = 0.5 m/s, Temperature of Rubbing Surfaces 40°C, Mating Material: Steel 100 Cr 6 700 Hv 10, No Lubricant)

Ultrason® Grade	Dynamic Friction Coefficient	Wear Intensity S (μm/km)	Mean Roughness RZ (μm)
S 2010	0.60	>1000	2.5
S 2010 G4 (20% glass fiber)	0.42	60	2.5
S 2010 G6 (30% glass fiber)	0.46	4	2.5

Index

B

C

D

H

M

N

O

P

R

S

T

U

V

W

X

Y

Z